Spectroscopy of Biological Molecules
Theory and Applications – Chemistry, Physics,
Biology, and Medicine

NATO ASI Series

Advanced Science Institutes Series

A series presenting the results of activities sponsored by the NATO Science Committee, which aims at the dissemination of advanced scientific and technological knowledge, with a view to strengthening links between scientific communities.

The series is published by an international board of publishers in conjunction with the NATO Scientific Affairs Division

A	Life Sciences	Plenum Publishing Corporation
B	Physics	London and New York
C	Mathematical and Physical Sciences	D. Reidel Publishing Company Dordrecht, Boston and Lancaster
D	Behavioural and Social Sciences	Martinus Nijhoff Publishers
E	Engineering and Materials Sciences	The Hague, Boston and Lancaster
F	Computer and Systems Sciences	Springer-Verlag
G	Ecological Sciences	Berlin, Heidelberg, New York and Tokyo

Series C: Mathematical and Physical Sciences Vol. 139

NATO Advanced Study Institute on Spectroscopy of Biological
" Molecules, 1983, Acquafredda di Maratea, Italy.

Spectroscopy of Biological Molecules

Theory and Applications – Chemistry, Physics, Biology, and Medicine

edited by

Camille Sandorfy

and

Theophile Theophanides

Department of Chemistry, University of Montreal,
Montreal, Quebec, Canada

D. Reidel Publishing Company

Dordrecht / Boston / Lancaster

Published in cooperation with NATO Scientific Affairs Division

Proceedings of the NATO Advanced Study Institute on
Spectroscopy of Biological Molecules
Acquafredda di Maratea , Italy
July 4 - 15, 1983

Library of Congress Cataloging in Publication Data

NATO Advanced Study Institute on Spectroscopy of Biological Molecules (1983 :
      Acquafredda di Maratea, Italy)
      Spectroscopy of biological molecules.

      (NATO ASI series. Series C, Mathematical and physical sciences ; vol. 139)
      "Proceedings of the NATO Advanced Study Institute on Spectroscopy of
Biological Molecules, Acquafredda di Maratea, Italy, July 4—15, 1983."-T.p. verso.
      Includes index.
          1.   Spectroscopy-Congresses.   2.   Biomolecules-Analysis-Congresses.
I.   Sandorfy, Camille, 1920-   . II.   Theophanides, Theo M.   III.   Title.
IV.   Series: NATO ASI series. Series C. Mathematical and physical sciences ; no. 139.
QP519.9.S36N38   1983   574.19'285   84-17862
ISBN 90-277-1849-0

Published by D. Reidel Publishing Company
P.O. Box 17, 3300 AA Dordrecht, Holland

Sold and distributed in the U.S.A. and Canada
by Kluwer Academic Publishers,
190 Old Derby Street, Hingham, MA 02043, U.S.A.

In all other countries, sold and distributed
by Kluwer Academic Publishers Group,
P.O. Box 322, 3300 AH Dordrecht, Holland

D. Reidel Publishing Company is a member of the Kluwer Academic Publishers Group

TABLE OF CONTENTS

PREFACE

This volume contains the proceedings of the NATO-Advanced
Study Institute on the "Spectroscopy of Biological Molecules",
which took place on July 4-15, 1983 in Acquafredda di Maratea,
Italy.

The institute concentrated on three main subjects: the
structure and dymanics of DNA, proteins, and visual and plant
pigments. Its timeliness has been linked to rapid advances
in certain spectroscopic techniques which yielded a considerable amount of new information on the structure and interactions of biologically important molecules. Among these
techniques Fourier transform infrared, resonance and surface
enhanced 'Raman spectroscopies, Raman microscopy and microprobing, time resolved techniques, two photon and ultrafast
electronic, and C-13, N-15 and P-31 NMR spectroscopies and
kinetic and static IR difference spectroscopy receiced a
great deal of attention at the Institute. In addition, an
entirely new technique, near-millimeter-wave spectroscopy
has been presented and discussed.

Two introductory quantum chemical lectures, one on the
structure of water in DNA, and another on the energy bands in
DNA and proteins set the stage for the experimentally oriented
lectures that followed. Fundamental knowledge on hydrogen
bonding was the topic of two other lectures.

Panel discussions were held on the structure and conformations of DNA, metal-DNA adducts and proteins and on visual
pigments.

Many scientists who normally attend different conferences
and never meet, met at Aquafredda di Maratea. We feel, that
at the end of our Institute a synthetic view emerged on the
powerful spectroscopic and theoretical methods which are now
available for the study of biological molecules.

The directors of this Advanced Study Institute would like
to express their profound gratitude to NATO's Scientific
Affairs Division for making the holding of the Institute
possible. They express their heartfelt thanks to all the
Lecturers and Participants for their valuable contributions
and for consenting a certain financial sacrifice so that

we could have the required number of lecturers in all the sections of our broad subject matter.

Last, but not least we thank our Italian friends for their help and kind hospitality and Nature for the blue sky and sea of Italy.

Montréal, Québec, December, 1983   C. Sandorfy
                                                    T. Theophanides

ENERGY BANDS IN DNA

Janos J. LADIK

Chair for Theoretical Chemistry, Friedrich-Alexander-
University Erlangen-Nürnberg, FRG and Laboratory of
National Foundation for Cancer Research at the Chair
for Theoretical Chemistry, University Erlangen-Nürnberg

ABSTRACT.- Ab initio SCF LCAO band structures of homopolynucleotides are presented. In the case of a cytosine stack the effect
of the surrounding water on the band structure is shown. It is
mentioned in the case of polyacetylene that using a good basis
set and taking into account the major part of correlation the
Hartree-Fock gap reduces to a value in reasonable agreement
with experiment. Therefore, one can expect that the gap for DNA
is also considerably smaller than minimal basis calculations
indicated and can raise the question whether DNA is an intrinsic
semiconductor.
    Finally, the effect of aperiodicity on the band structure
of DNA is discussed on the basis of calculations using the
negative factor counting (NFC) technique.

I. INTRODUCTION

Polymers play an important role as plastics. Highly conducting
polymers are in the last decade objects of extensive
experimental investigations being candidates for the discovery
of new physical phenomena and serious attempts are made for
their technical application (like batteries). Biopolymers like
nucleic acids (DNA and RNA), proteins, polysacharides, lipids
etc. have fundamental importance in life processes. To understand the different physical and chemical properties of polymers
(which underlie in the case of biopolymers also their biological
functions) one has to obtain a fair knowledge of their electronic
structure.
    To treat quantum mechanically DNA one has to proceed step-
wise:
1.) One starts with ab initio SCF LCAO crystal orbital (CO)

1

C. Sandorfy and T. Theophanides (eds.), Spectroscopy of Biological Molecules, 1–13.
© *1984 by D. Reidel Publishing Company.*

calculations /1/ taking a nucleotide base, a nucleotide base
pair or a whole nucleotide as unit cell (periodic polynucleotides).

2.) As next step one has to consider that DNA is aperiodic,
therefore one has to apply appropriate techniques /2/ to treat
this compositional disorder using the results of periodic
chain calculations as input (see below).

3.) After this one has to take into account the effect of the
surrounding water and ions on the band structure by
constructing an effective potential of the environment /3/.

4.) Finally, one should take into account (using a good basis
set) also the major part of the electronic correlation /4/.

5.) Having performed for a polymer steps 1.) - 4.) one is in a
good position to calculate different properties (electronic
and vibrational spectra, transport and magnetic properties
etc.) of it.

6.) In the case of biopolymers it is very important to treat
also interactions between polymer chains (for instance the
genetic regulation of a cell is mostly dependent on DNA-
protein interactions in a nucleoprotein).

Since the unit cells in DNA are fairly large, the calculations
performed until now belong mostly to the first step (band
structure calculations of homopolynucleotides. There are a few
attempts to treat aperiodicity in DNA and there exists a pilot
calculation to treat the effect of the surrounding water
molecules on the band structure of a cytosine stack. In this
short review the results of a part of these calculations
performed for DNA will be presented together with the necessary
methods. For the calculation of the electronic structure of
proteins which presents a still more formidable problem than that
of DNA, one has to perform the same steps. Until now only the band
structures of a few homopolypeptides have been computed and the
density of states of mixed glycine-alanine and glycine-serine
chains, respectively, have been determined (see the papers of
Seel and of Day, Suhai and Ladik at reference /2/). Since, however, these calculations though being similar are less advanced
than in the case of DNA, we do not discuss them in this paper.

No correlation calculations have been performed on biopolymers yet, but such computations have been successfully
executed in the cases of transpolyacetylene /5/ and polydiacetylenes.
The same holds for exciton spectra which have been successfully
computed applying intermediate exciton theory /6/ (including
correlation) for the before mentioned two chains /7/. There is
an early calculation on transport properties of periodic DNA
models using simple tight binding (Hückel) band structures /8/.
There exist pilot calculations also for the interactions between
hompolynucleotides and for polyglycine chains in different
conformations /9/. Finally one should mention that in the case
of a good basis set calculations of polyethylene $((CH_2)_x)$ taking
into account also the major part of the correlation even the

mechanical properties of this system could be computed with results
/10/ in good agreement with experiment.

The good results obtained for chains with small unit cells
give rise to hopes that with even larger computers and with the
improvement of the numerical techniques in the next few years
the calculations on the electronic structure of DNA and proteins
and on their properties will reach the same level of sophis-
tications as those on the simple chains.

II. METHODS

II.1. Ab initio Crystal Orbital Method

If there is a translational (or more generally any periodic)
symmetry in an infinite solid or polymer the infinite matrix
which one obtains in any LCAO theory can be brought with the help
of a simple unitary transformation into a block-diagonal form /1/.
The order of these blocks is (in the <u>ab initio</u> case) equal to the
number of basis functions in the unit cell. In this way the
original hypermatrix equation splits into N+1 matrix equations
if N+1 denotes the number of blocks (unit cells). Each such
equation has an index which denotes the serial number of the
matrix block to which the equation belongs. If $N \longrightarrow \infty$ this serial
number becomes continous. Physically it is the vector $\vec{k}$ (or one of
its components in the one-dimensional (1D) case) of the reciprocal
lattice which gives the crystal momentum /1/.

Here no attempt will be made to reproduce the derivation of
the expressions of the ab initio SCF LCAO CO method (for this see
/1/) only the final equations are written down (for the sake of
simplicity in the 1D case). Let us express a crystal orbital in
the form of a linear combination of Bloch orbitals

$$\Psi_n(k, \vec{r}) = 1/\sqrt{N+1} \sum_{j=-N/2}^{N/2} e^{ikja} \sum_{\jmath=1}^{m} c(k)_{m,\jmath} \chi_{\jmath}(\vec{r} - \vec{R}_{j\jmath}) \qquad (1)$$

Here N+1 is the number of unit cells, m is the number of basis
functions in the unit cell, a the elementary translation,
$\chi_s(\vec{r} - \vec{R}_{js})$ the sth atomic orbital (AO) centered at the Sth
atom (to which orbital s belongs) in the j-th unit cell (with
position vector $\vec{R}_{js}$) and n is the band index. The coefficients
$c(k)_{n,s}$ can be determined from the generalized eigenvalue equation

$$\underline{\underline{F}}(k) \; \underline{c}_n(k) = \varepsilon_n(k) \; \underline{\underline{S}}(k) \; \underline{c}_n(k). \qquad (2)$$

Here the overlap matrix $\underline{\underline{S}}(k)$ is the Fourier transform of the
matrix blocks containing the overlap integrals between basis
functions belonging to different cells,

$$\underline{\underline{S}}(k) = \sum_{q=-N/2}^{N/2} e^{ikqa} \underline{\underline{S}}(q); \qquad (3)$$

(for instance $\underline{\underline{S}}(o)$ contains all the overlap integrals within one cell, $S(\underline{1})$ the ones between the reference cell and its next neighbour and so on). Similarly one obtains /1/

$$\underline{\underline{F}}(k) = \sum_{q=-N/2}^{N/2} e^{ikqa} \underline{\underline{F}}(q) \qquad (4)$$

The elements of the matrices $\underline{\underline{F}}(q)$ are defined according to the detailed derivation /1/ as

$$\left[\underline{\underline{F}}(q)\right]_{r,s} = \langle \chi_r^0 | \hat{H}^N | \chi_s^q \rangle + \sum_{u,v=1}^{m} \sum_{q_1,q_2=-N/2}^{N/2} P(q_1-q_2)_{u,v} \qquad (5)$$

$$\times \left[\langle \chi_r^0(1)\chi_u^{q_1}(2)|\frac{1}{r_{12}}|\chi_s^q(1)\chi_v^{q_2}(2)\rangle - \frac{1}{2}\langle \chi_r^0(1)\chi_u^{q_1}(2)|\frac{1}{r_{12}}|\chi_v^{q_2}(1)\chi_s^q(2)\rangle \right]$$

Here the shorthand notation χ_s^q means $\chi_s(\vec{r} - \vec{R}_{sq})$, s $\in$ S, thus the superscripts are always cell and the subscripts basis function indices and $\hat{H}^N$ stands for the one-electron part of the Fock operator of the chain. Finally the elements of the charge-bond order matrices $\underline{\underline{P}}(q_1 - q_2)$ are defined as a generalization of the original definition given by Coulson as

$$P(q_1 - q_2)_{u,v} = 2a/2\pi \int_{-\pi/a}^{\pi/a} \sum_{h=1}^{n^*} c_{h,u}(k) \, c_{h,v}(k) \, e^{ika(q_1-q_2)} dk \qquad (6)$$

where n^* is the number of filled bands.

Inspecting equ.-s (2) - (6) we see that they represent nothing else but Roothaan's SCF LCAO equations for a closed shell system /11/ generalized for an infinite chain with periodic boundary conditions. One has to solve them at a number of points in k space simultaneously to be able to construct the matrices $\underline{\underline{P}}(q_1 - q_2)$ for the next iteration step. Using appropriate numerical integration techniques usually it is enough to have 6-8 k points between o and π/a $[\; \mathcal{E}(k) = \mathcal{E}(-k)]$ to obtain consistent results. The SCF procedure does not converge, however, in most cases so easily as in the corresponding molecules, especially if one uses a larger basis set. The main reason for this is that in practice one can go in the summations (3) and (4) only until a limited numbers of neighbours (actually to keep charge neutrality and obtain reasonably reliable results one has to go to different numbers of neighbours for different types of integrals /12/) and the error caused by this procedure is strongly amplified in cases when the matrix $\underline{\underline{S}}(k)$ has some very small eigenvalues /13/. (For the elimination of $S(k)$ in Equ. (2) one uses again Löwdin's symmetric orthogonalization procedure.)

Finally it should be pointed out that the formalism described here is valid not only for simple translation, but also for cases of repeated combined symmetry operations (for instance helix operation). As group theoretical considerations show it in this case 1) one has to put the nuclei into the right positions by going from one cell to the next and 2) one has to rotate correspondingly also the basis functions /14/.

II.2. Negative Factor Counting Technique for the Determination of States in a Disordered Chain

To obtain an orientation about the level distribution in aperiodic DNA and proteins one can apply the negative factor counting (NFC) method /2/ to determine the density of states in these disordered systems. According to this method if we write for the disordered chain a Hückel determinant which is tridiagonal due to the fact that only first neighbours' interactions are taken into account

$$
|\underline{\underline{H}}(\lambda)| = \begin{vmatrix} \alpha_1 - \lambda & \beta_2 & 0 & 0\ldots & 0 \\ \beta_2 & \alpha_2 - \lambda & \beta_3 & 0 & \\ 0 & \beta_3 & \alpha_3 - \lambda & \beta_4 & \\ \vdots & & & \ddots & \\ 0 & & & \beta_N & \alpha_N - \lambda \end{vmatrix} = 0,
$$

$$\tag{7}$$

this can be easily transformed into a didiagonal form with the help of successive Gaussian eliminations. Therefore, the determinant

$$
|\underline{\underline{H}}(\lambda)| = \prod_{i=1}^{N} (\lambda_i - \lambda) \tag{8}
$$

can be rewritten as

$$
|\underline{\underline{H}}(\lambda)| = \prod_{i=1}^{N} \varepsilon_i(\lambda), \tag{9}
$$

where the diagonal elements of the didiagonal determinant are given by the simple recursion relation

$$
\varepsilon_i(\lambda) = \alpha_1 - \lambda - \beta_i^2 / \varepsilon_{i-1}(\lambda), \quad i = 1,2,3,\ldots,N, \tag{10a}
$$

$$
\varepsilon_1(\lambda) = \alpha_1 - \lambda . \tag{10b}
$$

Comparing equ.-s (8) and (9) it is easy to see that for a given value, the number of eigenvalues smaller than λ

($\lambda_i \angle \lambda$) has to be equal to the number of negative $\varepsilon_i(\lambda)$ factors in equ. (9) /2/. (Calculations of all the eigenvalues ε_i of a long chain (N = 10^4 or 10^3) is impossible, but the computation of the $\varepsilon_i(\lambda)$ factors with the help of equ-s.(10a) and (10b) is very fast.) By giving λ different values throughout the spectrum and taking the difference of the number of negative $\varepsilon_i(\lambda)$'s belonging to consecutive λ values, one can ontain a histogram for the distribution of eigenvalues (density of states) of $\underline{\underline{H}}$ for any desired accuracy.

For actual calculations one has to make a band structure calculation for each component of the disordered chain assuming that it is periodically repeated. Then the values α_i (diagonal elements of $\underline{\underline{H}}$) can be determined from the positions of the bands of the components (the middle point or weighted middle points of the bands) and the off-diagonal elements β_i from the widths of the bands. It should be pointed out that the NFC in this simple form gives only the level distribution belonging to one band (for instance, valence band or conduction band) of the disordered chain.

In the case of disordered quasi-1D systems the NFC method can also be applied for the case of an arbitrary number of orbitals per site either in an ab initio form /2/ or in a semi-empirical, for instance, extended Hückel form /15/ (for the derivation see Day and Martino's paper in ref. /2/). In this case one has a matrix block instead of each diagonal element α_i and off-diagonal element β_{i+1}, respectively, in equ. (7) $\alpha_i - \lambda \rightarrow \underline{A}_i - \lambda \underline{S}_i$) $\beta_{i+1} \rightarrow \underline{B}_{i+1} - \lambda \underline{S}_{i+1}$). To construct these matrix blocks one has to perform ab initio MO calculations for the different units to obtain the diagonal blocks and cluster calculations for the off-diagonal ones. For instance in the case of a binary diordered chain in the first neighbours' interactions approximation one has to compute the AA, BB, BA and BB clusters. Having constructed the supermatrix of the disordered chain one can obtain   again very easily a histogram for the density of states for all the electrons (or valence electrons) of a long disordered chain in any desired accuracy. After obtaining the level distribution one can generate also a wave function for any particular energy level using the inverse iteration technique /16/.

II.3. Mean Field Treatment of the Effect of the Environment on the Electronic Strcuture of Chains

To take into account the effect of environment on the band structure of a periodic chain or on the level distribution of a disordered chain one can build into the one-electron part of the Fock operator of the chain the effective potential V_{eff} of the environment, $\hat{H}^N = \hat{H}^N + V_{eff}$. To generate V_{eff} one has to know the positions of the molecules (in the case of DNA water molecules and ions) in the environment of the chain. Having this information one lets to interact the systems building up the environment. If

one takes into accunt besides the electrostatic interactions also
their mutual polarizations (at least at the Hartree-Fock level),
one obtains the so-called mutually consistent charge distributions
of the constituent systems. To achieve this one can apply the
mutually consistent field (MCF) method /17/ developed in Erlangen.
(The same method can be used also for the calculation of interactions between infinite chains /18/). Having the MCF charge
distribution of each molecule and/or ion in the environment of a
unit cell one obtains for V_{eff} using the classical expression
/3/

$$V_{eff}(\vec{r}) = \sum_{q=-N}^{N} \sum_{i=1}^{M} \frac{\tilde{\varrho}_i^q(\vec{r}')}{|\vec{r} - \vec{r}'|} d\vec{r}' \tag{11}$$

Here N stands again for the number of unit cells, M denotes the
number of subsystems surrounding the reference cell and $\tilde{\varrho}_i^q(\vec{r})$
is the MCF electronic density of the i-th subsystem in the
environment of the q-th cell.

II.4. Calculation of Correlation in a Linear Chain

To calculate the correlation energy per unit cell for the ground
state of a polymer (either conductor or insulator) in principle
one can use any size consistent method (perturbation theory,
coupled cluster expansion, electron pair theories, etc.). To
keep the calculations in a manageable size one can use a simple
trigonometric series in k consisting of a few terms for the LCAO
coefficients occurring in the Bloch functions. In this way one can
put everything what is k-dependent before the integration and
therefore one has to perform the two-electron integrals over
the atomic orbitals only once. On the other hand, since the
calculations of the k-dependent prefactor is very fast one can
use a dense grid for the three independent k-values occurring
in the matrix elements

$$\langle IJ \| AB \rangle \equiv \langle \Psi_I(\vec{r}_1) \Psi_J(\vec{r}_2) | 1/r_{12}(1-\hat{P}_{12}) | \Psi_A(\vec{r}_1) \Psi_B(\vec{r}_2) \rangle \tag{12}$$

This type of matrix elements, in which the composite indices I,J..
denote a band index, a k-value and the spin occur in anyone of
the above mentioned methods suitable for the correlation
calculations.

As first step second order Moeller-Plesset perturbation
theory (MP2) /19/ can be applied which has provided (using a
4-31 G basis set) for a hydrogen chain about 70 per cent /4/ and
for polyacetylene (using a 4-31 G** basis) about 75 per cent of
the correlation energy.

Correlation calculations can be applied not only for the
more precise calculation of the total energy per unit cell (making
geometry optimization possible) but they can be used also for the

correction of the valence and conduction bands. Namely one can
define, following Takeuti /6/, (see also /5/) quasi particle energy
levels in the conduction and valence band, respectively, as

$$\mathcal{E}_c^{QP}(k_c) = E^{(N+1)}(k_c) - E^{(N)} = A(k_c) \tag{13a}$$

$$\mathcal{E}_v^{QP}(k_v) = E^{(N)} - E^{(N-1)}(k_v) = -I(k_v) . \tag{13b}$$

Here $E^{(N+1)}(k_c)$ is the total energy per unit cell with an extra
electron in the conduction band at level k_c, $E^{(N)}$ is the ground
state total energy per unit cell, and $E^{(N-1)}(k_v)$ stands for the
total energy per unit cell of the system with one electron
missing from the valence band level k_v. Further $A(k_c)$ is the
electron affinity (the energy gained by putting the extra
electron into level k_c of the conduction band) and $I(k_v)$ gives
the ionization potentital necessary to ionize an electron
from level k_v of the valence band.

One can approximate the total energies E (which are exactly
$E = E_{HF} + E_{corr.}$) by

$$E = E_{HF} + E_2 \tag{14}$$

$$\text{where } E_2 = - \sum_{I,J,A,B} \frac{\left| \langle \Psi_I(1) \Psi_J(2) | \frac{1}{r_{12}} (1-\hat{P}_{12}) | \Psi_A(1) \Psi_B(2) \rangle \right|^2}{\mathcal{E}_A + \mathcal{E}_B - \mathcal{E}_I - \mathcal{E}_J}$$

is the MP2 expression. Further taking into account Koopmans'
theorem,

$$\mathcal{E}_c^{HF}(k_c) = E_{HF}^{(N+1)}(k_c) - E_{HF}^{(N)} \tag{15a}$$

$$\mathcal{E}_v^{HF}(k_v) = E_{HF}^N - E_{HF}^{(N-1)}(k_v) \tag{15b}$$

one obtains from equ.-s (13)-(15) the expressions

$$\mathcal{E}_c^{QP}(k_c) = \mathcal{E}_c^{HF}(k_c) + \sum{}_c^e(k_c) + \sum{}_c^h(k_c) \tag{16a}$$

$$\mathcal{E}_v^{QP}(k_v) = \mathcal{E}_v^{HF}(k_v) + \sum{}_v^e(k_v) + \sum{}_v^h(k_v) . \tag{16b}$$

In these equations the electron- and hole self energies, $\sum^e$ and
$\sum^h$, respectively, belonging either to the conduction ($\sum_c$) or
to the valence band ($\sum_h$) express the change of the second order

perturbation theoretical correlation energy due to the $v(k_v) \rightarrow c(k_c)$
excitations. As the detailed derivation /5/ shows it in this
way one obtains a narrowing of both of the gap and of the
conduction and valence bands, respectively. The application of
this method to polyacetylene has reduced the STO-3G gap of 8.3 eV
in the 4-31 G** MP2 case to 3.0 eV while the experimental value
is ∼2.0 eV /5/. The reasons of the remaining ∼1 eV discrepancy
are well understood /5/

III. DISCUSSION OF THE RESULTS

III. 1. Periodic Polynucleotides

Ab initio SCF LCAO crystal orbital calculations (in which the
quasi momentum k is defined on the combined symmetry operation
translation + rotation which allows to reduce the unit cell from
a turn of the helix to a single chemical unit like A or A-T
etc.) of the four nucleotide base stacks /20,21/ using different
minimal basis sets /22,23/, of the poly(A-T) and poly(G-C) systems
/21/ modelling double stranded DNA and of the homopolynucletides
polycytidine /20/, polyadenilic and polythymidine /21/ have been
performed. Applying the geometry of B DNA determined by X-ray
diffraction /24/ these calculations have resulted in band
structures which show the following general features:

1) The conduction and valence bands of all the investigated
   systems have widths between 0.3 eV and 0.8 eV. These widths
   lie in the same range as those of the valence band of a TTF
   stack (0.3 eV) and of the conduction band of a TCNQ stack
   (1.2 eV) as our previous ab initio Hartree-Fock calculations
   have shown it /25/. Therefore, one would expect that by
   appropriate doping (which is by no means a trivial task)
   double stranded DNA B with periodic sequences A-T or G-C,
   respectively, could be made well conducting.

2) The gap of all the calculated DNA models /20,21/, as it is the
   case in all minimal basis ab initio calculations, is too
   large (around 11 eV). According to the experience, accumulated
   in our Laboratory in the case of polyacetylene /5/ if one
   applies a better basis set and introduces correlated one-
   electron bands with the aid of the electronic polaron model
   /5,6/ the gap decreases near to the experimental value. One
   has to point out, however, that if one compares the gaps
   (the energy difference of the lower limit of the conduction
   band and of the upper limit of the valence band) with
   experimental gaps, for the latter not the first electronic
   excitation energy should be used which due to exciton formation
   has a still smaller energy.

3) In the case of all the three homopolynucleotides polycytidine
   /20/, polyadenillic acid and polythymidine /21/ a charge

transfer of about 0.2e per unit cell has been found from the
sugar part of the nucleotide to the nucleotide base. The
inspection of the band structures of these systems shows that
the bands originating from the sugar-physophate chain and from
the nucleotide bases are still distinguisable /20,21/. The band
structures of the polynucleotides show that these composite
systems are insulators (completely filled valence bands).
Since a recent calculation on a cytosine stack using a 4-31 G
(double) basis for the valence electrons resulted in a
reduced gap of ∿6.8 eV /26/, one can expect that a 4-31 G**
+ MP2 calculations would result in analogy to the $(CH)_x$ case
/5/ in a quasi particle gap of ∿ 5 eV. Therefore one can
raise the question whether the system would not have a deeper
total energy, if one populated the bands in the course
of the calculation, if one took away 0.2e from the highest
filled sugar-phosphate type band and would put it into the
lowest unfilled band which is a base stack type band. If
this would be the case, one could expect that the periodic
sugar-phosphate chain of DNA and periodic segments of the base
stacks would show intrinsic conduction. This conduction would
be perturbed, however, by the aperiodic effective field
acting on this chain due to the different base pairs and - in
the case of nucleoproteins - due to the neighbourhood of
different amino acid residues at different parts of the chain.
This interesting problem certainly deserves further
investigations.

4) In the case of a cytosine stack - as mentioned in the
   Introduction - a model calculation has been performed to
   assess the effect of the water structure on its band structure
   /27/. For this purpose as first approximation the water
   structure around a cytidine unit determined by Clementi /28/
   has been applied /29/. For this water structure an effective
   potential of all the water molecules acting on the cytosine
   stack was generated applying the method described in II.3..
   The results of this pilot calculation show that a) the band
   widths hardly change due to the presence of water, but b)
   their positions are shifted to lower energies by values
   varying between 0.3 and 1.4 eV. These band shifts cause also
   a lowering of the total energy by ∿8 eV per cytosine
   molecule. These interesting first results certainly indicate
   that to obtain a more realistic description of the electronic
   structure of DNA one has to perform further similar calculations
   taking into account together with the water moelecules also
   the ions and the polypeptide chain surrounding a DNA double
   helix. Since, however, there is no X-ray on a nucleoprotein,
   one has to build up different plausible models for the
   geometrical structure of these polypeptide (first of all
   nucleohistone) chains.

III. 2. Some Remarks about Aperiodic DNA

Applying the NFC technique (see II.2.) in its simple form /2/ we
found that with increasing disorder the original bands of the
periodic components were split up into more and more peaks and
gaps in the density of states (DOS curve) /30/.

In the matrix block NFC calculations /31/ an aperiodic DNA
double helix of 1000 nucleotide base pairs has been treated. The
sequence of the base pairs has been generated with the help of
a Monte Carlo program. For the determination of the intra-chain
parameters for the valence bands ab initio band structure
calculations of the four nucleotide base stacks have been used.
On the other hand the interchain parameters have been estimated
on the basis of the strength of the interbase hydrogen bonds. We
have found again a strong aperiodicity effect on the DOS curves
especially in the case of strong disorder (50% A-T, 50% G-C) /31/.
It should be mentioned that in the last years a large number of
RNA sequences and DNA fragment sequences have been published in
the literature. These sequences are collected and stored in a
computer /32/. With the aid of a detailed mathematical statistical
analysis of these sequences one can hope that it will be possible
to find out some regularities and use them to construct so-called
representative sequences for different classes of biological
systems. (The calculations performed until now show for instance
that the sequences are not random, but possess a strong
preference to have the same unit repeated several times /33/.)
These effective sequences instead of the presently used random
sequences could be used then for the treatment of aperiodic DNA
with the NFC method.

Finally, one should point out that if one looks at the DOS
curves of disordered polymers (see for instance: /30/ and /31/ it is
obvious that they have a complicated structure with many gaps
between the occupied regions. Therefore, the question whether
such a polymer is a semiconductor or an insulator is by no means
a trivial one. The first problem which arises is whether the
Fermi level falls inside a continous region of filled levels or
is in one of the gaps. Even if the position of the Fermi level
would allow electronic conduction, one has to investiqate with the
help of a suitable method /34/ whether the states around it are
localized or delocalized. Only in the latter case can one expect
a non-vanishing conductivity.

ACKNOWLEDGMENT

The author would like to express his gratitude to Professors
C. Sàndorfy and T. Theophanides for the invitation and for the
cooperation of Professors E. Clementi , F. Martino and P. Otto
and of Drs. G. Corongiu, R.S. Day, M. Seel and S. Suhai in performing the investigations on DNA reviewed here. He is very much
indebted to them also for the many fruitful discussions. The

financial support of the "Fond der Chemischen Industrie" with
journals and books and of the "Kraftwerk Union AG" by supplying
free computer time is also gratefully acknowedged.

REFERENCES

/1/ Del Re, G., Ladik, J and Biczó, G., Phys. Rev. 155, (1967)
    997; André, J.-M., Gouverneur, L. and Leroy, G., Int. J.
    Quant. Chem. 1 (1967) 427 and 451; Ladik, J. in "Electronic
    Structure of Polymers and Molecular Crystals, ed.-s André,
    J.-M. and Ladik, J. (Plenum Press, New York-London) 1975,
    p. 23.
/2/ Dean, P., Proc. Roy Soc. London, Ser. A. 254 (1960) 507;
    ibid 260 (1961) 203; Dean, P., Rev. Mod. Phys. 44 (1972) 127;
    Seel, M., Chem. Phys. 143 (1979) 103; Day, R.S. and Martino, F.,
    Chem. Phys. Lett. 84 (1981) 86; Day, R.S., Suhai, S. and
    Ladik, J., Chem. Phys. 62 (1981) 165; For a recent review
    see: Ladik, J., Int. J. Quant. Chem. 23 (1983) 1073.
/3/ Ladik, J., Lecture at Int. Symp. Quantum Biology and Quantum
    Pharmacology, Palm Coast, Florida (1981) unpublished.
/4/ Ladik, J. in "Recent Advances in the Quantum Theory of Polymers"
    ed. André, J.-M., Brédas, J.-L., Delhalle, J., Ladik, J.,
    Leroy, G. and Moser, C. (Springer Verlag, Berlin-Heidelberg-
    New York) 1980, p. 155; Suhai, S. and Ladik, J. J. Phys. C:
    Solid State Phys. 15 (1982) 4327; Ladik, J., Int. J. Quant.
    Chem. 23 (1983) 1073.
/5/ Suhai, S., Phys. Rev. B 27 (1983) 3506.
/6/ Takeuti, Y., Progr. Theor. Phys. (Kyoto) 18, (1975) 421;
    Progr. Theor. Phys. (Kyoto) Suppl. 12 (1975) 75.
/7/ Suhai, S.,J. Chem. Phys. (submitted).
/8/ Suhai, S., J. Chem. Phys. 57 (1972) 5599.
/9/ Otto, P., Clementi, E., Ladik, J. and Martino, F., J. Chem.
    Phys. (submitted).
/10/ Suhai, S., J. Polym. Sci., Polym. Phys. 21 (1983) 00.
/11/ Roothaan, C.C.J., Rev. Mod. Phys. 23 (1957) 69.
/12/ Suhai, S. in "Quantenmechanische Untersuchungen an quasi-
    eindimensionalen Festkörpern", Habilitation Thesis, Erlangen,
    1983; J. Chem. Phys. (submitted).
/13/ Suhai, S., Bagus, P.S., and Ladik J., Chem. Phys. 68 (1982)
    467.
/14/ Ukrainski, I.I., Theor. Chim. Acta 28 (1975) 139; Blumen, A.
    and Merkel, C., phys. stat. sol. (b) 83 (1977) 425; See also:
    Ladik, J. in "Electronic Structure and Excited States of
    Polymers", ed.-s Nicolaides, C.A. and Beck, D.R. (D. Reidel
    Publishing Company, Dordrecht-Boston) 1978, p. 495.
/15/ Kertész, M. and Göndör, Gy., J. Phys. C 14 (1981) 1851.
/16/ Wilkinson, J.H., "The Algebraic Eigenvalue Problem"
    (Clarendon Press, Oxford) 1975.
/17/ Otto, P. and Ladik, J., Chem. Phys. 8 (1975) 192; ibid 19
    (1977) 209; Otto, P., Chem. Phys. 33 (1978) 407.

/18/ Ladik, J., Int. J. Quant. Chem. S9 (1975) 563.
/19/ Moeller, C. and Plesset, M.S., Phys. Rev. 46 (1934) 618.
/20/ Ladik, J. and Suhai, S., Int. J. Quant. Chem. QBS7 (1980)
     181.
/21/ Otto, P., Clementi, E. and Ladik, J., Chem. Phys. 78 (1983)
     4547.
/22/ Hehre, W., Stewart, R.F. and Pople, J.A., J. Chem. Phys.
     1 (1969) 2657.
/23/ The basis used in the integral package of IBMOL-5.
/24/ Arnott, S., Dover, S.P. and Wonacott, A.J., Acta Cryst. B28,
     2192 (1969).
/25/ Suhai, S. and Ladik, J., Phys. Lett. A77 (1980) 25.
/26/ Suhai, S., unpublished.
/27/ Otto, P., Ladik J., Corongiu, G., Suhai, S. and Förner, W.,
     J. Chem. Phys. 77 (1982) 5026.
/28/ For a review see: Clementi, E. "Computational Aspects for
     Large Chemical Systems" Lecture Notes in Chemstry, Vol. 19,
     (Springer-Verlag, Berlin-Heidelberg-New York) 1980;
     Corongiu, G. and Clementi, E., Biopolymers 20 (1981) 551.
/29/ At the time of the calculation no water structure around a
     cytosine stack was available. An inspection of the water
     structure around a cytidine unit shows, however, that this
     still contains the major part of the first hydration shell
     around a cytosine unit.
/30/ Seel, M. and Ladik, J. (unpublished); Reviewed by Ladik, J.
     in "Structure and Dynamics of Nucleic Acids and Proteins"
     ed.-s Clementi, E. and Sarma, R.H., (Adenine Press, New York)
     1983, p. 261.
/31/ Day, R.S. and Ladik, J., Int. J. Quant. Chem. 21 (1982) 917.
/32/ See for instance: Dayhoff, M.O., Schwartz, R.M., Chen. H.R.,
     Barker, W.C., Hunt, L.T. and Orsutt, B.C. "DNA", Vol. 1,
     (Mary Ann Liebert Inc.) 1981, p. 51.
/33/ Gentleman, J.F., Shadbolt-Forbes, M.A., Hawkins, J.W.,
     Ladik, J. and Forbes, W.F., Mathematical Scientist (submitted).
/34/ Gazdy, B., Day, R.S., Seel, M. Martino, F. and Ladik, J.,
     Chem. Phys. Lett. 88 (1982) 220.

H-BOND CHARGE-RELAY CHAINS IN MULTI-HEME CYTOCHROMES AND OTHER BIOMOLECULES

G. DEL RE

Cattedra di Chimica Teorica,Università di Napoli

via Mezzocannone 4, I - 80134 Napoli (Italy)

ABSTRACT. Biomolecules such as cytochromes function as systems capable of accepting or donating electrons, the redox centers being heme groups consisting of Fe-porphyrin macrocycles with imidazoles co-ordinated in the 5,6 positions.

When more than one heme group participate in the redox process, transfer of electrons from one heme group to another must take place.

The mechanism of this transfer has been the subject of much speculation: "through-space" and "through-bond" paths are the classes into which the suggested mechanisms fall. The latter class of paths are instances of H-bond relay chains, capable of transferring protons as well as electrons; examples are the "imidazole pump" and the "proton push-pull" chains.

The two basic properties of such chains are efficiency and transfer times. Both depend on the properties of H-bonds, which quantum mechanically can be treated as (usually asymmetric) potential energy double wells.

Charge-relay chains may consist of equivalent O.HO bonds or be more complicated. The quantities involved in a theoretical analysis of them as well as the chapters of theoretical chemistry involved are recalled. The chemical kinetics treatment is reported for the simpler proton-transfer mechanism. Next, the questions involved in a theoretical assessment of electron transfer through H-bridge chains involving peptide bonds are illustrated on simple models. Energy differences, energy barrier, transfer times, vibrational frequencies, stepwise vs. concerted mechanisms according to studies in collaboration with C.Minichino,A.Peluso and G.Villani are briefly discussed. The main conclusion is that relay chains whereby two imidazoles of two distinct hemes are connected through two peptide bonds,one water molecule, and one more peptide bond (and at least one of the peptide bonds is in -N=CR-OH form) are quite probable,efficient,and fast.

15

C. Sandorfy and T. Theophanides (eds.), Spectroscopy of Biological Molecules, 15–37.
© 1984 by D. Reidel Publishing Company.

Excitation by light or heat as well as other perturbations appear
to trigger in living systems biochemical processes whose main step
is electron transfer (E.T.) or proton transfer (P.T.). Examples
are some steps in photosynthesis, as well as a variety of redox
processes, e.g. those in multiheme cytochromes. In these notes
we intend to provide an introduction to the possible role of H-bridges in E.T. and P.T., thus providing a short review of a line
of research very actual and greatly dependent on advances in spectroscopy.

These notes are devoted to five point. First, we shall examine
a concrete example of electron-proton transfer in a biomolecule.
Then we shall recall some points regarding the double-well model
of the hydrogen bridge. Thirdly, we shall examine two model sequences of H-bonds. Next we shall present some computational results concerning the energies and vibrational modes of one of those
sequences. Finally, we shall indicate how those results and the
preceding considerations can be used to gain a deeper understanding
of the mechanisms under consideration.

1. Heme-heme intramolecular E.T. in cytochromes.

As a concrete example let us consider ET in a multiheme cytochrome.
A molecule of that type functions as an electron carrier, since it
receives electrons by oxidizing some partner and then yields the
electrons to some other element of a biochemical redox chain /1/.
Without entering the details, we shall here assume that we have
two heme groups separated by a number of amino-acid residues as
well as backbone peptide bonds. Possible pathways for the overall
ET process can be variations on two main themes, which borrowing
Hoffman's terminology/2/may be called "assisted throughspace" and
"through-bond" mechanisms.

a. Assisted through-space electron transfer mechanism.

A heme group consists of a Fe-porphyrin macrocycle with imidazole
rings coordinated in the positions 5 and 6.
If the electron to be transferred is initially on the macrocycle
A, its transfer to the macrocycle A' can be assigned a an electronic
push-pull or pull-push mechanism /3,4/ provided the energy levels
of A and A' are very close to one another. In both cases electron
transfer from A to A' is mediated by a species B which has a discrete set of electronic levels and is coupled to both A and A' -
i.e. is an 'electron bridge' connecting the two macrocycles.
The electron occupying the HOMO of A can reach the (empty) LUMO
of A' by two different schemes - as can be easily understood by assigning to B just two electronic levels, one occupied, the other
empty: a push-pull mechanism, one electron being released by A to
B and by B to A'; a pull-push mechanism, when first B releases an
electron to A', and then A yields an electron to B . Calculations
performed on a simplified model/5/in order to explain the dependence

of the through-space electron transfer on the energy of the HOMO
of B relative to the energy of the HOMO of A and of the LUMO of A'
give the expected result that the electronic transfer rate is greatest when the three levels in question have the same energy, if the
couplings of B and A' are small. On the other hand, if the coupling
is stronger (thus giving rise to a "level-shift"). The electron migration rate increases if the HOMO of B has a slightly higher energy that the states under consideration of A and A'. In general,
it must happen that, with due account of the coupling, the resulting
perturbed levels of A,B,A' are practically degenerate; or, if they
acquire some 'width' because of the perturbation, at least in part
the corresponding energy profiles occupy the same en rgy range.
Even with due account of electrostatic interactions, / 6/, this
condition places stringent requirements on the species B which should
form a bridge between two porphyrin macrocycles. the latter have
in fact fairly closely spaced energy levels; but their HOMO's and
LUMO's lie very low in energy with respect to some standard aromatic
like benzene, and, what is more important, their difference in energy should be small as compared to the first ionization potential
of benzene (ca. 9 ev.).

From the above consideration, it appears that, contrary to what
is considered almost obvious in many biochemical studies, the assisted through-space mechanism for electron transfer between porphyrin
macrocycles assuming any aromatic molecule as a bridge is in general quite unlikely. For instance, a benzene in a sandwich arrangement, as is well known, is an estremely stable electron system,
and even when it belongs to a larger conjugated system it shows
very little tendency to interact with other conjugated system in
the direction perpendicular to the ring. (As is well known, graphite
has a very weak conductivity in the direction perpendicular to the
'100) planes / 7/. Also typical organic conductors like poly-p-phe-
nylene, in addition to requiring doping, do not appear to conduct
transversally to the benzene ring).

A different form of assisted 'electronic' mechanism has been
suggested by some authors /8,9/. They have considered possible that
the protein backbone would provide the required electron relay, due
to the assumed existence of valence and conduction bands of that
backbone. They have suggested that, in spite of the high gap between the two bands a donor-acceptor ET could be assisted by the protein backbone if the donor (or the acceptor) level involved in ET
were in the energy gap, thus playing a role similar to that of impurities in semiconductors. An electron released to the conduction
band could then travel quite far, and travel from one residue to
another   along the protein backbone. This mechanism, however
attractive it may seem, is made unlikely by the fact that it seems
now established that the non-periodic perturbations resulting from
the residues in a protein chain disrupt the periodicity to such an
extent that the conduction band really consists of extremely narrow
sub bands or of isolated levels /10/, so that electron mobility
along the peptide chain is by far less probable than the analogy

with semiconductors would suggest. (Even the recently discovered
highly conducting polymers, the polyacetylenes, in spite of their
highly conjugated character require doping with very strong electron donors or acceptors in order to show a high conductivity.)
Thus, in agreement with others / 11/, we prefer to assign the main
peptide chain the rôle of a structure by which the groups participating in ET are more or less exposed or protected, and, more
important, held in the correct reciprocal positions.

b. Through-bond electron transfer.

We apply this mechanism assuming that the two hemes are initially
in the fully oxidized state. An electron taken up by the macrocycle of one of these prosthetic groups can move to another group by
a sequence of steps, the first one being transfer to one of the
imidazole rings co-ordinated to iron in positions 5 and 6. In particular, we can assume that in the oxidized form of the heme an imidazole ImH has lost a proton and become Im, so that the system is neutral
and has the structure of fig. 1b. The added electron passes to
the unprotonated imidazole which in turn takes up a proton giving the
now reduced structure of fig. 1a. At this point the transfer of
electrons to and from hemes can be attributed to two distinct mechanisms: imidazole pump and proton push-pull.

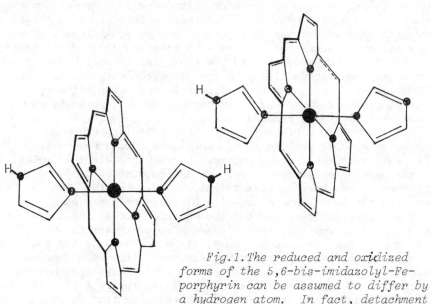

Fig.1.The reduced and oxidized
forms of the 5,6-bis-imidazolyl-Fe-
porphyrin can be assumed to differ by
a hydrogen atom. In fact, detachment
of a proton from the reduced form (a), followed by removal of an
electron, results in the oxidized form (b). The removal of the
electron is probably mediated by transfer from the imidazole ring
to the porphyrin macrocycle.

   The 'imidazole-pump mechanism' was proposed some years ago by
Urry and Eyring / 12/, but did not receive the attention it deser-
ved bacause of the lack of realism of some (otherwise unessential)
features of the original proposal. Some re-elaboration eliminating
those features provides a very elegant mechanism for the axial electron transfer through the iron atom of macrocycle. Consider
three heme groups A,B,C initially in the oxidized state and connected by a series of hydrogen bonds schematized by HX. Suppose
now that upon opening of the donor gate, an electron is initially
localized on the porphyrin ring of heme A. The imidazole pump mech
anism can then be represented by a series of steps as in fig.2:
   a) the three hemes (connected by hydrogen bridges) are all in the
oxidized state;then an electron enters the macrocycle A .
   b) the electron passes from the macrocycle to its unprotonated
imidazole ring. A fast electron rearrangement takes place, so that
the nitrogen which formed a covalent bond with iron now forms a
dative bond. The δ nitrogen of that ring takes a proton from
HX (with formation of an X^- ion) and
   c) X^- now yields its excess electron to the neighboring imidazole
of B, and takes up its proton. The ensuing fast electron rearrangement causes the other imidazole ring of B to abstract a proton
from the HX system connecting B to C, so that the situation of
fig.2c now holds. (Note again that the 'reduction' actually takes
place at X even if the essential role is played by the heme groups).
This electron transfer is strictly speaking the imidazole-pump step.
   d), e) these two steps are the analogs of steps b and c, but now
involve heme B and C. So that at the end we have the situation(not
shown), where C (actually, its X partner) is reduced.
   The 'proton -push-pull' mechanism / 13/ is already implied in
the above scheme, and need only be explicated in more detail when
the actual structure of HX is considered. Characteristic hydrogen
-bond chains involving peptide bonds and water molecules can be assumed to exist pairs of hemes. The two important elementary steps
are:
   i) the shift of the proton of a bond X...HY from the equilibrium
position adjacent to Y to that adjacent to X:XH...Y. This shift
of course implies that something happens either to X or to Y, so
that the latter situation is preferred. The switching process may
be assumed to start at one heme, which takes up one proton from one
hydrogen bridge in which it participates. This provides tha perturbation at X, which then results in capture of a proton;
   ii) X is normally a peptide link OCN near the oxidized heme; it
is in the tautomeric structure HO-C=N: which is stabilized by the
binding situation of the heme and by the presence of a neighboring
N...HN bridge. One link in the hydrogen bridge chain may be, as
has been said, a water molecule. If the reduction remains at a
given heme pair, then this molecule will turn into an hydroxyl group;
otherwise it will simply yield a proton on one side and capture one
on the other.

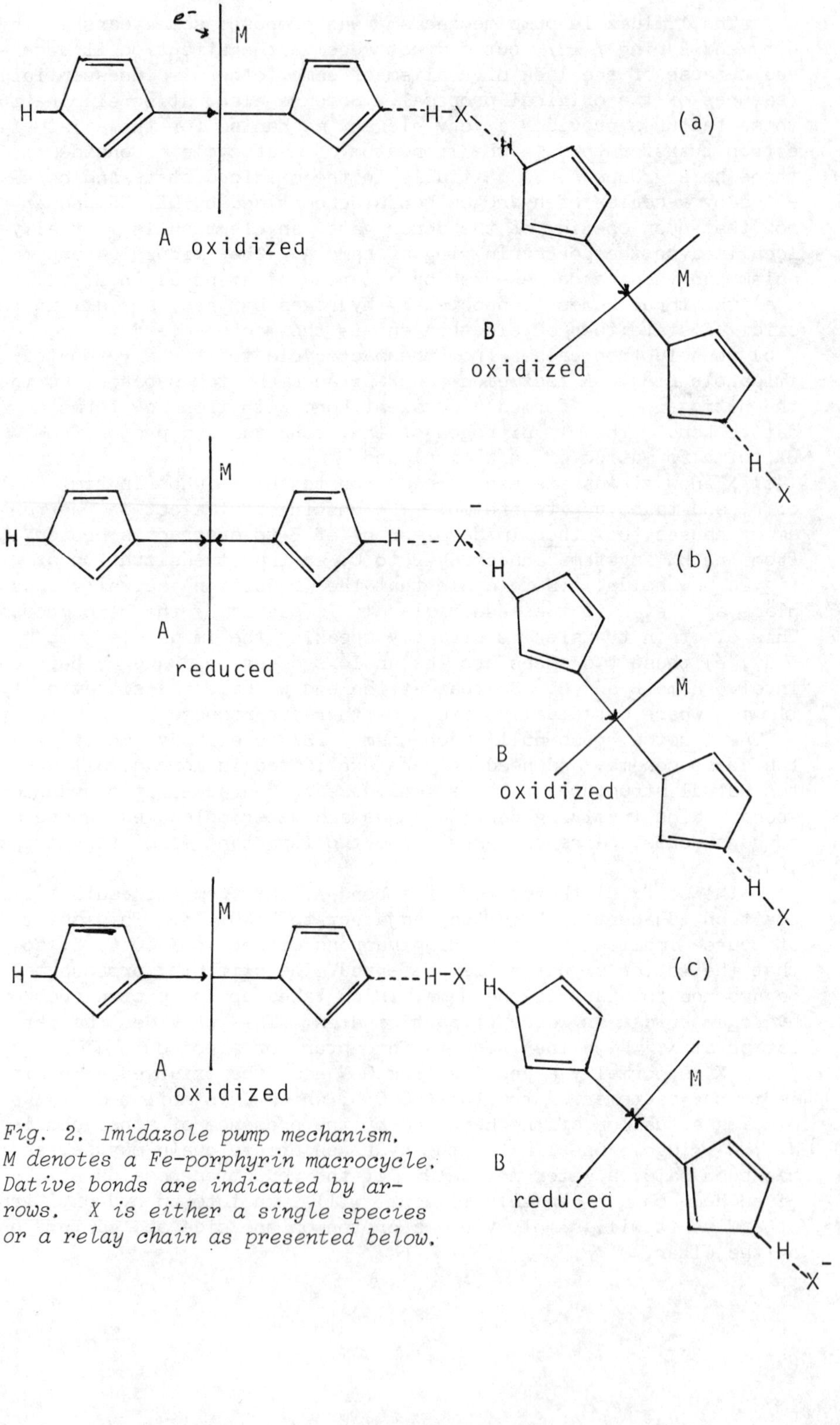

Fig. 2. Imidazole pump mechanism.
M denotes a Fe-porphyrin macrocycle.
Dative bonds are indicated by arrows. X is either a single species
or a relay chain as presented below.

2. The hydrogen bond

In order to gain a better understanding of the through-bond mechanism described above, let us discuss first of all the properties
of the H-bond. Much work by outstanding Authors has been devoted
to this molecular feature / 14/. We have studied it in connection
with the anomalous conduction in water, the model propsed for that
phenomenon being quite close to ideas advanced by Nagle,Mille and
Morowitz / 15/ concerning ion transport through biological membranes and other processes presumably involving ion transport.

It stands to reason that the hydrogen bond is associated,within
the frame of the Born-Oppenheimer approximation, with a potential
energy profile characterized by very steep walls near the heavy
nuclei (N or O, for our purposes) and a barrier up to one or two
tens of kJ/mole between two approximately equal minima.However
reasonable this intuitive picture may be misleading, for three reasons: (a) the energy profile under consideration may not be the
most significant one for the process under study; (b) the double
well is not necessarily a double well with a barrier, but is so
only when the distances of the heavy atoms are kept large by other
intramolecular or extramolecular constraints; (c) in most hydrogen
bridges the double well is normally asymmetric. This sounds obvious
in the case of NHO bonds, less obvious in the case of OHO or NHN
bonds. Actually, the asymmetry is less a consequence of the difference in the heavy atoms than of other factors, especially the
immediate environment of the H bridge.

The last point is most important for our purposes. When the
double well is symmetric, the probability that the proton will
occupy one of the wells is the same, and hence only the initial
conditions and the height of the barrier determine its actual probability of presence at any given instant of time. As has been
shown among others by Busch and De la Vega / 16/, even a slight
asymmetry causes the proton to collapse into the deeper well, regardless of its initial state. The reason for this behavior may
be understood from a two-state model: the mixing coefficient of
the two lowest localized states $|1>$ and $|2>$ is $2h/(\mathcal{E}_1-\mathcal{E}_2)$, with
$\underline{h}$ the coupling and $\mathcal{E}_1,\mathcal{E}_2$ the two energies; $\underline{h}$ tends to decrease rap-
idly when $|\mathcal{E}_1-\mathcal{E}_2|$ increases. This remark is very important, be-
cause it suggests that switching of H positions in H-bonds is effective when the double well passes in the course of time from shape
(a) to shape (c) of fig.3. This change in shape is only possible
if something happens to the partners. Examples are approach of another ion or reduction. The symmetric well (b) may or may not be
a stable situation.

The situation of fig.3 can be interpreted as stating that, although the symmetric well is clearly the step which allows transition
of the proton from say (a) to (c), actual proton or electron transfer takes place because the channel b→c is open, so that as protons
arrive to the right-hand well of (b) the system collapses into (c)
and there is ample time for the new system to yield its proton or

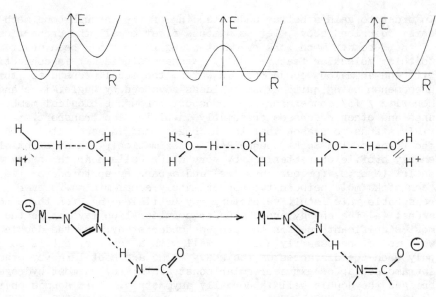

Fig. 3. Successive shapes of a double well during proton or electron transfer in a "charge-relay chain". Two typical examples are shown (M=porphyrin macrocycle).

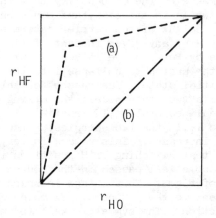

Fig. 4. Minimum-energy paths corresponding to stepwise (a) and concerted (b) mechanism for proton shift in the $NH_3 \ldots HF \ldots HOH_2^+$ chain. Not drawn are the contour lines representing the total B.O. energy for various values of the two coordinates. According to ref. 13, (a) is predicted by calculations when the NF and FO distances are kept equal to 3 A, (b) when those distances are 2.75 A.

electron to the surrounding medium.

Estimates of tunneling times for double wells of the $H_3O^+...H_2O$ type, where the barriers are of the order of 40 kJ/mole / 17/ give values between 10^{-12} and $10^{-10}s$/ 16/. (For recent accurate ab-initio studies of hydrogen bridges, see especially ref.18).

The mechanism shown in fig.3 immediately suggests a number of questions concerning chains of hydrogen bonds. In this section, we shall just remind the reader that a repetition of a switching process, such as the PT

$$H_3N...H-F...H-OH_2^+ \longrightarrow H_3N^{\pm}H...F-H...OH_2$$

studied by Elrod et al. / 13/, may follow either a stepwise or a concerted mechanism. This is not easy to decide; the minimum-energy path in the potential energy surface associated with changes in the HF and the HO distance (all other independent geometrical parameters being kept constant) may correspond to either possibility, depending on the NF and FO distances (fig.4). A criterion for deciding whether a PT process is concerted or stepwise is provided by a normal analysis: when there are either a mode involving simultaneous motion of both the mobile protons or two quasi-degenerate modes for their vibrations in their respective hydrogen bridges, then a concerted mechanism can be expected. This idea will be applied and briefly discussed below.

3. Types of charge-relay chains

Charge-relay through proton bridge chains in large system was suggested by Eigen / 19/ and Onsager / 20/. One fairly complete attempt to model such chains was provided by Nagle,Mille and Morowitz / 15, 21/ (which we shall refer to as NMM) and corresponds to the $(H_3O-H_2O)^+$ case of fig.3. The chain we have studied is different, and corresponds to the other example given in fig.3. As they illustrate different aspects of the same general problem, we shall briefly review both.

a. OH...O multiple chains.

The chains studied by NMM are sequences of OHO bonds as illustrated in fig.5, where five steps of proton transport from one solution to another are illustrated. When such H-bond chains are formed by side chains of proteins, they can function as 'wires', capable of relaying ions (or electrons) from one medium to another. NMM have also suggested that conformational changes of the protein may accompany such charge relay processes, thus providing a mechanism for conversion of chemical energy to mechanical energy.

The two most important characteristics of this possible mechanism are: (i) its efficiency; (ii) its rate. The efficiency is limited in principle because there will always be back leakage (even in the

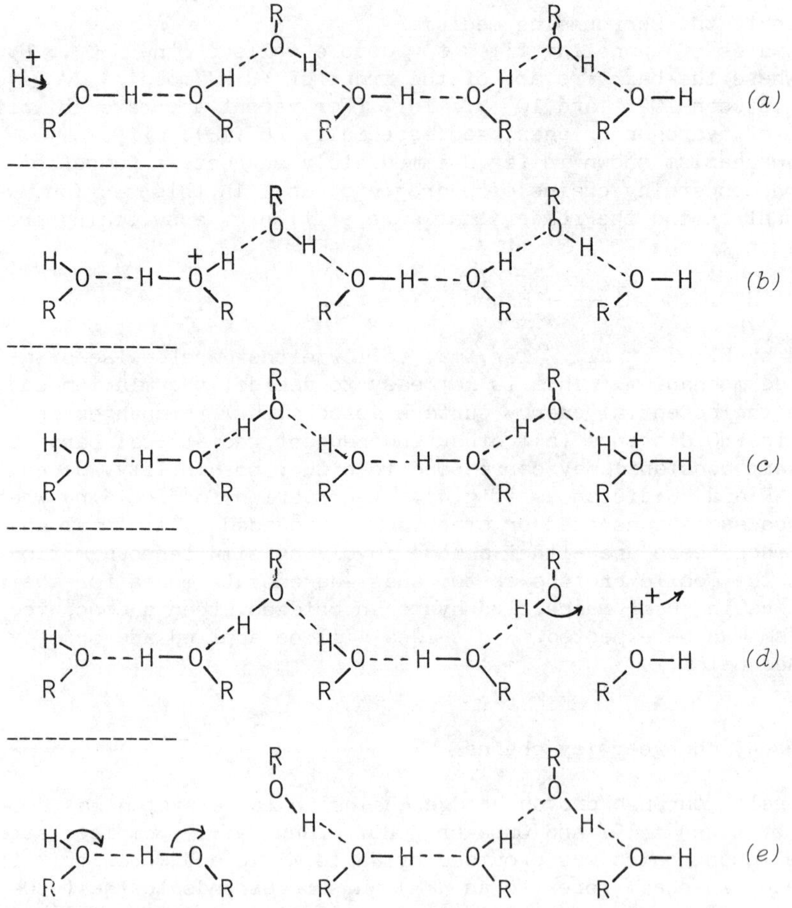

Fig. 5. A possible mechanism for proton transport along a chain which finally goes back to its initial state (according to NMM). It consists of left-to right switching of hydrogen bonds |steps (a,b,c)| followed by propagation to the left of a Bjerrum fault (proton vacancy). The latter involves 'rotation' of OH bonds. The R groups are amino-acid residues protruding from a single protein backbone, and ensure rigidity of the system.

presence of a driving force like a difference in pH) and intermediate states of the types (b) or (e) can perform local oscillations without ever getting to the appropriate end of the chain; in other words, the cycle (a) – (d) – (a) may be initiated without ever being completed. According to NMM, a free energy difference of ca. 12 kJ/mol between (d) and (a) is sufficient to ensure 99% efficiency.

As concerns rates, they depend of course on the free energy barriers to be overcome in the direct and reverse processes.

A brief review of the calculation of the rate of the forward process (proton transfer) will serve to illustrate the physical scope and limitations of theNMM model.

Suppose there are n-1 identical hydrogen bridges (and hence n proton sites); and let the various states be numbered according to the site to which the proton has hopped starting from the left--hand site 1. The process

$$\ldots H\text{-}\overset{+}{\underset{R}{O}}\text{-}H\ldots\overset{}{\underset{R}{O}}\text{-}H\ldots \qquad\longrightarrow\qquad \ldots H\text{-}\overset{}{\underset{R}{O}}\ldots H\text{-}\overset{+}{\underset{R}{O}}\text{-}H\ldots$$

j-th state   (j+1)-th state

requires a constant free energy change, $F_{j+1} - F_j = \epsilon$, for j=2 3,...,n-1. If a_+ is the rate of one-step forward hopping, and a_- the corresponding rate of backward hopping,the following relation must hold

$$z = a_-/a_+ = \exp\left(\epsilon/kT\right). \tag{1}$$

Let now $p_j(t)$ be the probability that the system is in the j-th state. As this probability can be interpreted as a concentration - because we assume that there is thermal equilibrium, the general kinetic equation is

$$\frac{dp_j}{dt} = a_- p_{j+1} + a_+ p_{j-1} - (a_- + a_+)p_j ; \quad 1<j<n \tag{2a}$$

for j=1 there is no hopping from state p_0, and for j=n there are two special rates: the rate b_- of hopping from the solution (state n+1) to the chain, and the rate $\bar{a}_+$ of hopping from the chain to the solution. Than eqns (2a) must be completed by

$$\frac{dp_1}{dt} = a_- p_2 - a_+ p_1 \tag{2b}$$

$$\frac{dp_n}{dt} = b_- p_{n+1} + a_+ p_{n-1} - (\bar{a}_+ + \bar{a}_-)p_n \tag{2c}$$

$$\frac{dp_{n+1}}{dt} = \bar{a}_+ p_n - b_- p_{n+1} \tag{2d}$$

where, of course

$$\frac{b_-}{\bar{a}_+} = e^{-(F_n - F_{n+1})/kT}$$

The special assumptions can be made that protons cannot go back from the (n+1)-th state (say, the right-hand solution) to the n-th state and hence $b_-=0$; that rate $\bar{a}_+$ is practically equal to a_+;

and that $a_+ + a_- = f$ is a constant equal to $10^{11}s^{-1}$ (from ice), so that from eqn 1, $a_+ = 10^{11}s^{-1}(1+z$

$$a_+ = 10^{11}s^{-1}(1+z) \qquad\qquad (3)$$

Eqns (2) can be solved by well-known Laplace-tranform techniques. NMM suggest an approximate form of solution which assumes that at all the chain sites quasi-equilibrium holds, so that

$$p_j(t) = C(t)/z^j \qquad ; \qquad (1<j<n) \qquad\qquad (4)$$

and, as the sum of all the p_j's, including p_{n+1} must be unity at any time, we obtain

$$C(t) = (1-p_{n+1}(t))\ \frac{z-1}{z^n-1}\ z^n \qquad\qquad (5)$$

Substitution of eqns(3) and (4) into eqn (2d) (with $b_-=0$, $\bar{a}_+ = a_+$) yields:

$$\frac{dp_{n+1}}{dt} = C(t)/z^n\cdot\bar{a}_+ = k_+(1-p_{n+1}) \qquad\qquad (6)$$

where k_+ is the rate of formation of the (n+1)-th state. Explicit computation with $\varepsilon = 1.2$ kJ/mol, n=20, kT=2.5 kJ/mol, gives $k_+ = 1.1\ 10^6s^{-1}$, an estimate in good agreement with the exact solution of eqns (2) for the same choice of parameters and initial conditions.

b. Intramolecular H-bond chains over peptide bonds.

The preceding application of the H-bond charge-relay idea indicates the kind of information one would like to have about such a chain and the natureand number of the parameters needed to obtain that information. It will be noted, however, that neither the shape of the double well nor the quantum aspects have been specifically intraduced; and the overall process studied has been assigned a stepwise mechanism to which statistic governed by thermal effects have been applied. Some appreciation of the features not taken into explicit account in the preceding subsection can be obtained by a study of the intramolecular chain already discussed in sec. 1.

The chains studied by NMM were suggested by the problem of describing charge relay between two solutions, e.g. through a cell membrane. The example of heme-heme ET is different, in that the two solutions are replaced by porphyrin macrocycles, and the 'driving force' is the electrostatic created by a charged macrocycle. Moreover, there is no reason why the backbone peptide bonds should not participate in the charge-relay process.

We shall specifically model the chain represented by X in fig.2 (cf.also the lower example in fig. 3) shown in full in fig. 6.

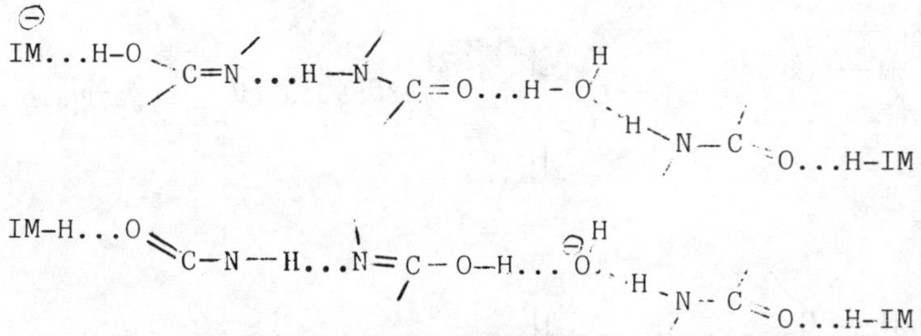

Fig. 6. A possible H-bond relay chain between the imidazoles of
two heme in a cytochrome c_3. The chain is shown with the reducing
electron on the first imidazole and on the bridging water molecule, which has lost a proton. The final stage, when the right-
hand imidazole has lost its proton, is not shown.

The construction of a suitable model is a difficult research
problem in itself. The chain shown in fig.6, which might function
as an 'electron wire' at least between certain heme pairs of multiheme cytochromes, has a very intriguing characteristic, namely that
it consists essentially of three peptide bonds and one water molecule; and two of the peptide bonds face each other by their N's,
so that one of them will always be in the enol form. Although
further discussions (as are being conducted with R. Haser and coll.
in Marseille/22/) are required to ascertain that this is actually
a valid mechanism in concrete cases, it is interesting to discuss
it from a general theoretical viewpoint. Indeed, we have studied
it on simpler schemes (fig.s 7,9) in order to obtain some quantitative information without exceedingly difficult computations.
    In our model N-methylacetamide has been used to model a peptide
unit. This is a better choice than formamide, as has been pointed
out by Del Bene/23/. Experimental studies have shown that this
molecule is planar/24/, and bond lengths confirm that there is a
high degree of resonance between the two forms

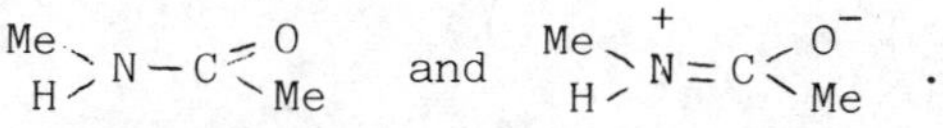

Because of this resonance, the NH group shoud behave like the NH
group of pyrrole, and hence its lone pair is not available for
hydrogen binding. This is why we have assumed that the right-hand
form can lose a proton to a neighboring species.
    In the most complete of our series of calculations, we have sim-
ulated the two imidazoles by OH and OH_2 groups (Fig. 7), trying
to preserve the most interesting feature of fig. 6, namely the
eneol-ketone transition of the peptide bond and the water molecule
serving as a relay. The simulation of the initial reduced species
by OH^- demands special caution because of the high stability of
the corresponding oxidized species (water), but many features of
the model can be considered as general ones without danger.

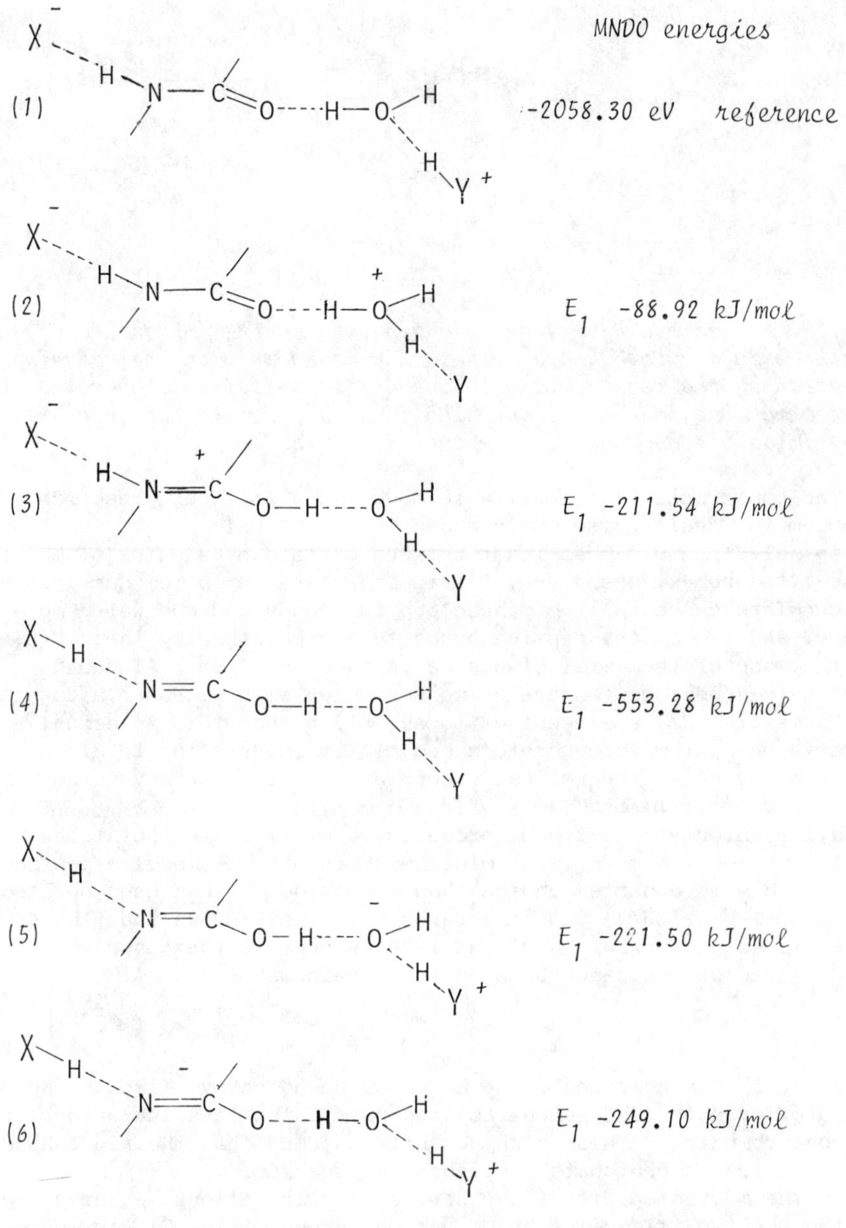

MNDO energies

(1)   −2058.30 eV   *reference*

(2) E_1 −88.92 kJ/mol

(3) E_1 −211.54 kJ/mol

(4) E_1 −553.28 kJ/mol

(5) E_1 −221.50 kJ/mol

(6) E_1 −249.10 kJ/mol

Fig. 7. Charge-relay chain involving a peptide bond. X and Y are the heme imidazoles in the chain discussed in sec. 1 (cf. fig. 6), but they have been simulated here (MNDO clcns.) by OH^- and OH_2, respectively. This explains why the chain where recombination has taken place (chain 4) is by far the most stable one. Reminder: to obtain kcal/mole divide by 4.184!

The process in which the species represented in fig.7 participate
is essentially a recombination process. The important feature of
recombination is that the electron - released, say, by removal of
a proton from water (HX) at the left end of the chain - is not removed at the other end, so that the free-energy change associated
with release to the oxidizing species need not be taken into acce-
unt. Let HP stand for the peptide group, P'H for its enol form,
X^- for OH^-; then the sequences of steps suggested by inspection of
the energies given in fig.7 is as follows:

Table 1.

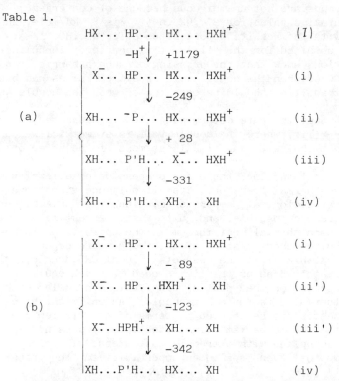

$$HX...\ HP...\ HX...\ HXH^+ \qquad (I)$$

$$-H^+ \downarrow\ +1179$$

$$X^{\bar{.}}...\ HP...\ HX...\ HXH^+ \qquad (i)$$

$$\downarrow\ -249$$

(a)

$$XH...\ ^-P...\ HX...\ HXH^+ \qquad (ii)$$

$$\downarrow\ +28$$

$$XH...\ P'H...\ X^{\bar{.}}...\ HXH^+ \qquad (iii)$$

$$\downarrow\ -331$$

$$XH...\ P'H...XH...\ XH \qquad (iv)$$

$$X^{\bar{.}}...\ HP...\ HX...\ HXH^+ \qquad (i)$$

$$\downarrow\ -89$$

$$X^{\bar{.}}..\ HP...HXH^+...\ XH \qquad (ii')$$

(b)

$$\downarrow\ -123$$

$$X^{\bar{.}}..HPH^{\bar{.}+}...\ XH...\ XH \qquad (iii')$$

$$\downarrow\ -342$$

$$XH...P'H...\ HX...\ XH \qquad (iv)$$

*Table 1. Electron transfer (a) and proton transfer (b) as pos-
sible paths for charge recombination after removal of a proton
from species I. Energy changes are given in kJ/mole.*

This table shows that, as expected, recombination is highly favored,
the first step being more likely when the negative charge travels,
as is reasonable. Of course, the differences in energy are significant as concerns our problem only if the barrier associated to
each bridge does not slow the transfer down so much that equilibrium
considerations are unrealistic. Investigation of the barriers has
resulted in no barrier for the $HO^-...HN$ bridge. This result is in
contrast with the existence of a barrier in $H_3O_2^-$ (40 to 75 kJ/mole
according to 4.31 G and PCILO calcns.,resp./25/) and a barrier of
$\sim$ 65 kJ/mole in $NH_4^+OH^-$ as predicted by MNDO. However, the very

fact that the MNDO method does not consistently give a zero barrier
for the NH...OH⁻ bond suggests that the properties of the peptide
bond are responsible at least for an important lowering of the barrier. Of course, a more accurate investigation will be necessary;
for the moment we can retain that the peptide bond shows a significant propensity to yield the proton linked to nitrogen to a negative partner; the resulting RN=CR'-O⁻ anion seems to be quite stable
at least in the H-bond chain under study here.

The results obtained so far concerning the barriers associated
to the other H-bonds are not unambiguous because of convergency
problems. A tentative value for P⁻..HX in (ii) is 82 kJ/mole, in
good agreement with the results reported above for O⁻..HO. A similar result is obtained for the (iii)⟶(iv) barrier. Tunneling-
-effect calculations show that the crossing of such a barrier requires less than 100 ps when the double well is symmetric: as has been
mentioned, the transition should be much faster when the double well
is unsymmetric.

We conclude that the rate determining step in sequence (a) of
table 1 is (ii)⟶(iii),where the double well is symmetric: we can
safely assume that the switching time is at most of the order of
10-100 ps.

In addition to tunneling, the other process which allows transfer is of course thermal transfer. Even if only one vibrational
ste is occupied at biological temperatures, transitions via higher
vibrational states are possible, and quite fast. An estimate of the
time needed to reach thermal equilibrium in the presence of tunneling effects requires consideration of the temperature dependent
density matrix $\exp(-\hat{H}/kT)$: in the case of an ideal OHO double well
it gives approximately 5 ps as the time needed to reach equilibrium.
Therefore, the assumption that proton transfer is controlled even
in our case by thermal effects is reasonable. Actually the hypothesis that each step can be treated in terms of equilibrium constants - as in the NMM mechanism illustrated above - does not apply
here, because of coupling with extremely fast steps. Therefore,we
can safely assume that even 5 ps is an upper limit for the transfer
time.

As concerns the efficiency of the transfer, the high difference
in energy (if MNDO calculations can be trusted) between the initial and the final state guarantees that it will be very high: once
a proton has crossed the (ii)⟶(iii) barrier the system collapses
into (iv).

As a last point concerning the system of fig.7, we have to consider the nature of the transfer mechanism. in fact, the mechanism
discussed so far is a stepwise one; but also a concerted one is possible in principle. A possible answer to this question requires
a brief mention of spectroscopic properties.

We have calculated the normal frequencies of chain (i) of fig.
7 using an empirical choice of the force constants. Most values
are easily assigned to stretching and bending vibrations of the various groups and agree with expected values. The most interesting

frequencies are $\nu_{41} = 3486$ cm^{-1}, $\nu_{38} = 2275$ cm^{-1}. The first value corresponds to in-phase vibrations of the two $O..._1O$ hydrogen bond (56% and 38% respectively);a frequency of 3416 cm^{-1} corresponds to the NH stretching vibration. The second value, also accompanied by a very close frequency (2067 cm^{-1}) corresponds to more complicated vibrations involving the three hydrogens of the bridges as well as other atoms.

The interest of these values, especially of the first pair lies in the inference that slightest coupling between the corresponding modes will give rise to a concerted motion. In particular, the vibration of the two OHO bridges appears to be simultaneous already at the normal level, while coupling with the NH...O bond should be highly effective due to the small energy denominator (~ 65 cm^{-1}).

We conclude that a concerted transfer mechanism in the system of fig.7 is quite reasonable in view of its vibrational characteristics. The next question is whether such a mechanism is also favored by energy considerations. An estimate of the barrier has been made by MNDO calculations by allowing the three protons involved in H bonds to move simultaneously along the corresponding bonds by equal distances. The shape of the overall double well is represented in fig.8. The barrier is of the order of 30 kJ/mole, a very low value. Thus, the concerted mechanism seems a valid alternative to the stepwise one in the model system of table 1. Nevertheless, we believe that this conclusion should not be generalized, because, as the chain to be considered becomes longer and the environment more varied, the degeneracies are likely to disappear or stay limited to local groups of bonds.

It remains to discuss models other than recombination. We have studied several models, but here we shall confine ourselves to the most pertinent one, the Im-P-HIm system shown in fig.9.

The stepwise mechanism proceeds from a minimum MNDO energy of -261.370×10^3 kJ/mole to a final energy of -261.269×10^3 kJ/mole both barriers being approximately 200 kJ/mole for the forward process and 100 kJ/mole for the reverse process. It thus seems that the forward electron transfer process involves a barrier system reaching ca. 250 kJ/mole above the initial state and overall energy differences of 100 kJ/mole in favor of the 'initial' state: a voltage difference of 1 V could make the forward process spontaneous. The given values are subject to revision due to the crudeness of the MNDO method; at any rate, when compared with the values of fig.7, they suggest that the intervention of water in the chain of fig.6 is very important, because it lowers the barriers involved in the transfer process. Inspection of fig.6 also shows that the present results strongly support the idea that electron transfer takes place starting from an imidazole which faces a peptide bond in its enol form. Objections based on the remark that the enol form of the peptide bond is not observed in NMR studies can be met on two grounds: (i) the peptide bonds in question are one or two out of hundreds; (ii) they could actually be just transient forms restored to the keto form after electron transfer (we owe this comment to D.Urry).

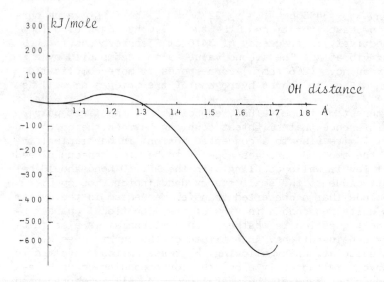

Fig. 8. Shape of the double well associated to the potential energy profile for the concerted mechanism of process (a) or (b) of Table 1. The hydrogens of the H-bridges are assumed to move to the left by equal fractions of the appropriate distances.

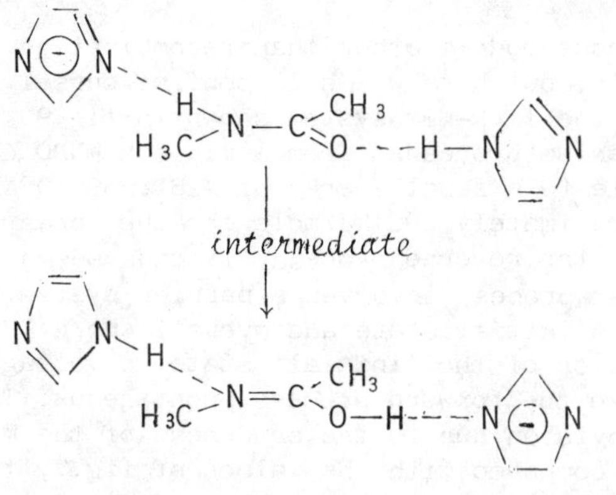

Fig. 9. Model system to study electron transfer from one heme imidazole to another via a peptide bond which passes from ketonic to enolic form (and viceversa). To clearly realize the limits of this model cfr. the full system of fig. 6, where there are peptide groups in either form at all stages, plus an interposed water molecule, which alone helps to lower the barriers.

c. Input and output data : Tables 2,3,4.

Table 2

(i) N-methyl-acetamide and water
distances. CC:1.53; CN:1.32, 1.47; CO:1.24; CH 1.04^a; NH: 1.02^b;
OH: 1.00^b.

angles. $\hat{CCO}$:121; $\hat{CCN}$:114; $\hat{OCN}$:125; $\hat{HCH}$(meth):110.9^{c,d} ;HOH:109.5.

(ii) imidazole^{e,f}
distances. $N_1 \cdot C_2$:1.35; $N_1 H_1$:1.04; $C_2 N_3$:1.31; $C_2 H_2$:1.08; $N_3 C_4$:1.38;
$C_4 H_4$:1.06^a; $C_4 C_5$:1.38; $C_5 H_5$:1.06; $C_5 N_1$:1.38.

angles. $N_1 \hat{C}_2 N_3$:112.4; $C_2 \hat{N}_3 C_4$:105.4; $N_3 \hat{C}_4 C_5$:109.8; $C_4 \hat{C}_5 N_1$:105.6;
$C_5 \hat{N}_1 C_2$;106.8; $H_1 \hat{N}_1 C_2$:126.7; $H_2 \hat{C}_2 N_3$:124.5; $H_4 \hat{C}_4 C_5$:129.1; $H_5 \hat{C}_5 N_1$:122.9.

(iii) hydrogen bridges^b
N...HO:2.74; O...HO:2.68; $\hat{CO}$...H:120; N(imid)...HN:2.9, N(imid)...
...HO:2.6.

Table 2. Geometrical input data. *Distances are in A, angles in de-*
grees. Except when other wise specifiesd, data are taken from: L.
Pauling: The Nature of the Chemical Bond (III ed). Ithaca, N.Y.:
Cornell U.P. 1960, p. 282.
Notes. A. assumed. b. From: S. Scheiner, C.W. Kern: J.Am.Chem. Soc.
99, 7042 (1977). c. From: Interatomic Distances. London: Chem.Soc.
1965. d. Methyl groups in eclypsed position. e. The imidazoles and
the NMA molecule of fig.'9 are assumed to lie on perpendicular planes.
f. As suggested by Scheiner and Kern, from: M.S.Lehmann, T.'F.'Koetzle,
W.C. Hamilton: Int.J.Pept.prot.Res. 4, 229 (1972).'

Table 3

(i)   amolecular force constants (mdyn/A)
K(C=O) = 8.5^b; K(C-N) = 5.5^b; K(C-CH$_3$) = 3.4^b; K(N-CH$_3$) = 3.7^b;
K(O-H) = 8.26^a; K(NH...) = 5.80^c; K(OH...) = 6.90^d;
H(HOH) = 0.98^a; H(O=C-N) = 0.35^b; H(CH$_3$-C=O) = 0.30^b ; H(N-C-CH$_3$)=
= 0.30^b; H(H-N-C) = 0.35^b; H(C-N-CH$_3$) = 0.20^b; H(CH$_3$-N-H) = 0.15^b

B.O.P.($^H_{Me}$ N-C) = 0.10^c;   B.O.P.($^{Me}_{Me}$ C=O) = 0.50^c; T(N-C) = 0.39^c;

F(H-O-H) = 0.40^a; F(N-C-O) = 1.50^b; F(CH$_3$-C=O) = 0.50^b

(ii) Intermolecular force constants (mdyn/A)
K(O-H...O) = 0.36^d; K(NH...C) = 0.13^c; H(N-H...O) = 0.04,

H(H...O–H) = 0.02^C; H(O...H–O) = 0.04^C; H(C–O...H) = 0.02^C; F(H–O...
...H) = 0.01^C; F(C=O...H) = 0.008^C

Table 3. Force constants adopted for normal mode calculations.
a. *L.H.Jones, R.R.Ryan: J.Chem.Phys.52, 2003 (1970)*
b. *T.Hiyazawa, T.Simanouti, S.Mizushima: J.Chem.Phys. 29,611 (1958)*
c. *K.Itoh, T.Simanouti: J.Mol.Struct. 41, 1554 (1964)*
d. *K.Nakamoto, S.Kiohida: J.Chem.Phys. 41,1558 (1964)*

Table 4

Computed experimental frequencies (cm^{-1}) for chain (1) of table 1
(Fig.7)

Theoretical	Experimental	Normal Mode
3974	3942	$1/\sqrt{2}(\Delta R(7-14) + \Delta R(7-13))$
3924	3832	$1/\sqrt{2}(\Delta R(7-14) - \Delta R(7-13))$
3897	3832	$\Delta R(4-5)$
3841	3832	$\Delta R(12-15)$
3486	3400	$0.8\,\Delta R(3-4) + 0.6\,\Delta R(6-7)$
3483	3400	$0.6\,\Delta R(3-4) - 0.8\,\Delta R(6-7)$
3417	3200–3300	$\Delta R(11-12)$
2275	–	$0.2\,\Delta\alpha(6,4,3)+ 0.2\,\Delta\alpha(7,6,4) +$ $0.70\,\Delta\alpha(13,7,6) - 0.6\,\Delta\alpha(14,7,6)$ $-0.4\,\Delta T(14,7,13)$
2271	–	$0.65\Delta\alpha(14,7,6) - 0.77\Delta\alpha(14,7,13)$
2067	–	$\Delta\alpha(5,4,3)$
1715	1690	$-.35\,\Delta R(1,2) -0.28\,\Delta\alpha(9,1,8) +$ $0.42\,\Delta\alpha(11,9,1) - 0.27\,\Delta\alpha(11,9,10)$ $-0.4\,\Delta T(10,9,1,8) - 0.64\,\Delta T(11,9,1,2)$
1569	1663	$0.55\,\Delta\alpha(7,6,4)+0.46\Delta\alpha(13,7,6)+$ $\Delta\alpha(14,7,6) + \Delta\alpha(14,7,13)$
1516	1550	$0.66\,\Delta\alpha(11,9,1) - 0.66\Delta\alpha(11.9.10)$
1276	1250–1300	$0.30\,\Delta R(1-9) - 0.25\,\Delta\alpha(8,1,2)+$ $0.25\,\Delta\alpha(9,1,2) + \big(\Delta\alpha(11,9,10)+$ $-\Delta\alpha(11,9,1)\big) - 0.56\,\Delta T(11.9,1,2)$
1213	–	$0.36\Delta\alpha(11,12,15) - 0.45\,\Delta B(9.1,10,11)+$ $-0.82\,\Delta T(11,9,1,2)$
972	1096	$-0.48\,\Delta R(1-8) + 0.78\,\Delta R(9-10)$ $0.37\,\Delta\alpha(8,1,2)$
953	881	$-0.30\,\Delta\alpha(9,1,2) + 0.42\,\Delta\alpha(15,12,11) +$ $-0.53\,\Delta B(9,1,10,11) + 0.55\,\Delta T(10,9,1,8)+$ $-0.42\,\Delta T(11,9,1,2)$
834	725	$-0.44\,\Delta R(1,2,8,9) - 0.88\,\Delta T(10,9,1.8)$
608	627	$-0.25\,\Delta R(1,8)-0.5\,\Delta\alpha(8,1,2)+0.45\Delta\alpha(9,1,2)$ $+0.65\,\Delta\alpha(3,2,1)$
466	–	$0.4\,\Delta\alpha(3,2,1) + 0.9\,\Delta\alpha(5,4,3)$
433	–	$-0.34\,\Delta\alpha(9,1,8) + 0.6\,\Delta\alpha(6,4,3) +$ $-0.5\,\Delta\alpha(15,12,11) + 0.41\Delta B(9,1,10,11)$

Theoretical	Experimental	Normal Mode (table 4,ctd.)
399	–	non assignable
347	–	$0.81 \, \Delta\alpha \, (6,4,5) + 0.59 \, \Delta\alpha \, (7,6,4)$
288	–	$-0.40 \, \Delta R(4-6) - 0.90 \, \Delta\alpha \, (7,6,4)$
278	–	$0.95 \, \Delta\alpha \, (15,12,11) - 0.25 \, \Delta B(9,1,10,11)$
271	–	$0.93 \, \Delta\alpha \, (15,12,11) - 0.2 \, \Delta B(9,1,10,11)+ -0.2 \, \Delta\alpha \, (7,6,4)$
255–238–184	–	
170–169–121		
84–81–32–27		

Table 4. Notes. 1.The numbering of the atoms is as follows

2. For the geometry and the input force constants of table 2
3. The predicted frequencies of chain 4 of fig.7 are the same as those of chain 1, except for: $\nu(N-C) = 1538$, $\nu(C=O) = 1700$, $\nu(O-H)= = 3540$
4.$\Delta R(i,j)$ is the variation of the internal stretching coordinate defined by the i-j bond; $\Delta\alpha(i,j,k)$ is the variation of the internal bending coordinate defined by the i-j-k valence angle; $\Delta T(i,j,k,r)$ is the variation of the internal torsional coordinate defined by dihedral angle i,j,k,l; ΔB is the out-of-plane bending internal coordinate.

4. Bibliographical note and acknowledgments.

The class of problems to which these lecture notes refer was
first suggested to us by R.Haser, head of the protein-structure
group of the CRMCC-CNRS, Marseille. He was trying to clarify the
functions of the four hemes of a cytochrome C_3 on the basis of
structural data and other evidence. The proposal of the H-bond
relay chain was elaborated by G.Del Re and C.Minichino (Naples)
and discussed extensively with R.Haser. Computations on model systems, including computations not mentioned here, have been carried
out in collaboration with A.Peluso (Naples) who prepared the program and selected the force constants. The information about thermal
transfer in the double well is taken from the thesis of G.Villani
(Naples). The complete papers are in preparation.

As concerns bibliography , we emphasize that the literature
on our subject is immense. A lot of work was based on the theories
of R.A.Marcus /26,27/ in regard to PT as well as ET. Jortner and
his collaborators (especially Bixon and Jortner /28/) have produced several formal analyses of ET in a quantum mechanical frame
although few applications have been made. The same topic has been
the subject of extensive work by Dogonadze and his collaborators
/29/. We have not covered those studies for lack of space, but
emphasize their importance. We also emphasize that the reference
given are but a very scanty selection of the relevant papers.

The original work to which reference is made in these notes was
undertaken in 1981-82 when the A. was a guest professor at the University of Marseille 3 (CRMCC-CNRS,Campus Luminy) and has been continued as a project sponsored by the Italian National Council of
Research (CNR). Support of the French Ministry of Education as
well as CNR is therefore gratefully acknowledged.

References.

1.   D.V. Der Vartanian, J. LeGall: Biochim. Biophys.Acta 346,79-
      -99 (1974)
2.   E.Heilbronner, A.Schmelzer: Helv.Chim.Acta 58,936 (1975)
3.   R.R. Dogonadze, J.Ulstrup, Yu.I.Kharkarts, J.Theoret.Biol.40,
      259-277 (1973)
4.   iidem: ibidem, p.279-283 (1973)
5.   L.J.Root,M.J.Ondrecher: Chem.Phys.Lett. 88, 538-42 (1982)
6.   G.Del Re: J.Chem.Soc. (Far.Trans.2) 77, 2167 (1981)
7.   A.R.Ubbelohde, F.A.Lewis: Graphite and its crystal componds.
      Oxford:Clarendon P., 1960
8.   A.S.Davydov: Phys.Stat.Sol. (b) 90, 457,464 (1978)
9.   E.G. Petrov: Int.J.Quantum Chem. 26, 137-152 (1979);
10.   M.Seel:Chem.Phys. 43, 103 (1979);J. Ladik, these Notes.
11.   C.E.Castro: J.Theor.Biol.33,475-490 (1971)
12.   D.W.Urry, H.Eyring:J.theoret.Biol. 8, 198-213 (1965)
13.   J.P.Elrod, R.D.Gandour, J.L.Hogg,M.Kise, G.M.Maggiora, R.L.

Schowen, K.S. Venkatasubban: Symp. Far. Soc. "Proton "Transfer" $\underline{10}$, 145-153 (1975)

14. cf. lectures by D.Hadži, this volume, p. 37 and 59.

15. J.M.F. Nagle, H.J.Morowitz, Proc. Natl.Acad.Sci. (USA) $\underline{75}$,298--302 (1978)

16. J.M.Busch, J.R.De La Vega: J.Am.Chem.Soc. $\underline{99}$,2397-2406 (1977)

17. A.Laforgue, C.Brucena-Grimbert, D.Laforgue-Kantzer,G.Del Re, V.Barone: J.Am.Chem.Soc. $\underline{86}$,4436 (1982)

18. P.Hobza, F.Mulder, C.Sandorfy: J.Am.Chem.Soc. $\underline{103}$,1360 (1981); $\underline{104}$, 925 (1982)

19. M.Eigen, L.De Maeyer: Proc.Roy.Soc.(London)A247,505-533 (1958)

20. M.S.Chen, L.Onsager, J.Bonner,J.Nagle: J.Chem.Phys. $\underline{60}$,405--419 (1974)

21. J.F.Nagle, M.Mille, H.J.Morowitz: J.Chem.Phys. $\underline{72}$,3959-3971 (1980)

22a. cf. R.Haser. M.Pierrot, M.Frey, M.Payan, J.P.Astier: in "Structural Aspects of Recognition and Assembly in Biological Macromolecules" (eds.Balaban $\underline{et}$ $\underline{al.}$) p.213-223 Rehovot (Philadelphia, PA.):I.S.S. 1981

22b. M.Pierrot, R.Haser, M.Frey,F.Payan,J.P.Astier:J.Biol.Chem.1983

23. J.Del Bene: J.Am.Chem.Soc. $\underline{100}$,1387 (1978)

24. L.Radom et.al., Aust.J.Chem.,$\underline{25}$, 1601 (1972)

25. cf.Ref.s 16-17

26. R.A.Marcus: J.Chem.Phys.$\underline{24}$,966 (1956);27.idem:JACS $\underline{72}$,891(1968)

28. M.Bixon, J.Jortner: Far.Disc.Chem.Soc. $\underline{74}$, 17-29 (1982)

29a. cf.R.R.Dogonadze, A.M.Kuznetsov: Prog.Surf.Sci. $\underline{6}$,1 (1975)

29b. A.M.Kuznetsov: Far.Disc.Chem.Soc. $\underline{74}$, 49-56 (1982)

HYDROGEN BONDING - THEORETICAL AND SPECTROSCOPIC ASPECTS

D. Hadži

LEK - Chemical and Pharmaceutical Works and
Boris Kidrič Institute of Chemistry. Ljubljana,
Yugoslavia

ABSTRACT. The electronic and vibrational theories of the hydrogen bond are reviewed. High level quantum mechanical treatments
of small systems which explain and reproduce well the experimental facts have led to the development of simpler models now used
in the treatment of biological and other large systems. The
application of H-bond theories to spectral phenomena of H-bonding
are also briefly reviewed.

In 1957, Coulson said (1) "... it would be very nice if by
the use of wave-mechanical methods we could calculate the energy
of a H-bond as a function of the positions of the atoms and molecules involved. But a moment's consideration shows that this
is not possible, and is unlikely to be possible for several years,
if ever ...". The H-bond theories developed in the meantime have
not only lead to the fullfilment of Coulson's desire but also to
the possibility of calculating most of the characteristics of
H-bonded systems. Since in a short review only the essential
achievements can be presented, the more interested reader should
consult for details the monographs (2) and the original articles
given in the References.

ELECTRON THEORY OF THE H-BOND

The interaction between a proton donor A-H and an electron
donor B (often termed proton acceptor) results in the formation
of a geometrically defined, reversible complex and a plethora of
changes in molecular characteristics with respect to the isolated
molecules. The task of the theory is to rationalize the origin

39

C. Sandorfy and T. Theophanides (eds.), Spectroscopy of Biological Molecules, 39–60.
© *1984 by D. Reidel Publishing Company.*

and extent of these changes in terms of the preexisting characteristics of isolated molecules and of the electronic rearrangements induced by the interaction.

Both the perturbation theory and the variational SCF MO
theories have been applied to the treatment of H-bond phenomena.
The former is not as universally applicable as the SCF MO theory
and most of theoretical work on H-bonds is done within the framework of the latter. Computations at any level of sophistication
were done on many simple H-bonded systems so that now the limitations and defficiencies of various semi-empirical and restricted
basis ab-initio methods are well known (3,4,5,10). The role of
the electron correlation energy has also been examined and it was
found that it is quite important in the calculation of force
constants (5) and of potential barriers in very short hydrogen
bonds (6, 24).

It is useful to start the electronic description of the H-
-bond by the analysis of the interaction energy. Several schemes
have been developed and perhaps the easiest to follow is that of
Morokuma (7-9).

Like in other schemes it starts with the antisymmetrized
wave functions of separate molecules $\tilde{A} \Psi_A^0$ and $\tilde{A} \Psi_B^0$ where $\tilde{A}$ is
the antisymmetrizer. The difference between SCF energy E_{AB} of the
complex, described by the wave functions $\tilde{A} \Psi_{AB}$, and the sum E_0 of
the expectation values of Ψ_A^0 and Ψ_B^0: INT $= E_{AB} - E_0$ represents
the interaction energy without the contribution of dispersion
forces. A series of model functions $\Psi_1, \Psi_2, \Psi_3, \Psi_4$ with
expectation values E_1, E_2, E_3, E_4 is introduced:

$$\Psi_1 = \tilde{A} \Psi_A^0 \cdot \tilde{A} \Psi_B^0, \qquad E_1$$

$$\Psi_2 = \tilde{A} \Psi_A \cdot \tilde{A} \Psi_B, \qquad E_2$$

$$\Psi_3 = \tilde{A} (\Psi_A^0 \cdot \Psi_B^0), \qquad E_3$$

$$\Psi_4 = \tilde{A} \Psi_{A \to B^x} \cdot \tilde{A} \Psi_{B \to A^x}, \qquad E_4$$

and the differences $E_i - E_j$ are defined as contributions of the
electrostatic, polarization, exchange and charge transfer energies.
Thus ES $= E_1 - E_0$ represents the Coulombic interaction between the
undistorted charges of A and B and PL $= E_2 - E_1$ is the contribution
due to mutual polarization. No electron exchange is allowed between A and B in the functions Ψ_1 and Ψ_2. This is introduced by
the antysymmetrized product function Ψ_3 and the exchange polarization energy EX is then $E_3 - E_1$. The charge trasfer CT is represented by Ψ_4. The difference between the total SCF energy and
the sum $E_1 + E_2 + E_3 + E_4$ represents the coupling energy MIX.
Thus the SCF interaction energy is:

INT = ES + PL + EX + CT + MIX

The dispersion energy is beyond the HF-SCF theory and has to be
calculated separately. It is quite important, particularly in
weaker bonds (10). The definition of the energy components through
the model wavefunctions is arbitrary and the values of the calculated energies depend on the quality of computation of the wavefunctions. For instance, the zero differential overlap schemes
do not yield EX at all and the apparent correctness of the computed interaction energy is due to the compensation of this defect
by the way certain Coulomb integrals are calculated. The incomplete basis sets exaggerate the electrostatic component and also CT
which is particularly notable with the STO-3G set (9). However,
these shortcommings affect rather the quantitative side of the
energy decomposition than the general trends which emerge from
the analysis of numerous examples by this and other schemes including the application of the perturbation theory. The main point
is that the major contribution to the stabilization of H-bonds
originates in the electrostatic attraction of the unperturbed
charge distribution. A substantial contribution comes also from
CT and, less,from PL. The latter two contributions are largely
counterbalanced by EX so that eventually ES is the dominating
stabilizing component.

Table 1. Examples of energy (kcal/mol) decomposition
         (data from Morokuma, ref. 2c, p. 46)

Complex	$R_e{}^a$(Å)	INT	ES	EX	PL	CT	MIX
$H_3N...NF$	2.68	-16.3	-25.6	16.0	-2.0	-4.1	-0.7
$H_2O...HF$	2.62	-13.4	-18.9	10.5	-1.6	-3.1	-0.4
$H_3N...HOH$	2.93	-9.0	-14.0	9.0	-1.1	-2.4	-0.4
$H_2O...HOH$	2.88	-7.8	-10.5	6.2	-0.6	-2.4	-0.5
$F...H...F^-$	2.29	-62.7^b	-91.2	65.4	-7.4	-34.9	-16.1

^{a}optimized intermolecular geometry, 4-31G basis set
^{b}The energy 21.6 kcal/mol necessary to stretch the H-F monomer
distance from 0.922 Å to 1.146 Å in the (FHF)$^-$ ion is substrac-
ted from INT (= -84.3 kcal/mol) obtained without H-F deformation.

Instructive is the dependence of these components upon distance
between A and B. Two examples are shown, one for a weaker complex
(acetamide-water, Fig. 1a) and one for extremely strong, symme-

trical H-bonding (FHF⁻, Fig. 1b). Whereas the (negative) ES, PL
and CT decrease with decreasing distance EX (positive) increases
thus leading to a minimum in the total energy of interaction at

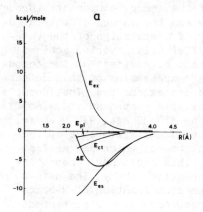 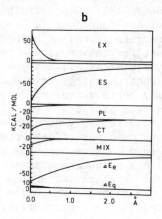

Fig. 1. Dependence of the energy components upon intermolecular
distance. a: acetamide-water complex (ref. 11), b: FHF⁻ (simplified after ref. 12).   E and E_e correspond to INT in the text.

equilibrium distance. The gradients are different so that at
large distances (weak complexes) ES is relatively more important
than CT and PL, but the latter and, particularly, CT rapidly gain
at shorter distances and in stronger complexes. Very important
is also the angular dependence of the energy components. This is
illustrated by the example of the water dimer (Fig. 2) which
shows that the curve for ES closely follows that of total energy.

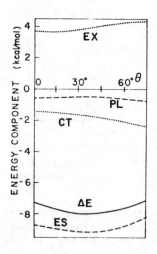

Fig. 2. Dependence of the
energy components of the
water dimer upon the torsional angle Θ between the
planes of the H_2O molecules
(from ref. 7).

The curves for CT and EX have opposite trends so that eventually the angular dependence of ES is decisive for the mutual orientation of the interacting molecules. This explains the success of the calculations of the geometry of H-bonded complexes considering only the electrostatic interactions (13,14,15). Clearly, in such calculations the AH...B distance has to be fixed since ES has no minimum. The importance of the electrostatic component lead to the development of electrostatic models of the H-bond which are useful for predicting the mutual orientation and interaction energy. This will be dealt with in the next paragraph.

   Although the interaction between the uperturbed charges dominates the energy and the mutual orientation most of the other H-bond phenomena are connected with the charge rearrangements. It is possible to calculate electronic densities according to each of the above defined wave functions (16), but it is the overall effect that is eventually comparable to various experimental data. From the many calculations the general trend can be schematically presented:

$$H_B \xrightarrow{\hspace{1cm}} B \ldots H \xrightarrow{\hspace{0.8cm}} A \xrightarrow{\hspace{0.5cm}} H_A$$

The charge transfer resulting from the donation of electrons from bonding orbitals of H_B to the nonbonding orbitals of H-AH and vice versa is a net donation to $H-AH_A$ (CT in Table 2) and it increases with the strength of H-bonding. There is an increase of

Table 2. Examples of charge distribution changes[a] and dimerization energies (6-31G basis set, 17)

$H_B-B\ldots HAH_A$	H_B	B	H	A	CT	E_D[b]
$H_3N\ldots H-NH_2$	0.026	-0.010	0.054	-0.041	0.016	2.94
$H_2O\ldots H-OH$	0.032	-0.007	0.042	-0.056	0.025	5.64
$H_3N\ldots H-OH$	0.054	-0.032	0.063	-0.064	0.022	6.48
$H_2O\ldots H-F$	0.048	-0.018	0.032	-0.062	0.030	9.20
$H_3N\ldots H-F$	0.087	0.054	0.044	-0.078	0.033	12.20
$(H_2O\ldots H-OH_2)^{+c}$	-0.084 -0.255	0.089	-0.0841	0.1236	-	-

[a] Negative values (in electrons) indicate the charge density increase relative to monomer values
[b] in kcal/mol, dimerization energy
[c] 4-31G basis set (18)

the electron density at B which occurs at the expense of H_B. The
bridging proton looses charge whereas A-H gains it mostly on A.
Some figures obtained on dimeric hydrides and collected in Table 2
illustrate this (17). However, in positive ions with very strong
H-bonds such as $H_5O_2^+$, $H_3F_2^+$, $HCl-HFH^+$ and $H_2Cl_2^+$ there is an increase of charge density at the central proton (18). The changes
in the charge distribution accentuate the polarity existing in
the separate molecules. For the formation of the H-bond, A has to
be more electronegative than hydrogen and B possesses one or more
lone electron pairs. π electrons in double bonds and aromatic
rings may also act as electron donors, but they give rise to weak
bonding only. The result of the charge redistribution is reflected in a larger dipole moment of the complex than given by the
vectorial sum of the dipole moments of the component molecules
(19). An important consequence of the increased charge on, for
instance, the proton donating oxygen in water is that it becomes
a better proton acceptor. On the other hand, the increased polarity of the electron donating water molecule makes it a better
proton donor. The result is the cooperative effect, i.e. the
H-bond in the linear water trimer is stronger than in the dimer
(20). The effect is attenuated in interrupted chains, e.g. the
2:1 water-imidazole complex (21).

The charge redistribution is graphically shown by the charge density difference plots obtained by subtracting the charge
density of the H-bonded complex from the sum of the charge densities of the isolated molecules with the nuclei at identical positions. An example of medium strong H-bond is shown in Fig. 3.

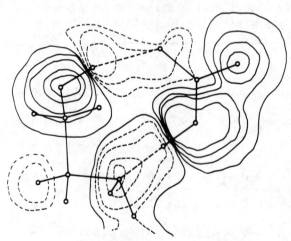

Fig. 3. Charge density difference plot of the ethanolamine-formic
acid complex. Full line: increased density (+), dotted: decreased
(-). Contours at: -0.01,-0.005,0,+0.005,+0.01,+0.02 e^-/a_o^3.

In general, besides the charge density increase on B, A and H_A, and a decrease on H and H_A there is an increase in the region of the H-A bond and the region between H and B so that in fact there is an alternation of increase and decrease in the region of the H-bond.

```
      d   i   d   i   d    i d i
      H - B ...... H - A - H
```

The charge density distribution can experimentally be obtained from X-ray or combined X-ray and neutron diffractions (22). Although the X-N deformation density maps are obtained from crystals and hence the density deformation has a component from the polarization by the surrounding molecules the bonding effects are dominant and the theoretically calculated charge distributions are at least qualitatively comparable to the experimental ones. Interesting results have been obtained with sodium hydrogen diacetate which is representative of symmetrical H-bonds. In the X-ray deformation density maps there is a continuous region of electron density extending along the H-bond from one oxygen to the other similarly as in covalent bonds. although the density region is more diffuse (23). This result corresponds to the orbital energy analysis of $H_5O_2^+$ (18) which shows a strong mixing and net energy lowering of the bonding orbitals of both monomers upon complex formation, and shifting of the proton into the central position. The mixing places more charge in the H-bond giving it some covalent character in the sense of a three center. four electron bond. However, this is pronounced only in the case of the extremely strong H-bonds.

What can be said about the nature of the "normal" H-bonds? This question has been often discussed (e.g. 2b,3.9). In fact, it is very difficult to delimitate H-bonding from other molecular interactions by comparing any single characteristic of the H-bond with the corresponding one in other types of complexes besides the fundamental fact that the proton is involved. Comparisons of the contribution of various interaction energy components of complexes in which the proton is not directly involved with typical H-bond complexes do not reveal essential differences. A good example are the ClF...HF and FCl-FN complexes of which the former only is H-bonded, and is slightly less stable (24). Its interaction energy of -2.5 kcal/mol decomposes (9) into: ES = -2.7, EX = 3.0, PL = -0.8, CT = -2.3 and MIX = 0.2 kcal/mol. The energy of FCl-FH (-3.2 kcal/mol) is decomposed into ES = -3.0. EX = 1.7, PL = -0.2. CT = - 1.9. and MIX = 0.1 kcal/mol. The directional properties of the H-bond, particularly the tendency of the proton to be on the A...B line are often said to be specific, but this results from the properties of the H-bonding molecules. For the other types of complexes definite mutual orientations are specifics on their own, e.g. the orientation of the I_2 molecule along

the sixfold axis of the benzene ring in the charge-transfer complex. What really makes the H-bond unique are the phenomena connected with the potential function governing the proton dynamics.

THE POTENTIAL ENERGY FUNCTION

     For the understanding of the vibrational spectra of H-bonded
systems and of the phenomena connected with proton transfer it is
necessary to consider the energy changes with proton and also
with the heavy atom displacements.

     The one-dimensional proton potential function obtained by
calculating the energy for stepwise displacements of the proton
from the equilibrium position at fixed positions of the heavy
nuclei is sufficient for many purposes. The comparison between
the functions of free A-H and weakly H-bonded groups shows a slight
displacement of the energy minimum and hence of the proton away
from A, and a flattening of the curve. Fig. 4a shows the example
of the water dimer. With stronger H-bonds, e.g. between an amine
and a carboxylic acid, Fig. 4b, a second energy minimum will appear.

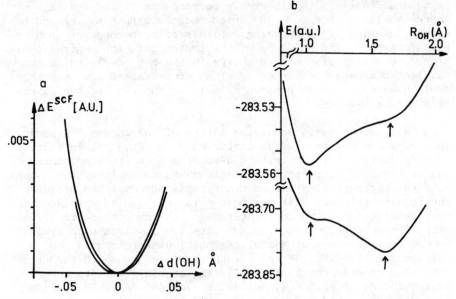

Fig. 4. Computed proton potential functions. a: water dimer (from
G.H.F. Diercksen, Theoret. Chim. Acta 21, 335 (1971)). b: methyl-
amine-formic acid dimer. Upper curve: computation "in vacuo",
lower curve: with surrounding charges (see text and ref. 32).

The occupation of the second minimum by the proton corresponds to the protonated form $A^-...H-B^+$ and the difference between the energies of both minima determines the equilibrium $AH...B \rightleftharpoons A^-...H-B^+$, if separated by a potential barrier from the first one. In homoconjugate ions, $(AHA)^+$ and $(BHB)^-$, the potential function will have either two equal minima separated by a potential barrier or a central single minimum. For instance, calculations on $(H_3N...H...NH_3)^+$ show at the equilibrium distance a barrier (25) whereas in $(F...H-F)^-$ and $(OH_2...H...OH_2)^+$ (26) there is only a single minimum. However, the barrier characteristics and, ultimately, the shape of the potential functions are very sensitive to the quality of computations and even correlation energy has an important influence on the calculated parameters of the potential (6).

The potential energy function depends strongly on medium influences. Extensive model calculations on the $H_5O_2^+$ potential function have shown how easily it is distorted into asymmetrical shapes under the influence of the electric field of other ions and by the mutual influence of neighbour H-bonds (27,28). In solids such effects are obvious and from X-ray and, particularly, neutron diffraction investigations it is known that the H-bond of chemically identical homoconjugate ions, e.g. hydrogen maleate, may be of the symmetrical (29) or asymmetrical (30) type depending on the overall crystal symmetry.

For biological mechanisms and those involving proton transfer in general the dependence of the potential function upon the medium is particularly important. For instance, in the gas phase most amines and carboxylic acids form neutral, H-bonded complexes whereas in polar solvents they are ionic. Model calculations show that by adding water molecules to the carboxyl and amine moieties the potential function is modified in the sense that the energy minimum corresponding to the ionic form $A^-...HB^+$ becomes deeper (Fig. 4b). The calculations can be also made by placing appropriate dipoles around the complex that will eventually stabilize the ionic form (31,32).

In double minimum potentials with small barriers the proton is tunneling between both minima with a rate that depends upon the barrier parameters, i.e. the barrier height and the distance between the minima. With physically acceptable barriers (i.e. 7 to 9 kcal/mol) the tunneling rates of 10^{10} to 10^{12} sec^{-1} are easily obtained (33,34) and this mechanism provides also the possibility for protonic conductivity in extended H-bonded systems such as ice (35), boric acid and some metal hydroxides (36). The mechanism involves defects such as H_3O^+ ions or water in the case of hydroxydes, and proton hopping between symmetrical or nearly symmetrical potential minima:

Since the molecules must rotate to accept a new proton the conductivity process involves activation energy which is not needed
for tunneling. The conductivity then depends on the rate of rotational diffusion of the defects. The importance of proton conductivity for biological mechanisms will be discussed in the second
lecture.

The potential of short, intramolecular H-bonds, e.g. in the
enol form of malonaldehyde, has been extensively treated in order
to answer the question whether there is a central, single minimum
or is the potential of the double minimum type. Recent computations (37) agree with experimental evidence (38) that in this and
related, more complex intramolecular H-bonded systems (39) the
potential is of the latter type allowing for rapid proton tunneling.

For the interpretation of dynamic properties of H-bonds,
particularly of vibrational spectra, it is necessary to use multi-
dimensional potential surfaces that govern the A...B stretching
and the deformation vibrations. The computation from first principles is possible only for the simplest systems e.g. HF_2^- (40).
However, the most interesting vibrational features are connected
with the coupled motions of the A-H...B and A-H...B coordinates,
i.e. the high and low frequency stretchings. The corresponding
potential surface of H_5O_2 has been computed by Janoschek et al.(27).
The coupling of the A-H and A...B stretchings generates a
large number of vibrational energy levels in the high frequency
region besides those predicted by the one dimensional potential
function.

Potential functions of electronically excited states are of
considerable interest in the possibility of photoinduced proton transfers and their possible role in mutations (41). Electronic spectra,
particularly luminescence, reveal that the acid and base properties
in the electronic excited states of H-bonding molecules may be drasti-
cally changed (42). However, there are few good quality computational
studies of potential functions in electronic excited states. Morokuma
and Iwata have applied their electron-hole potential theory to the
excited states of the formaldehyde-water complex (43) and the formic
acid dimer (44) calculating the potential curves for the intradimer
proton transfer in various excited states. Such calculations are
particularly demanding because of the necessary geometry optimization of the excited states. The importance of the latter was
recently shown on the formaldehyde-water system. Whereas previous
calculations on water-carbonyl group containing systems indicated

weakening and even absence of H-bonding in the excited state (45)
persistence resulted when the formaldehyde geometry was optimized (46).

MODEL TREATMENTS OF HYDROGEN BONDING

For the treatment of interactions in large systems such as
the proteins and nucleic acids even the less demanding MO schemes
are prohibitive in the supermolecule approach particularly when
the minimum energy arrangement is sought. More economic models
had to be developed. The model treatments can roughly be divided
into two groups. The first is the outgrowth of the older electrostatic, point charge models of the H-bond and is upgraded with
the modern possibillities to compute   reliable wave functions
and charge distributions for the separate molecules or at least
for the basic units of polymers and consider the coulombic part
only of the interaction energy. They are therefore less suited
for the treatment of the dynamic aspects of H-bondind including
the bulk properties of H-bonded liquids and solids in which the
proton potential function has the leading role. For this purpose
model potential functions have been developed. They are based
either on ab-initio calculated pair potentials or on suitably pa-
rametrized functions used in the treatment of spectra of diatomic
molecules. We shall successively consider both types of H-bond
models.

In the molecular wave function based electrostatic model of
H-bonding the starting point is the charge distribution function
of each molecule $\varsigma_i(\vec{r}, \vec{R})$ obtained from the molecule's wave
function. Here $\vec{r}$ denotes the position at which the density is
evaluated and $\vec{R}$ defines the location and orientation of the charge distribution. The charges originate both from nuclei and electrons. The electrostatic interaction energy E_{es} between molecules A and B may be written:

$$E_{es}(A,B) = \int V_A(\vec{r}_1, \vec{R}_\alpha) . \varsigma_B(\vec{r}_2, \vec{R}_\beta) \, dr_2$$

where $V_A(\vec{r}_1, R_\alpha)$ is the electrostatic potential of the molecule A:

$$V_A(\vec{r}_2, R_\alpha) = \int \frac{\varsigma_A(\vec{r}_1, \vec{R}_\alpha)}{|\vec{r}_1 - \vec{r}_2|} \, dr_1$$

The product of V_i with a point charge q placed at $\vec{r}$, q. $V(\vec{r})$, is
an energy and is often used for the representation of the potential (MEP) in the form of iso-energy curves, putting q = e = 1 a.u.
These MEP maps show the lowest energy path of approach of electrophilic or nucleophilic reagents to the molecule. For example, the
valleys of the negative potential mark the sites of protonation

and the values of V(r) near the nitrogen atom in amines roughly
correlate with proton affinities (47). However, in order to reproduce the trends in proton affinities in the series of alkyl-
amines MEP is not adequate and polarizability effects have to be
considered (48). The charge density distribution function ϱ (r)
may be expanded in terms of multipoles (usually limited to quadru-
poles) centered at appropriate points and the electrostatic interaction is then calculated as in classical electrostatics as
the total interaction energy between two sets of multipoles. This
procedure is used in the search of the lowest energy mutual
arrangement of interacting molecules as e.g. in the ethanolamine-
-formic acid complex (31). For biopolymers, e.g. DNA, the calculation of the wave function and MEP for the complete molecule
is not feasible and therefore procedures have been developed to
compute the MEP maps from a superposition of the MEP's of fragments (49). Such maps are very useful as a guide for tracing the
path of approach and placing of water molecules, ions and charged
entities.

The electrostatic model of H-bonding has been extensively
elaborated by Kollman (50,51) in the sense of searching simple
methods of representing the charge distributions which can qualitatively reproduce the magnitude and directional properties of
accurately computed MEP and be used as a guide in predicting the
relative energies and mutual arrangements of molecules in complexes. Significant correlations have been obtained between the
electrostatic potential predictions and the results of reliable
ab-initio calculations on a considerable series of representative
simple molecules. Simple expressions of the form

$$E_{int} = k \; POT \; A \; x \; POT \; B$$

result. Here, POT A and POT B are the electrostatic potentials
of the proton donor and electron donor computed at defined points
using contributions from partial charges calculated to reproduce
the accurately calculated MEP. This complements the earlier
approaches to reproduce the enthalpies of complex formation from
empirically derived parameters characteristic of the component
molecules (52), but has the advantage of being able to predict
also the geometries. Allen's model which is also based on good
quality ab-initio calculations offers even more extensive interpretive and predictive possibilities that include besides the interaction energy and directionality the internuclear separation,
the stretching force constants and infrared intensities (53). The
key quantities appearing in this model are the bond dipole of the
proton donor, the difference between the ionization potentials of
the electron donor and of the noble gas atom in its row and the
length of the H-bonding lone pair of the electron donor. The inclusion of the ionization potential accounts in this model also
for the charge transfer.

In the treatment of dynamic aspects of H-bonding, particularly of the vibrational spectra and special solid state properties such as ferroelectricity and proton conductivity empirical potential functions play an important role. The most used is the Lippincott-Schroeder (54) model function which has the form:

$$V_{A-H...B} = V_1 + V_2 + V_3 + V_4$$

V_1 and V_2 are Morse type functions:

$$V = D_o \left\{ 1 - \exp \left[-n \ (r-r_o)^2 \ / \ 2r \right] \right\}$$

parameterized in V_1 with the dissociation constant D_o, equilibrium internuclear distance r_o and n, a parameter related to the ionization potentials of A of the free A-H group, and in V_2 correspondingly for the H...B part. V_3 is an exponential term accounting for the Van der Waals repulsion between A and B, and V_4 accounts for electrostatic attraction between A and B. This function was successful in rationalizing the A-H frequency shifts and their temperature dependence in terms of the geometric and energy parameters (55). Recently, two back to back Morse type functions were used to interpret the vibrational spectra and the phase transitions of H-bonded ferroelectrics and of some other solids with short H-bonds (56).

H-bond potentials are essential in quantitative models used to explain and predict bulk properties of associated liquids, crystals and interactions between biopolymers, water and active smaller molecules. The potentials used in Monte Carlo type calculations are derived from high quality ab-initio calculations on small molecules (57) whereas for the molecular mechanics type calculations empirical potential functions (e.g. 58) are used. The energy is usually expressed as a sum of Van der Waals, electrostatic and Morse components and the parameters are chosen so as to fit the vibrational spectra of simple compounds, experimental energies of association etc. A recently developed potential for the O-H...O bond was successfully tested on the well known example of water dimerization (59).

HYDROGEN BOND EFFECTS ON SPECTRA

H-bonding affects any type of transitions and the differences between the spectra of monomeric and associated molecules are not only indicators of orbital energy changes and charge redistribution, but are also used to determine thermodynamic and kinetic parameters of association (for a compillation of data see ref.2b). We shall briefly review only the most utilized types of spectrometries - infrared and Raman, nuclear magnetic resonances (NMR) and electronic absorption. However, microwave spectrometry should

at least be mentioned as a source of precise data on the geometries of H-bonded complexes in the gas phase (60), of their dipole moments and deuterium quadrupole coupling constants. Electron
paramagnetic resonance is a tool for studying H-bonded radicals
equivalent to NMR for closed shell systems (61).

(i) Vibrational spectra

Infrared spectra (62) offer probably the most universal and
fastest way of demonstrating the presence of H-bonds and estima-
ting their strength by observing the difference $\Delta \nu$ between the
wavenumber of the free and bonded A-H stretching. Raman spectrometry is in principle equivalent to infrared, but in practice the
A-H bonds may be more difficult to observe because of weakness
$\Delta \nu$ may have any value between those observed with non-specific
medium effects up to 3000 cm^{-1} found with very short H-bonds as
in e.g. acid salts of carboxylic acids (63). The shift of the
A-H stretching band to lower wavenumbers is readily explained in
terms of the electron density redistribution which reduces the
electric field gradient around the proton. The computation of the
force constants with such precision as to match the observed frequencies is quite demanding (64), but the trend to the red shifting
is reproduced already with simpler methods. Since both the equilibrium positions of the nuclei as well as the vibrational energy
levels are determined by the potential surface, relationships
between the A-H distance lenghtening, A-B shortening, and the force
constants, correlations between the pertaining quantities (2b,62)
can be established and extended also to the H-bond energy. Most
interesting for practical purposes are the relations between $\Delta \nu_{A-H}$
and the A...B distance, and with the bonding energy, because they
permit the estimation of two important characteristics the direct
determination of which is much more difficult, from one easily
accesible datum, $\Delta \nu$. An example of the former is shown in Fig.5.
The linear relationship between $\Delta \nu$ and the bond enthalpy $-\Delta H^o$ has

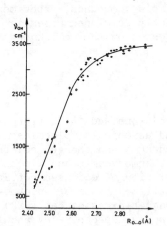

Fig. 5. Plot of infrared O-H frequency shift vs. O...O distances
(after ref. 65).

been proposed already in 1937 (66), but it has to be restricted
to series of H-bonds by the same proton donor to different electron donors. Different proton donors exhibit different slopes (67).
$\Delta \nu$ of standard proton donors are used to estimate the basicity
of proton acceptors (68). The relations between A-H and A...B distances are established from numerous crystal structure data (69).
Badger's rule set up in 1934 (70) for covalent bond force constants
has been revived by Allen's theoretical model (53) in the form:

$$K_{AB} (R - d_{AB})^3 = 1.86$$

to relate the ab-initio calculated force constants K_{AB} for the
A...B stretching with the A...B distance R. The row independent
parameters d_{AB} assume values of 1.00, 0.80 and 0.55 for groups
5, 6, and 7.

     Besides the red shift of νAH there is a strong increase in
infrared intensity. This intensity enhancement is, in fact, one
of the most conspicuous manifestations of the charge rearrangement
on H-bond formation and is connected with the change in the charge transferred on A-H elongation.

     A characteristic of the H-bond is the presence of non-linear
terms in the dipole moment (electrical anharmonicity) i.e. the
simple relationship $I \approx (\partial \mu / \partial q)^2$ is not valid. The role of electrical anharmonicity is experimentally demonstrated in the diffe-
rence between the infrared and Raman band shapes and in the anomalous ratios of the OH to OD infrared intensities (71).

     The third band characteristic that changes on H-bonding is
the band width. The width at half band height is roughly proportional to $\Delta \nu$ and may amount to several 100 cm^{-1}. It is usually
accompanied by the development of a complex band contour (Fig. 6).
The explanation of the band shaping mechanisms was for years one
of the most controversial subjects and a large amount of experimental work has been done in support of various theories. The modern theories (71,72) developed particularly consistently for
the gas and liquid phases incorporate much of the older hypothe-
ses in a more rigorous form permitting even the reproduction of
experimental spectra by using independent parameters. Common to
all theories is the anharmonic coupling between the high νAH and
the low frequency AH...B stretching. In other words, the potential
governing the A-H stretching is "modulated" by the AH...B motion.
The coupling of both motions gives rise to a number of combination
bands which may appear as a fine structure of the ν A-H band. The
interaction of the ν AH mode with combinations of internal modes
is an additional mechanism giving rise to subsidiary peaks and
also to regions of higher transmission commonly observed in the
spectra of crystals with short H-bonds. In the case of liquids
the band shaping mechanism includes also the interaction of the

slowly vibrating subsystem (A...B stretching) with the surrounding (73). In solids, interactions with low frequency phonons have
to be considered (74).

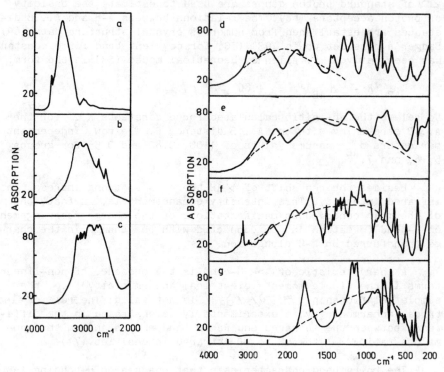

Fig. 6. Examples of ν A-H infrared bands with increasing H-bond
strength. a - liquid trifluoroethanol (l), b - trichloroacetic
acid (s), c - vinylimidazole (s), d - dichloroacetic acid - DMSO
complex (l), e - CsH_2PO_4 (s), f - 1-carboxyimidazole (s), g -
- $KH(CF_3COO)_2$ (s).   Dotted line in d-f indicates the probable
ν AH band without interference of other bands, (s) - solid,
(l) - liquid.

    A particularly striking broadening effect is observed in
aqueous solutions of strong acids and bases, as well as in aprotic
solutions of organic acids and bases with appropriate difference
in pH_a. The infrared spectra of many such systems show what is
called continous absorption, i.e. an absorption background extending from 2000 cm^{-1} into the far infrared on top of which more
or less broadened individual bands are superimposed (75). Common
to all these systems are symmetric or nearly symmetric double minima of potential energy separated by low barrier. Theoretical
calculations on simple representatives of such H-bonds, particularly on $H_3O_2^+$, have shown that such systems are highly polarizable

and that the potential shapes and hence the energy levels are
very susceptible to the effects of surrounding electric fields.
The numerous energy levels created by the anharmonic coupling of
the high- and low frequency H-bond vibrations and by tunnelling
become smeared out by the fluctuating electric fields thus creating an effectively continuous absorbtion (27). This type of spectra has been observed also in many biologically relevant systems
and has been taken as an indicator of H-bonds with double minimum
potentials with low barriers and proton tunnelling that warrant
proton translocation with low energy of activation (76).

H-bonding affects also the other vibrational modes. The in
plane deformation. A-H$\int$...B and the out-of-plane deformation,
A-H̄...B̃ shift to higher frequencies. The latter mode only is a
characteristic one and may be used for correlations with the
strength of H-bonding (62). The low frequency modes of the H-bond,
i.e. the AH...B stretching (77) and various bendings are difficult
to separate from other low frequency modes. particularly in solids.
Therefore they are not widely used for the characterization of
H-bonds. although they are important for the A-H stretching band
shaping by anharmonic interactions.

(ii) Nuclear magnetic resonance

The charge redistribution owing to H-bonding is reflected in
various NMR parameters of the directly involved nuclei and, to
various extents, also on the more distant ones. The dynamic effects of H-bonding affect the relaxation parameters. Most exploited in H-bond studies is the ^{1}H chemical shift which is to the
low field corresponding to the depletion of the electronic charge
on H-bonding. The shift is linearly related to the H-bond energy
for chemically related proton donors (78). The developments of
the multipulse techniques which allow high resolution measurements
on polycrystalline solids have made possible the determination of
the anisotropy of the shielding and it was shown that $\Delta\sigma$ (difference between the parallel and the perpendicular component) varies
with the strength of H-bonding (79). The influence of H-bonding
on the spin-spin coupling (J) comes in mainly through the increase of the A-H bond length and in FHF$^-$ the value of J_{HF} decreases
by 20 % (80). The study of H-bonded complexes with fast proton
exchange is complicated by the loss of the fine structure generated by spin-spin coupling owing to the averaging by fast proton
movements. At temperatures near 110 K, however, the exchange is
sufficiently slowed down and well resolved spectra are obtained
in mixed solvents (81). Thus it is possible to investigate the
separate spectra of ion pairs and neutral acid-base complexes.

Perhaps the most instructive probe of H-bonding is the ^{2}H
nucleous by virtue of its quadrupole moment. The quadrupole coupling constant DQCC and the assymmetry parameter η reflect the

changes in the electric field gradient which is created by all
nuclei and electrons in the molecule and also by its neighbours.
However, the effects of the latter are small in comparison with
the changes due to H-bonding charge rearrangements. Empirical
correlations of DQCC with the O...O distance (82), and with the
O-H force constant have been established (83) and theoretically
corroborated (84,85). The DQCC's span a large amplitude between
50 kHz (for symmetrical H-bonds) and 314 kHz (free HOD). The temperature dependence of the spin-lattice relaxation times gives
informations on ^{2}H motions and such measurements have revealed
rapid proton transfers in tropolone (86).

^{17}O and ^{14}N high resolution studies of H-bonded systems are
accumulating. The ^{17}O shifts of donor OH groups are downfield.
Even larger shifts occur when the hydroxylic oxygen is electron
donating, but the carbonyl oxygens display upfield shift (87,88).
^{14}N and ^{15}N resonances are shifted downfield and are much larger
than the ^{1}H shifts. Studies using ^{14}N and ^{17}O shifts of formamide,
N-methyl- and N,N-dimethyl formamide in various solvents indicate (89) that the latter are even more sensitive probes of H-bonding. The ^{15}N shifts have been studied in connection with H-bonding and protonation of amines (90) and nucleic base pairing (91).
^{31}P resonance changes as a function of the state of ionization
of the phosphate groups are extremely important as intracellular
probes of pH (92). ^{13}C though rarely directly involved in H-bonding is nevertheless an useful probe since the downfield shifts
on H-bonding of adjacent oxygens and reflecting the charge decrease on C are large (93). Moreover, the shielding anisotropy that
is measurable in solids is also informative (94) of H-bonding.

ELECTRONIC ABSORPTION SPECTRA

      The differences in the absorption spectra between the sepa-
rated proton donors and electron donors, and their complexes are
due to the respective effects of H-bonding on the ground and excited electronic states whereby the differences in the potential
functions of these states (Franck-Condon principle) are important.
The systematization of spectral shifts according to the types of
transitions was proposed long ago (95). As a rule the π - π^* transitions of proton donors show red shifts on H-bonding and the
classic example are the phenols. The shift is accompanied with an
intensification and increase of proton donating ability due to an
internal charge transfer from the oxygen $2p\pi$ electrons to the
aromatic ring. The n-π^* and n-σ^* transitions of proton acceptors
shift to the blue provided that the n-π transition is excited
from a n orbital localized at the acceptor atom. The blue shift
reflects the additional energy used up in weakening the H-bond
in the excited state. Such were the conclusions from MO calculations on formaldehyde-water and amide-water systems (43,45).

However, they were recently challenged by measurements of circular dichroism of α , β unsaturated ketones (96) and MO calculations on the formaldehyde-water complex (46). The experimental results suggest that there is no change in the adiabatic energy of excitation i.e. that the H-bond remains in the excited state. The blue shift occurs because the ground and excited state geometries of the proton acceptor are differently affected by H-bonding and hence the transition occurs between different vibrational states.This was confirmed by the MO calculations in which the excited state geometry of formaldehyde was optimized whereas in the previous theoretical work the geometry was frozen in the ground state. Although the blue shift of the n-π transition and weakening of the H-bond occurs also with heterocyclic systems e.g. pyridine, there may be exceptions (97). Although the fluorescence and phosphorence spectra are even more instructive of H-bonding and, particularly, the fluorescence quenching by H-bonding has many applications in the study of interactions in biological systems the subject is too complex to be treated briefly and the reader has to consult the original literature (42.98).

REFERENCES

1. Coulson, C.A., "The Hydrogen Bond", in: "Hydrogen Bonding" (Hadži, D., ed., Pergamon Press, 1959), p. 339.
2.a Schuster, P., Zundel, G. and Sandorfy, C., eds., "The Hydrogen Bond",North Holland, Amsterdam 1976.
  b Joesten, M.D. and Schaad, L.J., "Hydrogen Bonding", M. Dekker, New York 1974.
  c Ratajczak, H. and Orwille-Thomas, W.J., eds., "Molecular Interactions, Vol. 1, John Willey & Sons, Chichester 1980.
  d Kollman, P.A., "Hydrogen Bonding and Donor-Acceptor Interactions", in: "Applications of Electronic Structure Theory" (Schaefer, H., ed., Plenum Press, New York 1977), p. 109.
  e Pullman, B., ed., "Intermolecular Interactions: From Diatomics to Biopolymers", Willey, Chichester 1978.
3. Schuster, P., ref. 2a, Vol. 1, p. 26.
4. Kollman, P, McKelvey, J., Johansson, A. and Rothenberg, S. 1975, J. Am. Chem. Soc. 97, p. 955.
5. Lischka, H. 1979, Chem. Phys. Lett. 66, p. 108.
6. Meyer, W. 1973, J. Chem. Phys. 58, p. 1017.
   Stogard, A., Strich, A., Almlöf, J. and Roos, B. 1975, Chem. Phys. 8, p. 405.
7. Morokuma, K. 1977, Acc. Chem. Res. 10, p. 294.
8. Kitaura, K. and Morokuma, K. 1976, Int. J. Quant. Chem. 10, p. 325.
9. Morokuma, K., in ref. 2c, p. 46.
10. Kolos, W. 1979, Theor. Chim. Acta 51, p. 219.
11. Tomasi, J., in: "Chemical Application of Atomic and Molecular Electrostatic Potentials" (Politzer, P. and Truhlar, D.G.,eds., Plenum Press, New York 1981), p. 279.

12. Umeyama, H., Kitaura, K. and Morokuma, K. 1975, Chem. Phys. Lett. 36, p. 11.
13. Bonaccorsi, R., Petrongolo, C., Scrocco, E. and Tomasi, J. 1971, Theoret. Chim. Acta 20, p. 331.
14. Kollman, P. 1977, J. Am. Chem. Soc. 99, p. 4875.
15. Janoschek, R. 1982, Croat. Chem. Acta 55, p. 75.
16. Yamabe, S. and Morokuma, K. 1975, J. Am. Chem. Soc. 97,p. 4458.
17. Dill, J.D., Allen, L.C., Topp, W.C. and Pople, J.A. 1975, J. Am. Chem. Soc. 97, p. 7220.
18. Desmeules, P.J. and Allen, L.C. 1980, J. Chem. Phys. 72, p.473.
19. Sobczyk, L., Engelhardt, H. and Bunzl, K., in ref. 2a, Vol. 3, p. 937.
20. Del Bene, J. and Pople, J.A. 1970, J. Chem. Phys. 51, p. 4858.
21. Del Bene, J. 1980, J. Chem. Phys. 72, p. 3423.
22. Olovsson, I., in: "Electron and Magnetization Densities in Molecules and Crystals" (Becker, P., ed., Plenum Press, New York 1980), Olovsson, I. 1982, Croat. Chim. Acta 55, p. 171.
23. Stevens, E.D., Lehmann, M.S. and Coppens, P. 1977, J. Am. Chem. Soc. 99, p. 2829.
24. Janda, K.C., Klemperer. W. and Novich, S.E. 1976, J. Chem. Phys. 64, p. 2698.
25. Delpuech, J.J., Serratrice, G., Strich, A. and Veillard, A. 1971, Mol. Phys. 29, p. 849.
    Scheiner, S., Harding, L.B. 1981, J. Am. Chem. Soc. 103, p. 2169.
26. Kraemer, W.P., Diercksen, G.H.F. 1970, Chem. Phys. Lett. 5, p. 463.
27. Janoschek, R., Weidemann, E.G., Pfeiffer, H. and Zundel, G. 1972, J. Am. Chem. Soc. 94, p. 2387.
28. Weidemann, E.G., ref. 2a, p. 245.
29. Darlow, S.F. and Cochran, W. 1961, Acta Cryst. 14, p. 1250.
30. Hsu, H. and Schlemper, E.O. 1978, Acta Cryst. 34B, p. 930.
31. Šolmajer, T. and Hadži, D. 1983, Int. J. Quant. Chem. 23, p. 945.
32. Koller, J. and Hadži, D., to be published.
33. Brickmann, J. and Zimmerman, H. 1979, J. Chem. Phys. 50, p. 1608.
34. de la Vega, J.R. 1982, Acc. Chem. Res. 15, p. 185.
35. Yanagawa, Y. and Nagle, J.P. 1979, Chem. Phys. 43, p. 329.
36. Haas, K.-H. and Schindewolf, U. 1983, Ber. Bunsenges. Phys. Chem. 87, p. 346.
37. Catalan, J., Yanez, M. and Fernandez-Alonso, J.I. 1978, J. Am. Chem. Soc. 100, p. 6917.
    Fluder, E.M. and de la Vega, J.R. 1978, J. Am. Chem. Soc. 100, p. 5265.
38. Baughcum, S.L., Duerst, R.W., Rore, W.F. and Wilson, E.B. 1981, J. Am. Chem. Soc. 103, p. 6296.
39. Rosetti, R., Haddon, R.C. 1980, J. Chem. Phys. 73, p. 1546.
    Rosetti, R., Haddon, R.C. and Brus, L.E. 1980, J. Am. Chem. Soc. 102, p. 6913.

40. Lohr, L.L. and Sloboda, R.J. 1981, J. Phys. Chem. 85, p. 1332.
41. Rosen, P. 1975, Int. J. Quant. Chem. 9, p. 473.
42. Mataga, N. and Kubota, T., "Molecular Interactions and Electronic Spectra", M. Dekker, New York 1970, p. 346.
43. Iwata, S. and Morokuma, K. 1973, Chem. Phys. Lett. 19, p. 94.
44. Iwata, S. and Morokuma, K. 1977, Theor. Chim. Acta 44, p. 323.
45. Del Bene, J.E. 1978, J. Am. Chem. Soc. 100, p. 1395.
46. Taylor, P.R. 1982, J. Am. Chem. Soc. 104, p. 5248.
47. Kollman, P. and Rothenberg, S. 1977, J. Am. Chem. Soc. 99, p. 1333.
48. Bonaccorsi, R., Scrocco, E. and Tomasi, J. 1976, Theor. Chim. Acta 43, p. 63.
49. Pullman, A. and Pullman, B., "The Electrostatic Molecular Potential of the Nucleic Acids", in: "Chemical Application of Atomic and Molecular Electrostatic Potentials" (Pulitzer, P. and Truhlar, D.G., eds., Plenum Press, New York 1981).
50. Kollman, P.A., McKelvey, J., Johansson. A. and Rothenberg, S. 1975, J. Am. Chem. Soc. 97, p. 955.
51. Kollman, P.A. 1978, J. Am. Chem. Soc. 100, p. 2974.
52. Drago, R.S., Vogel, G.C. and Needham, T.E. 1971, J. Am. Chem. Soc. 93, p. 6014.
53. Allen, L.C. 1975, J. Am. Chem. Soc. 97, p. 6921.
54. Lippincott, E.R. and Schroeder, R. 1955, J. Chem. Phys. 23, p. 1131.
55. Lippincott, E.R., Finch, J.N. and Schroeder, R., in ref. 1, p. 361.
56. Robertson, G.N. and Lawrence, M.C. 1981, Chem. Phys. 62, p. 131.
    Lawrence, M.C. and Robertson, G.N. 1980, Ferroelectrics 25, p. 363.
57. Clementi, E., "Computational Aspects for Large Chemical Systems" (Lecture Notes in Chemistry), Springer Verlag, 1980.
58. Momany, F.A., McGuire, R.F., Burgess, A.W. and Scheraga, H.A. 1975, J. Phys. Chem. 79, p. 2361.
    Scheraga, H.A. 1979, Acc. Chem. Res. 12, p. 7.
    Oie, T., Maggiora, G.M., Cristoffersen, R.E. and Duchamp, D. J. 1981, Int. J. Quant. Chem. Q.B.S. 8, p. 1.
59. Taylor, R. 1980, J. Mol. Struct. 71, p. 311.
60. Bellot, E.M. and Wilson, E.B. 1975, Tetrahedron 31, p. 2896.
61. Blinc, R., in ref. 2a, Vol. 2, p. 831.
62. Hadži, D. and Bratos, S., in ref. 2a, Vol. 2, p. 565.
63. Hadži, D. and Orel, B. 1973. J. Mol. Struct. 18, p. 227.
64. Bouteiller. Y., Allavena, M. and Leclerq, J.M. 1980, J. Chem. Phys. 73, p. 2851.
65. Novak, A. 1974. Struct. Bonding 18, p. 177.
66. Badger, R.M. and Bauer. S.H. 1977. J. Chem. Phys. 5. p. 839.
67. ref. 2b, p. 208.
68. Kamlet, M.S., Solomonovici. A. and Taft, R.W. 1979, J. Am. Chem. Soc. 101, p. 3734.
69. Ollovsson. I. and Jönsson, P.G., in ref.2a, Vol. 2, p. 393.

70. Badger, R.M. 1934. J. Chem. Phys. 2. p. 128.
    Auvert, G. and Maréchal. Y. 1979. Chem. Phys. 40, p. 735.
71. Maréchal. Y.. in ref. 2c. p. 231.
72. Bratos. S., Lascombe. J. and Novak, A.. ref. 2c, p. 302.
73. Bratos. S. 1975. J. Chem. Phys. 63. p. 3499.
74. Delepiane, G., Piaggio. P., Rui. M.. Degli Antoni. L. and
    Zerbi. G. 1982. J. Mol. Struct. 80. p. 265.
75. Zundel, G., in ref. 2a. Vol. 2. p. 683.
76. Rastogi. P.P. and Zundel. G. 1981. Z. Naturforsch. 36c, p.961.
77. Novak, A. 1982. Croat. Chem. Acta 55. p. 147.
78. Schaefer, T. 1975. J. Phys. Chem. 79. p. 1888.
79. Bergelund. B. and Vaughan. R.W. 1980. J. Chem. Phys. 73. p.
    2037.
80. Fujiwara. M. and Martin. J.S. 1974. J. Am. Chem. Soc. 96,
    p. 7625.
81. Denisov, G.S. and Golubev. N.S. 1981. J. Mol. Struct. 75,
    p. 311.
82. Soda. G. and Chiba, T. 1969. J. Chem. Phys. 56. p. 439.
83. Blinc, R. and Hadži, D. 1966. Nature 212. p. 1307.
84. Ditchfield, R. 1976. J. Chem. Phys. 65. p. 3123.
85. Butler, L.G. and Brown, T.L. 1981. J. Am. Chem. Soc. 103. p.
    6541.
86. Jackman, L.M., Trewella. J.C. and Haddon. R.C. 1980. J. Am.
    Chem. Soc. 102, p. 2519.
87. St. Amour, T.E. and Fiat. D. 1980. Bull. Magn. Reson. 1. p.
    118.
88. St. Amour, T.E., Burgar, M.I.. Valentine, B. and Fiat. D.
    1981. J. Am. Chem. Soc. 103, p. 1128.
89. Burgar, M.I, St. Amour, T.E. and Fiat. D. 1981, J. Phys.
    Chem. 85. p. 502.
90. Duthaler, R.O. and Roberts. J.D. 1979. J. Magn. Res. 34, p.
    129.
    Allen. M. and Roberts. J.D. 1980. J. Org. Chem. 46, p. 130.
91. Dyllick-Brenziger. C., Sullivan, G.R.. Pang, P.P. and Roberts,
    J.D. 1980. Proc. Natl. Acad. Sci. U.S.A. 77. p. 5580.
92. Barton. J.K., Den Hollander. J.A.. Lee, T.M.. Mac Laughlin,
    A. and Shulman. R.G. 1980. Proc. Natl. Acad. Sci. U.S.A. 77,
    p. 2470.
    Roberts. J.K.M.. Ray. P.M.. Wade-Jardetsky, N. and Jardetsky,
    O. 1980. Nature 283. p. 870.
93. Maciel. G.E.. Dallas, J.L. and Miller, D.P. 98, J. Am. Chem.
    Soc. 98, p. 5074.
94. Schröter. B.. Rosenberger. H. and Hadži. D. 1983, J. Mol.
    Struct. 96, p. 301.
95. Baba, H. 1958. Bull. Chem. Soc. Japan 31. p. 1691.
96. Beecham. A.F.. Hurley. A.C.. Johnson, C.H. 1980, J. Chem. 33.
    p. 699.
97. Del Bene, J.E. 1980, Chem. Phys. 50. p. 1.
98. Lippert, E., in ref. 2a, Vol. 1. p. 1.

HYDROGEN BONDS IN BIOLOGICAL STRUCTURE AND MECHANISM

D. Hadži

LEK - Chemical and Pharmaceutical Works and
Boris Kidrič Institute of Chemistry, Ljubljana,
Yugoslavia

ABSTRACT. Hydrogen bonding in proteins and nucleic acids is reviewed. Examples are given of hydrogen bonds that are essential
for the stability of structures, in biomolecular recognition, and
in some special mechanisms.

Mechanisms in living systems are characterized by high specificity, high rates and by the involvement of rather low energies in the single steps of consecutive reactions. This is possible because of the high specificity of the structure of molecules
involved, both those of high and low molecular weight, and their
ability of undergoing conformational changes governed by subtle
changes in the surroundings. The specificity of structure is respon-
sable for the mutual recognition and subsequent association of
molecules that is fundamental to most biological mechanisms. The
primary structure of proteins, and nucleic acids, which are the
most important carriers of biological mechanisms, determines ultimately the higher order structures, but this largely occurs by
the expression of the former via non covalent interactions amongst
which the H-bonds play a particularly important role. They are
equally important for the productive associations between molecules. A certain degree of directionality of H-bonds, their energy
which is on the high side for non covalent interactions, but weak
relative to covalent bonds, and their fast kinetics make them
well suited for the mentioned roles. Moreover, the potential surfaces of the stronger H-bonds warrant proton transfers between
pairs of molecules and, in extended systems, for large distance
proton translocations which are likely to be involved in some important mechanisms. The fact that a living cell contains 70 %
water, 15 % protein and 7 % nucleic acids and that these as well

61

C. Sandorfy and T. Theophanides (eds.), Spectroscopy of Biological Molecules, 61–85.
© *1984 by D. Reidel Publishing Company.*

as most of the other consistituents are capable of forming a variety of H-bonds certainly supports the statements which are most
often found about the importance of H-bonding for biological processes in the introductions to papers dealing with H-bonding.
Indeed, the very existence of groups capable of H-bonding, but not
undergoing any is such a rarity that it merits special mentioning.
An example is the natural, cyclic decapeptide antamanide that has
six NH groups. Two of them are engaged in intrachain H-bonds of
the $4 \rightarrow 1$ type, two are used to complex a cation and two are just
unused (1). However, this and a couple of the other exceptions
cannot seriously shake the general rule that under normal conditions as many H-bonds will exist as there are groups capable of
forming them. It is, however, difficult to single out those H-
-bonds that play a key role in determining, for instance, the
conformation of a polypeptide in the sense of its specific function or those definitely intervening in the recognition of a ligand by a receptor because in such cases H-bonds are only one of
the many factors involved in the process. It is easier to demonstrate the role of H-bonding in enzymatic reactions and in the base pairing of nucleic acids. The treatment of water in the context
of biologically relevant H-bonds is very delicate. On one side it
represents the major cell constituent and, with its properties
largely depending on its H-bonding system, deeply involved in the
organization and functioning of living systems. On the other side,
individual water molecules can often be assigned a special role
in bridging other molecules as, for instance.the thyroxin hydroxyl
group to prealbumin in the hormone-carrier complex (2). There is
also something between the aspect of water as a universal bath
and dielectric medium, and the extreme of water molecules with
specific function. This is the water of hydration that contributes
to the tertiary and quaternary structures of biopolymers. As a
fourth aspect, water providing for hydrophobic effects may be
considered. The first mentioned aspect of H-bonding in water is
too wast to be treated presently and we shall restrict ourselves
to examples of water molecules with a specific role in structure
and mechanisms.

HYDROGEN BONDING IN PEPTIDES AND PROTEINS

     H-bonds are involved at all levels of protein organization
beyond the primary structure. The most abundant is the amide
NH...O=C H-bond represented by several types in intra- and interchain interactions. Side chain hydroxyl groups of e.g. serine and
tyrosine are engaged mostly in interchain interactions, but they
may interact also with backbone peptide groups (3). Side chain
amino and carboxyl groups are mostly ionized and engaged in the so
called salt bridges which are, in fact, also H-bonds (4). The
essential characteristics of the amide H-bond are derived from
experimental and theoretical studies of intermolecular bonding of

simple amides in crystals and solutions (for a review see ref.4).
The N...O distance in N-methylacetamide (2.82 Å) (5) is very
near to the arithmetic mean 2.87 Å of various interamide H-bonds
that vary between 2.7 Å and 3.1 Å (6,7). From the X-ray data the
preferred angle at the hydrogen atom in the NHO triangle seems
to be 160° although values between 120 and 180° are found. The
experimental values of H-bond self association energies of trans-
-amides are burdened by the tendency for multiple chain association and the cooperative effect predicted by theoretical calculations (8). The range between 3.7 and 7 kcal/mol is acceptable
for solutions in non-polar solvents. In polar solvents, particularly water, the energy is strongly reduced (9). Since the mutual
positioning of the C=O and NH groups in intrachain bonds is often
dictated by other conformation influencing factors, the recent
extensive ab-initio calculations of the energy dependence on geometrical parameters is interesting. The study was done on the
formamide dimer using the STO-3G basis set (10). Six parameters
were varied: three dihedral angles (rotations about the NH, OH,
and CO bonds), the ON distance, and the angles at the hydrogen
and oxygen atoms. The energy computation varying the N...O distance yielded a Morse type curve with a bottom near 4 kcal/mol. The
value of the angle at the hydrogen atom is important: the energy
falls off when the angle is below 180° and at 120° the H-bond
energy is lost. The energy is practically independent of the dihedral angles. The angle at the oxygen is also of little importance providing that it is not below 100° which indicates the flexibility of the H-bonds.

The intrachain H-bonds result from a compromise between the
H-bond energy, the backbone torsional energy (rotation around the
C-N and C-C' bonds defined by the angles (Ψ and ϕ) and the in-
teractions between the side chains (R. S).

The smallest intrapeptide H-bonded structure C_5 is believed to
stabilize the extended chain forms. It was investigated by infrared spectrometry in CCl_4 solutions of peptide models. e.g.
MeCO-L-Pro-NHMe (11). In more polar solvents dioxane and DMSO it
is replaced by solvent bonding. Although the ab-initio computed
N...O distance of 2.77 Å corresponds to the crystallographic criteria of H-bonding and the computed stabilization energy of the
C_5 conformation is quite substantial. doubts have been expressed
whether this really is a H-bond interaction (12). The concern
arises from the computed NHO angle to be 100.9° which is much below the usual amide H-bond angles. In protein crystal structures,

the C_5 bond is postulated to exist in the bifurcated C_5C_7 situation e.g. in chymotrypsin and thermolysin. Such bifurcated bonds were also characterized by NMR and infrared spectrometries in solutions of proline containing oligopeptides (13). The proposed structure is:

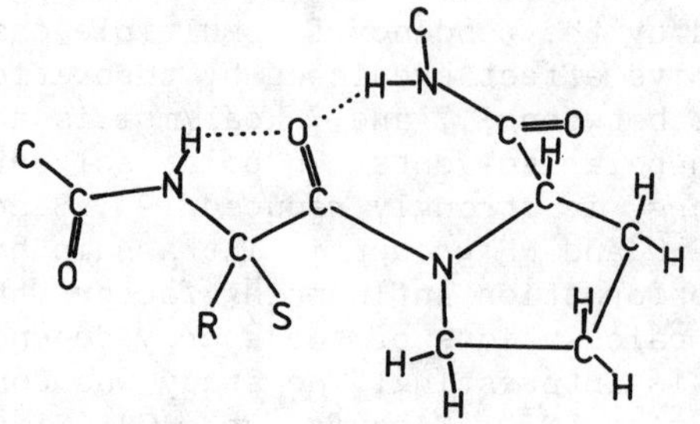

The C_7 H-bond appears as the key in hairpin bends at glycine residues, e.g. in insuline (14). The conformation of the glycine dipeptide was investigated recently by small ab-initio basis set calculations and the derived C_7 H-bond energy is about 3 kcal/mol (15). The C_{10} H-bond frequently occurs in bends that lead to chain reversals. The first empirical calculations (16) of three peptide units revealed three low energy conformations that contain short N...O distances and accordingly the types I, II and III of C_{10} H-bonds are now accepted. Types I and II are found in chain bends whereas type III leads to tight helical conformations:

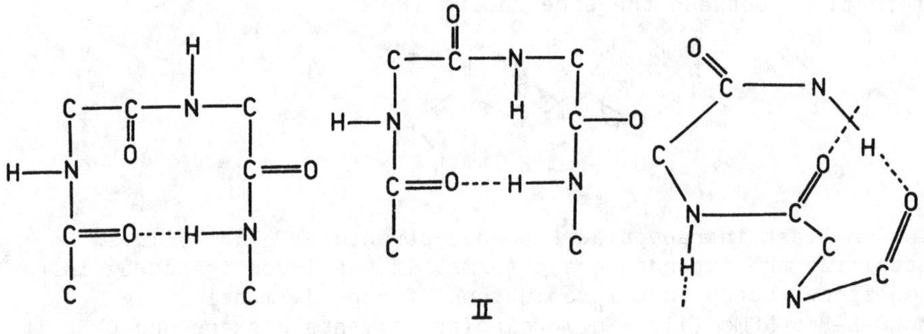

Fig. 1. C_{10} hydrogen bonds type I, II and III

A fourth type was postulated from ab-initio calculations of the glycine tripeptide, but the absolute energy of the corresponding

conformation is high (17). Type I is a candidate for the β-turn
in one of the pentapeptide enkephalins (18). β-turns with type II
H-bond are quite frequent in proteins though less so than type I
(19). Type II turns require for steric reasons the involvement of
glycine or a proline-glycine pairs. In view of the connection between structure and mechanical properties of elastomeric proteins
such as collagen and tropoelastin which are built of sequential
polypeptides, the repeating fragments as well as the corresponding
synthetic polypeptides have been throughly investigated and the
H-bonds were characterized also by NMR techniques. The H-bond is
quite stable in water and theoretical calculations on the
Val_1-Pro_2-Gly_3-Gly_4 show that the type II turn is more stable than
type I (20,21). Type II turns are also common in sequential polypeptides which appear as channel forming elements of biomembranes (22). Type III H-bonds giving rise to 3_{10} helical conformations are not very frequent because the absolute energy of the
corresponding conformation is rather high (17). The helical pieces
are therefore short (23,24).

The C_{13} H-bond appears in the most abundant secondary structure, the well known α-helix as defined in the classical work
of Pauling and Corey (25). Since the H-bond energy is not outstanding (Table 1) it must be the favourable arrangements of the
side chains and the torsions that stabilize the α-helix. In the
ideal α-helix there are 3.6 units per turn, but in many proteins
distorted helices appear that have between 3.4 and 3.9 units per
turn (26). From NMR work in solutions on a dodecapeptide analogue
of the repeat hexapeptide in tropoelastin C_{11} (γ bend) and C_{14}
H-bonds were proposed (27). The rare π helix (4.4$_{16}$) contains a
C_{16} H-bond (28).

The variety of H-bonds identified in bends by experimental
methods or revealed by molecular mechanics calculations (29) suggests the question about the importance of H-bonds for their stability. The first thing one would like to know is the actual energy of the various types of H-bonds. Since they are under constraint
of other factors contributing to the conformation, experimental
data on energies of simple amide associations are not relevant.
Theoretical calculations of H-bond energies require suitable reference systems. In the present problem this means systems that
allow the separation of the H-bond energy from external factors
intervening in the overall conformational energy. This has been
done in recent ab-initio calculations on glycine oligopeptides
(12,30) by constructing Morse type curves from differential energies between larger peptide and the constituent smaller units,
e.g. two tripeptide units minus a dipeptide unit at suitable conformations lead to the Morse curve of the α-helical C_{13} H-bond
in a tetrapeptide. The bottom of the curve is a good estimate of
the H-bond energy. The results of such computations are reproduced
in Table 1.

Table 1. Ab-initio computed values of CO...HN hydrogen bonds in
         glycine oligopeptides (30)

H-bond	Computed value (in kcal/mol)
C_7 in dipeptide	3
C_{10} in tripeptide	
Type I	2.82
Type II	4.27
Type III	3.45
Type IV	4.08
C_{13} in tetrapeptide (α-helix)	3.8

Since in the calculations the STO-3G basis set was used the absolute energies may be too small. However, the differences are certainly not large and hence they cannot be crucial for the preference of one or the other type of structure. In fact, the analysis of 135 bends in native structures of eight proteins has shown
that 43 % of bends do not contain C_{10} H-bonds (29). Molecular
mechanics calculations on pentapeptides have also demonstrated the
stability of bends without C_{10} H-bonds, but then the side chain-
-backbone H-bonds yield additional energy (29). The conformational energy analysis of N-acetyl-N'-methyl-amides of naturally
occuring amino acids has indeed revealed many low energy regions
with side chain-backbone H-bonds in derivatives of e.g. serine,
asparagine, aspartic acid and threonine which may explain the helix disrupting effects of these amino acids (3,31). The next
question is how resistant are these H-bonds to water. NMR work on
small peptides shows that C_5 and C_7 H-bonds are not stable in water (13,32). This is an additional argument to show that the role
of H-bonds in the stabilization of bends is not decisive.

The dominant role of H-bonds in the formation of β-sheets
is obvious from Fig. 2 representing the very regular $\alpha_L\alpha_R$ layer
structure occuring e.g. in silk fibroin. Infrared and ^1H NMR
studies of model peptides in CCl_4 solution yielded ΔH^o of 8.7-
-8.9 kcal/mol for a double interpeptide bond (33). Thus 4-4.5
kcal/mol may be taken for the energy of a single interpeptide
H-bond. β-sheets occur also in globular proteins with an average
of 15 % (34), but the length of the chains is usually shorter than
in fibrous proteins. The sheets contain up to six strands. The CO
and NH groups located at the borders of a sheet may be used either
in rolling up to yield barrel like structures or in interpeptide
H-bonding, e.g. in the formation of the tertiary structure of
silk fibroin (Fig. 3). This has been termed (35) $\alpha_L\alpha_R\beta$ structure because of the sharp β-hairpins. The structural hairpins
are stabilized by interpeptide H-bonds. These bends are possible

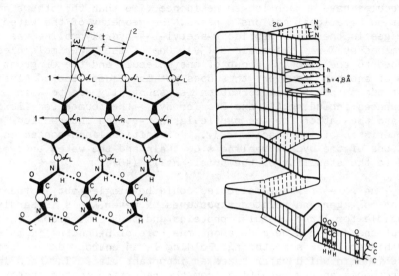

Fig. 2. β-sheet $\alpha_L \alpha_R$ structure. Circles denoted α represent side chains located at opposite sites of the pleated sheet (from ref. 35).

Fig.3. Helical structure of silk fibroin, vertical shading: $\alpha_L \alpha_R$ structures. Oblique dashed lines: interpeptide H-bonds in β-hairpins.

when no bulky side chains exist at the borders of the pins. Serine side chains of the C_p fragment of fibroin mediate H-bonds that held together the layers in the quaternary structure (35).This nicely explains the mechanical properties of silk fibers. In another example of fibrous protein. collagen. the situation concerning interchain interactions is more complex. The triple helix of collagen is made up of three polypeptide chains coiled around a common axis. The super coilling is facillitated by glycine residues occupying every third position. The other abundant amino acid is hydroxyproline and it was said that two interchain H-bonds involving water molecules are important for the triple helix formation and hence for the mechanical properties (36). The interest in the connection between the latter and the structural factors, particularly interactions involving the side chain of hydroxyproline,triggered many experimental investigations on model polypeptides and of molecular mechanics calculations (37). The latter have shown that the collagen like triple stranded structure is equally stable with $(Gly-Pro-Pro)_n$ polymers as it is with $(Gly-Pro-Hyp)_n$ and that the OH group does not contribute to the stabilization of the lowest energy structure. This refers to the absence of water. In hydrated collagen, however, the water molecules connecting CO and NH groups of adjacent chains seem to be indeed specifically bonded. NMR measurements (38) show that these

molecules have a much longer residence time than the others and
that no rotational motions appears. The geometry of the water
bridges in the model peptide N-acetyl-L-4-hydroxyproline was determined by X-ray difractions (39). One of the water molecules is
bonded to the N-acetyl group of one molecule and the OH group of
Hyp of another. However, this does not invalidate the conclusions
that the triple strand formation is essentially due to amino acid
sequences including Gly and Pro (or Hyp) - the former for flexibi-
ty and the latter for the particular non-bonded interactions characteristic of imino acids (37). The additional non-bonded interactions of the hydroxyproline side chain and the water bridges
add to the stability of the triple helix (40).

      The role of water molecules could be singled out in the examp-
le of collagen and related structures. However, this is hardly
possible for large globular proteins and thus some global considerations only can be made about the role of H-bonding in folding
and higher level structuring. Folding is an entropy driven process (41) in which water takes an important place. The gain in
hydrophobic energy resulting from the reduction in the number of
non-polar contacts with water compensates for the loss of chain
conformational entropy. This is supplemented by the energy of formation of H-bonds in the interior, including possible interior
water and unsaturated polar groups. and by H-bonding with water
of the surface polar groups. The analysis (42) of H-bonds in 15
globular proteins shows that amino-acids with non polar-side chains
particularly valine. alanine and leucine, tend to be burried in the
interior (50-70 %), those with one polar group such as serine and
threonine much less (about 25 %) whereas those with two polar
groups, e.g. aspartic and glutamic acids, are less than 10 % burried.
There seems to be a tendency to gain on surface polar group-water
interactions and hydrophobic stabilization of the interior.
Chothia (42) counts the interior H-bonds amongst hydrophobic groups,
but this is debatable (43). The role both of surface and interior
water molecules in the structure and dynamics of globular proteins
has been recently reviewed (44). The interior H-bonds certainly
do contribute to the stabilization of the tri-dimensional structure. This is illustrated by the example of zinc insulin dimer
where the X-ray structure analysis shows that the burried H-bonds
are consistently shorter than those exposed to the solvent. In the
β-sheet of the dimer two out of four H-bonds are shorter and
straighter indicating co-operativity of H-bonds (45).

      The higher order structuring may be considered as a process
of mutual recognition of pieces of secondary structure. What is
the role of H-bonding in this process may be gathered from the
fact that all H-bonds become saturated on folding (46) and, in general, that denatured proteins will refold to yield the original
H-bond network. For the association of protein oligomers which is
essentially a similar process of recognition between certain regions

of monomers, at least one example of specific H-bond interaction
can be given. It is offered by hemoglobin mutants in which amino
acids with non polar side chains replace those with H-bonding side
chains (47). In hemoglobin Phylly the replacement of Tyr 35β by
Phe leads to an increased dissociation into monomers. Tyr 35β in
the helix C of the β chain projects into the interior of the te-
tramer forming a H-bond with Asp 126α. Similarly, in hemoglobin A
there is a H-bond at the interface between Asn 102β and Asp 94α
which disappears in the Kansas mutant having instead a Thr at
102β. This reduces ΔG° by 3 kcal/mol. By this example we have
touched on the extremely important topic of molecular recognition.
Since nucleic acids besides the proteins are largely engaged in
such processes we have to treat H-bonding in the former before
dealing with molecular recognition in general.

HYDROGEN BONDING AND STRUCTURE OF NUCLEIC ACIDS

    H-bonding is involved as one of the factors determining the
structure of natural nucleic acids as well as of their synthetic
analogs. It enters with different weights at various levels of
the organization. In the formation of base pairs it is believed
not only to be the main component of stabilization energy, but
also the principal genetic information determinant. In the higher
order structuring the contribution of H-bonding is much less evident, except if we include the nucleic acid-water interactions.
Most of the investigations concerning H-bonding, both experimental and theoretical aim at the characterization of base pairs
with respect to preferential stability between complementary bases as the basis of template selection. H-bonding in mismatched
pairs and between tautomeric forms are equally interesting in the
search for mutation mechanisms.

    The NH...O and NH...N bonds between the bases in crystals
obtained by co-crystallization of various bases appear, besides
in the Watson-Crick pairs guanine-cytosine and adenine-uracil,
also in the Hoogsten form (Fig. 4). In the triple stranded polynucleotides such as poly(dT), poly(dA), poly(dT) both types of
H-bonds occur together (47a). The consideration of the doubly and
triply bonded pairs reveals no cooperative effect. H-bonding between one type of base in its own crystals leads to centrosymmetric
dimers, but acyl derivatives form cyclic tetramers and continuous
helices (for crystallographic data see ref. 48, 49).

    One of the most interesting problems related to genetic in-
formation is the relative stability of various base pairs. In the
determination of thermodynamic parameters in solution infrared,
^{1}H and ^{15}N NMR and UV spectroscopy were mainly used (49). Whereas
in organic solvents H-bonding with the formation of base pairs
prevails (50), stacking is dominant in aqueous solvents (51).

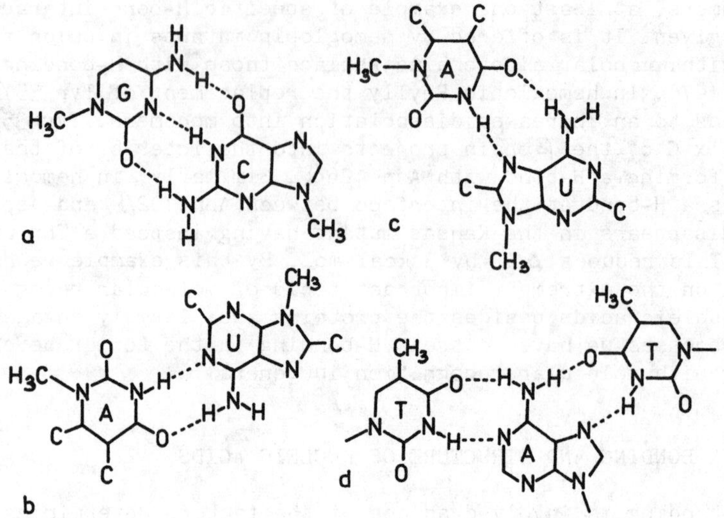

Fig. 4. Nucleic base pairs: a - Watson-Crick type, b - Hoogsten
type, c - both types in poly(dT). poly(dA). poly(dT) triplex.

However, a small proportion of water in $CHCl_3$ does not disrupt
H-bonding between base pairs (52). Theoretical calculations of
association energies in $CHCl_3$ and water are in agreement with the
experimental results (53). Recently, field ionization mass spectrometry has been used to determine the H-bond enthalpies of base
pairs (54). The advantage of this method is that the derived
enthalpies refer to interactions in vacuo. Some values obtained
by both methods are given in Table 2.

Table 2. Binding enthalpies of base pairs (kcal/mol)

Base pair	ΔH^a	ΔH^b
Adenine - uracil	14.5	6.2
Uracil - uracil	9.5	4.3
Adenine - thymine	13.0	-
Thymine - thymine	9.0	-
Guanine - cytosine	21.0	10-11.5
Cytosine - cytosine	16.0	6.3

a: data from (54)
b: data from (50)

The theoretically computed energies, though slightly lower on the
whole (for a compilation of data see ref. 53) than those obtained
by mass spectrometry, reproduce well the relative order. Solution
energies are considerably lower because of the influence of the
dielectric properties of the solvent on H-bond energies. The important result is that the Watson-Crick configurations are usually
more stable for each pair, but the differences between complementary and non-complementary base pair energies may be not large
enough to attribute the recognition potency to the energy characteristics of the H-bonds alone and that other factors are likely
to contribute to the information content of the organized nucleic
bases. It is more difficult to single out the role of interbase
H-bonding in the structure of duplex. Besides interbase H-bonding,
the influence of water on those bonds, the conformational energy
contributions from the phosphate-sugar backbone, the base stacking
energy and the influence of water on the backbone have to be taken in account. Experimental data on helix-coil transition enthalpies embrace all these factors. The observed Δ H-values for this
process range between 7-9.5 kcal/mol per base pair (56). Depending on the guanine-cytosine content the Δ H values show a spread
of 2-2.5 kcal/mol. Data from high resolution melting experiments
indicate a difference of 2.44 kcal/mol between the (CG)-(GC) and
(AT)-(TA) base pairs with Δ H being about 50 % of the in vacuo
experimental H-bond energy, i.e. 10.04 and -7.6 kcal/mol (57).
The recent infrared and calorimetric data on oligonucleotides with
non-complementary GU base pairs show that the "wobble" pairs have
also quite high Δ H values (58). To extract from this overall
ΔH the share of H-bonding it is necessary to calculate the other
components. The hydration energies have been calculated by the
Monte Carlo method using analytical interatomic potentials based
on ab-initio calculations of water-base interactions (59). The
stacking energies and the influence of water on stacking are also
calculated by quantum methods (60). From an analysis using such
data Rein (61) concluded that the H-bond energy contribution is
only between 1.5 and 2.5 kcal/mol per base pair. Since hydration
energies are nearly by an order of magnitude higher than the base
pair H-bond energies and the stacking energies are of the same
order as the H-bond energies such values can be taken only as
orientative. Thus only high level ab-initio calculations on helix
fragments should be able to set such analyses on solid grounds (62).

The strong influence of hydration on the relative stabilities
on the A, B and Z type DNA helices has triggered out the investigations on the details of water-backbone interactions. Infrared
studies have shown (63) that the free phosphate oxygens are the
first to bind two water molecules on the average. This is followed
by the hydration of P-O-C and C-O-C groups of phosphates and sugars, and the N and O atoms on the edges of base pairs bind water
least strongly. The A helix is stable with 4-5 water molecules per
base pair, tied to N and O atoms. The A to B helix transition

occurs after the groves are covered with a monolayer of water.
The spectroscopic observations are now complemented by X-ray diffraction studies of crystallized oligomeric DNA helices of both
A and B types which show a very regular pattern of hydrating water molecules (64). A particular role in the different properties
of RNA with respect to DNA is assigned to the ribose 2'OH group
that is H-bonded via one water molecule to the nearby phosphate
group (65).

Löwdin's hypothesis on point mutations (66) which is based
on the proton tunneling in H-bonds connecting complementary bases
was followed by numerous quantum mechanical calculations of proton
potential functions and the energy of tautomeric forms of the
bases. The transfer of a single proton leads to the formation of
ions, e.g. G-HC $\rightarrow$ GH^{+}-C^{-} and the separation of the double helix
at this stage would result in the impossibility of finding an
appropriate pair in the code copying process. However, the computed proton potential curves show that this process is energywise
highly unlikely (67). The double proton transfer leads to tautomeric forms G -C (Fig. 5) acceptable in the code copying process,

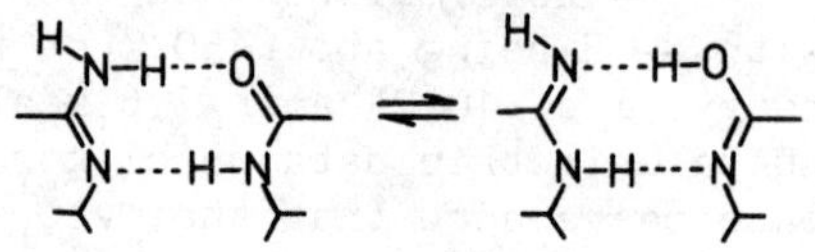

Fig. 5. Tautomerization by double proton transfer (schematic).

but would yield a change of the code. From the computational
point of view, this process is very demanding. To obtain a reliable potential surface, the full geometry optimization for every
step in the proton movement along the reaction coordinate is necessary. Moreover, the influence of medium should be considered
and, finally, the possibility of accomodation of the tautomeric
forms in the helix. The calculation of the relative energies of
the tautomeric forms with geometry optimization requires itself
some computational effort (68). The simultaneous proton transfer
has been studied on carboxylic dimers as the simplest model (69).
This has demonstrated the necessity of a full geometry optimization in order to obtain a potential function with a barrier comparable to that expected from NMR determinations of proton relaxation times (70). The existing computations of the proton potential functions for base pairs are not at this level and hence the
question of the likelihood of ground state proton tunneling between the bases is still pending. The double proton tunneling in
electronic excited states is also being considered (71). CNDO/S
calculations on the adenine-thymine base pair show that the barrier

in charge transfer excited $\pi - \pi^*$ states is lower than in the ground state. However, the life time in this state is short and thus the probability of the double proton transfer may be smaller than in the ground state. Concerning the possibility of accomodating tautomeric mispairs in the DNA helix, molecular mechanics calculations show that this exists for the GA. GC and AA pairs (72).

HYDROGEN BONDS IN MOLECULAR RECOGNITION

An important step in most biological mechanisms is the mutual recognition of molecules resulting in non-covalent complex formation and in the lowering of free energy with respect to the separate molecules. It may be called also interaction or just binding and the recognition efficiency equated to ΔG^o. The term recognition implies specificity and the classical examples are the antigen-antibody interaction, and the enzyme-substrate (or inhibitor) initial binding. The specific interaction occurs between biopolymers as well as between polymers and small ligands. The interaction may contribute to the organization of parts of a living system, e.g. protein-lipid-water interaction in the membrane formation, or it may trigger a chain of events as in hormone-receptor binding. In the latter type of interaction, the specificity is higher, though not quite exclusive. Similarly to small molecule interactions, the energy may be decomposed into contributions of electrostatic, charge transfer and van der Waals forces with the addition of hydrophobic interactions. Amongst the predominantly electrostatic type forces the H-bonding should be specified. However, it is only one amongst other factors and its share is usually difficult to define. As emphasized by Weber (73) a fine degree of molecular recognition by a protein can be achieved only if differences in binding result from sources none of which is truly dominant. Nevertheless, there are some examples of recognition in which particular H-bonds can be considered as important for specificity even if not dominant in the interaction energy. Firm evidence for the decisive role of certain H-bonds in recognition would require the combination of X-ray structure determination, spectroscopic methods, theoretical calcualtions and, eventually, chemical modifications of the recognition site. There are few cases where at least X-ray and NMR data were combined with theoretical calculations. These are mostly from the group of enzyme-substrate interactions. Therefore we shall examine first some examples of this type.

The task of the enzyme-substrate recognition process is to position the substrate so that the reactive bond is near the catalytic group in a conformation that will be similar to the transition state. Part of the binding energy is probably used in the change of conformation. Since the catalytic group is usually burried in a more or less hydrophobic cleft of the enzyme the intro-

duction of the substrate into the cleft and close to the reactive
site is already part of the recognition process. The future position of the substrate is usually occupied by one or more water
molecules which are replaced by the substrate. Thus, in fact, the
process is quite complex and involves a series of H-bond breakings
and formations until the final position is reached. An example is
the sliding of the substrate with its terminal carboxyl group into the pocket of carboxypeptidase A in which three arginine residues with their guanidine groups are important. On inserting of
the substrate a H-bond is formed between Arg 145 and the substrates carboxylate whereupon the guanidine group moves by 2 Å by
rotation from its initial position (74). A molecular orbital model calculation (75) on three arginyl residues and formic acid
has demonstrated the possible mechanism whereby the acid is first
bonded by a double H-bond to Arg 71 and then by a series of H-bond
breakings and formation of new ones ends up bonded to Arg 145 by
a double H-bond.

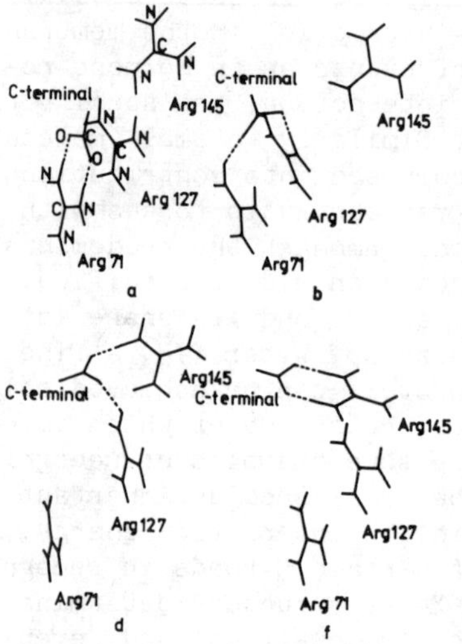

Fig. 6. Stages (a to f) in the
sliding mechanism of a peptide
terminal carboxylate (represented
by formic acid) over arginine residues 71. 127 and 145 towards the
reactive site of carboxypeptidase A
following to MO model calculations
(from ref. 75).

X-ray diffractions indicate that Tyr 198 and Phe 279 may also be
involved in the early stage of substrate binding (74). Analogous
sliding and recognition mechanisms could be envisaged also for
several other enzymes (75) with arginine residues at the bottom
of the pocket. The fact that nine out of ten glycolytic enzymes

contain arginine residues at their active sites demonstrates the
importance in recognition of the H-bonds mediated by this residue
to the carboxylate or phosphate groups of the substrates (76).

Most of the H-bonding schemes for the enzyme-substrate interaction are derived from X-ray structure determinations which reveal also other short contacts indicating hydrophobic interactions.
The schemes proposed (77) for a bacterial serine protease complex
(Fig. 7a) with oligopeptides, of penicillopepsin with a hexapeptide substrate (Fig. 7b), and of thermolysin (78) (Fig. 8)

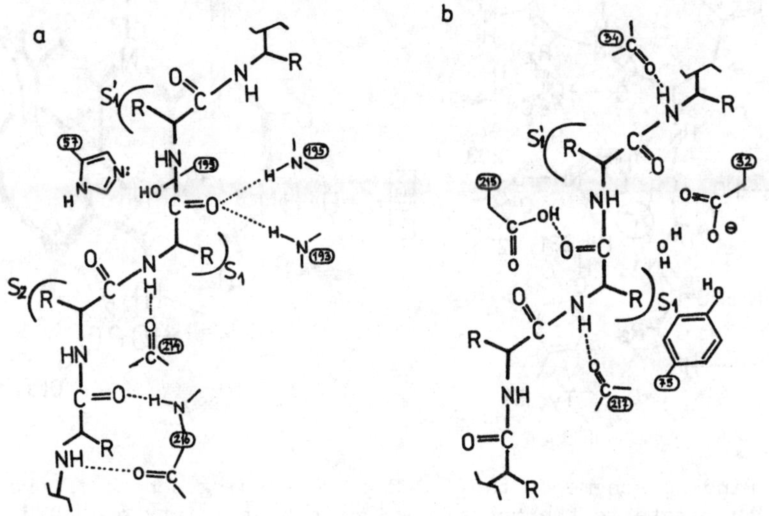

Fig. 7. Binding schemes of the peptide substrate to (a) bacterial
serine protease and (b) to penicillopepsin (after ref. 77). S in
Figs. 7 and 8 indicate hydrophobic interactions.

with inhibitors are reproduced. A combined H-bond and stacking interaction of the nicotinamide moiety of NADPH and flavin in glutathione reductase is sketched in Fig. 9 (79). Although the H-bonds
in these and many other schemes look quite impressive, one may
be wondering how much H-bonds contribute to the energy of interaction besides other stabilizing factors. From the analysis of
contacts in the lysozyme-N-acetylglucosamine hexamer complex, resulting from molecular mechanics calculations, about one fourth
of the total energy of stabilization is attributed to the H-bonding scheme (80). The interesting question of stereoselectivity
of enzymes was recently approached by molecular mechanics calculations on the interaction of α -chymotrypsin with L- and D-N-ace-
tyltryptophanamide (81). It was concluded that the enzyme binds

both configurations equally in the noncovalent initial complex.
The discrimination appears in the tetrahedral intermediate L-
-enantiomer which makes a better contact of its $H_2N-C-O-$ with
His 57. This stabilizes the L-enantiomer intermediate and also
facilitates the protonation of the leaving group. Another enantio-
mer discriminating H-bond is between the N-acetyl NH and C=O of
Ser-214.

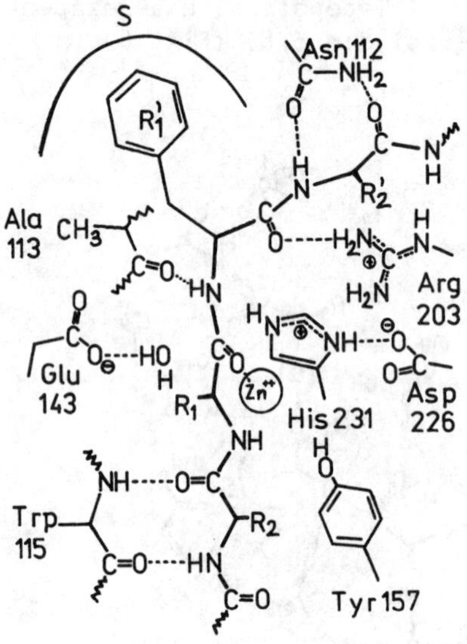

Fig. 8. Binding scheme of the
peptide substrate to thermolysin
(after ref. 78)

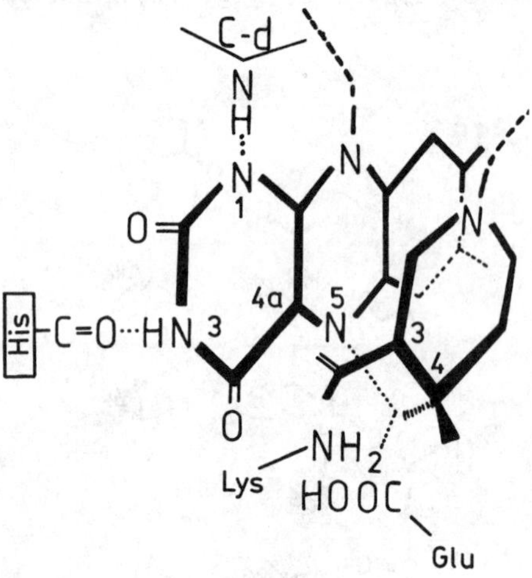

Fig. 9. Binding scheme of the
nicotinamide moiety of NADPH,
and flavin to glutathione reductase. C-d: central domain backbone (after ref. 79).

        Turning to other types of recognition we have to leave the
firm support of X-ray structural data and thus the indications of
the role of H-bonding become weaker. In the antibody-hapten interaction at least the architecture of the combining site is known
and the actual information on interactions come from NMR studies (82).
Binding to arginine residues of phosphate containing haptens to the
F_v fragment of mouse myeloma protein 315 is not only obvious from
NMR spectra, but is crucial: glyoxalation of Arg 95 abolishes binding. Similarly, NO_2 groups from chloronitrobenzene haptens bind
to asparagine and tyrosine residues of the F_v fragment of the immu-
noglobulin. Besides H-bonding, hydrophobic interactions of the
aromatic side chains of phenylalanine and tyrosine are specificity
determining.

The vast and for therapeutical purposes most important area of hormone, neurotransmitter and drug receptor interactions with specific ligands is in an inferior position because nothing is directly known about the structure of the binding sites of receptors and all conjectures about their architecture are based on the structure of ligands. Even the extracting of characteristic structural features from series of ligands having similar biological properties is quite difficult. For instance, drugs of the phenylethylamine type, opioids and histaminics have in common an amine group which is necessary for binding to the respective receptors. Hence it is likely that the receptors have a carboxyl or phosphate group at the recognition site. However, there is no positive evidence whether the H-bond between the amine and acidic site is of the neutral N...HO or ionic $NH^+...O^-$ type. Quantitative structure-activity relations in a series of β-adrenergic ligands indicate (83) that receptor affinity correlates with the electrostatic potential near the nitrogen, but not with proton affinity. This indicates that the neutral H-bonding is more likely for this type of ligands. Probably the best existing model for the ligand-receptor interaction is the tyroxine-prealbumin complex. Prealbumin is a plasma protein with high affinity for the thyroid hormone and its analogs. High resolution X-ray diffraction work (2, 84) has revealed the structure of the complex and the interaction was simulated by molecular mechanics calculations (85). Besides the hydrophobic interactions involving the aromatic residue and the iodine atoms of thyroxine the bonding is governed by the H-bond interaction between the thyroxine phenolic hydroxyl and a water molecule at the bottom of the prealbumine channel accomodating the hormone, and ionic H-bonds between the thyroxine side chain and the lysine, and glutamic residues at the entrance to the channel.

DNA can also be considered a receptor since it is the target of several antibiotic and antitumor drugs. However, they function mostly by intercalation. On the other hand, H-bonding is expected to be important in nucleic acid-coat protein interaction, in the recognition of specific sites on the outer surface of the helices by various enzymes involved in the replication process, and by regulatory proteins involved in the expression of the genetic formation. The knowledge about these important interactions is mostly within the general terms of H-bond, hydrophobic and electrostatic interactions (86). However, some more detailed informations are available.

A possible role of tyrosine residues of a lac repressor of Escherichia coli with DNA has been invoked on hand of photo-CIDNP--^{1}H NMR experiments (87). The interactions between the outer surfaces of the NA helices and basic proteins like histones and polyamines like spermidine are important for the higher order structures of nucleic acids. Histones are rich in arginine and there-

fore the guanidinium side chains are prone to interact with the
phosphate backbone of NA. With the four protons of the guanidinium moiety a variety of H-bonds with phosphate groups are possible and were indeed observed on model compounds (for references
see 88). A NMR study of arginine rich oligopeptides suggest that
the asymmetric bonding, i.e. involving the "cis" protons should
be most stable (88). NMR investigations of the interaction of
spermine with polynucleotides and with t-RNA indicate that a specific interaction occurs with the 2'OH group of ribose and the
adjacent phosphate residue which may be important for enzyme specificity towards DNA and RNA (89).

H-BONDS IN SPECIAL MECHANISMS

We shall now briefly treat some special mechanisms in which
proton transfers governed by peculiar potential surfaces are involved. Strong polarizations and charge transfers may be transmitted by extended H-bond systems of this type. Many examples can
be found amongst catalytic processes where particular amino acid
formations provide the source of protons for the general acid ca-
talysis and also the proper spatial disposition of charges necessary both to create the appropriate potential for the proton movement and for polarizing the substrate.

In the large group of serine proteases. the amino acid triad
aspartate, histidine and serine play the key role in the catalytic mechanisms of hydrolysis of internal peptide bonds in a polypeptide substrate. The triad (Fig. 10) is supposed (90) to supply

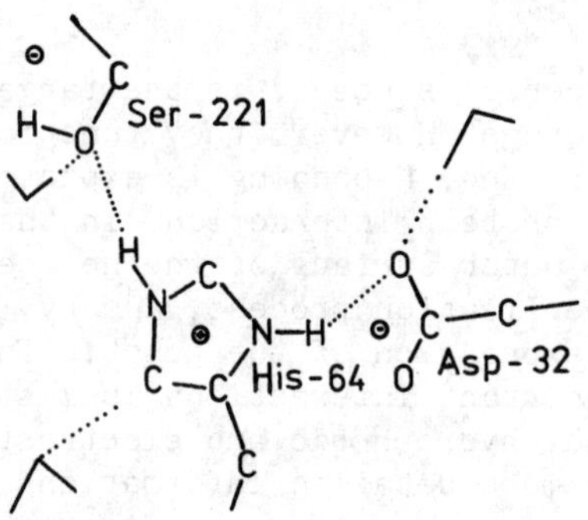

Fig. 10. The catalytic triad of
serine proteases surrounding water
molecules indicated by open triangles (after ref. 94).

and withdraw protons through concerted movements in the H-bond
system to the serine side chain to facilitate the nucleophilic
attack at the carbonyl carbon of the substrate leading to the acyl
serine intermediate and to assist the scission of the serine acyl
bond by activating a water molecule. In the acylation step, histi-

dine is supposed to accept the proton from serine and redeem it
to the leaving alcohol or amine. In the deacylation step, a proton is accepted from the water molecule to be returned to the
serine after the scission of the serine-acyl bond. The process
should be assisted by proton transfers between aspartate and histidine. Both the serine-histidine and histidine-aspartate H-bonds
were investigated by NMR spectrometry (91,92) and by theoretical
calculations on model systems (93,94). The significance of the
serine-histidine H-bond was questioned because of the large N...O
distance found by X-ray work on bacterial serine proteases (77).
However, infrared work on model systems shows that such a bond
can be strong (95) and it is possible that in the enzyme the questioned distance is modified upon substrate binding. Both experimental evidence (92) and model calculations (94) make the proton transfer between histidine and aspartate unlikely. Thus the
"proton shuttle" does not appear to be complete. Nevertheless,
the H-bond system in the triad is essential in providing the charge
movements necessary for the catalytic mechanism. The role of the
surrounding water and of other amino acid residues is essential
in providing the proper polar influences on the stability of the
active site and for the catalytic mechanism (94). The aspartate
proteases have a different charge relay system. Based on intermediate resolution (2.0 Å) X-ray structure determination (77) of
penicillopepsin the mechanism in Fig. 11 was proposed. The proton

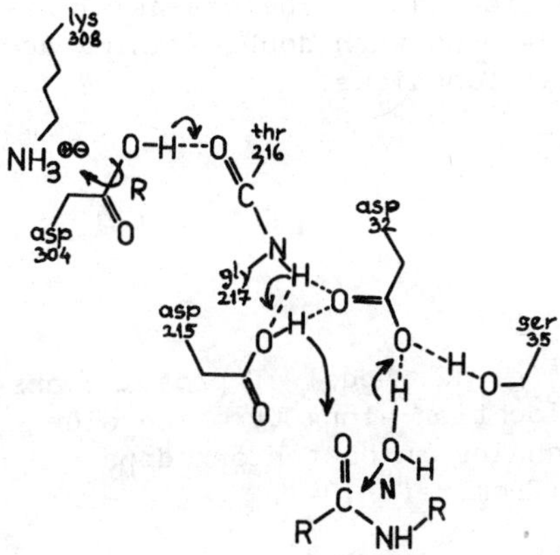

Fig. 11. Proposed initial stage
mechanism of aspartate proteases.
Arrows indicate proton transfers.
R indicates the rotation of
Asp 304 to accept charge from
Lys 308. and N the nucleophilic
attack of the deprotonated water molecule on the substrate
(after ref. 77).

shared between Asp 32 and Asp 215 seems to provide the strong
electrophile needed in the mechanism of aspartate proteases. The
very low pK_a of Asp 32 ($\simeq$ 1.3) is connected with the presence of
three H-bonds from neighbouring protein groups and one from a water molecule. The proposed mechanism does not involve any covalent
intermediate and may be analogous to that of metalloproteases if

the role of Zn^{2+} is equated to that of the electrophilic proton.
Tyr 75 of the aspartate protease would donate a proton to the
cleaved peptide similarly to Tyr 248 of carboxypeptidase A (77).
To be just to the variety of H-bonds participating in charge re-
lays we add one example of -SH...N bonding. This is papain where
cysteine-histidine-asparagine form the catalytic triad. The stabilization of the ion pair Cys^--His^+ is provided by the electric
field of the central α helix of papain the N-terminal of which is
close to the catalytic site (96). These few examples chosen amongst
many others should suffice to give a rough idea of the role of
H-bonds in catalytic mechanisms.

One of the most exciting problems in bioenergetics is the
transport of charges across membranes. The electric conductivity
of proteins was explored both for the possibility of electron and
proton conduction. Quantum mechanical calculations of the energy
bands in simple, periodic polypeptides have shown that the gap is
too large (97) although it may be reduced by the sorroundings (98).
Experiments on conductivity of bovine serum albumin and lysozyme
were interpreted in favour of protonic conduction (99). The main
argument is the lower conductivity of D_2O -hydrated protein than
of H_2O hydrated ones. Molecular mechanisms for proton transport
along H-bonded chains formed by hydroxyl groups of amino acids
side chains were proposed by Nagle and Morowitz (100) and by Knapp
et al. (101). In the latter case, serine side chains were explicitly considered. The general mechanism is analogous to that proposed for conductivity in water and ice (102). The essential requirement for such mechanisms are H-bonds with double minima and
low barriers in the proton potential functions.

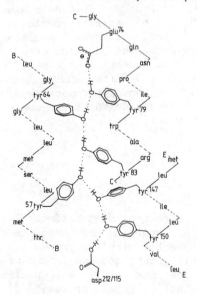

Fig. 12. Model of proton translocation along tyrosine side
chains in bacteriorhodopsin
(from ref. 104).

Potential energy surfaces for proton transfers in the simplest model, the protonated chain of water oligomers $H^+(H_2O)_n$ with n up to 5, were computed (103) from ab-initio wave functions (4-31G basis set). For O...O distances of 2.95 Å the activation energy of 16 kcal/mol was found. It is smaller for shorter distances. but increases with distortions of the chain and is independent of n. However. it is interesting that single proton transfers in the pentamer require the same activation energy as do simultaneous double proton transfers. This means that once the activation energy for the transfer of one proton is provided it will be sufficient to move all the protons in the chain. A realistic model (Fig. 12) of the proton translocating chain in bacteriorhodopsin was constructed by Merz and Zundel (104). It is based on tyrosine side chains in accord with the primary structure of the protein (105). The terminal aspartate and glutamate residues make the whole chain structurally symmetrical when one of these residues becomes protonated. The proton would be provided by the chromophore by photon activation. In this event specific, strong H-bonding probably of the double minimum type is again involved (106). Since the mechanism will be more extensively treated at this Institute we shall not discuss it here.

REFERENCES

1. Karle, I.L. 1978, Int. J. Quant. Chem. 5, p. 91.
2. Blake, C.C.F. and Oatley, S.J., 1977, Nature 268, p. 115.
3. Lewis, P.N., Momany, F.A. and Scheraga, H.A. 1973, Israel J. Chem. 11, p. 121.
4. Hadži, D. and Detoni S., in: "The Chemistry of Acid Derivatives", Suppl. B (Patai, S., ed., Wiley, Chichester 1979), p. 241.
5. Katz, J.L. and Post, B. 1960, Acta Cryst. 13, p. 624.
6. Donohue, J., in: "Structural Chemistry and Molecular Biology" (Rich, A. and Davidson, N., eds., Freeman, London 1968), p. 458.
7. Hagler, A.T., Huler, E. and Lifson, S. 1974, J. Am. Chem. Soc. 96, p. 5319.
8. Hinton, J.F. and Harpool, R.D. 1977, J. Am. Chem. Soc. 99, p. 349.
9. Hibberd, G.E. and Alexander, A.E. 1962, J. Phys. Chem. 66, p. 1854.
10. Peters, D. and Peters, J. 1980, J. Mol. Struct. 68, p. 255.
11. Boussard, G., Cung, M.T., Marraud, M. and Neel, J. 1974, J. Chem. Phys. 71, p. 1159.
12. Peters, D. and Peters, J. 1980, J. Mol. Struct. 64, p. 103.
13. Boussard, G., Marrana, M. and Aubry, A. 1979, Biopolymers 18, p. 1297.
14. Birkstoft, J.J. and Blow, D.M. 1972, J. Mol. Biol. 68, p. 187.
15. Peters, D. and Peters, J. 1980, J. Mol. Struct. 62, p. 229.

16. Venkatachalam. C.M. 1968, Biopolymers 6, p. 1425.
17. Peters, D. and Peters. J. 1980, J. Mol. Struct. 68, p. 243.
18. Prange, T. and Pascard, C. 1979. Acta Cryst. 35B, p. 1812.
19. Zimmerman, S.S., Pottle, M.S., Nemethy. G. and Scheraga, H.A. 1977, Macromolecules 10, p. 1.
20. Urry, D.W. 1978, Int. J. Quant. Chem. Q.B.S. 5, p. 51.
21. Abu Khaled. Md., Rennyopolakrishnan. V. and Urry, D.W. 1976, J. Am. Chem. Soc. 98, p. 7547.
22. Urry, D.W. and Lony, M.M. 1976, CRC Crit. Rev. Biochem. 4, p. 1.
23. Prasad, B.V.V., Shamala, N., Nagaraj, R., Chandrasekharan, R. and Balaram, P. 1979, Biopolymers 18. p. 1635.
24. Hendrickson, W.A., Love, W.E. and Karle, J. 1973, J. Mol. Biol. 74, p. 331.
25. Pauling, L. and Corey. R.B. 1951, Proc. Natl. Acad. Sci. USA 37, p.729.
26. Blundell, T.L. and Johnson, L.N.. "Protein Crystallography" (Academic Press, New York 1976). p. 35.
27. Abu Khaled, Md., Sugano, H. and Urry, D.W. 1979, Biochim. Biophys. Acta 577. p. 273.
28. Watson, H.C. 1969, Progr. Stereochem. 4, p. 299.
29. Lewis, P.N., Momany, F.A. and Scharaga, H.A. 1973, Biochem. Biophys. Acta 303, p. 211.
30. Peters, D. and Peters, J.   1980, J. Mol. Struct. 69, p. 259.
31. Lewis, P.N. and Scheraga H.A. 1971. Arch. Biochem. Biophys. 144, p. 576.
32. Takita. T., Oki, F., Maeda, K. and Umezawa, H. 1962, J. Antibiot. Ser. A, 15, p. 46.
33. Mizuno, K., Nishio, S. and Shindo, Y. 1979, Biopolymers 18. p. 693.
34. Chou, P.Y. and Fasman, G.D. 1974, Biochemistry 13, p. 211.
35. Lim, V.I. and Steinberg, S.V. 1981, FEBS Lett. 131, p. 203.
36. Ramachandran, G.N., Bansal, M. and Bhatnagar, R.S. 1973, Biochim. Biophys. Acta 322, p. 166.
37. Miller, M.H., Nemethy, G. and Scheraga, H.A. 1980, Macromolecules 13, p. 470, 914.
38. Grigera, J.R. and Berendsen, H.J.C. 1979, Biopolymers 18, p. 47.
39. Hospital, M., Courseille. C.. Leroy, F. and Roques, B.P. 1979, Biopolymers 18, p. 1141.
40. Bhatnagar, R.S. and Rapaka, R.S.. in: "Biomolecular Structure and Function" (Agris, P.F.. ed., Academic Press, New York 1978), p. 429.
41. Afinsen, C.B. and Scheraga, H.A. 1975, Adv. Protein Chem. 29, p. 205.
42. Chothia, C. 1975, Nature 254. p. 304.
43. Finney, J.L.1978. J. Mol. Biol. 119, p. 415.
44. Finney, J.L., Goodfellow, J.M. and Poole, P.L. 1982, in: "Structural Molecular Biology: Methods and Applications" (D.B. Davies, W. Saenger, and S.S. Danyluk, Eds., Plenum, New York), p. 387.

45. Sakabe, N, Sazaki, K. and Sakabe, K. 1981, Acta Cryst. A37 (Suppl.), C 50.
46. Ptitsyn, O.B. 1981, FEBS Lett. 131, p. 197.
47. Perutz, M.F. and Lehman, H. 1968, Nature 219, p. 902. Muirhead, P.H., Cox, J.M. and Goaman, L.C.O. 1968, Nature 219, p. 29.
47a. Arnott, S., Bond, P.J., Selsing, E. and Smith, P.J. 1976, Nucl. Acid. Res. 3, p. 2459.
48. Voet, D. and Rich, A. 1970. Progr. Nucl. Acid Res. Mol. Biol. 10, p. 183.
49. Ts'o, P.O.P., in: "Basic Principles in Nucleic Acid Chemistry" (Ts'o, P.O.P., ed., Academic Press, New York 1974), p. 517.
50. Kyogoku, Y., Lord, R.C. and Rich, A. 1969, Biochim. Biophys. Acta 179, p. 10.
51. Jardetzky. O. and Roberts, G.C.K.. "NMR in Molecular Biology" (Academic Press, New York 1981), p. 190.
52. D'Albis, A., Wickens, M.P. and Gratzer, W.P. 1975, Biopolymers 14, p. 1425.
53. Egan, B.T., Nir, S., Rein, R. and Mac Elroy, R. 1978, Int. J. Quant. Chem. Q.B.S. 5. p. 433.
54. Yanson, I.K., Teplitzky. A.B. and Sukhodub, L.F. 1978. Biopolymers 18, p. 1149.
55. Rein. R., in: "Intermolecular Interactions" (Pullman, B., ed., J. Wiley and Sons. Chichester 1978), p. 307.
56. Wada, A., Yabuki, S. and Husimi. Y. 1980, CRC Crit. Rev. Biochem. 9, p. 87.
57. Gotoh, O. and Tagashira, Y. 1981. Biopolymers 20. p. 1033.
58. Stulz, J. and Ackerman, Th. 1983, Ber. Bunsenges. Phys. Chem. 87, p. 443, 447.
59. Clementi, E. and Corongiu, G. 1980, J. Chem. Phys. 72, p. 3979.
60. Langlet, J., Giessner-Prettre, G., Pullman, B., Claverie, P. and Piazzola, D. 1980. Int. J. Quant. Chem. 18, p. 42.
61. Shibata, M., Kieber-Emmors, T. and Rein. R. 1983, Int. J. Quant. Chem. 23, p. 1283.
62. Clementi, E. and Corongiu, G. 1982, Int. J. Quantum Chem. Q.B.S. 9, p. 213.
63. Falk, M., Hartman, K.A.,Jr. and Lord, R.C. 1963, J. Am. Chem. Soc. 85, p. 391.
64. Dickerson, R.E., Drew, H.R., Conner, B.N., Wing, R.M., Fratini, A.V. and Kopka, M.L. 1982, Science 216, p. 475.
65. Bolton, P.H. and Kearns, D.R. 1979, J. Am. Chem. Soc. 101, p. 479.
66. Löwdin, P.O. 1964, Rev. Mod. Phys. 35, p. 724.
67. Clementi, E., Mehl, J. and von Niessen, W. 1971, J. Chem. Phys. 54, p. 508.
68. Sygula, A. and Buda, A. 1983, J. Mol. Struct. 92, p. 267. Palmer, M.H., Wheeler, J.R., Kwiatkowski, J.S. and Lesyng, B. ibid, p. 283. Mandragon, A. and Ortega Blake, I. 1982, Int. J. Quant. Chem. 23, p. 89 and references therein.

69. Nagaoka, S.. Hirota, N.. Matsushita. T. and Nishimoto. K. 1982, Chem. Phys. Lett. 92, p. 498.

70. Nagaoka. S., Terao, T.. Imashiro, F.. Saika, A. and Hirota,N. 1981, Chem. Phys. Lett. 80, p. 580.

71. Maranon, J., Grinberg. H. and Nudelman, N.S. 1982, Int. J. Quant. Chem. 22, p. 69.

72. Kothekar, V., Bolis. G. and Rein, R. 1983. Int. J. Quant. Chem. 23, p. 1295.

73. Weber, G. 1975, Adv. Protein Chem. 29, p. 61.

74. Quiocho, F.A. and Lipscomb. W.N. 1971, Adv. Protein Chem. 25, p. 1.

75. Nakagawa. S. and Umeyama. H. 1978. J. Am. Chem. Soc. 100, p. 7716.

76. Riordan. H.F., McElvany. K.D. and Borders, Jr, C.R. 1977. Science 195. p. 884.

77. James, M.N.G. 1980. Canad. J. Biochem. 58. p. 251.

78. Kester. W.R. and Matthews. B.W. 1977. Biochemistry 16, p. 2506.

79. Pai. E.E., Schirmer, R.H. and Schulz, G.E., in: "Structure of Complexes between Biopolymers and Low Molecular Weight Molecules" (Bartmann. W. and Snatzke. G., eds.. Wiley, Chichester 1982). p. 39.

80. Scheraga, H.A.. Pincus. M.R. and Burke. K.A., in: ibid, p.53.

81. Wippf, G., Dearing. A.. Weiner. P.K.. Blanley. J.M. and Kollman, P.A. 1983. J. Am. Chem. Soc. 105. p. 997.

82. Dower, S.K. and Dwek. R.A., in: "Biomolecular Structure and Function" (Agris, P.F.. ed.. Academic Press, New York 1978), p. 295.

83. Šolmajer. T., Lukovits. I. and Hadži. D. 1982. J. Med. Chem. 25, p. 1413.

84. Blake, C.F., Period. Biol.. in print.

85. Blaney, J.M., Weiner, P.K., Dearing. A., Kollman, P.A., Jorgensen, E.C., Oatley, S.J., Burridge, J.M. and Blake. C.F. 1982. J. Am. Chem. Soc. 104. p. 6324.

86. Schimmel, P. 1979. Adv. Enzymol. 49, p. 187.

87. Buck, F.. Rüterjans, H., Kaptein, R. and Beyreutter, K. 1980, Proc. Natl. Acad. Sci. USA 77, p. 5145.

88. Klevan, L. and Crothers, D.M. 1979. Biopolymers 18, p. 1029.

89. Bolton, P.H. and Kearns, D.R. 1978. Nucl. Acid Res. 5, p. 1315.

90. Blow, D.M., Birktoft, J.J. and Hartley, B.S. 1969, Nature 221, p. 337.

91. Robbilard. G. and Schulman. R.G. 1974, J. Mol. Biol. 86, p. 519.

92. Bachovchin, W.W. and Roberts. J.D. 1978, J. Am. Chem. Soc. 93, p. 8149.

93. Umeyama, H. and Nakagawa, S. 1979, Chem. Pharm. Bull. 27, p. 1524.

94. Naray-Szabo, G.. Kapur, A., Mezey, P.G. and Polgar, L. 1982, J. Mol. Struct. THEOCHEM 90. p. 137.

95. Zundel, G. 1978, J. Mol. Struct. 45. p. 55.

96. Van Duynen, P.Th., Tholl, B.Th. and Hol, W.G.J. 1979, Biophys. Chem. 9, p. 273.
97. Ladik, J., Suhai, S. and Seel, M. 1978, Int. J. Quant. Chem. Q.B.S. 5, p. 35.
98. Kertesz, M., Koller, J. and Ažman. A. 1977, Nature 266, p. 278.
99. Behi, J., Bone, S., Morgan, H. and Pethig, R. 1982, Int. J. Quant. Chem. Q.B.S. 9, p. 367.
100. Nagle, J.F. and Morowitz, H.J. 1978, Proc. Natl. Acad. Sci. USA 75, p. 298.
101. Knapp, E.W., Schulten, K. and Schulten, Z. 1980, Chem. Phys. 46, p. 215.
102. Onsager. L., in: "The Neurosciences" (Schmitt, F.O., ed., Academic Press, New York 1975), p. 493.
103. Scheiner, S. 1981, J. Am. Chem. Soc. 103, p. 315.
104. Merz, H. and Zundel, G. 1981. Biochem. Biophys. Res. Commun. 101, p. 540.
105. Ovchinikov, Yu. A. 1982. FEBS Lett. 148, p. 179.
106. Leclercq, J.-M., Dupuis, P. and Sandorfy. C. 1982. Croat. Chem. Acta 55, p. 105.

RAMAN STUDIES ON DINUCLEOSIDE MONOPHOSPHATES

E.D.Schmid and V.Gramlich

Institut fuer Physikalische Chemie,
Universitaet Freiburg, Albertstrasse 23a,
78 Freiburg, F.R.Germany

ABSTRACT

Intermolecular interactions of dinucleoside
monophosphates in aqueous solution have been
investigated by Raman spectroscopy. The measurements
show that the sequence isomers have different
relative conformations owing to different intra- and
intermolecular interactions. The investigations show
that the dinucleotides ApG, GpA, UpG, GpC and GpG
can form ordered structures under the conditions used.

INTRODUCTION

Dinucleoside monophosphates have been studied
extensively as model compounds in order to obtain
information about the structure and conformation of
polynucleotides and nucleic acids. The conformational
and structural behaviour of the dimers has been
investigated by various methods, but only a few
dinucleoside phosphates have so far been investigated
by Raman spectroscopy (1,2). The Raman investigations
published so far reveal that valuable information on
conformational problems of dinucleoside
monophosphates may be obtained by this method. We
therefore tend to investigate the spectral properties
of all dinucleoside monophosphates. In this paper we
discuss the sequence and temperature dependence of
structure and conformation of dinucleotides containing

C. Sandorfy and T. Theophanides (eds.), Spectroscopy of Biological Molecules, 87–89.
© *1984 by D. Reidel Publishing Company.*

guanine in neutral aqueous solution in order to establish the capability and type of selfassociation of these dinucleotides.

RESULTS AND DISCUSSION

All compounds were prepared as approximately 0,02 M solutions by dissolving the required amount in deionized water containing 0,05 M $KClO_4$ as an internal standard.

Sequence dependence: Our measurements show that the sequence isomers have different relative conformations owing to different intra- and inter-molecular interactions. The most striking differences are found between ApG and GpA. In neutral H_2O solutions at 18°C ApG shows more lines than GpA. The appearance of new lines in the 650-900 cm^{-1} region together with the low relative intensities of the ring modes at 728 (RI (ApG) = 42, RI=35 (GpA)) and 1478 cm^{-1} suggest that GpA forms a highly ordered secondery structure with strong intermolecular association. This is consistent with the fact that GpA forms a gel under the experimental conditions used in our investigation. The fact that the PO_2^- band in ApG is replaced by a few other lines in GpA seems to show that the PO_2^- group must also be involved in the intermolecular association, probably by hydrogen bonding.

Temperatur dependence: It can be shown that most of the spectral differences in ApG and GpA are removed on heating to 70°C. In both ApG and GpA, structure formation is accompanied by an intensity decrease of the band at 1485 cm^{-1}. This line results from a superposition of a strong guanine ring mode and an adenine ring vibration with low intensity (3). The guanine ring mode is believed to involve the N-7 = C-8 band stretch and the intensity of this vibration depends strongly on any molecular interaction with N-7 (4). Thus, Mansy et al. (5) have interpreted a decrease in the intensity of this band as due to hydrogen bonding at the N-7 position of guanine. The intensity decrease of this line can now be used as a measure of the thermal stability of the ordered structures. The plot of the relative peak heights of the 1485 cm^{-1} line as a function of temperature correspond to typical cooperative transitions with half-conversion temperatures of about 18°C for ApG and $37^{o}C$ for GpA.

In Table 1 the self-associating dinucleotides, the postulated type of base pairing and the half-conversion temperatures of the ordered structures are shown. We also studied the Raman spectral properties of all the other eleven 3',5'-ribodinucleoside mono-phosphates but we could not find self-association with intermolecular hydrogen bonding in these cases under the conditions used (nucleotide concentration o.o2 M; salt concentration o.1 M; pH = 7). obviously only guanine-containing dinucleotides are able to form ordered structures under these conditions. Details of the Raman spectra of dinucleotides are discussed by Schmid et al.(6).

Table 1: Postulated types of Base pairing and half-conversion temperatures of self-associating dinucleotides

Dinucleotide	Postulated type of base pairing	Half-conversion temperature
ApG	G-G (dimers)	18 ^{o}C
GpA	G-G (tetramers)	37
UpG	U-U and G-G (dimers) (possibly U-G)	15
GpC	G-C (dimers)	25
GpG	G-G (tetramers)	60

References:

(1) S.C. Erfurth, E.J. Kiser and W.L. Peticolas, Proc. Natl.Acad.Sci.USA 69,938 (1972)
(2) B. Prescott, R. Gamache, J. Livramento and G.J. Thomas,Jr., Biopolymers 13,1821 (1974)
(3) S.C. Erfurth and W.L. Peticolas, Biopolymers 14,1259 (1975)
(4) S. Mansy and W.L. Peticolas, Biochemistry 15,2650 (1976)
(5) S. Mansy, S.K. Engstrom and W.L. Peticolas, Biochem.Biophys. Res. Commun. 68,1242 (1976)
(6) V. Gramlich, R. Herbeck, P. Schlenker and E.D. Schmid, J. Raman Spectrosc. 7,101 (1978)
V. Gramlich, H. Klump and E.D. Schmid, J. Raman Spectrosc. 14,333 (1983)

MOLECULAR INTERACTION BETWEEN NUCLEIC ACIDS AND ALKYLATING
AGENTS BY RAMAN SPECTROSCOPY

A. Bertoluzza

Cattedra di Chimica e Propedeutica Biochimica, Centro
Studi Interfacoltà di Spettroscopia Raman,
Istituto Chimico
"G. Ciamician" University of Bologna, Italy.

INTRODUCTION

Reactions between electrophiles and nucleic acid bases are of great
significance and interest from a biological point of view, because
of their involvement in mutagenesis and carcinogenesis. It has been
suggested that DNA replication will lead to pairing errors, i.e.
to mutation, if the hydrogen bond interaction between DNA bases
are different from those proposed by Watson and Crick and involve
tautomers of bases (1).
Although this hypothesis has been subsequently extended to cover
the kinetics of DNA replication, the study of structural modifications of bases, capable of producing non   Watson-Crick complementary base pairs, is of paramount importance within the field of
mutation and carcinogenesis (2).
For such a study, Raman spectroscopy investigation has proved to
be a very useful tool.
Within the broader scope of a study of the correlations between the
structural modifications induced in the bases of nucleic acids by
electrophiles (alkylating agents) and their carcinogenic (mutagenic)
power, this work will single out the following points for discussion and illustration:
1) the molecular decomposition mechanism of three methylating agents,
   i.e. dimethylsulfate (DMS), methylmetanesulfonate (MMS), N-methyl-
   N-nitrosourea (MNU) which are characterized by different carcinogenic power;
2) the spectroscopic study of methylated and methylated-protonated

91

C. Sandorfy and T. Theophanides (eds.), Spectroscopy of Biological Molecules, 91–111.
© 1984 by D. Reidel Publishing Company.

derivatives of adenine;
3) the spectroscopic characterization of the product resulting
   from reaction of methylating agents (dimethylsulphate)   and
   nucleosides (adenosine), nucleotides (adenosine monophosphate)
   and polynucleotides (poli dA).

1. MOLECULAR REACTIVITY OF METHYLATING AGENTS.

The reactivity of the methylating agents examined, which have been
reported to differ greatly in their mutagenicity, shows the following trend: DMS<MMS<MNU (3).

1A. Dimethylsulphate $(CH_3O)_2SO_2$.

Fig. I shows the Raman spectra of liquid DMS and of its buffered
solution (after complete reaction) together with the spectrum of
the buffer solution blank (0.4 M NH_4HCO_3; pH 7.2) (4).

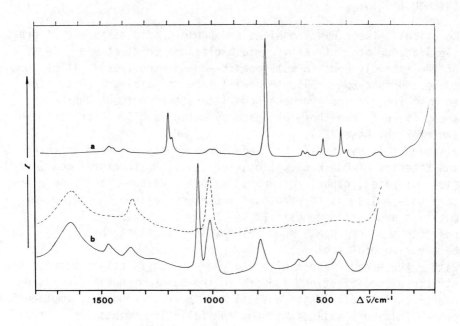

Fig. I - Raman spectra of: a) liquid dimethylsulphate (DMS)
b) DMS in buffered solution after complete transformation;
buffer solution (dotted line).

A comparison of the spectra is shown in Fig. 2, where it is apparent that the spectrum of the buffered DMS is very different from the spectrum of liquid DMS and more related to the spectra of the ion $CH_3OSO_3^-$ methylsulphate and of methanol CH_3OH.

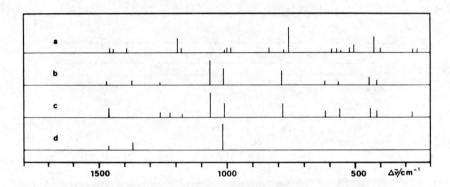

Fig. 2 - Comparison of the Raman spectra of: a) liquid DMS; b) DMS
in buffered solution after complete transformation; c)
aqueous solution of $CH_3OSO_3^-$ (alkaline methylsulphate);
d) methanol in buffered solution.

In particular, as seen in Table I, in the buffered solution spectrum the bands at 1000 and 985 cm^{-1}, assigned to in phase and out-
of-phase stretching O-C modes respectively, are missing, while a
new strong band appears at 787 cm^{-1}. This new band can be assigned
to a stretching S-O mode from the $CH_3OSO_3^-$ ion.
Similar changes are also observed in the spectral region above
1000 cm^{-1}, where the bands arising from O=S=O stretching appear
(at 1195 cm^{-1} the symmetric mode, at 1392 cm^{-1} the asymmetric mode).
These bands are missing in the spectrum of the aqueous buffered
DMS solution and are replaced   by two new bands, a strong one at
1065 cm^{-1} and a weak one at 1260 cm^{-1}, which can be attributed to
the symmetric and asymmetric stretching mode of the SO_3 group in
$CH_3OSO_3^-$.
Other minor modifications can be observed in the spectra.
In addition, the O-CH_3 stretching at~1000 cm^{-1}, characteristic of
the $CH_3OSO_3^-$ ion is not clearly visible in the spectrum owing to
the presence of a strong band of methanol CH_3OH and buffer at the
same wavelength.
Thus the Raman results make it possible to detect the hydrolysis reaction undergone by DMS in an aqueous buffered solution.

Table I. Raman spectra of liquid DMS, DMS in aqueous buffered solution (after complete transformation, $CH_3OSO_3^-$ (alkaline aqueous solution), CH_3OH (aqueous solution) and buffer (aqueous solution of NH_4HCO_3).

DMS (liquid) (5)	DMS (aqueous buffered solution)	$CH_3OSO_3^-$ (alkaline aqueous solution) (6)	CH_3OH (aqueous solution)	Buffer
$\tilde{\nu}/cm^{-1}$	$\Delta\tilde{\nu}/cm^{-1}$	$\Delta\tilde{\nu}/cm^{-1}$	$\Delta\tilde{\nu}/cm^{-1}$	$\Delta\tilde{\nu}/cm^{-1}$
110 w,b				
253 w				
270 w,sh		273 w,b		
401 w				
	415 w,sh	413 m ρSO_3		
426 s sciss O-S-O				
	445 m	438 m ρSO_3		
505 m wag O=S=O				
522 w,sh δSOC				
555 vw,sh δSOC	565 w	559 m $\delta s\ SO_3$		
573 w sciss O=S=O				
592 w rock O=S=O				
	618 w	615 m $\delta as\ SO_3$		635 w
				675 w
759 vs ν_sO-S-O				
778 vw,sh	787 s	781 s νSO		
835 w ν_{as}O-S-O				
985 w,sh νO-C				
1000 w νO-C				
1010 vw,sh	1013 s	1006 s,b	1020 vs νCO	1020 vs
	1065 vs	1063 vs $\nu_s SO_3$		1070 w
1178 w,sh rock CH_3		1174 vw,b		
1195 s ν_sO=S=O				
		1221 w,b $\nu_{as}SO_3$		
	1260 vw,b	1257 w,b $\nu_{as}SO_3$		
	1370 w			1365 s,b
1392 w ν_{as}O=S=O				
1445 w,sh δCH_3				
1460 w δCH_3	1470 w	1462 m,b	1465 m δCH_3	
	(*)		(*)	(*)
2855 w	2850 s	2845 m	2845 vs $\nu_s CH_3$	
	2920 w,sh			
2970 s $\nu_s CH_3$	2965 s	2958 s	2955 vs $\nu_{as}CH_3$	
	2985 w			
3035 w,sh $\nu_{as}CH_3$		3036 s,b		
3055 w				
	(*)		(*)	(*)

(*) broad strong water band

Such reaction leads to the formation of the corresponding $CH_3OSO_3^-$ ion and of methanol

$$
\begin{array}{c}
CH_3 \\
| \\
O \\
| \\
O=S=O \ + \ 2 \ H_2O \\
| \\
O \\
| \\
CH_3 \ + \ buffer
\end{array}
\longrightarrow
\left[
\begin{array}{c}
CH_3 \\
| \\
O \\
| \\
O=S=O \\
:| \\
O
\end{array}
\right]^-
\ + \ H_3O^+ \ + \ CH_3OH
\qquad
+ \ buffer
$$

It has also been observed that the same reaction occurs for an aqueous DMS solution in the absence of the buffer, and under these conditions DMS acquires, in addition to methylating character, a protonating character, due to the formation of H_3O^+. Consequently it is necessary to take into account both the formation of methylated and protonate species when studying the interaction between DMS and biological molecules (in the present case, the basis of the nucleic acids).

1B. Methylmethanesulphonate $(CH_3O)CH_3SO_2$.

The Raman spectra of liquid MMS, its buffered aqueous solution after complete transformation and the buffer solution are shown in fig. 3 and listed in Table II (4).
In this case, also, some interesting changes are observed.
The weak 814 cm^{-1} band, due to the S-O stretching mode, disappears. At the same time, the sharp 722 cm^{-1} stretching C-S band of liquid MMS is largely affected and shifted in the buffered solution to 787 cm^{-1}. As in the case of DMS, the sharp 1165 cm^{-1} symmetric stretching O=S=O band and the weak 1355 cm^{-1} asymmetric O=S=O band are both modified giving rise to a symmetric stretching SO_3^- strong band at~1050 cm^{-1} and to a asymmetric stretching SO_3^- weak band at about 1200 cm^{-1}.
Other smaller modifications are observed in the Raman spectra.
Therefore for MMS it is reasonable to postulate the same type of hydrolysis reaction as for DMS the case

$$
\begin{array}{c}
CH_3 \\
| \\
O=S=O \ + \ H_2O \\
| \\
O \\
| \\
CH_3
\end{array}
\quad + \ buffer
\longrightarrow
\left[
\begin{array}{c}
CH_3 \\
| \\
O=S=O \\
:| \\
O
\end{array}
\right]^-
\ + \ H_3O^+ \ + \ CH_3OH
\qquad
+ \ buffer
$$

Table II. Raman spectra of liquid MMS, MMS in aqueous buffered solution (after complete transformation), $CH_3SO_3^-$ (alkaline aqueous solution), CH_3OH (aqueous solution) and buffer (aqueous solution of NH_4HCO_3).

$CH_3^ISO_2OCH_3^{II}$ MMS (liquid) (7)	MMS (aqueous buffered solution)	$CH_3SO_3^-$ (alkaline aqueous solution) (8)	CH_3OH (aqueous solution)	Buffer
$\tilde{\Delta\nu}/cm^{-1}$	$\tilde{\Delta\nu}/cm^{-1}$	$\tilde{\Delta\nu}/cm^{-1}$	$\tilde{\Delta\nu}/cm^{-1}$	$\tilde{\Delta\nu}/cm^{-1}$
270 w				
330 w,sh				
343 m	350 m	348 m,b $\delta_{as}CS$		
446 m				
530 m δSO_2	530 w,sh	533 w δSO_3		
554 w δSO_2	560 m	560 m δSO_3		635 w
	680 vw			675 w
722 vs νC-S				
	787 s	789 s ν_s C-S		
814 w νS-OCH$_3$				
977 w	972 w	970 vw		
998 w $^{wCH}_3$	1002 w,sh			
	1022 m		1020 vs νCO	1020 vs
	1050 vs	1054 vs $\nu_s SO_3$		
				1070 w
1165 s $\nu_s SO_2$				
	1185 vw,sh			
	1210 w,b	1200 w,b $\nu_{as}SO_3$		
	1280 vw			
1332 vw,sh $\delta_s CH_3^I$				
1355 w $\nu_{as}SO_2$				
		1370 w,sh		1365 s,b
		1385 w		
1418 w $\delta_{as}CH_3^I$	1430 w	1429 m $\delta_{as}CH_3$		
1464 w,sh $\delta_{as}CH_3^{II}$	1475 w		1465 m δCH_3	
	(*)		(*)	(*)
2655 vw				
2815 vw,sh				
2850 w $\nu_s CH_3^{II}$	2850 m		2845 vs $\nu_s CH_3$	
2865 vw,sh				
	2920 vw,sh			
2942 s $\nu_s CH_3^I$	2945 vs	2944 s $\nu_s CH_3$		
2966 m $\nu_{as} CH_3^{II}$			2955 vs $\nu_{as} CH_3$	
	2990 m			
3028 m $\nu_{as} CH_3^I$	3030 w,sh	3024 m $\nu_{as} CH_3$		
	(*)		(*)	(*)

(*) broad strong water band

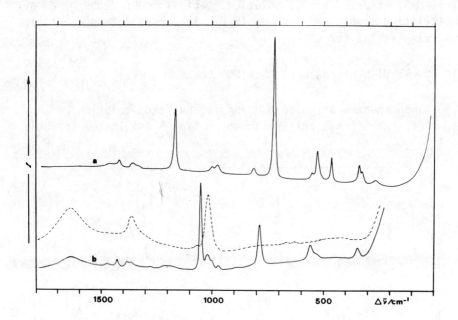

Fig. 3 - Raman spectra of: a) liquid methylmethanesulphonate MMS;
b) MMS in buffered solution after complete transformation; buffer
solution (dotted line).

This assumption is supported by the good agreement, as shown in
Fig. 4 between Raman spectrum of the MMS (buffered solution) alkaline methansulphonate, methanol and buffer solution.

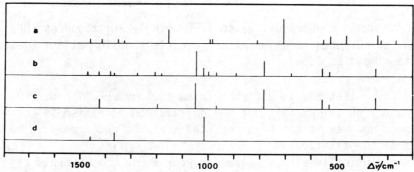

Fig. 4 - Comparison of the Raman spectra of a) liquid MMS; b) MMS
in buffered solution after complete transformation; c) aqueous so
lution of $CH_3SO_3^-$ (alkaline methanesulphonate); d) methanol in buf
fered solution.

This MMS buffered aqueous solution reacts through methanol as a methylating agent, and through $(H_3O)^+$, as a protonating agent, in the same way as for DMS.

1C. N-methyl-N-nitrosourea $CH_3N(NO)CONH_2$.

The Raman spectra of solid MNU and its buffered solution (after complete transformation) are shown in fig. 5 and listed in Table III

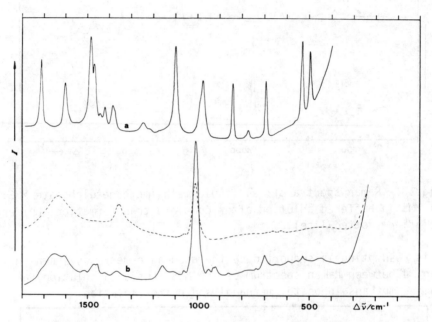

Fig. 5 - Raman spectra of: a) solid N-methyl-N-nitrosourea (MNU); b) MNU in buffered solution after complete transformation; buffer solution (dotted line).

Is is known that MNU is unstable in an aqueous solution buffered at neutral pH values (9), but the degradation mechanism is not well defined. However it has been reported that MNU decomposes to form methanol quantitatively.
This conclusion is also confirmed by our Raman measurements (fig. 6). The aqueous buffered MNU solution shows evolution of gas after stirring for 3 hours and the Raman spectrum is reproduceable only after 24 hours.

Table III. Raman spectra of solid MNU, MNU in aqueous buffered solution (after complete transformation), CH_3OH (aqueous solution) and buffer (aqueous solution of NH_4HCO_3).

MNU (solid) $\tilde{\nu}/cm^{-1}$	MNU (aqueous buffered solution) $\tilde{\nu}/cm^{-1}$	CH_3OH (aqueous solution) $\tilde{\nu}/cm^{-1}$	Buffer $\tilde{\nu}/cm^{-1}$
215 s			
291 s			
	348 vw		
370 w			
395 w			
498 s			
	526 w		
535 s			
	555 vw		
	590 vw		
	645 vw		635 w
			675 w
695 s	700 m		
	735 vw		
775 w	780 vw		
843 s			
	860 vw		
	920 w		
	948 vw		
978 s	980 vw,sh		
995 m,sh	1006 vs	1020 vs	1020 vs
	1060 vw		1070 w
1102 vs	1110 vw		
	1157 m		
1215 vw,sh			
1248 w			
	1315 vw,sh		
	1365 w		1365 w
1383 m			
1422 m	1420 vw		
1442 w	1450 m		
1465 m,sh	1470 m	1465 m	
1485 vs			
	1510 vw		
1602 s	1602 w		
	(*)	(*)	(*)
1710 s			
	1720 vw,sh		
	1765 vw,sh		
	2845 s	2845 vs	
2905 vw			
2955 s	2955 s	2955 vs	
2998 m			
	(*)	(*)	(*)

(*) broad strong water band

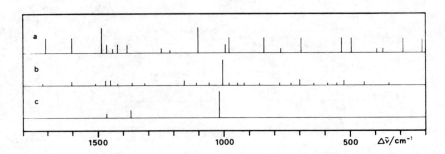

Fig. 6 - Comparison of the Raman spectra of a) solid MNU; b) MNU
in buffered solution after complete transformation; c) methanol
in buffered solution.

2. STRUCTURE AND REACTIVITY METHYLATED AND METHYLATED-PROTONATED
ADENINE DERIVATIVES.

In this work we discuss the Raman (and infrared) spectra of the
adenine $C_5H_5N_5$ and its derivatives, 1-methyladenine $C_5H_4N_5CH_3$ and
1-methyladenine hydrochloride $C_5H_4N_5CH_3 \cdot HCl$. Some molecular aspects
deserve attention. They are:
- a shift of the tautomerism from free base in the derivative
  compounds;
- different reactivity in the new tautomeric forms;
- H-bond coupling in the derivative compounds different from that
  established by Watson and Crick (1) for the free base.

2A. Adenine $C_5H_5N_5$.

For adenine, eight tautomeric forms can in principle be envisaged.
Four are of "aminic" nature and can be referred to the more stable
aminic form with H attached to N in 9 position

shifting the H atom to positions 1 or 3 or 7. The other four tautomers are of "iminic" nature

A number of theoretical calculations indicate that N(9)H the aminic form   is the most stable (10) and this conclusion is supported by I.R., Raman and NMR spectra (11-18).

2B. 1-methyladenine.

For 1-methyladenine previous infrared spectra (19-20) have been
reported and they suggest an iminic structure. Also for 1-methylade-
nosine (which is isoelectronic with 1-methylinosine (21)) an iminic
form has been proposed (26).
A theoretical investigation also suggests that the imino structure
is intrinsically more stable than the amino one (22).
On the contrary, preliminary NMR data (23) are consistent with a
predominant amino structure.
Raman and X-ray spectra have not been examined. Our Raman and infrared spectra (24) are shown in figg. 7 and 8 and the results limited to the spectra region 1700-1250 cm^{-1} listed in table IV and
compared with those of inosine and 1-methylinosine, the latter is
known to have a structure similar to the iminic form of adenine.
Some main features are shown in the spectra of 1-methyladenine:
the appearance of a band around 1645 cm^{-1}, which is absent in the
spectra of adenine and may correspond to the localized C=O stretching mode at about 1690 in the Raman spectra of inosine and me-
thylinosine; the disappearance of the band at about 1300 cm^{-1}, which
in the spectrum of adenine is due to the stretching $C-NH_2$ mode; the
disappearance of the δNH and γNH deformation modes of imidazole
ring, which respectively appear at about 1250 and 870 cm^{-1} in the
spectra of adenine and are absent in its sodium salt (13); the appearance of the δNH_2 deformation mode at about 1700 cm^{-1}.
These data suggest the hyphothesis of a form of 1-methyladenine
derived from the iminic one through intermolecular hydrogen-bond

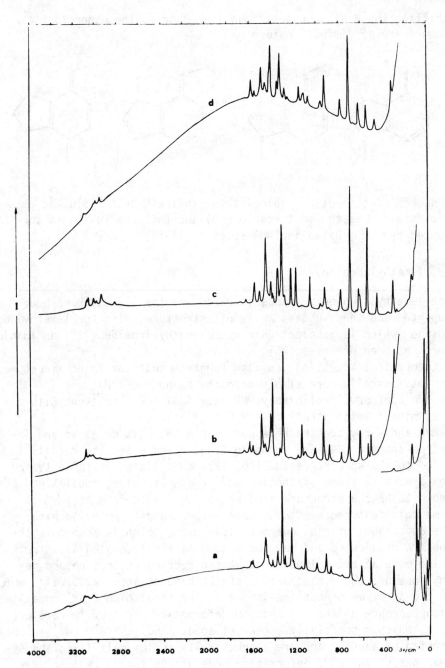

Fig. 7 - Raman spectra of: a) adenine, adenine hydrochloride,
c) I-methyladenine, d) I-methyladenine hydrochloride.

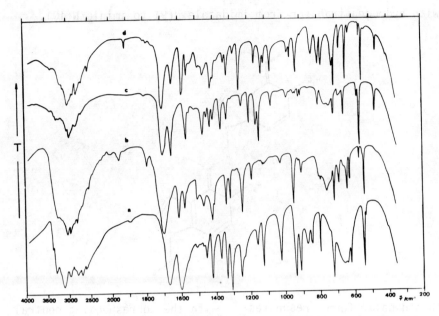

Fig. 8 - I.R. spectra of: a) adenine, b) adenine hydrochloride, c) 1-methyladenine, d) 1-methyladenine hydrochloride.

Table IV. Raman and infrared spectra of adenine, 1-methyladenine and 1-methyladenine hydrochloride.

(11)	Adenine (crystalline) I.R. $\tilde{\nu}$/cm	Adenine Raman $\Delta\tilde{\nu}$/cm	Inosine (H$_2$O solution) (2) Raman $\Delta\tilde{\nu}$/cm	1-methylinosine (H$_2$O solution) (2) Raman $\Delta\tilde{\nu}$/cm	1-methyladenine (crystalline) I.R. $\tilde{\nu}$/cm	1-methyladenine Raman $\Delta\tilde{\nu}$/cm	1-methyladenine hydrochloride (cryst.) I.R. $\tilde{\nu}$/cm	Raman $\Delta\tilde{\nu}$/cm
			νC=O 1690 (2)	νC=O 1687 (1)	δNH$_2$ {1700 vs / 1684 m,sh}	1695 vw	δNH$_2$ 1695 vs	
νNH$_2$	1670 vs	1675 vw,b			νC=N 1647 vs	1645 w	νC=N 1640 s	1640 w
8 apy	1600 vs	1595 w	1594 (3)	1587 (3)	1560 s	1565 m	1580 vs	1585 m
8 bpy	1555 w,sh		1554 (10)	1555 (10)	1540 w,sh		1553 m	1550 s
			1518 (4)	1514 (8)	1510 vw	1520 w	1492 w	1495 s
R$_{Im}$	1505 w	1483 m						
			1472 (7)	1480 (2)	1460 s	1463 s	1466 w	1458 m
19 bpy	1450 m	1442 vw			1440 m	1435 vw,sh		
19 apy	1420 s	1418 w	1422 (3)	1434 (3)	1420 s	1420 m	1420 s	1423 m,sh
R$_{Im}$	1417 s							
					1403 m	1405 w	1407 s	1410 s
						1395 vw		
14py	1368 m	1372 w	1382 (4) / 1351 (4)	1386 (3)	1360 s	1358 m	1344 m	1350 m
R$_{Im}$	1333 s	1333 sh	1322 (5)	1336 (7)	1323 s	1325 s	1324 m	1324 s
νC-NH$_2$	1300 vs	1307 w			νC-NH$_2$ -	-	νC-NH$_2$ -	-
δNH$_{Im}$	1250 s	1250 s	1272 (1)	1273 (0)	δNH$_{Im}$ -		δNH$_{Im}$ 1285 m	1280 m

with hydrogen transfer from imidazole N(9) to iminic N(10).

This dipolar form   resonates   with the corresponding neutral
form, postulated by ultraviolet and NMR data of 1-methyladenine
in solution (25)

In the solid state the ionic form may be prevalent. This hypothe
sis seems to be supported by investigations at present in progress
in our laboratory (27), which indicate that a proton transfer bet
ween nitrogen acid-base pairs occurs when there is a large $|\Delta pK_a|$
difference (i.e. the difference between the acid pK_a and pK_a of
the acid conjugate to the base).

2C. 1-methyladenine hydrochloride.

No I.R., Raman, X-ray or NMR spectra have been reported in the
literature for 1-methyladenine hydrochloride. If the structure
proposed for 1-methyladenine is that above, for its hydrochloride
a resonance between the structures

can be envisaged.

The close similarity of the I.R. and Raman spectra of 1-methyladenine hydrochloride and the analogous derivative of inosine (21) and adenosine (26) justifies the postulation of a predominanting structure with $C=N_{10}$ double bond.

The appearance in the spectra of the bands due to the stretching mode $C=\overset{+}{N}_{10}$ at about 1650 cm^{-1}, to the δNH imidazole deformation at about 1280 and the absence in the spectra of the stretching νC-NH$_2$ band is significant.

All the specie discussed show a change of tautomery toward the i-minic form with double character between carbon (6) and nitrogen (10), which is the least stable at room temperature for adenine.

This change of tautomery involves a different molecular reactivity due to a different distribution of the π bond in the molecular skeleton.

But, what is most important for the purpose of mutagenesis and carcinogenesis, is that the products of the methylation of the adenine are subjected to an interaction of the hydrogen bond between the bases which is different from the canonic interaction postulated by Watson and Crick (1).

In particular we consider the most simple case of the adenine hydrochloride derivative (2). While the adenine has a preferential Watson - Crick coupling with thymine

A = T

the adenine hydrochloride shows a preferential pairing with cytosine

$$A^+ = C$$

In the same way the methylated and protonated methylate derivatives of adenine give rise to unpairing which is at the origin
of metagenesis and cancerogenesia phenomena.

3. RAMAN SPECTRA OF THE PRODUCTS FROM THE METHYLATION OF
   ADENOSINE, ADENOSINE MONOPHOSPHATE AND POLYADENYLIC ACID.

The methylation of adenosine and alkaline salts of adenosine-5'-
monophosphate and 5'-polyadenylic acid (poly dA) has been obtai
ned by leaving the derivatives of adenine to react in a buffered
medium (NH_4HCO_3) at pH 7,2 at 3°C for three hours (28).
The spectra of the mixtures are shown in figg.9,10,11.
Fig. 9 shows the spectrum of the adenosina, dimethylsulphate and
buffered mixture, and the spectrum is compared with the adenosina
and dimethylsulphate spectra which are both buffered. In the same
way, figures 10 and 11 show respectively the adenosine monophosphate, and poly dA spectra.
The changes in the spectra of methyl derivatives relative to the
region 1700-1300 are listed in Table V and compared with those of
literature on adenosine monophosphate at pH = 7,5 and 3,5 (29).
Considering also the behaviour of the relative intensity of the
bands, one observes a substantial agreement among the spectra of
the mixtures, in buffered solution, of adenosine-dimethylsulphate,
adenosinemonosulphate-dimethylsulphate, poly dA-dimethylsulphate
with the spectra of I-methyladenosine phosphate in acqueous solution at 7.5 and 3.5 pH.

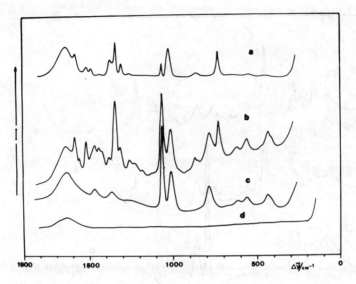

Fig. 9 - Raman spectra acqueous buffered solution of: a) adenosine; b) adenosine + dimethylsulphate; c) dimethylsulphate;
(Raman spectra of liquid water, d)).

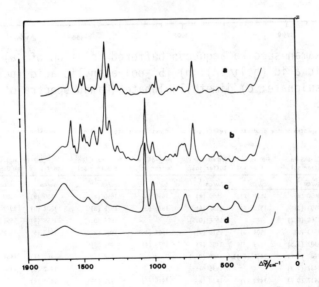

Fig. 10 - Raman spectra acqueous buffered solution of: a) adeno
sine-5'-monophosphate; b) adenosine-5'-monophosphate + dimethylsulphate; c) dimethylsulphate; (Raman spectra of liquid water, d)).

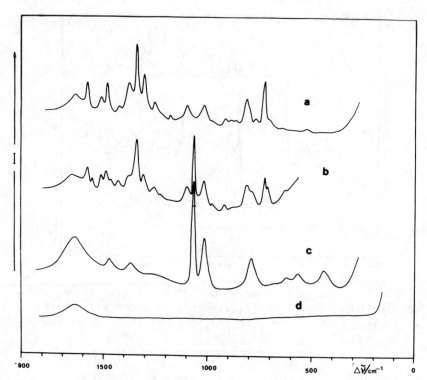

Fig. 11 - Raman spectra acqueous buffered solution of: a) 5'-polyadenylic acid (poly dA); b) 5'-polyadenylic acid (poly dA) + dimethylsulphate; c) dimethylsulphate;(Raman spectra of liquid water, d)).

Table **V**. Raman spectra of the products of reaction between dimethylsulphate and adenosine, adenosine monophosphate and poly-dA.

ADS in buffer sol. $\tilde{\nu}/cm^{-1}$	ADS+DMS in buffer sol. $\Delta\tilde{\nu}/cm^{-1}$	AMP in buffer sol. $\Delta\tilde{\nu}/cm^{-1}$	AMP+DMS in buffer sol. $\Delta\tilde{\nu}/cm^{-1}$	Poly-dA in buffer sol. $\Delta\tilde{\nu}/cm^{-1}$	Poly-dA + DMS in buffer sol. $\Delta\tilde{\nu}/cm^{-1}$	1-CH$_3$AMP in pH 7.5 $\Delta\tilde{\nu}/cm^{-1}$	H$_2$O sol (26) pH 3.5 $\Delta\tilde{\nu}/cm^{-1}$
1650 (*)	1650 (*)	1650 (*)	1650 (*)	1650 (*)	1650 (*)	1680 (⁂)	1680 (⁂)
1583 (3)	1583 (3)	1576 (4)	1580 (4)	1572 (3)	1575 (3)	1582 (3)	1579 (3)
	1560 (0.7)		1556 (0.5)		1555 (0.7)	1559 (1)	1556 (1)
1510 (2)	1513 (3)	1506 (3)	1510 (4)	1505 (1.7)	1508 (2.2)	1512 (5)	1512 (5)
1485 (2)	1487 (1.5)	1479 (3)	1483 (2)	1475 (4)	1482 (3)		
	1460 (1.7)	1455 (0.4)	1460 (0.7)		1462 (1)	1467 (1)	1464 (1)
1430 (0.5)	1432 (1.3)		1435 (1)			1432 (3)	1430 (3)
	1417 (0.7)	1420 (1)	1420 (2)	1415 (1)	1420 (1)	1417 (4)	1414 (5)
1370 (5)	1377 (3)	1375 (4)	1377 (3)	1370 (4.5)	1378 (3)		
1337 (10)	1337 (10)	1333 (10)	1337 (10)	1335 (10)	1335 (10)	1337 (10)	1336 (10)

*) Water band; ⁂) Partially or completely obscured by solvent scattering.

The spectra of I-methyladenosine monophosphate at two dif
ferent pH values are almost identical and therefore in these pH
conditions the presence of an amine protonate form has been proposed (26).

The agreement among the Raman data relative to our systems with
those of the literature relative to I-methyladenosine monophosphate suggests the principal presence in our case, too, of a me
thyl product of the base coinciding with that of the structure men
tioned above.
It is interesting to observe how the structure mentioned above
corresponds to one of the two resonant forms proposed by us for
I-methyladenine hydrochloride, and how this resonance can be affected in the passage from the solid state to that in solution.
On the basis of the observations set out above it is therefore
possible to establish that the principal product of methylation
of the mixtures adenosine-dimethylsulphate, adenosinemonophosphate-
dimethylsulphate, and polydA-dimethylsulphate in a buffered solution at about 7 pH, consists of the protonate and methylate deri-
vate in position I on the base.
In this case too the same considerations are valid as were previously made in the case of the derivatives of the base, that is, a molecular reactivity due to the distribution of the bonds π in the mo
lecular framework, and an interaction of hydrogen bonds caused by
the base different from the canonical one postulated by Watson and
Crick.
This latter aspect, as has already been said, plays an important
role in the phenomenon of mutagenesis and carcinogenesis. Measurements are at present being made relative to the components of
the nucleic acids including bases different from adenine, and mea
surements of the interaction of the hydrogen bond of the derivatives of bases likely to show a coupling different from the preferential one of Watson and Crick.

This programme of research, which has been partly effected in col
laboration with the School of Chemistry of Bradford University
(Prof. D.A. Long) and financed by CNR, the Overall Project
"Controllo della Crescita Neoplastica" (Neoplastic Growth Control),
pursues the purpose of making a contribution of a molecular spectroscopic nature to the phenomenon of mutagenesis and carcinogene
sis in the products of reaction between alkylating substances and
nucleic acids.

REFERENCES

1) Watson, J.D. and Crick, F.H.C. 1953, Nature 171, pp. 964-967.
2) Topal, M.D. and Fresco, J.R. 1976, Nature 263, pp. 285-293.
3) Sun, L. and Singer, B. 1975, Biochem. 14, pp. 1795-1802.
4) Bertoluzza, A., Fagnano, C., Morelli, M.A. and Tosi, R.
   "Molecular interactions between electrophil and nucleic acids.
   Molecular reactivity of three typical methylating agents by
   Raman spectroscopy" in print on J. Raman Spectroscopy.
5) Christe, K.L. and Curtis, E.C. 1972, Spectrochim. Acta 28A,
   pp. 1889-1898.
6) Siebert, H. 1957, Z. Anorg. Allg. Chem. 289, pp. 15-28.
7) Freeman, D.E. and Hambly, A.N. 1957, Aust. J. Chem. 10, pp.
   239-49.
8) Simon, A. and Kriegsmann, H. 1956, Chem. Ber. 89, pp. 1718-
   1726.
9) Snyder, J.K. and Stock, L.M. 1980, J. Org. Chem. 45, pp. 1990-
   1999 and therein reported.
10) Mezey, P.G. and Ladik, J.J. 1979, Theor. Chim. Acta 52, pp.
    129-145 and therein reported.
11) Angell, C.L. 1961, J. Chem. Soc. 1, pp. 504-515.
12) Lord, R.C. and Thomas Jr., G. J. 1967, Spectrochim. Acta 23A,
    pp. 2551-2591.
13) Lautie, A. and Novak, A. 1974, J. Chim. Phys. Physicom. Biol.
    71, pp. 415-420.
14) Hock, M. J. R. 1981, Chem. Phys. Letters 78, pp. 592-595.
15) Kos, N.J., Van der Plas, H.C. and Van Veldhuizen, B. 1979,
    J. Org. Chem. 44, pp. 3140-3143.
16) Guo, J.M. and Li, N.C. 1977, Tung XU Shu Li Hsueh Pao 3,
    pp. 143-150.
17) Richard, R.E. and Thomas, N.A. 1974, J. Chem. Soc., Perkin
    Trans 2, pp. 368-374.
18) Jones, A.J., Grant, D.M., Winkley, M.W. and Robins, R.K. 1970,
    J. Am. Chem. Soc. 92, pp. 4079-4087.

19) Kanaskova, J.D., Sukhorukov, B.I., Peutin, J.A. and Komaroskaya, G.V. 1970, Izv. Acad. Nauk. S.S.S.R., Ser. Khim. 8, pp. 1735-1742.

20) Montgomery, J.A. and Thomas, H.J. 1965, J. Org. Chem. 30, pp. 3235-3236.

21) Mansy, S. and Tobias, R.S. 1974, J. Am. Chem. Soc. 96, pp. 6874-6884.

22) Pullman, B., Berthod, H. and Dreyfus, M. 1969, Theor. Chim. Acta 15, 265-268.

23) Townsend, L.B. 1973, Synth. Proced. Nucleic Acid Chem. 2, pp. 267-398.

24) Bertoluzza, A., Fagnano, C., Morelli, M.A., Tosi, R. and Long D.A. "Molecular interactions between electrophyles and nucleic acids. Raman and infrared spectra of adenine, adenine.HCl,(adenine)$_2$.HCl and of the corresponding I-methyl derivates in relation to tautomerism and mutagenic phenomena of bases of nucleic acids" in print on J. Raman Spectroscpy

25) Dreyfus, M., Dodin, G., Bensande, O. and Dubois, J.E. 1977, J. Am. Chem. Soc. 99, pp. 7027-7037.

26) Samir Mansy, Peticolas, W.L. and Tobias, R.S. 1979, Spectrochim. Acta 35A, pp. 315-329.

27) Bertoluzza, A., Marinangeli, A.M., Simoni, R. and Tinti, A. 1982, Ital. J. of Biochem. 6, pp. 466-468.

28) Unpublished results.

29) Sun, L. and Singer, B. 1975, Biochemistry 14, pp. 1795-1802.

RAMAN AND FT-IR SPECTROSCOPY OF DRUG-BIOLOGICAL TARGET
INTERACTIONS IN VITRO AND IN VIVO

M. Manfait[a], T. Theophanides[b], A.J.P. Alix[a]
and P. Jeannesson[c]

[a]Université de Reims, U.E.R. Sciences,
Laboratoire de Recherches Optiques, B.P. 347,
51062 Reims Cedex, France.

[b]Université de Montréal, Département de Chimie,
C.P. 6210, Montréal, Québec, Canada H3C 3V1.

[c]Université de Reims, U.E.R. Pharmacie,
Laboratoire de Biochimie,
51096 Reims Cedex, France.

1. INTRODUCTION

In the last decade, chemotherapy has been used alone or in association with surgery, radiotherapy and lately immunotherapy in
cancer treatment. This method has increased continuously and led
to a great improvement in the cure of tumours, in particular,
when the tumour cells have been spread all over the body.

Nearly fifteen years ago, the first antitumor compounds were
selected randomly by a large screening program and in many cases
without precise knowledge of the mode of action. Today the search
and design of new potential antitumor drugs have required metho-
dologies which permit elucidation in the first step of their in
vitro mechanism of action with biological targets (nuclear DNA,
membrane, protein, etc...). Moreover, little or nothing is known
about the in vivo molecular interactions of these drugs within
intact cancer cells, which undoubtly is the most accurate way of
reproducing and studying the biological events leading to pharma-
cological results.

For this purpose, laser Raman spectroscopy has proved to be a
113

C. Sandorfy and T. Theophanides (eds.), Spectroscopy of Biological Molecules, 113–136.
© 1984 by D. Reidel Publishing Company.

powerful and very sensitive method for structural studies of biological systems as well as for determining the intermolecular
interactions. This is in order to characterize either the chemical
parts or the atoms which are really implicated in the interactions.

Raman spectroscopy is compared favorably with current spectroscopic techniques. The real advantages of Raman spectroscopy as a
powerful technique used in the study of biological systems, can
be summarized as follows :

(i)   Raman spectroscopy is a non destructive technique and it is
        applicable without difficulties to samples in aqueous solu-
        tions of biological systems ;
(ii)   only a few μl of solution are necessary to record the Raman
        spectra and
(iii)   under resonance conditions, with the appropriate wavelength
        excitation versus the electronic absorption band of the
        molecular moiety (chromophore), Raman spectroscopy permits
        us to obtain a selective chromophore information ; today the
        use of the resonance effect is the best way for drug target
        interaction studies. In many cases, one has the choice to
        force resonance conditions on the drug alone or on the
        target.

In this respect we like to point out that Raman spectroscopy can
now be only applied in a sophisticated manner routinely, if one
has :

(i)   a tunable laser available in the UV-visible range (with
        continued or pulsed excitation) ;
(ii)   monochannel or multichannel spectrographs and spectrometers
        with a good rejection rate of background light ;
(iii)   in monochannel mode one uses a classical photomultiplier
        and in multichannel mode a linear photodiode array generally
        coupled with a light intensifier and
(iv)   a computerized system to drive the laser source, the spec-
        trometer and to enable several operations to be carried out
        on the stored spectra (difference spectra, smoothing, decon-
        volution, normalization, etc...).

Moreover, in the case of a study conduced at a micrometric scale
the extension of the usual Raman spectroscopy technique to the
NEW micro-Raman technique needs, in particular, the coupling of a
microscope (visible, UV) to a modified spectrometer (see the
M.O.L.E. of Pr Delhaye). It appears that the Raman and the micro-
Raman techniques are suitable for a study in vivo, either on a
small population of living cells (ca 5000) or on a single living
cell, in order to study the drug-target interactions at the intracellular level (membrane, nuclear DNA, protein, etc...).

Although the Fourier transformation formulae were known for more than one century, its application to infrared spectroscopy and to biological molecules is quite recent. Its application to several spectroscopic techniques has been generalized as the FT methods which are suitable to study even cellular systems.

A brief historical introduction to the previous pionnering works in this field is described first concerning the in vitro systems.

(i) Raman

. A few years ago, Chinsky et al. (1) studied the complex actinomycin-DNA. Actinomycin D (phenoxazine linked to two different peptide chains) is an antibiotic seen to be effective against many gram-positive and gram-negative organisms. The actinomycin chromophore gives resonance-enhanced Raman bands with laser excitation at 457.9 nm. Some intensity bands are sensitive to interaction with DNA which consists in an insertion of the actinomycin into the DNA groove (interaction phenoxazine ring-guanine). Later, the same authors refined the above proposed model of interaction by a UV-resonance Raman study of the complex (2).

. The above authors (3), worked on the complex ethidium bromide (trypanocidal drug) - DNA, showing by Raman the existence of a specific interaction in this system.

. Martin et al. (4), by conventional Raman spectroscopy investigated the distamycin-DNA and netropsin-DNA complexes. Both drugs are antiviral agents as well as oligopeptide antibiotics. The obtained data provided a binding model, where the pyrrole ring (methyl-groups) interacts with DNA and the peptide N-H groups are strongly hydrogen bonded with DNA.

. Fairclough et al. (5) using favorable preresonance Raman conditions (viz., bithiazole ring) in their study of bleomycin (antitumoral glycopeptide drug) bound to calf thymus DNA, showed a strong interaction located at the thymine bases.

. Some authors (6-7) worked on the drug methotrexate binding to L. Casei dihydrofolate reductase (DHR) and concluded by comparison of the Raman spectra of the native enzyme and the drug-DHR complex, that significant conformational changes occured in the DHR upon complexation. Also, the pteridine ring of methotrexate was shown to be protonated (8).

. T. O'Connor et al. (9) studied the complex cis-platinum-DNA by Raman difference spectroscopy and showed evidence of the formation of C-DNA upon metalation.

. Our experimental attempts to describe the drug-DNA inter-

actions by use of classical and/or resonance Raman spectroscopy
techniques have been made on adriamycin (10-11), cis-platinum
(12), new anticancer inorganic rings ; hexaziridinocyclotriphos-
phazene $(NPAz_2)_3$ (13) and pentaziridinocyclodiphosphathiazene
$(NPAz_2)_2NSOAz$, $Az=NC_2H_4$ (14) and on gallium chloride (15).

(ii) FT-IR

. Fourier Transform Infrared Spectroscopy applied to the
study of drug interactions has been used only recently, as in the
case of cis-platinum-calf thymus DNA complex (16) and in the case
of adriamycin-DNA complex (17).

The in vivo studies are now in progress. Interactions
between a major antitumor drug adriamycin with living cancer
cells, i.e. K 562, have been studied by resonance Raman spectroscopy (18), micro-Raman spectroscopy on a single cell (19) and by
FT-IR spectroscopy (17).

2. METHODOLOGY (See Scheme I)

The basis of a vibrational study in the drug-target interactions
requires for their understanding two major steps :

(i)   the recording of spectroscopic data on the free drug and
       bound drug in order to determine by a careful comparison of
       the treated data what part of the molecule or which atoms
       are implicated in the mechanism of interaction and

(ii) the spectroscopic data as above performed on the target
     (free and bound).

By using the classical techniques (Raman and FT-IR) steps (i)
and/or (ii) are simultaneously obtained (exception is done for
very low concentration of drug vs target, i.e. DNA). The key
technic used in both cases (i) and (ii) is the difference method.

The following concrete example (drug = hexaziridinocyclotriphos-
phazene) illustrates clearly the above procedure (step (i)). The
comparison is achieved by computer-subtracting variable amounts
of one spectrum from another. Previously, the various spectra are
normalized to the same relative Raman intensity, by using an
internal standard (here $\nu_s(C10_4^-) = 934$ cm^{-1}). The intensity of
the standard band scattering measures the combined effect of such
experimental factors as counting time, optical alignment and
laser power, (see Figure 1).

Comparison of the complete spectra of the solvent and the solution
(drug-DNA in the solvent) shows that a region around 1850 cm^{-1}
can be considered to have nearly "zero Raman intensity" at least

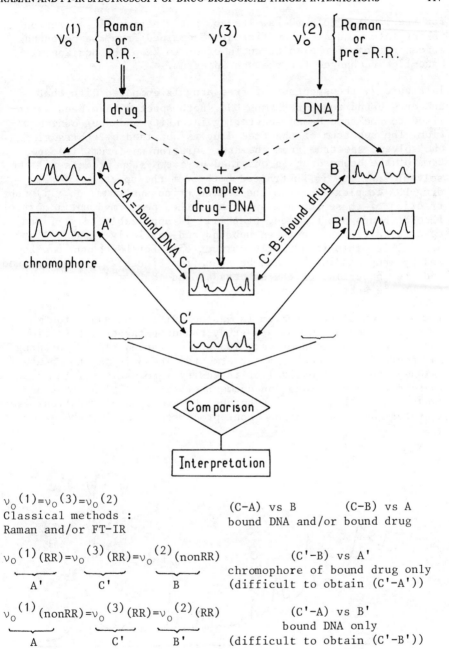

$$\nu_0(1)=\nu_0(3)=\nu_0(2)$$
Classical methods :
Raman and/or FT-IR

(C–A) vs B   (C–B) vs A
bound DNA and/or bound drug

$$\underbrace{\nu_0(1)\,(RR)}_{A'}=\underbrace{\nu_0(3)\,(RR)}_{C'}=\underbrace{\nu_0(2)\,(nonRR)}_{B}$$

(C'–B) vs A'
chromophore of bound drug only
(difficult to obtain (C'–A'))

$$\underbrace{\nu_0(1)\,(nonRR)}_{A}=\underbrace{\nu_0(3)\,(RR)}_{C'}=\underbrace{\nu_0(2)\,(RR)}_{B'}$$

(C'–A) vs B'
bound DNA only
(difficult to obtain (C'–B'))

Scheme I : Methodology of drug-DNA interaction studies

for the lower wave numbers. The solvent spectrum is subtracted
taking into account a coefficient determined from the compound
concentration of the solution in order to keep the horizontal
base-line unchanged for the new spectrum.

In Figure 2, the spectrum of free drug is compared with that of
the drug bound to calf thymus DNA. Both spectra have been normalized to the same Raman scattering intensity and drug concentration. The spectrum of the free drug is obtained by subtracting
the solvent spectrum from the drug solution spectrum. The spectrum of the bound drug is obtained by a two-step process (i) the
solvent spectrum is subtracted both from the spectrum of the
drug-DNA complex and from the DNA solution spectrum and (ii) subtraction of these two new spectra is then performed until most of
the DNA bands are removed. Step (ii) is applicable only if the
spectral difference between unbound and partially bound DNA are
negligible. This is strictly correct only for DNA bands unaffected by drug binding. As can be seen from Figure 2, the background
and the base-line of the spectra around 1850 cm^{-1} are not affected by the subtraction process.

The approach of the problem is rather different and often more
difficult when one wishes to study the drug-target interactions
under such conditions of very low concentration ($10^{-5}M$) of drug,
as required in chemotherapy treatments. The only way to counter-
balance the very low level of intensity information of the drug
is to use the resonance enhancement effect ($\times 10^4$) by the resonance Raman spectroscopy technique. However, only vibrations and
therefore only information about processes involving the chromo-

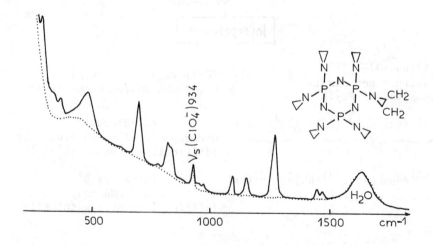

Figure 1.   Raman spectra of $(NPAz_2)_3$ in 10 mM $NaClO_4$ aqueous
            solution (−) and of the solvent (···)

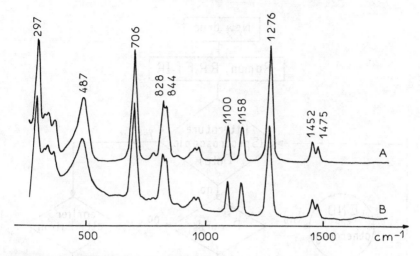

Figure 2. Raman spectra of unbound (A) and bound (B) $(NPAz_2)_3$
          after subtraction of free calf thymus DNA and
          $NaClO_4$ (10 mM) in aqueous solution. The Raman
          spectra are normalized to the same scattering
          intensity and drug concentration.

phore of the drug itself can be obtained. Anyhow, if one is interested by the mechanism of interaction at the target level, it is
then necessary to increase the drug concentration to see the modifications induced on the target. The subtracting technique may
be of use for the comparison of the spectra of the free and bound
systems. Similar data on the target are obtained by an other
appropriate choice of the wavelength (ν_0 which is near one of the
electronic absorption bands of the target) (see Figure 3).

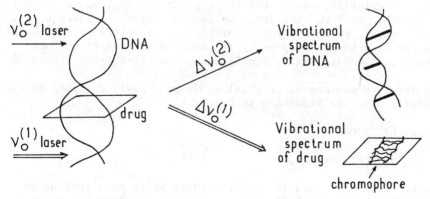

Figure 3. Wavelength laser excitations for visualization of
          the part of the drug–DNA complex by resonance
          Raman effect.

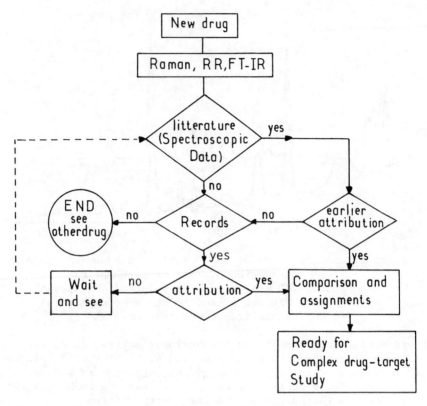

Figure 4. Drug study flowchart

Regarding the special case of some drugs such as, cis-platinum, gallium chloride, etc., we like to point out that as it is quite impossible to "see" the drug, in aqueous solution, consequently one has only to show evidence of the modifications of the target in the drug-target complex by the above described techniques (Raman, resonance Raman, FT-IR)(see below cis-DDP-DNA Figure 7).

In order to perform an original study on a newly designed or discovered drug see flowchart in Figure 4.

3. APPLICATIONS

3.1. In vitro studies

After having outlined the general ideas which permitted us to apply Raman and/or FT-IR spectroscopy to the study of the drug-target interactions in biological systems, we will now present some concrete examples and results obtained by us using these techniques which appear to embrace a wider range of applications.

3.1.1. The cis-platinum-DNA system

The antitumor drug cis-dichlorodiammine-platinum(II), cis-DDP cis-Pt(NH$_3$)$_2$Cl$_2$ (cis-platinum) was discovered by Rosenberg et al. in 1964 (20-21). The major problem was then to determine the mechanism of antitumor activity of cis-platinum as compared to the trans-isomer which is quite inactive against tumors (see Figure 5).

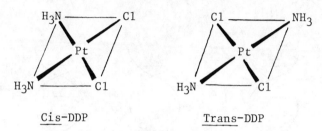

Cis-DDP   Trans-DDP

Figure 5. Cis- and trans-isomers of dichlorodiammineplatinum(II)

Results on these two isomers have led to studies which show, for example, that the drug cis-DDP reacts preferentially with the DNA bases (in particular with the G-C pairs of DNA) and that interstrand cross links are important (see Figure 6). Difference Raman spectra in which the solvent bands are subtracted, showing the changes occured in the carbonyl vibration band by complexation with low doses of the drug (Pt/Phosphate ratio $\cong$ 0.06) and other important effects which are characteristic of changes taking place in the secondary structure of DNA are reported in Figure 7. The major changes on DNA upon platination are described in Table 1.

DNA	DNA+cis-platinum	DNA+trans-isomer
684(G)	Small decrease	No change
838(OPO)	Shift to 834 cm^{-1}	Shift to 834 cm^{-1}
1238(T)	Small decrease	Increase
	1412(G$^+$)[a]	
1490(G,A)	Large decrease	Small decrease and shift to 1510 cm^{-1} (shoulder)
	1540(G$^+$)	1540(G$^+$) Slightly visible
1578(A,G)	Shift to 1584 cm^{-1}	No change
1628(C=O of G,C)	Shift to 1596 cm^{-1}	No change
1662(C=O of T)	Small decrease	Large decrease

[a]G$^+$: platined G.

Table 1. Modifications of some DNA Raman bands upon platination of the guanine N7 site

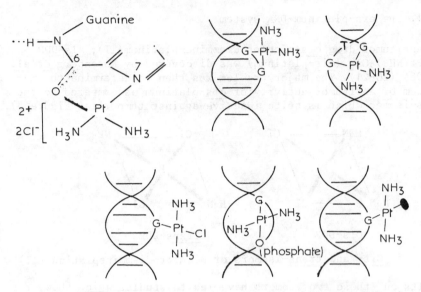

Figure 6. Types of interactions : cis-DDP and trans-DDP with DNA

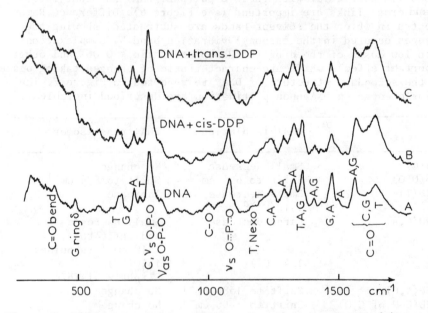

Figure 7. Difference Raman spectra of salmon sperm DNA (A),
cis-DDP + DNA (B) and trans-isomer + DNA (C). Concentration of DNA was 8.5 mg/ml in the presence of 8 mM NaCl.
The Pt/Phosphate ratio was 0.06 ; pH = 6.5 at 25°C.
Excitation wavelength 488 nm. The Raman spectrum of the
solvent has been subtracted in all three cases.

The results allow us to propose a model of drug action illustrated in Figure 8.

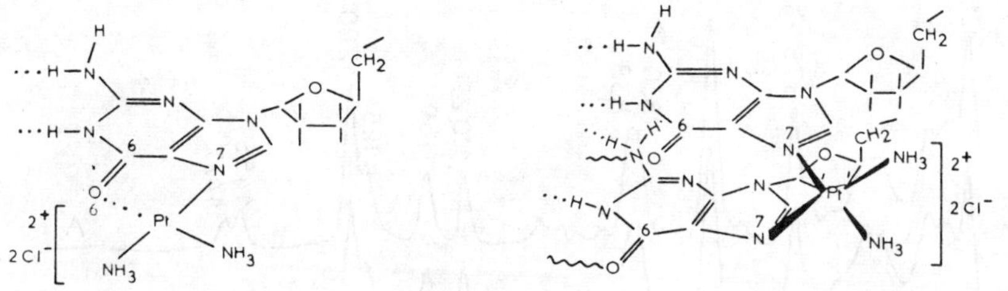

Figure 8. Models of binding of cis-platinum to guanine bases

The cis-platinum, thus binds in a specific manner to DNA requiring adjacent guanines or guanines brought together after local melting to form a spefific (G)N7-Pt-N7(G) binding which, at low doses of Pt/Phosphate does not seem to cause a great disruption of the secondary structure of DNA. In addition, the drug apparently affects stacking. The modification of the guanine carbonyl band could be explained either by a direct weak interaction of the drug with the carbonyl (Pt...O=C) after blocking one N7 site of guanine or by an indirect effect on the guanine bases by forming a PtG$_2$ adducts.

3.1.2. The SOAz – DNA system

Among all the new antitumor inorganic ring systems of the type cyclodiphosphathiazene (NPAz)$_2$(NSOX) designed by Labarre et al. (22-23), the most successful compound, in all the cases investigated, was SOAz, namely pentaziridinocyclodiphosphathiazene (NPAz)$_2$NSOAz (Az = aziridino = NC$_2$H$_4$) (see Figure 9).

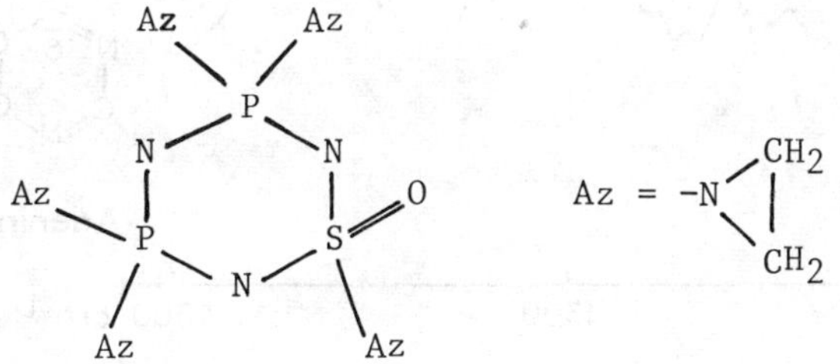

Figure 9. The antitumor drug pentaziridinocyclodiphosphathiazene namely SOAz

By using the difference Raman spectra, one can discriminate both the modifications induced in SOAz by its linkage to DNA and vice et versa.

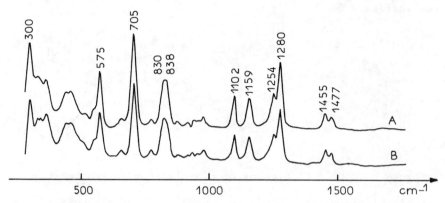

Figure 10. Difference Raman spectra of free SOAz (A) and bound
           SOAz (B)

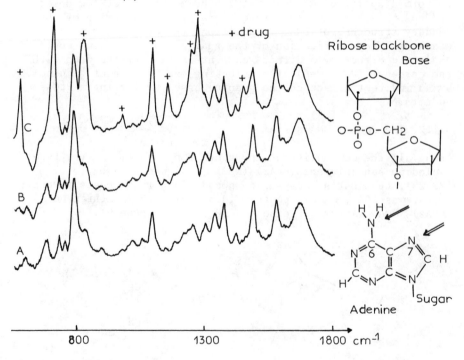

Figure 11. Difference spectra of free DNA (A), bound DNA (B) and
           SOAz-DNA complex (C).
           r = drug/base = 1, final DNA concentration = 2.8 mg/ml.
           Excitation wavelength 488 nm

Taking into account the previous frequency assignments made on
MYKO 63 (cyclophosphazene drug $N_3P_3Az_6$) (24), the assignments of
the main group frequencies of SOAz were obtained (25). The comparison between the bound and free SOAz difference spectra is shown
in Figure 10. The main alterations in intensity for the bands
attributed to a N_3P_2S ring vibration mode 705 cm^{-1} and to the
breathing mode of the aziridinyl ring linked to the phosphorus at
1280 cm^{-1} are clearly seen. The decrease in intensity of the
latter band is due to the mechanism of interaction of the aziridino ligands in the complex. The aziridinyl ring is opened to
give the corresponding $-NH-CH_2-CH_2-^+$ carbocation. Moreover, it
seems that complexation occurs essentially through the four Az
groups grafted on the two phosphorus atoms of the SOAz, the Az
group on sulfur being not involved in complexation. The spectra
(solvent subtracted) of free DNA (A), SOAz–DNA complex (C) and
bound DNA (B), obtained by subtracting the free drug spectrum
(see Figure 10 A) from the spectrum of the complex are shown in
Figure 11.

The greatest intensity decrease of the sugar–phosphate bands at
900 cm^{-1} (mixing of $\nu(C-C_{sug})$ with $\nu(C-O_{sug})$) and at 1016 cm^{-1}
($\nu(C-C-O)$ sugar-phosphate backbone) prove clearly that SOAz interacts with DNA at the ribose-phosphate backbone level, probably
either by grafting of its $-NH-CH_2-CH_2^+$ carbocations on the oxygens of the (C-C-O) links or by a weak disruption of the secondary
structure of DNA (some complementary effects support these pre-
sumptions). Bands characteristic of adenine (1181, 1304, 1341 cm^{-1})
which are slighty affected upon complexation are indicative of
second order interactions between drug and adenine with the N7
and the external NH_2 groups (dialkylation). Finally the non alteration of the specific guanine band (1492 cm^{-1}) on binding leads
to the conclusion that there is no interaction between SOAz and
guanine.

3.1.3. The adriamycin – DNA system

(i) Raman studies

The major anticancer drug adriamycin (ADM) (26) is obtained by a
biosynthetic method and is the 14-hydroxyl derivative of dauno-
mycin (isolated from cultures of streptomyces peucetius, which
is a var. caesius). ADM has the anthracycline structure linked to
an amino sugar (see Figure 12). It is believed that the antitumor
activity of the drug is due to its interference with DNA (27).
It has been shown that sugar residue do affect the nature of intercalation and the stability of DNA-ADM complex. All preliminary
works are consistent with intercalation of ADM into the helix of
DNA (28-29). Resonance Raman spectroscopy is a powerful technique
in order to determine the specific interaction of the chromophore

(quinonoid part of ADM) with the nucleic acid.

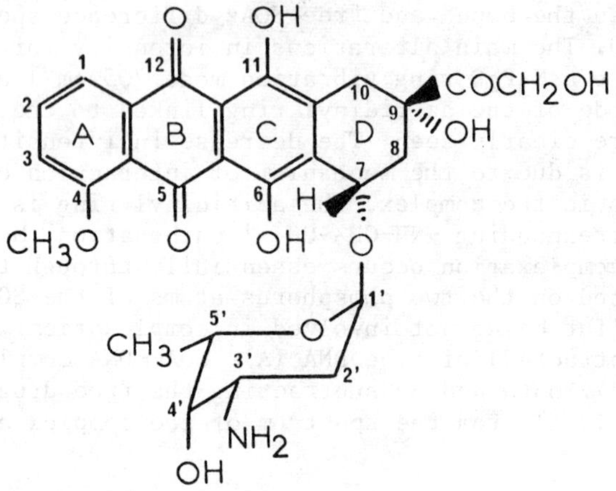

Figure 12. Structure of adriamycin

The Raman and resonance Raman spectra of calf thymus DNA and
DNA-ADM complexes formed with low concentrations of ADM which have
biological significance are shown in Figure 13. Evidence that all
ADM molecules are intercalated follows from the same appearance
of the resonance Raman spectra of the ADM chromophore regardless
of the ratio r = nucleotide/ADM varying from 1000 to 50 ; however,
it is seen that even for a small value of r $\cong$ 250 (see Figure 13C)
the Raman spectrum of DNA is hidden under the strong resonance
Raman bands of the chromophore. Then, in order to avoid difficulties in analysing the above spectra and to point out the interactions taking place between DNA and ADM, we have taken the UV-
Raman spectra in pre-resonance conditions vs DNA (see Figure 14).
Under these experimental conditions it is observed only the band
DNA spectrum of ADM-DNA complex (see Figure 15). The resonance
spectra of free ADM (quinonoid anthracycline chromophore) and
that of bound ADM to DNA obtained by subtraction from the Raman
spectrum of the complex (r = 500) are reported in Figure 16 and
show substantial changes on the bound drug.

Model building studies on the modifications of the DNA structure
which allow intercalation of a drug with an anthracycline structure such as ADM have been extensively reported (X-ray diffraction, crystal structures studies on drugs, fluorescence studies,
etc...). (see References given in (11)).

On the bases of our spectroscopic studies, we have proposed a
model describing the process of intercalation of ADM. The phenolic
vibrations (466, 1214-1246, 1306 cm^{-1}) of the chromophore of ADM

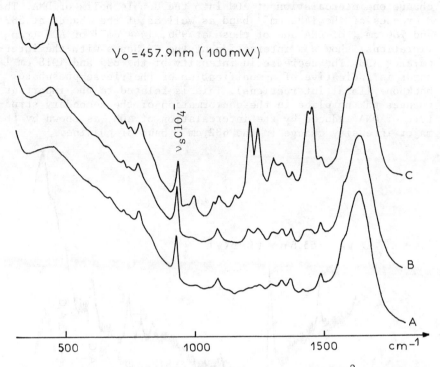

Figure 13. Raman spectrum of calf thymus DNA in 10^{-2}M NaClO$_4$
aqueous solution (A) ; Resonance Raman spectra of
ADM–DNA complexes in 10^{-2}M NaClO$_4$ aqueous solution
r(nucleotide/drug) = 1000 (B), r = 250 (C). Final DNA
conc. = 4 mg/ml. All intensities are normalized

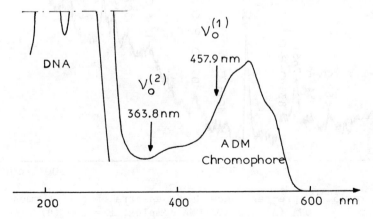

Figure 14. Absorption spectrum of ADM–DNA complex
showing the frequencies of the laser
excitation wavelengths

change on intercalation of ADM into the double helix of DNA. The decreases of the 1482 cm^{-1} band as well as of the bands at 682 and 786 cm^{-1} of DNA and of those at 996, 1444 cm^{-1} of ADM (ring vibrations) show a π interaction of the GC bases with the intercalator ADM. The decrease in intensity of the 835 and 1018 cm^{-1} bands is indicative of a modification of the ribose-phosphate backbone (local interactions). This is related to the important changes taking place in the conformation of the secondary structure of DNA induced by the intercalation of ADM, as shown by the major intensity change in the 682 cm^{-1} band of guanine.

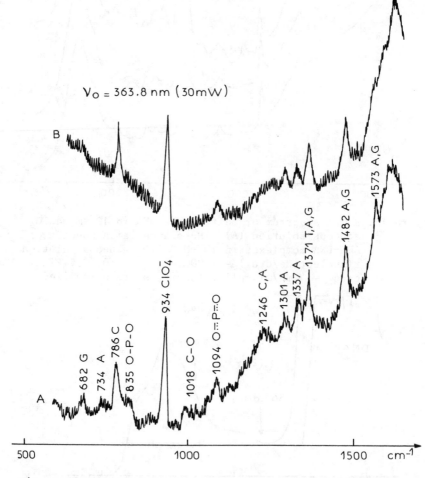

Figure 15. UV pre-resonance Raman spectra of calf thymus
DNA (A) and ADM-DNA complex, r = 200 (B).
Final DNA conc. = 1.8 mg/ml

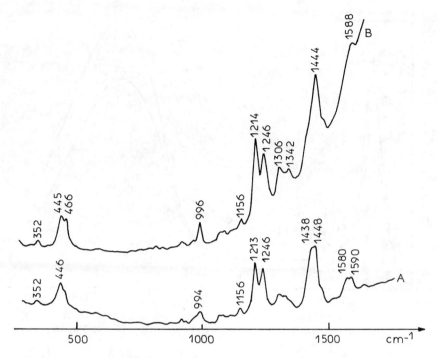

Figure 16. Difference resonance Raman spectra of bound ADM
to DNA, r = 500 (A) and free ADM 2.6 x 10^{-4}M in
aqueous solution (B). Excitation wavelength 457.9 nm

(ii) FT-IR studies

Fourier Transform Infrared Spectroscopy has been employed to study the adriamycin molecule and its interactions with the pharma-
cological target DNA. The spectrum of free DNA in 10^{-2}M NaClO$_4$
solution (see Figure 17A) and that of the solvent in aqueous
solution (see Figure 17B) have been recorded with a DIGILAB
FTS-15/C Fourier Transform Michelson infrared interferometer. The
power of the present FT-IR spectrometer equipped with a computer
is shown when the difference method is applied. Both zooming
spectra, show DNA (Figure 18A) and DNA-adriamycin, r = 200 (Figure 18B) spectra. They reveal that contrary to the Raman spectra
in the region 1300–900 cm^{-1}, the intense IR bands of the sugar
and the phosphate vibrations of DNA are observed. Intensity changes in the region 1500–1800 cm^{-1} confirm the intercalation of the
drug into the G–C planes of the double helix. On the other hand,
the decreasing of the 1087 cm^{-1} band ($\nu_s PO_2^-$) shows a drug phosphate interaction, indicating that probably it is the amino-sugar
moiety of drug adriamycin which interacts with the backbone of
DNA.

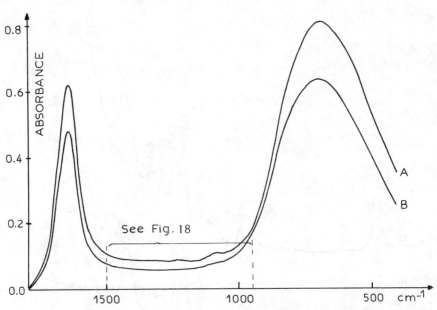

Figure 17. FT-IR spectra of calf thymus DNA (5 mg/ml in H_2O,
$10^{-2}M$ $NaClO_4$) (A) and solvent H_2O, $10^{-2}M$ $NaClO_4$ (B)

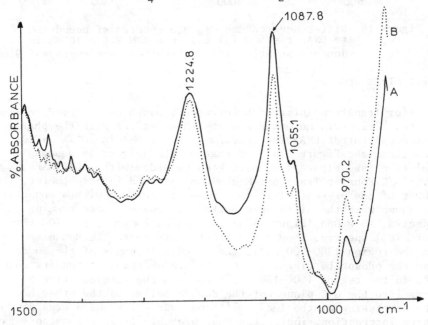

Figure 18. Difference FT-IR spectra of free DNA (A) and
ADM -DNA complex (r = 200) (B) after subtraction
of the solvent FT-IR spectrum

3.2 In vivo studies on cancer cells

Taking into account the obtained results on the nature of the interaction of adriamycin with the DNA by resonance Raman spectroscopy, it was thus of interest to develop a method of investigation, at the intracellular level, of the drug-target interactions,
either on a small cellular population or even on a single cell.

3.2.1 Raman studies on a small cell population (18)

The human erythroleukemia K 562 cell line has been used. Cells
were grown in Dulbecco's medium with 10 % heat-inactivated fetal
calf serum supplemented with 2 mM L-glutamine in 5 % CO_2 in air
at 37°C. It is noticed that the viability of the cells at the be-
ginning of each experiment always exceeded 98 %. The treated
cancer cells are obtained by incubation of K 562 cells in exponential growth with 50 μg/ml of adriamycin for 2h.

Raman spectra of treated and untreated cells are obtained with a
classical computerized spectrometer as described above. However,
two major problems must be solved. It is necessary (i) to immobi-
lize the cells in a matrix which will keep without change the
cell viability during the experimentation ; so, it has been found
that an agarose solution at 0.5 % in PBS giving rise to a gel
medium was adequate, (ii) to find the relationship between the
irradiance of the laser beam and the exposition time of the biological system, in order to preserve the cell viability. This
latter problem, may become more important in case of treated
cells. In fact, when the laser excitation wavelength takes a
place into the absorption band of the drug, the increase of the
temperature leads rapidly to the death of the cells. In such conditions, it appears that the exposition time must be strongly
reduced by a factor 10 or 20 compared to that of untreated cells
(18). However, the use of the agarose gel medium as cell matrix,
which has a heat-capacity and conductive properties, greatly
reduces the thermal effect of the laser beam.

These major obstacles being circumvented, the Raman spectra of
treated or untreated cancer cells can be obtained. Raman spectra
of untreated K 562 cells in agarose gel and of adriamycin-treated
K 562 cells are shown in Figures 19A and 19B respectively. We
like to point out that the Figure 19B does not show the crude
spectrum, but a given spectrum after (i) subtraction of a mathematical baseline and (ii) a Savitsky-Golay smoothing (30).
Actually, this spectrum represents the resonance Raman spectrum
of the bound intracellular adriamycin under 457.9 nm laser excitation wavelength. The comparison with the spectrum of free ADM
(see Figure 16B) reveals substantial differences showing that
the ADM chromophore changes when it interacts with the intracellular target, which is the nuclear DNA (27).

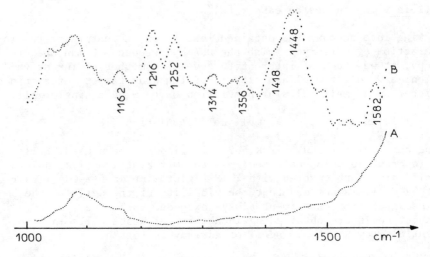

Figure 19. Raman spectra of A) untreated K 562 cells in
          agarose gel ; B) ADM-treated K 562 cells (resonance
          Raman spectrum of the bound intracellular ADM).

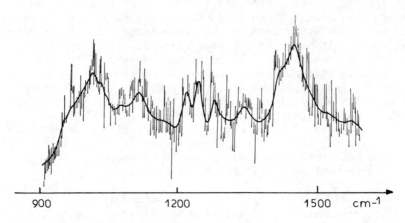

Figure 20. Resonance Raman spectrum of ADM into one single
          cell nucleus.

These modifications are in agreement with the ones previously
observed in the interaction ADM-native DNA. These results would
suggest that ADM should interact in vivo and in vitro within
the same process.

Finally, we like to notice that when treated or untreated cells
are fixed by glutaraldehyde (a fixative agent used in electron
and scanning electron microscopy) there is no more problems

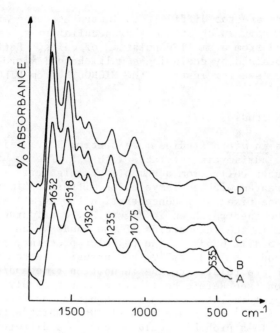

Figure 21. FT-IR spectra of A) untreated K 562 cells ;
B) treated cells with adriamycin 50 μg/ml,
incubation time 2mn ; C) 5mn ; D) 3h.

regarding the cell viability versus the power irradiance of the
laser beam itself, but only one about the possibility of burning
the chemical sample. The technique of fixation by glutaraldehyde,
which prevents sample alterations, leads to results similar to
those obtained from living cells and is useful when one requires
a lenghty time of data acquisition.

3.2.2. Raman studies on a single living cell (19)

Another advanced approach is to study the biological system described as above, at the cellular level and more precisely to
experiment on one single living and/or fixed cell. This could be
performed only by use of the micro-Raman techniques. The first
attempts for recording a resonance Raman spectrum of bound ADM
to nuclear DNA, on the Raman microprobe M.O.L.E. (31), are illustrated in Figure 20. For this purpose, it was found necessary to
put the treated K 562 cells in the physiological NaCl medium and
to leave them deposited on the bottom of a petri dish. The laser
beam is focused on the nucleus of a single chosen cell through a
water immersion objective (magnification X100). The rough crude
spectrum given in Figure 20 was obtained by multiscanning data
acquisition. However the record, in monochannel mode, implies to
choose a new cell at each scan in order to prevent any dammage.

These spectra are too difficult to interpret, even after smooting, they present some interest and show peculiarities similar to the ones obtained from a small population of cells. Better results should be obtained by designing a multichannel micro-Raman spectrometer (see for instance the MICRODYL from DILOR).

3.2.3. FT-IR studies (17)

The previous in vitro studies (see section 3.1.3 (ii)) indicated how the drug adriamycin interacts with DNA. The in vivo studies in which cancer cells from a human erythroleukemia K 562 cell line were treated with adriamycin for different times of incubation and one fixed drug concentration (50 µg/ml) are illustrated by the spectra shown in Figure 21. Considerable changes are observed in the intensities of the bands which are mainly the amide I and II bands of the cellular proteins, particularly the amide band I at 1632 cm^{-1}. The intensity increase of these bands, which are related to the incubation time and/or the drug concentration (see Reference 17), are most probably due to an adriamycin-protein interaction. It should be noticed that the 1632 and 1518 cm^{-1} bands are also slightly shifted to higher frequencies. This probably could be due to a different intracellular concentration of adriamycin entering the cell and interacting at the cell surface with the proteins (32). Finally, it must be pointed out that it remains difficult to interpret at present the complete FT-IR spectra. However, they are promising and FT-IR can be useful too in studying such complex biological systems.

REFERENCES

1. Chinsky, L., Turpin, P.Y., Duquesne, M. and Brahms, J. 1975, Biochem. Biophys. Res. Commun. 65, 1440.

2. Chinsky, L. and Turpin, P.Y. 1978, Nucleic Acids Res. 5, 2969.

3. Chinsky, L., Turpin, P.Y. and Duquesne, M. 1976, Proc. Int. Conf. Raman Spectrosc. 5th, E.D. Schmid, J. Brandmueller and W. Kiefer, Eds., Hans Ferdinand Schulz Verlag, Freiburg/Br., Germany, pp. 196-197.

4. Martin, J.C., Wartell, R.M. and O'Shea, D.C. 1978, Proc. Nat. Acad. Sci. 75, 5483.

5. Fairclough, D.P., Fawcett, V., Long, D.A., Taylor, L.H. and Turner, R.L. 1978, Proc. Sixth Int. Conf. Raman Spectrosc., E.D. Schmid, R.S. Krishnan, W. Kiefer and H.W. Schrotter, Eds., Heyden, London, Philadelphia, Rheine, pp. 90-91.

6. Durig, J.R., Dunlap, R.B. and Gerson, D.J. 1980, J. Raman Spectrosc. 9, 266.

7. Ozaki, Y., King, R.W. and Carey, P.R. 1981, Biochemistry 14, 3219.

8. Saperstein, D.D., Rein, A.J., Poe, M. and Leaty, M.F. 1978, J. Am. Chem. Soc. 100, 4296.

9. O'Connor, T., Mansy, S., Bina, M., Mc Millin, D.R., Bruck, M.A. and Tobias, R.S. 1982, Biophys. Chem. 15, 53.

10. Manfait, M., Alix, A.J.P., Jeannesson, P., Jardillier, J.-C. and Theophanides, T. 1980, Proc. VIIth Int. Conf. Raman Spectrosc., W.F. Murphy, Ed., North-Holland, Amsterdam and New-York, pp. 636-637.

11. Manfait, M., Alix, A.J.P., Jeannesson, P., Jardillier, J.-C. and Theophanides, T. 1982, Nucleic Acids Res. 10, 3803.

12. Alix, A.J.P., Bernard, L., Manfait, M., Ganguli, P.K. and Theophanides, T. 1981, Inorg. Chim. Acta 55, 147.

13. Manfait, M., Alix, A.J.P., Butour, J., Labarre, J.-F. and Sournies, F. 1981, J. Mol. Struct. 71, 39.

14. Manfait, M. and Labarre, J.-F. 1981, Adv. Mol. Relax. Interac. Proces. 21, 117.

15. Manfait, M. and Collery, P. 1983, Proc. on XIIIth Int. Congr. Chemother. August 28th to Sept. 2nd, Vienna, Austria.

16. Theophanides, T. 1981, Applied Spectrosc. 35, 461.

17. Manfait, M. and Theophanides, T. 1983, to be published.

18. Jeannesson, P., Manfait, M. and Jardillier, J.-C. 1983, Anal. Biochem. 129, 305.

19. Manfait, M., Jeannesson, P., Jardillier, J.-C. and Dhamelincourt, P. 1982, XXIXth Annual Conf. Spectrosc. Soc. of Canada, Sept. 26-29 1982, St. Jovite, Québec, Canada.

20. Rosenberg, B., Van Camp, L. and Krigas, T. 1965, Nature, London 205, 698.

21. Rosenberg, B., Van Camp, L., Trosko, J.E. and Mansour, V.H. 1969, Nature, London 222, 385.

22. Labarre, J.-F., Sournies, F., Van de Grampel, J.C. and Van der Huizen, A.A. ANVAR French Patent No. 79-17336, July 4 1979 (World extension : July 4 1980).

23. Labarre, J.-F., Sournies, F., Cros, S., François, G., Van de Grampel, J.C. and Van der Huizen, A.A. 1981, Cancer Lett. 12, 245.

24. Manfait, M., Alix, A.J.P., Labarre, J.-F. and Sournies, F. 1982, J. Raman Spectrosc. 12, 212.

25. Manfait, M., Alix, A.J.P., Lahana, R. and Labarre, J.-F. 1982, J. Raman Spectrosc. 13, 44.

26. Arcamone, F., Cassinelli, G., Fantini, G., Grein, A., Orezzi, P. Pol, C. and Spalla, C. 1969, Biotechnol. Bioeng. 11, 1101.

27. DiMarco, A. 1975, Cancer Chemother. Rep. 3, 91.

28. Duvernay, Jr, V.H., Pachter, J.A. and Crooke, T.S. 1979, Biochemistry 18, 4024.

29. Johnston, F.P. Jorgenson, K.F., Linn, C.C. and Van de Sande, J.H. 1978, Chromosoma (Berl.) 68, 115.

30. Savitsky, A. and Golay, M.J.E. 1964, Anal. Chem. 36, 1627.

31. Delhaye, M. and Dhamelincourt, P. 1975, J. Raman Spectrosc. 3, 33.

32. Tritton, T.R. and Yee, G. 1982, Science 217, 248.

SPECTROSCOPIC PROPERTIES OF METAL-NUCLEOTIDE AND METAL-NUCLEIC ACID INTERACTIONS

T. THEOPHANIDES and H.A. TAJMIR-RIAHI

Département de Chimie, Université de Montréal
C.P. 6210, Succ. A, Montreal, Québec, CANADA   H3C 3V1

ABSTRACT
     The Fourier Transform Infrared Spectra of a series of
mononucleotides and their transition and non-transition metal
complexes are studied and discussed. The potential coordination
sites of the mononucleotides are analysed with respect to the
various metals employed. Several complexes which were structurally known have been investigated in order to relate their
crystal structures to the infrared spectra and vice versa. The
binding sites of the bases and phosphate when coordinated to
the metals produce modification in the infrared spectra of the
mononucleotides which often can be diagnostic of the binding
sites involved in coordination. It has been found that the
purine base is coordinated to the metals through the N7 atom of
guanine. In addition the phosphate group is also bound to the
metal either directly or indirectly through a coordinated water
molecule forming hydrogen bonding with the negative oxygen atoms
of the phosphate group. The N7 binding produces considerable
changes in the skeletal vibrations of the imidazole ring, while
the direct phosphate binding produces changes in the phosphate
vibrations of the PO_3^{2-} group, which are diagnostic of direct or
indirect metal coordination.

C. Sandorfy and T. Theophanides (eds.), Spectroscopy of Biological Molecules, 137–152.
© 1984 by D. Reidel Publishing Company.

1.1 INTRODUCTION

In the last two decades coordination chemistry of metal ions
with nucleic acids and their components has been expanded enor-
mously. The interest in this area of research is based on the
importance of the metal ion role in a large variety of biological
processes, such as, genetic expression (replication, transcription or translation), metaloenzyme activity and other metal-
protein or metal-nucleic acid processes (1). Furthermore, the
discovery (2) of the antitumor activity of certain platinum
compounds in recent years led to an extensive investigation of
the metal–DNA interaction (3), and the cross-linking of DNA
molecules by certain platinum compounds. Several review articles
(1,4-7) cover the structural analysis and spectroscopic properties of metal-nucleoside and metal-nucleotide complexes studied
until now.

Mononucleotides, such as, guanosine-5'-monophosphate,
adenosine-5'-monophosphate, deoxyguanosine-5'-monophosphate,
uridine-5'-monophosphate and cytidine-5'-monophosphate are the
constituents of nucleic acids and the targets of metal-ligand
bonding (Fig. 1).

A mononucleotide contains three potential coordination
parts which can function as ligands:
(a) the base heteroatoms of the purine and pyrimidine bases and
    their exocyclic nitrogens or carbonyl oxygens),
(b) the phosphate oxygen atoms and
(c) the sugar oxygen atoms.

Each of these coordination sites is capable of acting as
a donor ligand under specific chemical conditions. Hard acids,
such as, alkaline-earth and alkali metal ions prefer phosphate
binding, whereas soft acids (heavy transition and non-transition metal cations) predominantly bind to the nitrogen atoms
of the base moiety (7). Recently (6), several complexes of
transition and non-transition metals with mononucleotides have
been reviewed and structurally analysed. Most of the structurally known (8-22) metal-nucleotide complexes with their
binding sites are given in Table 1. The 5'-GMP and its metal
complexes are among the most extensively studied DNA components
and this is mainly due to the strong basic character of the
guanine moiety which makes it the most plausible target of
metal-binding in DNA. There is also a strong indication that
the effectiveness of some platinum(II) anti-cancer drugs is
mainly related to the primary attack of the Pt(II) on the N7-
site of the guanine moiety (3). In the present report we limit
ourselves to the structural and spectroscopic properties of
some metal-GMP and metal-DNA complexes.

1.2 5'-GMP-METAL COMPLEXES AND BONDING SITES

The 5'-GMP molecule (Fig. 1) binds to metal cations
through the N7-atom of its imidazole ring, the phosphate group
and the sugar moiety (6). At neutral chemical conditions the

N7-site seems to be the preferred metal-ligand target. The
O6 atom of the pyrimidine ring is also a potential coordination
site of the molecule. Crystallographic evidence so far, is
overhelming for N7-metal interaction, but less abondant for
O6-metal interaction (6). The phosphate group does interact
with the metal especially the hard metals and there are several examples (8, 13-14, 16-18, 22). It is of interest to
study the coordination properties of the 5'-GMP molecule in
the presence of hard and soft metal ions and to learn more about
its complexing abilities including the sugar oxygen atoms.
N7-Bonding. X-ray structural data show (9-12) that divalent
Mn, Co, Ni and Cd are N7-bound (Fig. 2) in an octahedral
environment, with five water molecules and the N7-atom of the
purine ring system, $(H_2O)_5M(5'-GMP)$. The phosphate and the
carbonyl groups are hydrogen bonded to the coordinated water
molecules. Spectroscopic evidence shows (23) that Pt(II) and
Mg(II) are also N7-bound in the corresponding 5'-GMP compounds.
In figure 2 is also illustrated the crystal structure of free
5'-GMP acid.
Phosphate-Bonding. The Cu(II) ion is directly linked to the
phosphate group through the negative oxygens and to three
water molecules in addition to the N7-atom of the base moiety
(8) in a polymeric $(H_2O)_3Cu_3(5'-GMP)_3.5H_2O$ coordination
compound (Fig. 2). The complexes of Mg(II) and Cd(II) synthesized in acidic media also showed phosphate-binding, in
addition to N7-binding (23-25).
Spectroscopic Studies. The Raman and NMR techniques have been
widely used to distinguish between coordination sites in mono-
and poly-nucleotides (4), however, with nucleotides reliable
correlation between structural information and spectroscopic
properties is still missing. Fourier Transform Infrared Spectroscopy (FT-IR) is a newly developed technique and has been
the least used to characterize the coordination site in metal-
nucleotide complexes (23-27). Therefore, we wish to analyse
the infrared spectra of several structurally known and some
newly synthesized metal-GMP compounds whose structures have
not yet been resolved with the aim to detect the characteristic features of each structural type of the metal complexes
reviewed here and hopefully to establish a correlation between
the spectral changes and the binding sites used by the 5'-GMP
molecule. The FT-IR spectra of the free $5'-GMPNa_2$ and its
metal complexes have been recorded in the region $4000-400$ cm^{-1}.
The molecular structure of the 5'-GMP molecule with the
numbering of the atoms is illustrated in Figs. 1 and 2.
     In the spectral region of $4000-2700$ cm^{-1} the asymmetric
and symmetric stretching vibrations of the NH_2, NH, OH, H_2O,
CH and CH_2 functions of the free 5'-GMP molecule have been
recorded (27-30). They showed no considerable changes in the
spectra of the metal complexes. The small shifts of the bands
and the intensity changes which are observed in this region of

Fig. 1. Common mononucleotides

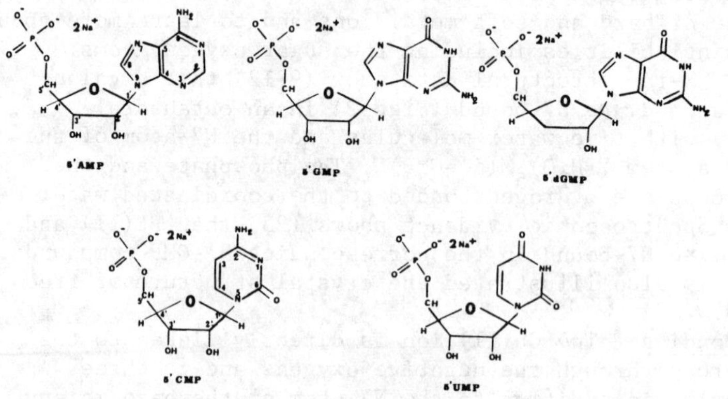

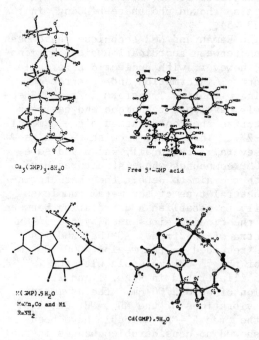

Fig. 2. Structurally know GMP metal complexes.

TABLE 1. Structurally known nucleotide complexes of some metal ions

COMPOUND	COORDINATION SITE	REFERENCES
$Cu_3(GMP)_3 \cdot 8H_2O$	N_7 and O atom of PO_3^{-2} and $2H_2O$	8
$Ni(GMP) \cdot 5H_2O$	N_7 and $\underline{5}$ H_2O	9
$Co(GMP) \cdot 5H_2O$	N_7 and $\underline{5}$ H_2O	10
$Mn(GMP) \cdot 5H_2O$	N_7 and $\underline{5}$ H_2O	11
$Cd(GMP) \cdot 5H_2O$	N_7 and $\underline{5}$ H_2O	12
$Co_2(UMP)_2 \cdot 4H_2O$	O atoms of H_2O and PO_3^{-2}	13
$Cd(IMP) \cdot 3H_2O$	N_7, O atoms of OH groups and PO_3^{-2}	14
$Ni(AMP) \cdot 5H_2O$	N_7 and $\underline{5}$ H_2O	15
$Zn(IMP) \cdot 3H_2O$	N_7 and O atoms of PO_3^{-2}	16
$Cu(IMP) \cdot 3H_2O$	N_7 and O atoms of PO_3^{-2}	17
$Cd(CMP)_2 \cdot H_2O$	N_3 and O atoms of PO_3^{-2}	18
$Co(CMP)_2 \cdot H_2O$	N_3 and O atoms of PO_3^{-2}	18
$Pt(tn)(Me-5'GMP)_2^{++}$	N_7 of the purine ring only	19
$Pt(IMP)_2(NH_3)_2^{++}$	N_7 of the purine ring only	20
$Pt(IMP)_2(tn)^{++}$	N_7 of the purine ring only	21
$Pt(CMP)_2enCl_2$	N_3 and O atoms of PO_3^{-2}	22

the spectra could be mainly due to the rearrangement of the
strong hydrogen bonding network of the free nucleotide upon
metalation. The presence of such a strong hydrogen bonding
network both in the free ligand structure (31,32) and in the
metal complexes (8-12) is shown in their crystal structures.
However, it is difficult to draw any conclusions on the nature
of the metal-ligand bonding and the coordination sites involved
from the absorption bands in this region.

In the 1800-400 cm^{-1} region, the IR spectra of 5'-GMP
showed considerable changes upon complexation. These spectral
changes are discussed in terms of metal molecule interaction
and structural changes: (a) The purine and pyrimidine vibrational frequencies (1800-1180 cm^{-1}) are analyzed first and then
(b) the phosphate and sugar stretching vibrational frequencies
(1100-900 cm^{-1}). The structurally known N7-bound bivalent metal
ions of Mn, Co, Ni and Cd to 5'-GMP showed marked similarities
in the purine and pyrimidine region of the spectra while Cu
showed major differences in the phosphate region of the spectra,
which has been interpreted as a direct Cu-phosphate bonding
compared to the indirect metal-phosphate interaction through
hydrogen bonded water molecules of several other transition metal
cations (Fig. 2 and Table 2). Cis- and trans-Pt(NH$_3$)$_2$Cl$_2$ and
K$_2$PtCl$_4$ complexes of 5'-GMP showed distinct spectral similarities
with those of the structurally known N7-bound complexes mentioned
above, whereas the Mg(II)-GMP compounds exhibited major resem-
blance with the corresponding Cd(II) compound bound to phosphate.
The main features of the spectra relevant to this discussion
with possible assignments are given in Table 2 and Fig. 3.
(a)   The purine vibrational frequencies (1800-1180 cm^{-1})
In the spectrum of the free 5'-GMPNa$_2$ there are five
absorption bands at 1692 cm^{-1} (bs), 1662 cm^{-1} (sh), 1597 cm^{-1} (s),
1575 cm^{-1} (sh) and 1535 cm^{-1} (m) assignable tentatively to the
C6=O, the scissoring mode of NH$_2$, C4-N3 and C5-C4 and C4-N9
stretching vibrations, respectively (28-30). These absorption
bands showed considerable changes in the spectra of the structurally known N7-bound compounds and those of the corresponding
Pt(II) and Mg(II) complexes (Table 2 and Fig. 3). The C6=O
stretching vibration of the free base at 1692 cm^{-1} lost intensity, but did not show any major shifting ($\Delta\bar{\nu}$ = 6 cm^{-1}) and
this is mainly due to the rearrangement of the carbonyl hydrogen
bonding network upon N7-metalation. In the crystal structure of
the free GMP acid the carbonyl group is shielded from hydrogen
bonding (31), while in the free 5'-GMPNa$_2$ (32) the carbonyl
group is hydrogen bonded to water molecules, just as those of
the structurally known (8-12) N7-bound metal complexes, where
the C6=O group is hydrogen bonded to a coordinated water molecule. The crystal structure of Pt(tn)(Me-5'-GMP)$_2^+$ showed strong
hydrogen bonding of the carbonyl group via a water molecule
which is also linked to a phosphate group (P — O$^-$···H-O-H···O6)
(19). This hydrogen bonding network causes the broadening of

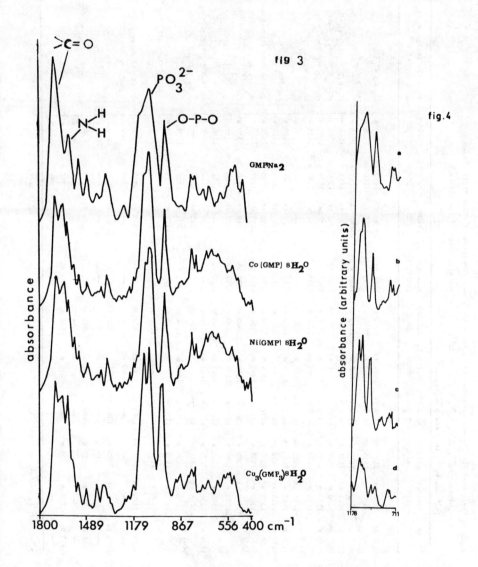

Fig. 3.   FT-IR Spectra of 5'-GMP-Metal Complexes

Fig. 4.   FT-IR Spectra of 5'-GMPNa$_2$ (a), Co(GMP).8H$_2$O (b),
          Cu$_3$(GMP)$_3$.8H$_2$O (c) and Mg(GMP).6H$_2$O (d) in the
          phosphate region.

Table 2. IR-in bands (cm⁻¹) and Possible Assignments for 5'-GMP Complexes of structurally known M_7-bonded Mn^{+2}, Co^{+2}, Ni^{+2}, Cu^{+2}, and Cd^{+2} in comparison with those of the corresponding Pt^{+2} and Mg^{+2}, in a and b regions.

5'-GMP	Mn(GMP) 8H₂O	Co(GMP) 8H₂O	Ni(GMP) 8H₂O	Cu₃(GMP)₃ 8H₂O	Cd(GMP) 8H₂O	cis-Pt(NH₃)₂(GMP)₂Cl₂	Pt(GMP)₂ Cl₂	Mg(GMP) 10H₂O	Mg(GMP) 6H₂O	Possible Assignments (28-32)
a										
1692 bs	1690 vs	1690 vs	1690 vs	1694 vs	1694 vs	1694 vs	1693 s	1697 bs	1695 s	$\nu C_6=O$, $\nu C_6=C_5$
1662 sh	1655 sh	1656 s	1656 sh	1663 sh	1656 s	1653 sh	1655 sh	1660 sh	1655 sh	δNH_2, νC_2-N_2
1597 sh	1645	1640 s	1640 s	1648 s	1643 sh	1643 s	1643 vs	1645 m	1638 m	νC_4-N_3, νC_4-C_5, νC_5-N_1
1575 sh	1611 s	1611 s	1611 s	1611 s	1610 s	1599 s	1597 vs	1611 m	1616 sh	νC_5-C_4, νC_3-N_3, νC_2-N_2
1555 m	1570 m	1570 m	1560 m	1551 sh	1550 sh	1550 m	1543 m	1550 sh	1560 sh	νC_4-N_9, $\nu C_6=O$, νC_4-N_3
1479	1533 m	1535 m	1535 m	1539 m	1533 m	1543 m	1577 m	1556 m	1535 m	δC_8-H, νC_8-N_7
	1476 m	1478 m	1479 m	1489 sh	1479 m	1503 m	1501 s	1483 m	1483 m	
1415 m	1465 m	1465 m	1466 sh	1450 w					1466 sh	$\delta CH + \delta CH_2$
1410 sh	1425 vw	1423 vw	1423 vw	1416 w	1420 sh	1423 w	1422 vw		1420 w	
1358 s	1410 vw	1412 vw	1412 w		1412 m	1412 vw	1410 vw	1414 w	1410 sh	purine ring vibration
	1389 s	1391 m	1391 m	1574 m	1379 m	1385 m	1585 vw	1383 m	1575 m	
1333 sh	1352 s	1352 m	1354 m	1350 sh	1358 m	1348 m	1550 m	1357 m	1367 m	νC_8-N_9, νN_7-C_8
1290 vw	1290 vw	1300 s	1300 s	1300 m	1300 sh	1300 sh	1310 sh	1330 sh	1330 sh	
1280 m	1280 m	1280 m	1290 m	1290 m	1290 m	1290 vw	1290 m			
1254 sh	1260 m	1260 m	1260 m	1260 vw	1265 m	1270 vw		1254 m	1250 m	νC_8-N_7, νN_1-C_6, νN_3-C_2
1234 s	1236 m	1255 m	1236 m	1240 m	1236 m					νC_2-N_2, νC_2-N_1, νC_6-NH_2
1216 m	1216 sh	1215 sh	1216 sh		1216 sh	1218 m	1215 sh	1220 sh		νC_8-N_7, δC_8-H
1204 m	1205 s	1206 s	1208 s	1208 s	1204 m	1205 s	1205 m	1206 m	1208 m	
1180 m	1177 m	1177 m	1175 m	1178 m	1175 m	1175 m	1177 w	1175 sh	1182 m	νC_8-N_7, νN_9-sugar, νC_8-N_9
1155 m	1155 s	1155 s	1155 s		1150 m					
b										
1125 sh	1109 sh	1107 sh	1109 sh	1111 vs	1112 sh	1109 vs	1110 sh	1107 bs	1102 bs	νCO of the sugar ring
				1040 sh						
1070 bs	1084 vs	1080 vs	1080 vs	1076 vs	1080 bs	1084 s	1069 bs	1080 sh	1076 sh	νdeg. PO_3^{-2}
				1001 sh					1052 s	
972 s	981 s	982 s	980 s	991 va	979 va	974 a	976 a	993 s	976 s	νsym. PO_3^{-2}
				975 sh					940 s	
805 s	802 s	802 s	802 s	815 s	802 s	800 a	800 m	792 m	802 m	$\nu P-O$
				790 w						

s, strong; m, medium; b, broad; sh, shoulder; w, weak; v, very; ν, stretching; δ, bending.

the C6=O stretching which shifts slightly to a lower frequency. The scissoring motion of the NH_2 group of the free base at 1662 cm^{-1} gained intensity and split into two components in the spectra of metal complexes (Fig. 3 and Table 2). The crystal structure of the free 5'-GMPNa$_2$ shows (31,32) strong $O\cdots H-N-H \cdots O$ hydrogen bonding to the phosphate group, whereas such a hydrogen bonding system would be weakened in the crystal structures of the metal complexes (8-12). Releasing or weakening the NH_2 group from hydrogen bonding may cause the sharpening and shifting of the NH_2 stretching vibration at 3400 cm^{-1} to a higher frequency, while the bending mode of NH_2 group at 1662 cm^{-1} shifts to a lower frequency. It should be noted that metalation or deprotonation of the NH_2 group brings considerable changes to that absorption frequency. The band at 1597 cm^{-1} in the spectrum of the free 5'-GMPNa$_2$ attributed to a skeletal vibration, gained intensity and shifted to 1600-1610 cm^{-1} in the spectra of the metal complexes. The shift is due to the inductive effect of the metal ion electrophile bound at N7, which causes less electron delocalization and finally there is less contribution to the skeletal vibrations of the ring system.

A band with medium intensity at 1479 cm^{-1} of the free 5'-GMPNa$_2$ assignable to the vibration involving primarily the five membered ring C8-H bending and the N7-C8 stretching frequencies (33), split into two absorption bands at 1476 cm^{-1} and a shoulder at 1465 cm^{-1} in the spectra of N7-metal complexes. This absorption band also shifted to 1500 cm^{-1} in the spectra of Pt(II) complexes (Fig. 3 and Table 2). It has been found (33) that the Raman line corresponding to that absorption at 1480 cm^{-1} disappeared upon protonation or methylation and shifted on deuteration of H8, whereas heavy metalation particularly platination does affect that Raman line but not to the same extent (7,34). It has been experimentally found (35) that the $C=N^{+}$ groups have in general higher frequencies than the parent C=N groups, therefore protonation (36) or metalation (37) could cause an increase to the infrared absorption of the C=N frequency of the purine ring system in these series of metal-GMP complexes.

Three absorption bands with medium intensities at 1333, 1254, and 1204 cm^{-1} in the spectrum of the free nucleobase assigned to N7-C8, N9-C8, N7-C5 stretching vibrations and to the C8-H bending mode (28,38), each one appeared as a doublet at about 1350-1290 cm^{-1}, 1280-1260 cm^{-1} and 1216-1205 cm^{-1} in the spectra of the metal-GMP complexes (Table 2). A medium absorption band at 1180 cm^{-1} in the spectra of the free ligand assigned to N7-C8 (28) showed intensity changes in the spectra of the metal complexes. The observed shifts upon N7-coordination can be explained by a distortion of the imidazole ring, where these vibrations are mostly localized and this is why N7-metalation could change the electron distribution in the ring system.

A shoulder absorption band at 1125 cm^{-1} in the infrared spectrum of the free base was attributed to the C-O stretching vibration of the suger ring (29) shifted to a lower frequency upon N7-metalation (Table 2). This shift would be either due to a rearrangement of the strong hydrogen bonding of the sugar moiety, or to the conformational changes around the ribose-phosphate bond upon direct or indirect metal-phosphate binding.

(b)   The phosphate and sugar vibrations of mononucleotides (1100-400 cm^{-1})

The mononucleotides, 5'-GMP, 5'-AMP, 5'-IMP, 5'-dGMP, 5'-UMP and 5'-CMP have a phosphate group at the C-5' position of the ribose ring. The disodium salts of these nucleobases contain a $-PO_3^-$ group capable of acting as a polydentate ligand or hydrogen bonding agent through the negative oxygen atoms.

The Raman spectra of the phosphate group in mononucleotides, polynucleotides and DNA have been studied (39-41). There are calculations on the normal vibrational modes of the phosphate group and the ribose in DNA (40,41), but the IR spectra of the nucleotide-phosphate group has not yet been fully investigated (29,42). For the phosphate anion $-PO_3^{2-}$ there are two characteristic bands at 1100 cm^{-1}, degenerate stretching and at 900 symmetric stretching (42).

The monoalkyl phosphate anion ($CH_3-OPO_3^{2-}$) shows infrared absorption bands (42), which can be compared with those of the already mentioned mononucleotides in this region of the spectra (see Table 3). There are also two other absorption bands with medium intensities in the IR spectra of the PO_3^{2-} anion at about 600-350 cm^{-1} (42), which are assigned to the $-PO_3^{2-}$ symmetric and $-PO_3^-$ degenerate deformation (Table 3). Six fundamental frequencies are expected for the $-PO_3^-$ ion of C_{3v} symmetry, three of these belong to the symmetric vibrations of A_1 type and the other three to the doubly degenerate vibrations of E type. The $CH_3-OPO_3^{2-}$ no longer has C_{3v} symmetry if we consider the hydrogen atoms due to the lack of a C_3 axis, but it possesses a lower C_s symmetry. Therefore, the degeneracy of the absorption band at 1100 cm^{-1} is removed and splits into a doublet at 1115 and 1090 cm^{-1} (Table 3). In the FT-IR spectra of the solid mononucleotides studied here such splitting has not been observed but most likely it has been overlapped by the strong and broad absorption at about 1090 cm^{-1}. Upon coordination of the $-PO_3^{2-}$ group the degeneracy of the E vibration frequencies is also removed.

The Raman spectra of mononucleotides (39), such as, 5'-AMP and 5'-GMP showed bands, at 850, 882, and 980 cm^{-1} which are assigned to the $-PO_3^-$ symmetric stretchings and the bands at 1123 and 1170 cm^{-1} are related to asymmetric stretchings. Protonation and metalation caused a large perturbation in the Raman intensities of these frequencies (43).

The Cu(II) ion is coordinated directly to the phosphate group, as well as to the N7-atom in $Cu_3(GMP)_3.8H_2O$ (8), whereas the corresponding complexes of Co(II), Ni(II), Mn(II) and

TABLE 3. FT-IR Absorption bands (cm^{-1}) of the phosphate group in mononucleotides ROPO$_3^{2-}$ in the region, 1120-450 cm^{-1} compared to the corresponding absorption bands of CH$_3$OPO$_3^{-2}$

5'-CMPNa$_2$	5'-UMPNa$_2$	5'-IMPNa$_2$	5'-AMPNa$_2$	5'-GMPNa$_2$	CH$_3$-OPO$_3^{-2}$ (42)
-	-	-	-	-	1115 PO_3^{-2} deg. (A')
1080 bs	1086 bs	1092 bs	1091 bs	1070 bs	1090 PO_3^{-2} deg. (A'')
-	-	1072	-	-	-
-	-	-	-	-	1056 C-O str. (A')
984 vs	979 vs	976 vs	976 vs	975 s	983 PO_3^{-2} sym. str.
812 s	816 s	825 m	797 s	805 m	750 P-O str. (A')
580 m	565 m	527 m	540 m	540 m	PO_3^{-2} sym. def. (A)
-	-	-	-	-	-
470 w	460 w	480 w	470 m	460 m	PO_3^{-2} deg. def. (E)

s, strong; b, broad; m, medium; w, weak; v, very.

Cd(II) ions are indirectly bound to the phosphate group via
two hydrogen bonded water molecules (9-12), in addition to the
linkage through the N7-atom of the base. The crystal structure
of 5'-GMPNa$_2$ showed (32) that there is no direct interaction
between the Na$^+$ ions and the phosphate group. Comparing the
IR spectra of the sodium salt with that of the corresponding
divalent Mn, Co, Ni, Cu and Cd complexes in the region 1070-
970 cm^{-1} and at 805 cm^{-1}, the spectra of the free base showed
drastic changes in the Cu-GMP compound, upon direct phosphate
coordination (Fig. 4). The bands at 1070, 972 and 805 cm^{-1}
split each into two components, while in the corresponding
Mn(II), Co(II), Ni(II) and Cd(II) complexes, where the phosphate group is not directly bound to these metal cations
(9-12), there have been no significant changes in this region
of the spectra (Fig. 4 and Table 2) compared to the ligand.
      The Cu(II) and Zn(II) ions are also directly bound to the
-PO$_3^{2-}$ group of the 5'-IMP molecule (16,17), as well as to the
N7-atom of the purine base residue, whereas in the corresponding Ni(II) and Co(II) complexes the phosphate anion is not
directly bound to these metal ions (44). On comparing the IR
spectra of these metal-nucleotide complexes, the absorption
bands at 1092 and 976 cm^{-1} in the spectra of the free 5'-IMPNa$_2$
showed splitting and shifting for the copper and zinc compounds,
while the corresponding cobalt and nickel complexes did not
show appreciable changes in this region of the spectra. The
Co(II) ion is coordinated only to the phosphate group of the
5'-UMPNa$_2$ molecule in Co$_2$(UMP)$_2$.4H$_2$O (13). On comparing the
infrared spectra of free 5'-UMPNa$_2$ with that of the Co(II)
compound, each of the bands at 1086, 979 and 816 cm^{-1} is split
into a doublet in the spectra of the coordination compounds.
The Cd(II) and Co(II) ions are directly bound to the phosphate
moiety of 5'-CMPNa$_2$, as well as to the N3 of the pyrimidine
base residue (18). The IR spectra of these two complexes
showed splitting and shifting of the bands at 1080, 984 and
812 cm^{-1} compared to the free nucleobase in the same region of
the spectra. In (H$_2$O)$_5$Ni(AMP).H$_2$O, the only crystal structure
known (15) for the 5'-AMP nucleotide, the Ni(II) ion is bound
to the N7-atom of the base moiety and to five water molecules
(two of these water molecules are hydrogen bonded to the
phosphate group). The infrared spectrum of 5'-AMPNa$_2$ in the
phosphate region showed no considerable changes upon this
indirect metal-phosphate interaction in the (H$_2$O)$_5$NiAMP.H$_2$O
compound.
      It could be concluded from the observations made on the
analysis of the IR spectra of the structurally known (8-18)
examples discussed here, that the absorption bands at about
1090-970 cm^{-1} and 800 cm^{-1} in the spectra of mononucleotides
are certainly due to the -PO$_3^{2-}$ vibrational frequencies, and they
are assigned in Table 3. They show large perturbations upon
phosphate coordination. Due to the skeletal and breathing

modes of the pyrimidine and purine vibrational frequencies
together with the NH_2 out-of-plane wagging (28) vibrations in
the region 700-400 cm^{-1} of the spectra of mononucleotide complexes, it is very difficult to distinguish any changes occurred
in the sym. def. (A) and deg. def. (E) modes upon metal-phosphate
bonding or hydrogen bonding. The IR spectra of the Pt(II) and
Mg(II) compounds of the 5'-GMPNa$_2$ molecule studied earlier (23)
showed marked similarities with those of the corresponding hydrogen bonded phosphate group (indirect bonding through water
molecules) in the Mn(II), Co(II), and Ni(II) complexes (Table 3
and Fig. 3). Their spectra were markedly dissimilar with those
directly coordinated phosphate groups in the case of the Cu-GMP
compound. Therefore, it is concluded that the phosphate group
of the 5'-GMP is indirectly linked to Pt(II) and Mg(II) <u>via</u>
hydrogen bonded water molecules. It should be noted that the IR
spectra of $(H_2O)_5Mg(GMP).H_2O$ compound recrystallized in acidic
media (pH=4) exhibited (23) a direct Mg-phosphate interaction
showing splitting and shifting of the absorption bands at 1070,
972 and 805 cm^{-1} related to the phosphate vibrational frequencies, whereas the spectra of the $(H_2O)_5Mg(GMP).5H_2O$ compound
obtained from basic solutions (pH=7-9) showed distinct spectral
similarities with the spectra of $(H_2O)_5Ni(GMP).H_2O$ which is known
to have a hydrogen-bonded phosphate group through the coordinated
water molecules (Fig. 4 and Table 2).

The ribose vibrational frequencies appear almost as sharp
bands (29,33) with medium intensities in the region of 1400-
400 cm^{-1} of the spectra. Most of these absorptions are obscured
by the strong and broad absorptions of the phosphate group in
the region 1100-800 cm^{-1} and the base vibrational frequencies in
the other regions of the spectra. The phosphate-ribose stretching vibrations show weak absorption bands at 900-790 in the
spectra of the 5'-GMPNa$_2$ molecule. These absorption bands show
similar behavior upon direct or indirect metal-phosphate bonding
(Fig. 3). The changes here can be explained by the inductive
effect of the phosphate electron distribution on the ribose-
phosphate bond or by conformational changes about the phospho-
diester bond upon $-PO_3^{2-}$ complexation (45) or H-bonding.

Finally, the guanine ring breathing mode at 680 cm^{-1} (28)
in the spectrum of 5'-GMPNa$_2$ was shifted to 700 cm^{-1} in the
spectra of the N7-metal complexes (Fig. 3). The bending vibration and the out-of-plane deformation modes of NH_2 and HN (28,
30) at 700-400 cm^{-1} in the ligand spectra do not shift considerably upon N7-metalation. The in-plane binding deformation of
the HN is masked in the 1610 cm^{-1} by the skeletal vibrations of
the ring.

1.3 CONCLUDING REMARKS

On the basis of spectroscopic studies on the N7-bound metal
complexes discussed here, the following conclusions can be drawn:

Metalation of the N7-site of 5'-GMP brings considerable spectral changes in the vibrations of the purine ring, which are
strongly coupled and which are often difficult to be rationalized in the free ligand. Metal coordination, however, can
change the electron distribution in the ring system and cause
alteration in the vibrational frequencies. Therefore, metalation could be a useful probe to expand our understanding of the
nucleotide vibrational problem. Some of the most meaningful and
common changes observed in the vibrational spectra of 5'-GMPNa$_2$
upon metal complexation are discussed together with the characteristic features of the metal N7-bound guanine ring and are
summarized as follows:
(1) modification of the band at 1480 cm^{-1};
(2) splitting and shifting of the bands at 1333, 1254, 1204 and
    1180 cm^{-1}, and
(3) shifting of the band at 680 cm^{-1} to about 700 cm^{-1} assigned
    to the breathing mode of the ring system.
    Furthermore, it is found that the characteristic features
of metal-phosphate bonding are:
(1) splitting and shifting of the broad and strong bands at
    about 1070 and at 972 cm^{-1}, upon direct metal-phosphate
    bonding
(2) the lack of splitting and major shifting is indicative of
    an indirect metal-phosphate interaction (most likely through
    a hydrogen bonded water molecule) and
(3) shifting of the band at 680 cm^{-1} to lower frequencies upon
    metal-phosphate interaction is indicative of a direct metal-
    phosphate bonding as well as N7-interaction.

REFERENCES
1. T.G. Spiro, "Nucleic Acid-Metal Ion Interaction",
    Ed. Wiley, New York (1980).
2. B. Rosenberg, Naturwissenschaften, 60, 339 (1973).
3. L.G. Marzilli and T.J. Kistenmacher, Accts. Chem. Res.,
    10, 146 (1977).
4. D.G. Hodgson, Progr. Inorg. Chem., 23, 211 (1977).
5. L.G. Marzilli, Progr. Inorg. Chem., 23, 255 (1977).
6. R.G. Gellert and R. Bau, in "Metal Ions in Biological
    System" H. Sigal, Ed., Marcel Dekker, New York, Vol. 8,
    p. 1 (1979).
7. T. Theophanides (ed.), "Infrared and Raman Spectroscopy
    of Biological Molecules", 205-223, 1979, Reidel Publishing
    Company, Dordrecht, Holland.
8. K. Aoki, G.R. Clark, and D.J. Orbell, Biochim. Biophys.
    Acta, 425, 369 (1976).
9. P. de Meester, D.M.L. Goodgame, A.C. Skapski, and B.T.
    Smith, Biochim. Biophys. Acta, 340, 113 (1974).
10. P. de Meester, D.M.L. Goodgame, T.J. Jones, and A.C.
    Skapski, C.R. Acad. Sci. Paris, 279, 667 (1974).

11. P. de Meester, D.M.L. Goodgame, T.J. Jones, and A.C.
     Skapski, Biochim. J., 139, 791 (1974).
12. K. Aoki, Acta Cryst., B32, 1454 (1976).
13. B.A. Cartwright, D.M.L. Goodgame, I. Jeeves, and A.C.
     Skapski, Biochim. Biophys. Acta, 477, 195 (1977).
14. D.M.L. Goodgame, I. Jeeves, C.D. Reynolds, and A.C.
     Skapski, Nucl. Acid Res., 2, 1375 (1975).
15. A.D. Colins, P. de Meester, D.M.L. Goodgame, and A.C.
     Skapski, Biochim. Biophys. Acta, 402, 1 (1975).
16. P. de Meester, D.M.L. Goodgame, T.J. Jones, and A.C.
     Skapski, Biochim. Biophys. Acta, 353, 392 (1974).
17. K. Aoki, Chem. Comm., 600 (1977).
18. G.R. Clark and J.D. Orbell, Chem. Comm., 697 (1975).
19. L.G. Marzilli, P. Chaliopil, C.C. Chiang, and T.J.
     Kistenmacher, J. Amer. Chem. Soc., 102, 2480 (1980).
20. D.M.L. Goodgame, I. Jeeves, F.L. Phillip, and A.C.
     Skapski, Biochim. Biophys. Acta, 378, 153 (1975) and
     J. Amer. Chem. Soc., 101, 1143 (1979).
21. T.J. Kistenmacher, C.C. Chiang, P. Chalilpolyil, and
     L.G. Marzilli, Biochim. Biophys. Res. Comm., 84, 70
     (1978).
22. S. Louie and R. Bau, J. Amer, Chem. Soc., 99, 3874 (1977).
23. H.A. Tajmir-Riahi and T. Theophanides, Can. J. Chem., 61,
     1813 (1983).
24. H.A. Tajmir-Riahi and T. Theophanides, Inorg. Chim. Acta,
     80, 183 (1983).
25. H.A. Tajmir-Riahi and T. Theophanides, Inorg. Chim. Acta,
     80, 223 (1983).
26. T. Theophanides, Appl. Spectrosc., 35, 461 (1981).
27. A. Pasini and R. Mena, Inorg. Chim. Acta, 56, L17-L19
     (1981).
28. M. Tsuboi, S. Tokahoshi and I. Horada, Physicochemical
     Properties of Nucleic Acid, Vol. 2, J. Duchense Ed.,
     Academic Press, New York (1973).
29. M. Tsuboi, Infrared and Raman Spectroscopy in basic
     principles in nucleic acid chemistry. Vol. 1.
     Academic Press, 1974, p. 399.
30. R.C. Lord and G.J. Thomas, Spectrochim. Acta, 23A, 2551
     (1967).
31. J. Emerson and M. Sandaralingam, Acta Cryst., B36, 1510
     (1980).
32. S.K. Katti, T.P. Seshardi, and M.A. Viswamitra, Acta
     Cryst., B37, 1825 (1981).
33. M.J. Lane and G.J. Thomas, Jr., Biochemistry, 18, 3839
     (1979) and S. Mansy and W.L. Peticolas, Biochemistry, 15,
     2650 (1976).
34. G.Y.H. Chu, S. Mansy, R.E. Duncan, and R.S. Tobias,
     J. Amer. Chem. Soc., 100, 593 and 607 (1978).
35. C. Sandorfy, "The Chemistry of Carbon-Nitrogen Double
     Bonds", S. Patai, Ed., p. 42, John Wiley, New York (1970).

36. J.I. Bullock, M.F.C. Ladd, D.C. Povey, and R.A. Tajmir-
     Riahi, Acta Cryst., B36, 2013 (1979).
37. F.A. Hart, M. Ul-Huque, and C.N. Caughlan, Chem. Comm.,
     1240 (1970) and J. Inorg. Nucl. Chem., 31, 145 (1969).
38. L. Chinsky, P.Y. Turrin, M. Duquesne, and J. Brahms,
     Biopolymers, 17, 1347 (1978).
39. L. Rimai, T. Cole, J.L. Parson, J.T. Hechmontt, and
     E.B. Carew, Biophys. J., 9, 320 (1969).
40. E.B. Brown and W.L. Peticolas, Biopolymers, 14, 1259
     (1975).
41. K.C. Lu, E.W. Prohofsky, and L.L. Vazndt, Biopolymers,
     16, 2491 (1977).
42. T. Shimanouchi, M. Tsuboi, and T. Kyogoku, Adv. Chem.
     Phys., 7, 435 (1964).
43. S. Mansy and R.S. Tobias, J. Amer. Chem. Soc., 96, 6894
     (1974).
44. K. Aoki, Acta Cryst., B32, 1454 (1976).
45. P. Papagiannakopoulos, G. Makrigiannis, and T. Theophanides,
     Inorg. Chim. Acta, 46, 263 (1980).

Abbreviations:

$5'$-GMPNa$_2$ = Guanosine-$5'$-Monophosphate Disodium Salt
$5'$-dGMPNa$_2$ = Deoxy-Guanosine-$5'$-Monophosphate Disodium Salt
Me-$5'$-GMP = N$_7$-Methyl-Guanosine-$5'$-Monophosphate
$5'$-AMPNa$_2$ = Adenosine-$5'$-Monophosphate Disodium Salt
$5'$-IMPNa$_2$ = Inosine-$5'$-Monophosphate Disodium Salt
$5'$-UMPNa$_2$ = Uridine-$5'$-Monophosphate Disodium Salt
$5'$-CMPNa$_2$ = Cybidine-$5'$-Monophosphate Disodium Salt
tn = Trimethylenediamine

CONFORMATION AND DYNAMICS OF NUCLEIC ACIDS AND PROTEINS FROM LASER RAMAN SPECTROSCOPY

Warner L. Peticolas
Department of Chemistry and Institute of Molecular
Biology, University of Oregon, Eugene, Oregon 97403.

INTRODUCTION

Nucleic acids and proteins may be studied with both classical or resonance Raman spectroscopies. In the former a visible laser line is selected, commonly one of the argon lines at 514.5 nm or 488.0 nm. One sees in these spectra Raman bands which are due to the vibrations of the backbone of the biological micromolecules as well as the side chains. Much information can be obtained on the average conformation of these biological macromolecules from analysis of the vibrational frequencies arising from the backbone portion of the chain. This is because the frequencies from the backbone are closely related to the torsional angles in which the principle chain or backbone of the biological polymer exist. On the other hand, with resonance Raman spectroscopy one can study individual side chains such as tyrosine or tryptophane. One can also design chromophoric substrates to make enzyme substrate complexes and study the conformation or the structure of the chromophore at the change in the enzymatic catalytic event. Examples of all of these types of spectroscopies and conformational information both static and dynamic which can be obtained will be discussed below. Suffice it to say however, that at the present time classical and resonance Raman spectroscopy presents powerful techniques for the study of the conformation of biological macromolecules both in vivo and vitro. Furthermore, by comparison of the vivo and vitro studies with studies on crystal fibers and similar samples one can make definite correlations between the structure in a living cell and the structure in the crystal. This ability to go from living cells to solutions to crystals and fibers represents one of the most powerful possible uses

153

C. Sandorfy and T. Theophanides (eds.), Spectroscopy of Biological Molecules, 153–170.
© 1984 by D. Reidel Publishing Company.

of the laser Raman technique.

II.   The Raman Spectroscopy of Nucleic Acids.

Early work on the Raman spectroscopy of mononucleotides
was done by Malt (1) but it was not until the pioneering investigation of Lord and Thomas (2) that the laser Raman spectroscopy
of mononucleotides was investigated in a systematic way.   This
pioneering paper is still useful for the assignment and identification of bands arising primarily from the nucleic bases.

After the publication of the Raman spectra of the nucleic
bases in aqueous solution at various pH's, a study was launched
on study of Raman spectroscopy of polynucleotidesand nucleic
acids in several laboratories included our own.(3-11) The early
work in our laboratory emphasized the correlations that existed
between the x-ray determined conformation of nucleic acids and
the Raman spectroscopy of the corresponding nucleic acid fibers.
It was found that there are certain Raman marker bands which
clearly show the existence of either the A, B, or C forms of
DNA.   These Raman bands were quantitatively assigned to conformationally dependent vibrations of the sugar-phosphate backbone
chain using a Wilson-GF calculation.   More recently work by
Pohl and Stockberger (12) in Germany and Lord and his collaborators at MIT have shown the existence of Raman marker bands characteristic of the Z form of DNA.(13)

Figures 1, 2, 3, and 4, show Raman spectra of the A, B,
C, and Z forms of DNA. A list of the characteristic marker
bands and their frequencies is given in Table 1.   Although the
results listed in these tables are taken from the original observations and assignments, it should be pointed out that these
observations and assignments have been repeated and confirmed
in a number of laboratories, so there is little doubt as to
their reliablity.(14)   The normal coordinate calculations have
also been more extensively refined and completed by other workers
which have made a more reliable quantitative assignment of these
bands possible.(15) The Raman marker bands which are marker
bands for the A-B-C type conformation occur in characteristic
frequencies and possess characteristic intensities. For the
A-genus polynucleotide structure, a strong sharp band in the
region 807-815 cm^{-1} is always found and calculations show that
this is in indicative of the C3 prime-indo conformation of ribose
rings.   For the B and C forms, weaker, broader bands are found
at 835 and 875 cm^{-1}, respectively.   The 835 cm^{-1} band has been
assigned to asymmetric -O-P-O- stretching vibration of the sugar-
phosphate chain when the dioxy ribos rings are in the C3 prime-
exo or C2 prime-endo conformation.

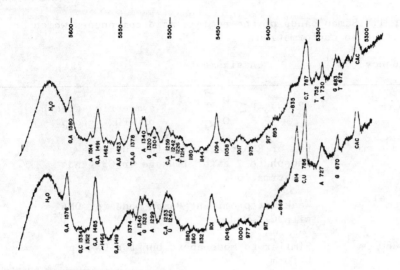

Figure 1.   Top: Raman spectra of a 20 mg/ml solution of calf-
thymus DNA at pH 7.0.   Bottom: Raman spectrum of transfer RNA
under similar conditions.   Note that the principle differences
in the Raman spectra of DNA and RNA are due to the substitution
of the thymine vibrations in DNA for the uracil vibrations in
RNA and the presence of the 835 cm^{-1} band in DNA and the 814 cm^{-1}
band in RNA.   These spectra are taken from paper of Small and
Peticolas (4).

In the past workers have generally associated the distinc-
tive Raman bands with the A, B, C, or Z, structure.   However,
care must be taken because we now know that the conformation
of DNA can be much more heterogeneous than previously thought.
For example, strictly speaking the term A-genus refers to a
double helical polymer with a certain number of residues per
turn, a certain pitch, a certain furanose ring pucker, etc.
As we will discuss later on in this article it is possible to
change the furanose ring pucker without changing the position
of any of the other structural parameters.   This means then
that strictly speaking we must not say that the band at 811
is A-genius marker band as we have done in the past, but say
that it is really a marker band for C3'-endo conformation of
the furanose-phosphate ester linkage.   In the past it is believed
that such a ring pucker of the furanose ring could only occur
in A-DNA or RNA but as we will see below this is no longer believed to be the case.   Similarly the 835 cm^{-1} band now must
not be regarded specifically as a marker band for the B-form
but rather the marker band for C2'-endo or C3 prime-exo of furanose ring pucker.   This is because it is possible to have ring

Table 1.   Raman bands of the nucleic acid components which are
sensitive to conformation.

Frequency (cm^{-1}) Assignment

1100 cm^{-1} $(R-O-)_2-P \Big\langle {}^{O \nearrow}_{O \searrow}$ symmetric stretch
                                                              of $-PO_2-$ group

811 ± 4 cm^{-1} DNA or RNA chain mode for furanose-
                          phosphate   ester linkage in C3'-endo
                          conformation.

835 cm^{-1} Weak but prominent Raman mode for DNA
                          with C2'-endo

870–880 cm^{-1} Similar to mode above but for DNA in
                          C-form

682 cm^{-1} Guanine ring breathing for C2'-endo-anti[a]

665 cm^{-1} Same as above but for C3'-endo anti[a]

625 cm^{-1} Same as above but for C3'-endo syn[a]

(a) Taken from Nishimura, Y., Tsuboi, M., Nakano, T., Higuchi, S.,
Sato, T., Shida, T., Uesugi, S., Ohtuka, Eiko, Ikehara,
Nucleic Acids Res. 11(5), pp 1579–88 (1983).

puckers which belong to one canonical form of DNA whereas the
rest of the conformational parameters belong to another conformational form. Thus throughout this paper we wish to emphasis
the great possibility for flexibility and fluctuations in the
conformations of DNA. Perhaps these conformational deformations
make it easy for enzymes to recognize where they are along the
DNA when performing such important biological functions such
as replication or transcription.

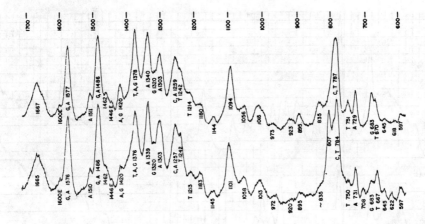

Figure 2. Raman spectra of fiber of the sodium salt of calf-thymus DNA at 98 (upper) and 75% (lower) relative humidity. The spectra are completely reversible upon reversing the humidity. The fiber in the top spectrum, 98% humidity has been shown from x-ray defraction measurements to be in the B-form while the fiber in the bottom spectra has been shown definitively to be in the A-form; thus we see that the transition from A to B causes a change in the 807 cm^{-1} band to the 835 cm^{-1} band. Thus these two bands become the marker bands for the A- and B-form insofar as the backbone is concerned. However, as will be discussed in the text, it is possible to have a backbone change with no concurrent change in the base positions.

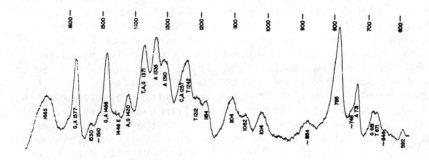

Figure 3. The Raman spectrum of a lithium salt of calf-thymus DNA at 32% relative humidity. This spectrum has subsequently been shown to be arised from the C-form of DNA. The spectra in figues 2 & 3 are due to Erfurth et al.

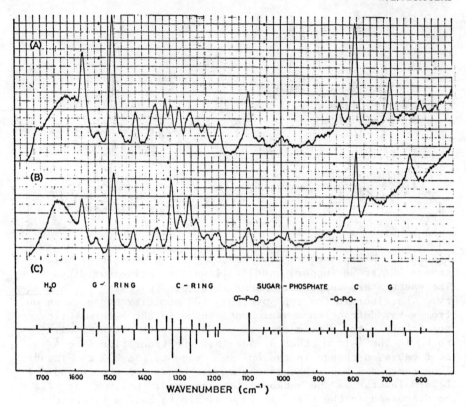

Figure 4. The Raman spectrum of low salt (upper) and high salt
(lower) poly(dG-dC). This spectrum was originally taken by
Pohl and Stockberger (12). Subsequently this spectral change
has been shown to be due to the formation of a zigzag double
helical structure called the Z-structure (13). It should be
noted that the Raman marker bands which occur in the 620-690 cm^{-1}
range and which are sensitive to the conformation of the G-C
sequences of Z-DNA arise from a quanine ring vibration and hence
are of a completely different nature from the backbone vibrations. The ring vibrations are active in the resonance Raman
effect which can be obtained at low concentrations. This opens
up the possibility of studying B-Z conformational transitions
at low concentrations using the resonance Raman effect.

The fact that DNA in solution has always been found to
show essentially the same Raman spectrum as DNA in the fiber
that 95% relative humidity is probably the most rigorous experimental proof that order DNA in solution is of the same B-genus
conformation as that in the fiber. Similarly the existence
of A marker bands in the Raman spectra of all ribo nucleic acids
structures clearly shows that ordered ribo nucleic acids belong
to the A-genus. For ribo nucleic acids there is a slight shift
in frequency of the marker band from 807 cm^{-1} A-DNA to 814 cm^{-1}
in RNA. However, intensity characteristics are identical. There
are a number of reasons reviews (see references (16) and (17).

It should be noted that the labels A and B strictly refer
to the conformational characteristics of DNA fibers at 75% and
95% relative humidity respectively. Thus the degree of conformation of any nucleic acid in solution can only be determined
by quantitative comparison of the marker band intensity relative
to the corresponding intensities in fibers of A-DNA and B-DNA.
In order to compare the intensities between the DNA and RNA,
most investigators have chosen the PO_2^- asymmetric stretching
as an internal reference band. On the basis of quantitative
measurements in a number of laboratories, it is now well established that the double helical DNA in solution is identical
with DNA fibers at 95% relative humidity whereas ordered RNA
in solution is essentially identical with DNA fibers at 75%
relative humidity. Thus until very recently the conformation
of nucleic acids in solution seemed a rather straight forward
somewhat uninteresting result. All DNA double stranded helices
possessed C2 prime-endo (or the equivalent C3 prime-exo) ring
pucker and existed in a rather well known Crick-Watson B-form
double helix, whereas all ribo nucleic acids in a double helical
structure existed in the A-genus form with a C3 prime-endo ring
pucker. More recenlty however, this whole field been thrown
into some state of confusion by the proposal that other forms
of nucleic acids may be important biologically.

A number of years ago Pohl and Stockberger (12) showed
that high salt poly (dG-dC) showed different Raman bands than
the low salt form and recent studies by Lord and coworkers have
shown that the Raman spectra of this material are identical
to crystals of oligo d(G-C) showing that at high salt alternating
d(G-C) polymers exist in the form labeled by Rich and coworkers
as the Z form.(13) However, these are not the only nucleic
acid polymers with alternating purine-pyrimidine sequences which
show secondary structure different from that of B-DNA. In particular poly (dA-dT) poly (dA-dT) fibers show a Raman spectrum
which indicates the existence of both C3 prime-endo C2 prime-endo
puckered furanose rings.(18) Actually, the deoxytetramer,
(dA-dT)$_2$ has been shown to have alternating C3 prime-endo-C2
prime-endo pucker furnose rings by x-ray defraction studies

on single crystals.(19) This finding of furanose conformational
heterogeneity has led to the interesting hypothesis that alternating (dA-dT) polymers in solution will possess alternating ring
pucker. It was this x-ray observation and hypothesis that led
to the Raman spectra of a study on this material.(18)

     Thus, we see from very recent work the necessity of a clear
establishment of a correlation between the intensity and frequency of Raman bands on molecules whose structure is known without
any doubt. Recently in this laboratory G.A. Thomas and the
present author began the study of the Raman spectroscopy of
crystals of three dinucleotides whose sugar-phosphate conformation is precisely known from x-ray defraction measurements.(20)
The three dinucleotides studied were uridylyl (3 prime-5 prime)
adenylmonophsophate (UpA), guanylyl (3 prime-5 prime) cytidine
monophsophatenonahydrate (GpC) and sodium thymidylyl (5 prime-3
prime) thymidylate 5 prime-hydrate (pTpT). The first two
of these dinucleotides have been shown by x-ray crystalography
to belong to the A-genus species with all the backbone coordinates in conformation almost identical to that of canonical
A-form DNA. Similarly it has been found that the Raman spectroscopy of these materials shows typical intensity and frequency
for the A-genus Raman marker band. On the other hand, it was
found that pTpT   which had been shown by x-ray analysis
to belong to the B-genus (C2 prime endo furanose ring conformation) showed a conformationally dependent B-genus Raman marker
band at 833 cm^{-1}. The interesting observation is that whereas
the A-genus Raman marker band found in the first two   dinucleotides was almost identical intensity with that found in
A form DNA and crystals and solutions of ordered RNA, the B
form marker band at 833 cm^{-1} found in the crystal was almost
ten times more intense than that found in ordinary B-DNA fibers
or double helices in solution. The 833 cm^{-1} band found in the
pTpT crystal is conformationally dependent and does not exist
in solution, showing that in solution the dinucleotide is much
less ordered than it is in the crystal.(20)

     On the bases of these observations it has been suggested
that B-DNA is much more flexible in its double helical conformation than A-DNA. Thus, B-DNA has a weak marker band even in
the double helical form in fibers or solutions due to a fairly
broad distribution of torsinal angles of the backbone chain
about the average B form torsinal angles. On the other hand,
RNA must possess torsinal angles which are almost identical
to those of the model compounds in the crystal. Finally, one
observation has been made which is very difficult to understand
in single stranded disordered DNA and RNA the marker bands are
too weak to be observed. This is in spite of the fact that
one observes easily the Raman bands due to the base vibrations.
The absence of either the C2 prime or C3 prime marker bands

for dinucleotides at room temperature as well as for polynucleotides at a temperature where the chains are no longer double
helical and largely disordered must be interpreted as a very
broad distribution of the backbone dihedral or torsinal angles.
This observation seems to be at variance with the observation
of NMR that dinucleotides exist in a equilibrium between the
C3 prime and C2 prime endo ring pucker.(21) One possible explanation of this apparent discrepency may be found in the observation that it is the dihedral angles along the backbone which
are primarily responsible for the intensity and frequencies
of the Raman bands whereas it is the dihedral angle around the
sugar themselves which apparently dominate the NMR results.
The two are rigorously related if no change in bond angle is
permitted. However, small change in bond angle would permit
a considerable more flexibility in the backbone dihedral angles
while keeping the torsinal angles of the sugar itself relatively
rigid.

Figures illustrating the above points are shown. For example figure 5 shows the Raman spectrum of aqueous solution
of UpA and its crystal grown from the above solution. Similarly
figure 6 shows the Raman spectrum of pTpT in aqueous solution
of pH 7 and in crystals grows from ethanol solution. The appearance of a very strong band at 833 cm^{-1} which occurs only in
crystal form can be easily noted. Figure 7 shows the Raman
spectra of poly (dA) poly (dT) as a function of temperature
in pH 7 showing that at low temperatures both 814 and 841 bands
are obtained. This shows the first evidence of ring pucker
heterogeity in aqueous solutions.

It can be seen from the above work that the Raman spectrum
presents a powerful technique for studying the conformation
of the conformation and diversity of conformation of nucleic
acids in solution. Quantitative results can be obtained. For
example, the 811 cm^{-1} band generally occurs at 814 cm^{-1} in RNA
structures and may be used as a measure of the A-type conformation. The 1100 cm^{-1} band is used as an internal standard of
reference, and the intensity ratio, I(811)/I(1100), is taken
as a measure of the A-type conformation. The value of 1.65
± 0.05 is generally taken to be 100% A-type configuration both
for DNA and RNA structures (7), (8).

Additional information can be obtained from the intensities
of the Raman bands from the nucleic bases. These intensity
changes are due to base-base stacking interactions which change
both the intensity of the ultraviolet absorption bands and the
intensities of the bands which arise from a preresonance effect
with the absorption band. Both the absorption band and the
resonant Raman ring vibrational modes tend to decrease in intensity upon stacking--an effect called hypochromism, Table 2 lists

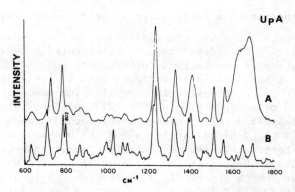

Figure 5. (A) Raman spectrum of aqueous solution of UpA at 20°C, pH 3 (HCl acidified), 15 mg/ml. A 514.5 nm line (100 mW) of argon laser was used as excitation source. Spectrum is 20 scans at 1 cm^{-1}/s scan rate. (B) Raman spectrum of crystals of UpA grown from the above solution. Ref (20).

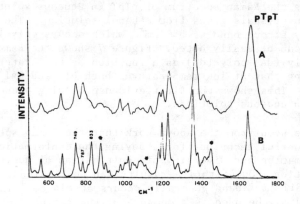

Figure 6. Raman spectra of pTpT. The bottom spectra marked, (B) is taken in crystals grown from 50% ethanol solution while the top spectrum marked (A) is taken in aqueous solution. Note the very strong conformational dependent marker band at 833 cm^{-1}. The asterisks show the presence of the ethanol bands. The 833 cm^{-1} band is 3 to 7 times stronger in this model of b-type DNA than it is in fibers of B-type DNA (see text). Ref (20).

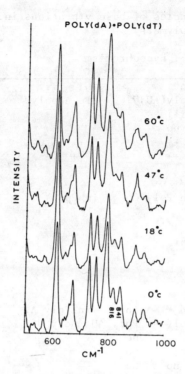

Figure 7. The Raman spectra of poly(dA) poly(dT) (20 mg/mL) in 0.2 M NaCl and 0.02 M sodium cacodylate at pH 7.0. Note that 0°C both the b- and a-form marker bands at 841 cm^{-1} and 816 cm^{-1} respectively are present. On the other hand there are no other little changes in the intensity or frequency of the nucleic acid based vibrations. This was interpreted as evidence for re-puckering of the furanose rings without a change in base orientation. This interpretation has been strengthened by the recent fiber x-ray defraction measurements (18), (21).

Table 2. Raman frequencies of base vibrations in double helical ribonucleic acid homopolymers

	Hypochromism		Preresonance
Uracil residues	in Poly(AU)[29]	in Plu A.polyU[18]	(likely excited state)
1680(s)(C=O strs.)	++		
1634(m)(C=O strs.)	no		
1400(m)	++	++	A(265 nm)
1235(s)	++	++	A(265 nm)
785(s)(ring breath.)			

Cytosine residue	in Poly C	in GpC	
1657(m)			
1607(m)			
1528(m)	+	+	A(268 nm)
1292(s)	+		A(268 nm)
1240(s)			
782(s)(ring breath.)			B(230 nm)

Adenine residue	in Poly A	in Poly AU	in Poly A. poly U	
1580(s)		no		A(276 nm)
1510(m)	++	no	+	
1484(m)	no	no	−	A(276 nm)
1379(m)	+	++		
1340(m)		no		
1310(s)	++	+	++	
1255(w)	+	no		
729(s)(ring breath.)	+	++	++	C(210 nm)

Guanine residue	in 5'GMP	in GpC	
1582(s)	+		A(276 nm)
1487(s)	++	−	A(276 nm)
1375(m)	+		
1328(m)	++	−	A(276 nm)
670(s)	--	−	Far

Table 3. Raman spectral changes observed on melting of ribonucleic acids.

Frequencies (cm^{-1})		Intensity changes observed upon melting (if any)	Assignment
H_2O pH7	D_2O pH7		
670	670	Large decrease	G
725	720	Large decrease	A
785	780	Increase due to shift in 814 cm^{-1} band	U,C OPO symmetric, if chain is disordered
814	814	Disappears completely, shifts to 785 cm^{-1}	OPO symmetric stretch
867	860	No change	Ribose
915	915	No change	Ribose
975	990	Decrease	Ribose
1003		No change	A,U,C
1047	1045	Small decrease	Ribose–phosphate
1100 (5)	1100 (4)	No change	PO_2–symmetric stretch
1182 (2)	1185 (1)	Small increase	A.G.C.
1240 (6)		Large increase	U
1248 (5)		Small increase	A,C
	1310 (7)	Small increase	C
1320 (7)	1320 (7)	Small increase	A,G
1340 (7)	1340 (7)	Increase	A
1375 (5b)	1370 (3B)	No change	A,G
1420 (2)		No change	G,A
1460 (sh)	1460 (sh)	No change	Ribose
1484 (10)	1480 (8)	Small increase	G,A
1527 (2)	1526 (3)	Small increase	C,G
1575 (8)	1578 (10)	Small increase	G,A
	1658 (4)		C=O in U,G,C
1692 (4)	1688 (4)		

W. L. PETICOLAS

Table 4. Raman spectral changes observed on melting of calf
thymus DNA[30]

cm^{-1} in H_2O pH = 7	cm^{-1} in D_2O pH = 7	Changes observed upon melting (if any)	Assignment
672	656	Small increase	T
683	677	Decreases	G
729	716	Large increase	A
750	734	Shifts to lower cm^{-1}	T
	765	Large increase in intensity	
	785	No change	PO_2 diester symmetric stretch
786		Small increase in intensity	PO_2 diester symmetric stretch
835	828	Shifts to lower cm^{-1}	PO_2 diester anti-symmetric stretch
879	867	Decreases	Deoxyribose–phosphate
	893	Decreases	Deoxyribose
920		Decreases	Deoxyribose
	966	Decreases	Deoxyribose
	1013	Decreases	C–O stretch
1015		No change	C–O stretch
1051	1047	Decreases and shifts to a higher cm^{-1}	C–O stretch
1094	1091	No change	PO_2^- diestersymmetric stretch
1144		Disappears	Deoxyribose–phosphate
1186		Moderate increase	T
1214		Moderate increase	T
1225		Moderate increase	A
1240		Large increase	T
1259	1260	Small increase	C,A
1303	1300	Moderate increase	A
1340	1343	No change	A
1378	1375	Moderate increase	T,A
1421	1418	Small increase	A,G
1643		Decreases	Deoxyribose–phosphate
	1484	Shifts to lower cm^{-1} deuteration of C-8 proton on A and G	
1491		Moderate increase	G,A
	1501	No change in intensity shifts to lower cm^{-1}	A

the hypochromic Raman bands; the electronic excited states from which these bands derive much or most of their intensity are also listed.

Finally Table 3 and 4 shows the changes experimentally observed upon the melting of double helical RNA or DNA.

Very recently Ziegler et al. (21) at the University of Oregon have used the multiple frequency doubled YAG laser to obtain the Raman spectra of the mononucleotides in the far U.V. The ultraviolet resonance Raman spectra of aqueous solutions of the 5'-monophosphates GMP, CMP, AMP and UMP have been obtained with 266 nm and 213 nm excitation. These two wavelengths are resonant with the first and second (or third) strong transition of these nucleotides. The resulting spectra are shown below. Large differential intensity changes are observed. For GMP the change from 266 to 213 nm excitation results in a large increase in intensity for a 1670 cm^{-1} line and a large decrease in intensity for a line at 1490 cm^{-1}. Binding of <u>cis</u>-diaminodicholoplatinum (CPD) to GMP results in a considerable decrease in the relative intensity of the 1670 cm^{-1} band and a band at 1344 cm^{-1} as well as some change in structure near 1535 cm^{-1}. For CMP, excitation at 213 nm results in apparent elimination of bands at 1243 and 1294 cm^{-1} and a decrease in intensity of a band at 1650 relative to 266 nm excitation. For AMP, the shift of the excitation energy toward the vacuum ultraviolet results in a large increase in the intensity of a 1580 cm^{-1} line and similar decreases in the intensity of lines at 1484 and 1338 cm^{-1}. For UMP, excitation at 213 nm results in virtual elimination of lines at 1690, 1630 and 1232 cm^{-1} which are quite strong at 266 nm. Attempts to rationalize these intensity changes on the basis of CNDO calculated bond order changes have been made.

Table 8. Raman Frequencies Commonly Observed in Classical Raman
Spectra of Proteins.

Frequency (cm^{-1})	Assignment
20–50	Strong low frequency modes of globular proteins may involve over-all breathing motion but usually over-damped in solutions
510 525 540 C–S–S–C stretching	Gauche-gauche-gauche gauche-gauche-trans trans-gauche-gauche
624	ring mode of phenylalanine
630–670	C–S stretching gauche
700–745	C–S stretching trans
760	strong sharp tryptophan ring mode
830–850	Doublet of tyrosine. $\dfrac{I(830)}{I(850)}$ H-bond donor or H-bond acceptor
930–940	C–C or C–N strong in α-helical regions
950–960	C–C or C–N strong in discordered regions
1004	strong sharp phenylalanine breathing
1016	tryptophan ring mode strong sharp
1220–1300	Amide III
1361	tryptophan ring mode
1440	CH_2 bending in side chains
1556	tryptophan
1584	trpt, phen
1605	phenylalanine
1613	trp, tyr, phe
1640–1680	C = O Amide I vibration
2560–2580	SH stretch
2800–3000	C–H stretch

III.   Classical Raman Spectroscopy and the Secondary Structure
of Polypeptides and Proteins.

The Raman spectrum of a protein is a sum of the Raman bands
due to the side chain residues of the amino acids and the backbone polypeptide chains. This can be seen in Table 8 where
a list of the bands commonly found in proteins is given. As
may be seen the aromatic residues -phenylalanine, tyrptophan
and tyrosine- give prominent bands due to strong preresonance
with their $\pi \to \pi^*$ transitions. However for the purpose of
discussing the secondary structure, the amide I and amide III
bands are most useful. Table 8 gives a list of the amide frequencies before and after deuterium exchange. The amide I vibration which is primarily a $C_1 = O$ stretching vibration changes
only slightly (about 30 cm^{-1}) upon H $\to$ D interchange. However,
the amide III vibration which is a combination C-N stretch and
N-H wag changes remarkably from 1200-1300 cm^{-1} to 900-1000 cm^{-1}
upon deuterium exchange.

It is also possible to follow the H-D exchange of the six
tryptophanes in lysozyme through following an interesting change
in the Raman spectrum which occurs upon H-D substitution. As
may be seen in Table 8 there is a strong tryptophane band at
1361 cm^{-1} which occurs in the native amino acid. This mode
disappears upon H-D exchange. Another mode in 1386 cm^{-1} appears
in the deuterum exchanged tryptophane. Recently in our laboratory we found it possible to determine the relative accessibility
of the six tryptophanes in lysozyme by virtue of their different
rates of exchange.

REFERENCES

1. Malt, R. A., Biochim. Biophys. Acta, 120, pp. 461 (1966).
2. Lord, R. C., Thomas, G. J., Jr., Spectrochim. Acta, Part A
    23A, pp 969 (1967).
3. Fanconi, B., Tomlinson, B., Nafic, L. A., Small, W. and
    Peticolas, W. L., J. Chem. Phys. 51, pp. 3993-4005 (1969).
4. Small, E. W., Peticolas, W. L., Biopolymers 10, pp 69-88
    (1971); 10, pp. 1377-1416 (1971).
5. Erfurth, S. C., Kiser, E. J., Peticolas, W. L., Proc. Natl.
    Acad. Sci. U.S.A. 69, pp 938-941 (1972).
6. LaFleur, L., Rice, J., Thomas, G. J., Jr., Biopolymers
    11, pp. 2423 (1972).
7. Brown, K. B., Kiser, E. J., Peticolas, W. L. Biopolymers
    11, pp. 1855 (1972).
8. Thomas, G. J., Jr., Hartman, K. A. Biochim. Biophys. Acta
    312 pp 311 (1973)
9. Prescott, B., Gamache, R., Livramentao, J., Thomas, G.
    J., Jr. Biopolymers 13, pp. 1821-1845 (1974).
10a. Erfurth, S. C., Peticolas, W. L. Biopolymers 14, pp. 247-264
    (1975).
10b. Brown, E. B., Peticolas, W. L. Biopolymers 14, pp. 1259-1271
    (1975).
11. Erfurth, S. C., Bond, P. J., Peticolas, W. L., Biopolymers
    14, pp. 1245-1257 (1975).
12. Pohl, F. M., Ranade, A. and Stockberger, M., Biochimi.
    Biophys. Acta, 335, pp. 85-92, (1973)
13. Thamann, T. J., Lord, R. C., Wang, A. H. T., Rich, A.,
    Nucleic Acid Res. 9, pp. 5443-57 (1981).
14. Goodwin, D. C., Brahms, J. Nucleic Acid Res. 5, 835-850,
    (1978).
15. Lu, D. C., Prohofsky, E. W., Van Zandt, L. L. Biopolymers
    16, pp. 2491-2506, (1975).
16. Hartman, D. A., Lord, R. C., Thomas, G. J., Jr., In "Phy-
    sico-chemical Properties of Nucleic Acids", Duchesne, J.,
    Ed., Academic Press: New York, Vol. 2, pp. 92-143, (1973).
17. Peticolas, W. L., Tsuboi, N., In "Infrared and Raman Spec-
    troscopy of Biological Moelcules", Theophanidies, T. M.,
    Editor, Reidel: Dordrecht Holland, pp. 153-165, (1979).
18. Thomas, G. A., Peticolas, W. L., J. Am. Chem. Soc., 105,
    pp. 993-996, (1983).
19. Klug, A., Jack, A., Viswamitra, M. A., Kennard, O., Shak-
    kend, Z., Steitz, T., J. Mol. Biol. 131, pp. 669-680, (1979).
20. Thomas, G. A., Peticolas, W. L. J. Am. Chem. Soc. 105,
    pp. 986-992, (1983).
21. Zeigler, L. D., Stromen, D. P. Hudson, B. S., and Peticolas,
    W. L. In "Raman Spectroscopy Linear and Nonlinear", Las-
    combe, J. and Huong, P. V., Ed. John Wiley & Sons:Great
    Britain, pp. 707-708, (1982).

STRUCTURAL TRANSITIONS IN DNA (A,B,Z) STUDIED BY IR SPECTROSCOPY

E. Taillandier   J. Liquier   J. Taboury   M. Ghomi

Laboratoire de Spectroscopie Biomoléculaire,
Université Paris-Nord, 74, rue Marcel Cachin
93000 Bobigny (France)

ABSTRACT

   This review is concerned by the possibility given by infrared spectroscopy to study the DNA secondary structures and the
transitions which may be observed between them. The infrared
linear dichroism method applied to oriented films gives the
opportunity to follow the B → A and B → Z conformational transitions in double stranded polynucleotides . Some infrared
absorptions are also related unambigously to the different fa-
milies of right-handed and left-handed helixes. This paper
describes the spectroscopic features and their eventual applications.

INTRODUCTION

   The present review will be focused on the extent to which
infrared spectroscopy is useful for elucidating the secondary
structures of DNA. The discovery of the left-handed helix
(Z-DNA) in crystals of short sequences of purine pyrimidine nucleotides (1,2) as well as recent results concerning the right-
handed helix of dodecanucleotides (3) have demonstrated the
polymorphism of DNA molecules. A number of new aspects of the
relationship between secondary structure and function are developing after the identification of Z DNA in material of
biological origin (4) (5) (6) (7) (8). In view of this situation, it is required to characterize a variety of DNA secondary
conformations depending on its environment (cations, proteins,
drugs) and its base sequence.

C. Sandorfy and T. Theophanides (eds.), Spectroscopy of Biological Molecules, 171–189.
© 1984 by D. Reidel Publishing Company.

The conformational transitions of the double stranded helix of DNA are reflected by changes in dichroic ratios of the infrared bands. The infrared spectra of the different forms of the same polynucleotide exhibit also quite large differences in vibrational frequencies and intensities.

Infrared spectroscopy imposes no requirement of sample crystallizability. Spectra of satisfactory quality are obtained not only from amorphous solids but also from aqueous gels, hydrated films and solutions. Figure 1 shows the spectrum of a nucleohistone hydrated sample.

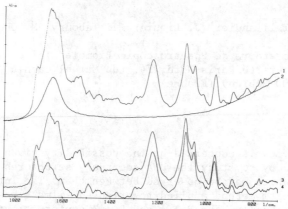

Figure 1.   1 hydrated nucleoprotein   2 water contribution   3 corrected spectrum of nucleoprotein 4 DNA spectrum in the complex.

The adsorbed water presents interfering absorptions in the 1640 cm^{-1} region and in the $700-600$ cm^{-1} region. As may be seen on figure 1, these strong and broad bands can be removed by substraction of a water spectrum recalculated and scaled by a computer coupled to the IR spectrophotometer.

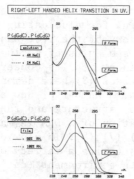

Figure 2.   B → Z transition   in U.V.

Because liquid water, the usual medium in biology, does
almost not present any absorption band in the U.V. region, U.V.
spectroscopy has been favoured by the biochemists. Figure 2
shows the U.V. spectrum of poly(dG-dC).poly(dG-dC). The number
of absorption bands of DNA is much smaller in the ultraviolet
than in the infrared. The I.R. spectrum contains more informations than the U.V. spectrum where most absorption bands caused
by base residues are superimposed on one another.

Recent advances in DNA synthesis methods have made it
possible to carry out single crystal X-ray analyses of double
stranded DNA molecules of predetermined sequence from 4 to 12
base pairs. X-ray diffraction is playing the principal role in
structure studies of nucleic acids. Infrared spectra however,
will be useful to determine the relative amounts of the different conformations of DNA when they coexist with different
crystallinities in the same complex sample or in systems available in smaller quantities.

Polymorphism of DNA

The DNA molecule has long been considered as existing only
in well defined right-handed double helical canonical forms A
and B, (9) : the right-handed B family with anti glycosidic
bonds and C_2'-endo sugar pucker, the right-handed A family with
anti glycosidic bonds and C_3'-endo sugar pucker. A third family
of nucleic acid double helixes is now known. The hexamer fragments d(CpG)$_3$ can crystallize with a left-handed sugar phosphate
backbone (1) (10). The left-handed form (called Z-DNA due to the
zig-zag character of the backbone) has alternating syn/anti glycosidic bonds and alternating C_3'-endo/C_2'-endo sugar pucker respectively for the guanine/cytosine base. There are two different
environments of phosphate groups in GpC and CpG sequences. Small
variations in the conformations of the different crystals of alternating G,C oligo-nucleotides suggest that there exists a
family of left-handed structures. Pohl and Jovin (11) were the
first to provide evidence for two conformations of alternating
poly(dG-dC).poly(dG-dC) in solution. The circular dichroism
of the high salt form was reversed when compared to that of the
low salt form. Recently sequence dependent structural variation
was found in the B-form of a CGCGAATTCGCG crystal (3). In contrast the structure uniformity of the A helix does not seem to
be influenced by local variations in the base sequences (12).
From the very first studies of DNA (13) it was apparent that
water is an important factor in helix structure. In films of
DNA the B → A transition (or the B → Z transition for poly d(G-C)
.poly d(G-C) is observed by infrared spectroscopy when humidity
is lowered (14) (15).

INFRARED DICHROISM

B → A conformational transition

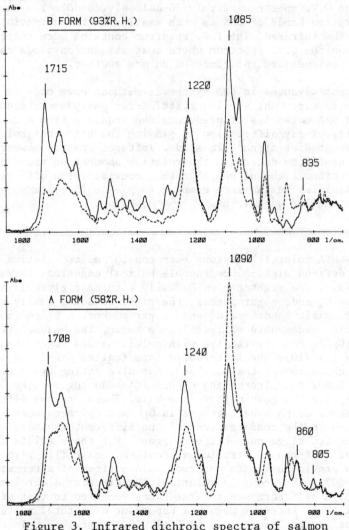

Figure 3. Infrared dichroic spectra of salmon
sperm DNA (1 Na^+/PO_4^{2-}) —— E perpendicular
--- E parallel

Two characteristic IR spectra are observed when stretched
films of salmon sperm DNA are dried (figure 3). The B spectrum
is obtained above 81 percent relative humidity (R.H.) and the
A spectrum when the film is dried to 66% R.H. without added salt.
Further drying below 58% R.H. leads to increasing disorder and
deterioration of the quality of the dichroic ratios of the IR bands.

To observe the dichroism of DNA films, two spectra have to be
recorded with the electric field direction of the incident IR
beam parallel and perpendicular to the polymer orientation axis
(a wire grid polarizer is used oriented at ± 45° to the orientation of the gratings). The differences between these two
spectra contain informations about the directions of the dipole
moment derivatives of the molecule and eventually about the bond
directions themselves. Fraser (16) (17) was the first to study
oriented films of DNA and has found a weak perpendicular dichroism for the in plane stretching vibrations of the bases.
Tsuboi (18) has analyzed the change in the IR spectrum of NaDNA
using different humidities inducing the transition from the
helical to random coil forms. Pilet and Brahms (19) have introduced the infrared linear dichroism for the study of the B → A
transitions as a function of the relative humidity of the DNA.
Infrared polarized spectra (figure 3) show major changes with
humidity in three spectral regions : at 1710 cm^{-1} considered to
be caused by base pairing, around 1230 cm^{-1} and 1090 cm^{-1} absorptions which are mainly due to the PO_2^- nearly antisymmetric
stretching vibration and to a PO_2^- stretching vibration of more
symmetric character coupled with a $C_5'-O_5'$ stretching vibration.
The B form is characterized by perpendicular bands at 1715 cm^{-1}
and 1085 cm^{-1} and by a non dichroic band at 1220 cm^{-1}. The A
form is characterized by an inversion of the polarization of the
1085 cm^{-1} band which becomes parallel at 1090 cm^{-1} and by two
perpendicular bands at 1240 cm^{-1} and 1708 cm^{-1}. The B → A conformational transition is thus accompanied by a substantial
rotation of the phosphate groups.

    To evaluate the dichroic ratio R of a band we measure the
absorption of the maximum of the band   for both parallel and
perpendicular spectra. The dichroic ratio $R(\perp///) = A\perp///$ allows
to calculate the angle Θ which the transition moment forms with
the axis of the oriented polymer. A correction is necessary because of the imperfect orientation of the film (16) (20) :

$$R(\frac{\perp}{A}{//}) = \frac{g + \sin^2\Theta}{2\cos^2\Theta + g} \qquad (1)$$

where g is a parameter which characterizes the axial orientation
in the film :

$$g = \frac{F}{N - \frac{3}{2} F} \qquad F = \int_0^{\pi/2} \sin^2\gamma f(\gamma) d\gamma \text{ and } N = \int_0^{\pi/2} f(\gamma) d\gamma$$

where $f(\gamma)$ is the distribution function of the chains (normalized if N = 1). The parameter g is first calculated using the
dichroic ratio R of the band at 1710 cm^{-1} or an average made
on eleven bands attributed to in plane vibration of the bases.
Once g is known, the Θ values for other dichroic bands in the
same spectrum are deduced by use of the formula (1). Figure 4

shows by example the variation of Θ_{1090}, angle of the transition
moment with respect to the helix axis deduced from the band at
1090 cm^{-1}, versus the hydration of the sample. It varies from
49° to 68° during the A → B transition in the case of a salmon
sperm DNA. If we make the assumption that bonds and transition
moments are colinear and that only one internal coordinate contributes to a mode, it is possible to determine the orientation
of the bonds involved in the corresponding infrared band. The
PO_2^- antisymmetric stretching vibration would have its transition moment along the direction of the 02 03 line of this group
and that of the PO_2^- symmetric stretching vibration would be along
the bi sector of the angle 02 PO3. With the preceding assumption
Θ_{1090} cm^{-1} represents the inclination of the OPO bisector to the
helix axis (49° in the A DNA and 68° in the B form) and Θ_{1220}cm^{-1}
the inclination of the 02 03 line versus the helix axis (55° in
the B form and 65° in the A form). Fuller et al (21) and more
recently Premillat et al (22) took the infrared data into account in their construction of models for A and B structures of
DNA.

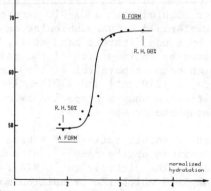

Figure 4. B → A conformational transition in salmon
sperm DNA followed by the orientation of the transition dipole moment of the IR band at 1090 cm^{-1} as a
function of the water content of the sample.

Protein-DNA interactions

The orientation of the dipole transition moment of the absorption band at 1090 cm^{-1} with respect to the double helix axis,
Θ_{1090}, is a sensitive conformational probe to identifie the geometry of the DNA helix. We have used this probe to study protein
DNA interaction (23) (24) (25). If we consider for example
figure 5 the nucleoproteic complex reconstituted with the fragment 1-89 of histone H2A (which contains all the residues in α
helix), Θ_{1090} values indicate the coexistence of the two A and
B geometries of the DNA in low R.H. conditions. The relative
amounts of these conformations depends on the DNA/protein input

ratio. The values of Θ_{1090} allow us, to obtain the percentage
of B form and the number of DNA base pairs stabilized at low
R.H. in a B geometry by one histone fragment. Figure 6 are shown
the results concerning different fragments of another histone,
H2B. The N terminal 1-59 fragment mainly in a random conformation is totally unefficient with respect to the stabilization
of the DNA in a B form as is shown by the values of Θ_{1090} which
remain around 49° whatever the amount of protein present. On the
contrary the 1-83 fragment which contains the totality of the
α helixes of the histone blocks the DNA in the B geometry as the
whole histone. Taking in account all the results obtained with
the different histones, it is possible to conclude that in a
nucleosomal core particle, the α helical parts of the four inner
histones interact with the DNA and stabilize the DNA in a B geometry (see table 1).

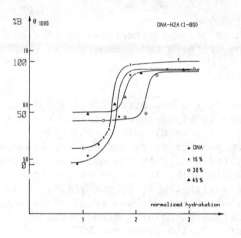

Fig. 5. Partial stabili-
zation of B DNA by his-
tone H2A (1-89) fragment.

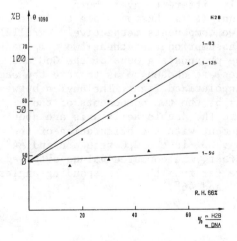

Fig. 6. Determination of
the relative amounts of A
and B DNA geometries in
nucleoproteins.

HISTONE FRAGMENT		Number of DNA base pairs blocked in B conformation per Histone fragment
H2A	1-56 1_____47 56	5
	1-89 1_____47 66 78 88	20
	1-129 1_____47 66 78 88_____129	21
	73-129 73 76 88_____129	14
H2B	1-59 1_____59	6
	1-83 1_____65 83	24
	1-125 1_____65 84_____125	24
H4	1-53 1_____53	0
	1-67 1_____55 67	10
	1-84 1_____55 67 78 84	18
	1-102 1_____55 67 78 90 102	22
	85-102 85_____102	2

Table 1. Role of the α helical structures of histones in the stabilization of the B DNA geometry.

B → Z conformational transition

The two bands mainly due to the PO_2^- stretching vibrations at 1220 cm^{-1} and 1090 cm^{-1} are weakly perpendicular in the polarized infrared spectrum of poly(dG-dC).poly(dG-dC)MgCl$_2$ in the left handed conformation. Two geometries Z_I and Z_{II} are proposed by Xray diffraction studies, and differ by the orientation of the phosphate group versus the helix axis. The observed dichroic ratios are in better agreement with the Z_I geometry (26). Assuming that all the bands are Lorentzian we have recalculated the IR spectra for the two directions of polarization between 1800 cm^{-1} and 900 cm^{-1} (figure 7). The symmetric PO_2^- stretching vibration is found at slightly different wavenumbers for the two directions of polarization. It is thus possible to deconvoluate this absorption into two components respectively parallel and perpendicular. With the assumption that these two components reflect the vibrations of the phosphate groups of the GpC and CpG residues, their intensities are chosen so as to have the same extinction coefficients in unpolarized light. The angles between the transition dipole moments of the two components of the 1090 cm^{-1} band with respect to the double helix axis are about 70° and 40°. This is in agreement with the orientation of the OPO bissector in the Z_I form : 60° in the CpG sequence and 44° in the GpC sequence. In contrast, the computed values obtained from the dichroic IR data are far from the corresponding angles of the Z_{II} form which are 47° and 46°.

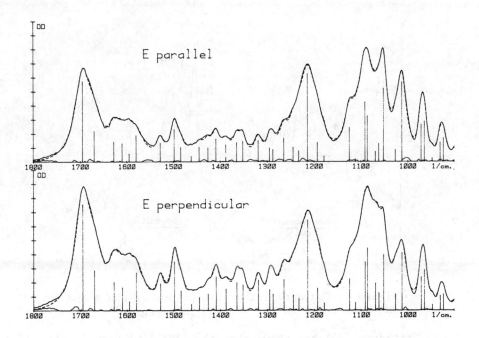

Figure 7. Polarized IR spectra of poly(dG-dC).poly
(dG-dC) MgCl$_2$ in Z form. ——— Experimental spectra.
--- Calculated spectra using Lorentzian bands ; their
wavenumbers and absorption coefficients are represented
by vertical lines.

IR BANDS CHARACTERISTIC OF THE DIFFERENT DNA FAMILY CONFORMATIONS

A and B conformations

Unambiguous assignments of some infrared bands to related
DNA forms have been made. As can be seen figure 8 an infrared
band near 835 cm^{-1} has been found in the B form of DNA. On the
other hand the A form is characterized by bands at 860 cm^{-1} and
805 cm^{-1} (27). The dehydration of the DNA films is accompanied
by a continuous decrease of the 835 cm^{-1} band and an increase
of the 860 cm^{-1} band indicating the B → A transition. The B → A
transition has been followed figure 9 by the measurement of the
ratio $A_{860}/A_{835} + A_{860}$ at regular time intervals (as we can
notice it is a steep and cooperative transition occuring over a
narrow range of humidity). This ratio can be used to evaluate
the amounts of A and B geometries even in unoriented samples(28).

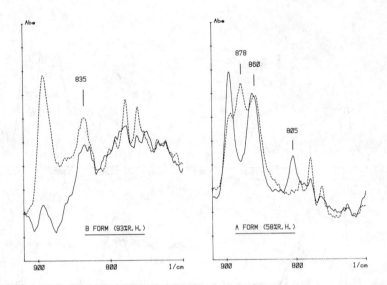

Figure 8. Characteristic absorptions of the A and B
DNA geometries in the 800–900 cm^{-1} region.

A Fritzsche (29) has shown that the 1185 cm^{-1} band is also
characteristic of the A form. This band has been assigned tentatively to vibrations of the sugar-phosphate backbone with a
high contribution from the sugar moiety.

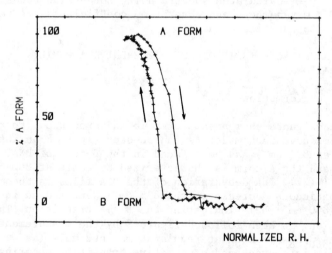

Figure 9. B → A conformational transition in salmon
sperm DNA followed by the relative intensities of
characteristic absorptions at 835 cm^{-1} and 860 cm^{-1}.

Z conformations

    Different IR spectra of poly(dG-dC).poly(dG-dC) films are
obtained with decreasing hydration when an excess of NaCl or
MgCl$_2$ is present in the sample. The modifications of the IR
spectrum observed at about 90% R.H. with NaCl or MgCl$_2$ (r > 0,5
MgCl$_2$ per nucleotide) have been correlated with the B → Z transition. A second reversible transition is observed in the case
of MgCl$_2$ and not in the case of NaCl, when the relative humidity is lowered to 30% for high magnesium chloride content
(r > 5). At low R.H. drastic modifications of the IR spectrum occur in the region between 1300 cm^{-1} and 1050 cm^{-1} (see figure 10
the infrared spectrum of the low R.H. form called Z$_o$).

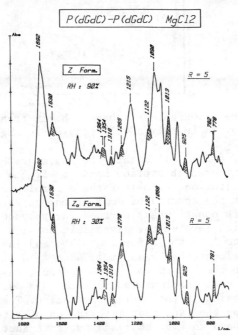

Figure 10. Infrared spectrum of poly(dG-dC).poly(dG-dC)
film (MgCl$_2$).

The IR study of the poly(dG-dC).poly(dG-dC) conformational
transitions as a function of the different values of the relative humidity and of the salt content in the films is summarized
figure 11.

    The most predominant differences between the spectra of the
Z form and of the B form consist in the presence in the Z form
spectrum of the following absorption bands : 1409 cm^{-1}, 1364 cm^{-1},
1354 cm^{-1}, 1318 cm^{-1}, 1265 cm^{-1}, 1124 cm^{-1}, 1013 cm^{-1}, 925 cm^{-1}

182

E. TAILLANDIER ET AL.

and by a doublet at 782-778 cm^{-1} (shaded bands in figure 10).
These bands are associated to the vibrations of the guanine and
of the deoxyribose in a syn conformation. The strong shifts of
the C = O stretching mode from 1710 cm^{-1} to 1690 cm^{-1} and of the
1660 cm^{-1} band of the cytosine to 1630 cm^{-1} are indicative of
the important rearrangement of the base stacking in the Z form.

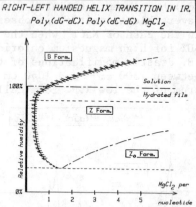

Figure 11. Right-left handed helix transitions observed by IR in poly(dG-dC).poly(dG-dC) MgCl$_2$ film.

Figures 12 and 13 illustrate the basis of the assignments
of the Z-DNA infrared absorption bands : effects of deuteration
or of CH$_3$ substitution, values of the dichroic ratios and by
comparison with the spectra of smaller molecules as the cytidine
or the GMP which constitute a part of the polynucleotide. Two
absorptions which involve the glycosidic torsion are observed at
1374 and 1364 cm^{-1} in the B form, at 1364 and 1354 cm^{-1} in the
Z form. We attribute the bands at 1374 cm^{-1} in the B form and
at 1354 cm^{-1} in the Z form to the dG residue, respectively in an
anti and a syn conformation. The 1364 cm^{-1} is due to the dC residue which remains in the anti conformation. These assumptions
are supported by the fact that methylation of cytosine shifts
the 1364 cm^{-1} band to 1356 cm^{-1} and the fact that in the
poly(dG-m^{5}dC).poly(dG-m^{5}dC) Z spectrum only one absorption is
detected at 1354 cm^{-1} with an intensity double of those of the
1364 cm^{-1} and 1354 cm^{-1} bands.

In heavy water, the B form of poly(dG-dC).poly(dG-dC)
presents an absorption at 778 cm^{-1}, the Z form two absorptions
at 784 cm^{-1} and 778 cm^{-1}. The latter is shifted to 768 cm^{-1} by
C$_5$ methylation of cytosine. We can thus attribute this 778 cm^{-1}
band, which is parallely polarized, to an out of plane vibration
of cytosine. The 784 cm^{-1} band in the Z form is due to an out of
plane vibration of the guanine as it is observed at the same
position as in guanidine. The presence of the doublet 784-778cm^{-1}
is characteristic of the Z form ; only one absorption is observed

at 778 cm^{-1} in the B form.

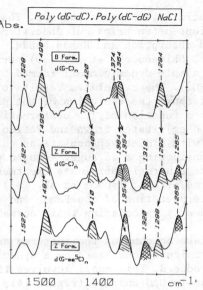

Figure 12. Influence of the B → Z transition on the 1374 cm^{-1} band. Effect of the 5-methylation of the cytosine on the Z form IR spectrum.

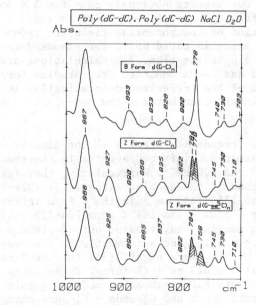

Figure 13. Deuterated spectra of B and Z forms of poly(dG-dC).poly(dG-dC). Influence of the 5-methylation of the cytosine.

NORMAL COORDINATE TREATMENT

The nucleic acid is a complex polymer. The calculation of
the normal vibrations often helps our understanding of the
nature of infrared absorption and Raman bands. In the past and
for the first time, Shimanouchi et al. (30) performed a calculation based on a simple dynamic model formed by a dimethyl-
phosphate $(C_3'-PO_4-C_5')$. The calculated results obtained from
this primary model show the evolution of the phosphate group
vibration modes as a function of its conformational degrees of
freedom (tt, tg and gg). Later Brown and Peticolas (31) using
a repeating model with a diethylphosphate as chemical unit and
Lu et al. (32) considering a di sugar-monophosphate dynamic
model tried to make a normal coordinate treatment of the right-
handed A and B forms of DNA. Tsuboi et al. (33) carried out
separately the interpretation of the Raman and infrared spectra
of the base residues by calculating their normal vibrations.

In order to take account of the infrared and Raman peaks,
we performed some calculations based on a dynamic model in which
the bases are introduced. This model involves the two independent sequences, namely CpG and GpC (figure 14), which take
account of the structural particularity of the "wrinkled B" (34)
and Z (1) forms of poly(dG-dC).poly(dG-dC). In these calculations
the hydrogens are neglected. So, each of the considered sequences
contains 35 atoms. One expects obviously 3N - 6 = 3 x 35 - 6 = 99
normal vibrations from CpG or GpC sequences. To respect the
harmonic approximation of the potential field the redundant internal coordinates are eliminated using the standard method of
B.B̃ matrix-product diagonalization. The calculations are based on
the Wilson-GF method and use a modified Urey-Bradley force field
which takes account of the π-electron delocalization in the base
planes and of the interaction between the non-planar modes of
the bases.

The calculated frequencies agree well the experimental
peak-positions in Raman and infrared spectra in the spectral
region between 1800 and 600 cm^{-1}. Particularly, they take account of the hehaviour of the doublet situated at 1374-1364 cm^{-1}
in the B form and at 1364-1354 cm^{-1} in the Z form infrared
spectra. The B form 778 cm^{-1} and the Z form 784-778 cm^{-1} bands
are assigned to the guanine and cytosine out-of-plane modes. Finally, these results are in good agreement with the recent at-
tribution proposed by Nishimura et al. (35) for the Raman bands
at 685 cm^{-1} (B form) and 625 cm^{-1} (Z form). These bands are
assigned to the ring-breathing vibration mode of guanine altered
simultaneously by anti → syn and C_2'-endo → C_3'-endo changes
respectively for the base and sugar conformations, when the
B → Z transition occurs (tables 2 and 3).

Experimental frequencies IR	polarization L.D. exp.	* Calculated frequencies (cm^{-1}) and assignments		
		Brown and Peticolas	Lu et al.	This work
1220 (s)	n.d.	1210 O=P=O asym.st.	1209 O=P=O asym. st.	1224 G planar modes
				1217 O=P=O asym.st.
				1216 C planar modes
1185 (sh)	n.d.		1202 sugar-phosphate	1197 G planar modes
			1160 sugar-phosphate	
1115 (sh)	⊥	1108 C-O st. + P-O st.	1112 C-O st.	1126 G planar modes + sugar
				1106 $C_4'-C_5'$ st. + sugar
1086 (s)	⊥	1093 O=P=O sym.st.	1094 O=P=O sym.st	1089 O=P=O sym.st.
1065 (s)	⊥			1066 sugar
1050 (s)	⊥	1057 C-O st. + P-O st.	1058 C-O st.	1051 sugar
				1036 sugar + C planar modes
1025 (m)	⊥	1016 C-C st. + C-O st.		1023 $C_4'-C_5'$ st. + sugar
				1010 sugar + $C_3'-O_3'$ st.
1000 (sh)	⊥		999 sugar-phosphate	980-999 G planar modes + $C_3'-O_3'$ st. + sugar
966 (m)	⊥	942 C-C st. + C-O st.	959 sugar	964 $C_5'-O_5'$ st. +sugar
890 (m)	//		886 sugar-phosphate	880 sugar + C planar modes
865 (m)	//			861 G out-of-plane modes
835 (w)	//	839 O-P-O asym.st.+ $C_3'-C_4'$ st. + $C_5'-O_5'$-P bend.	834 sugar-phosphate	844-835 G planar modes + O-P-O asym. st + sugar
		774 O-P-O sym.st.+ C-O-P bend. + C-C st.	789 O-P-O sym.st.	797-773 O-P-O syn.st. + sugar +$C_3'-O_3'$-P bend + $C_3'-O_3'$ st.
778 (m)	//			776 G out-of-plane modes
740 (w)	⊥			757 G planar modes
730 (w)	n.d.			740 C planar modes
703 (w)	n.d.			701 sugar + O-P-O sym.st.
682 **				680 G planar modes

G : guanine C : cytosine (s) : strong (m) : middle
(w) : weak (sh) : shoulder // : parallel to DNA axis
⊥ : perpendicular to DNA axis n.d. : non-dichroic
* Only those calculated frequencies corresponding to the important infrared bands are reported in this table.
** Raman bands from spectra of Tsuboi et al.

Table 2. Comparison between the experimental and calculated frequencies of the Z form of poly(dG-dC). poly(dG-dC) (1250-600 cm^{-1})

186   E. TAILLANDIER ET AL.

Experimental frequencies	Polarization L.D. exp.	Calculated frequencies (cm^{-1})	
		Z(I) form	Z(II) form
1215 (s)	n.d.	1229 O=P=O asym. st.	1228 O=P=O asym. st.
		1220 G planar modes	1219 G planar modes
		1213 C planar modes	1209 C planar modes
1194 (sh)	n.d.	1194 G planar modes	1197 G planar modes
1124 (m)	n.d.	1128-1110 C_4'-C_5' st. + sugar + G planar modes	1130-1110 C_4'-C_5' st. + sugar + G planar modes
		1093 G planar modes + C_1'-N_9 st. + sugar	1095 G planar modes + C_1'-N_9 st. + sugar
1090 (s)	⊥	1076 O=P=O sym. st. + sugar	1074 O=P=O sym. st. + sugar
1070 (sh)	⊥	1071 O=P=O sym. st.	1070 sugar + O=P=O sym. st.
1053 (s)	//	1051 sugar+C planar modes	1055 sugar+C planar modes
1013 (s)	//	1016-1010 sugar + C_4'-C_5' st +C_3'-O_3' st.	1018-1008 sugar + C_4'-C_5' st. + c_3'-O_3' st.
1000 (sh)	//	995-977 G planar modes + C_3'-O_3' st.+ sugar	997-980 G planar modes + C_3'-O_3' st.+ sugar
966 (s)	n.d.	960 C_5'-O_5' st. + sugar	960 C_5'-O_5' st.
925 (m)	//	934-924 sugar +G planar modes +C planar modes	933-922 sugar +C_3'-O_3' st. + C planar modes
885 (sh)	//	883-873 sugar +C planar modes +G planar modes	893-874 sugar + G planar modes
869 (m)	//	860 G out-of-plane modes	866-860 G planar modes and out-of-plane modes
845 (sh)	//	845 O-P-O asym.st.+sugar	
835 (m)	//		833-831 O-P-O asym.st.+sugar
810 (sh)	n.d.	819-808 O-P-O asym.st.+ sugar	821-812 O-P-O asym.st. + sugar
784 (m)	//	783 G out-of-plane modes	783 G out-of-plane modes
778 (m)	//	778 C out-of-plane modes + O-P-O sym. st.	777 C out-of-plane modes
745 (m)	n.d.	743 C out-of-plane modes	742 C out-of-plane modes
730 (sh)	n.d.	728 G and C planar modes + sugar	725-721 C planar modes + sugar
710 (m)	n.d.	717 G planar modes + sugar	
625 **			625 G planar modes

G : guanine   C : cytosine   (s) : strong   (m) : middle
(w) : weak   (sh) : shoulder   // : parallel to DNA axis
   : perpendicular to DNA axis   n.d. : non-dichroic
* Only those calculated frequencies corresponding to the important infrared bands are reported in this table.
** Raman bands from spectra of Tsuboi et al.

Table 3. Comparison between the experimental and calculated frequencies of the B form of poly(dG-dC). poly(dG-dC) (1250-600 cm^{-1})

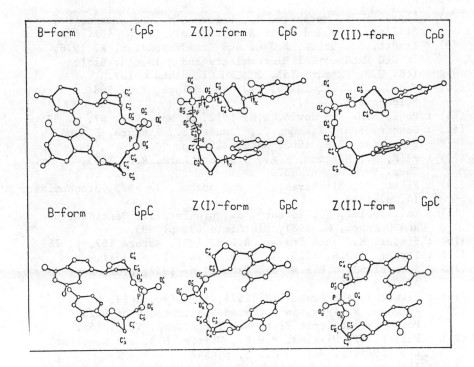

Figure 14. Conformation of CpG and GpC sequences
of the wrinkled B, Z(I) and Z(II) forms.

REFERENCES

(1) : Wang, A.H.J., Quigley, G.J., Kolpak, F.J., Crawford, J.L.,
      Van Boom, J.H., Van der Marel, G. and Rich, A. 1979,
      Nature 282, p. 680.
(2) : Crawford, J.L., Kolpak, F.J., Wang, A.H.J., Quigley, G.J.,
      Van Boom, J.H., Van der Marel, G. and Rich, A. 1980, Proc.Natl
      Acad Sci USA 77, p. 4016.
(3) : Wing, R.M., Drew, H.R., Takano, T., Broka, C., Tanaka, S.
      Itakura, K., Dickerson, R.E. 1980, Nature 287, p. 755.
(4) : Nordheim, A., Pardue, M.L., Lafer, E.M., Möller, A.,
      Stollar, B.D. and Rich, A. 1981, Nature 294, p. 417.
(5) : Viegas-Pequignot, E., Derbin, C., Lemeunier, F. and
      Taillandier, E. 1982, Ann Genet 25, p. 218.
(6) : Lemeunier, F., Derbin, C., Malfoy, B., Leng, M. and
      Taillandier, E. 1982, Exp Cell Res 141, p. 508.
(7) : Viegas-Pequignot, E., Derbin, C., Malfoy, B.,
      Taillandier, E., Leng, M., Dutrillaux, B. 1983, Proc
      Natl Acad Sci (in press).

(8) : Lipps, H.J., Nordheim, A., Lafer, E.M., Ammermann, D., Stollar, B.D. and Rich, A. 1983, cell 32, p. 435.

(9) : Arnott, S., Smith, P.J.C. and Chandrasekaran, R. 1976, in CRC Handbook of Biochemistry and Molecular Biology (ed. G.D. Fasman) Vol. 2 (CRC Cleveland Ohio).

(10) : Wang, A.H.J., Quigley, G.J. and Kolpak, F.J. 1981, Science 211, p. 171.

(11) : Pohl, F.M. and Jovin, T.M. 1972, J. Mol. Mol. 67, p. 375.

(12) : Conner, B.N., Takano, T., Tanaka, S., Itakura, K. and Dickerson, R.E. 1982, Nature 295, p. 294.

(13) : Falk, M., Hartman, K.A., J.R. and Lord, R.C. 1963, J. Am. Chem. Soc. 85, p. 387.

(14) : Pilet, J., Blicharsky, J. and Brahms, J. 1975, Biochemistry 14, p. 1869.

(15) : Taillandier, E., Taboury, J., Liquier, J., Sautière, p. and Couppez, M. 1981, Biochimie 63, p. 895.

(16) : Fraser, M.J. and Fraser, R.D.B. 1951, Nature 167, p. 759.

(17) : Fraser, R.D.B. 1953, J. Chem. Phys. 21, p. 1511.

(18) : Sutherland, G.B.B.M. and Tsuboi, M. 1957, Proc. Roy. Soc. A239, p. 446.

(19) : Pilet, J. and Brahms, J. 1973, Biopolymers 12, p. 387.

(20) : Zbinden, R. 1964, in Infrared Spectroscopy of High Polymers, Academic Press New York, Chap. V, p. 167.

(21) : Fuller, W., Wilkins, M.H.F., Wilson, H.R. and Hamilton, L.D. 1965, J. Mol. Biol. 3, p. 547.

(22) : Albiser, G. and Premilat, S. 1982, Nucleic Acids Res. 10, p. 4027.

(23) : Liquier, J., Taboury, J., Taillandier, E. and Brahms, J. 1977, Biochem. 16, p. 3262.

(24) : Taillandier, E., Taboury, J., Liquier, J., Gadenne, M.C., Champagne, M. and Brahms, J. 1979, Biopolymers 18, p. 1877.

(25) : Couppez, M., Sautière, P., Brahmachari, S.K., Brahms, J., Liquier, J., and Taillandier, E. 1980, Biochem. 19, p. 3358.

(26) : Pilet, J. and Leng, M. 1982, Proc. Nat. Acad. Sci. USA 79, p. 26.

(27) : Brahms, S., Brahms, J. and Pilet, J. 1974, Israel J. Chem. 12, p. 153.

(28) : Liquier, J. Gadenne, M.C., Taillandier, E., Defer, N., Favatier, F. and Kruh, J. 1979, Nucleic Acids Res 6, p. 1479.

(29) : Pohl, W. and Fritzsche, H. 1980, Nucleic Acids Res. 8, p. 2527

(30) : Shimanouchi, T., Tsuboi, M. and Kyogoku, Y. 1964, Adv. Chem. Phys. 7, p. 435.

(31) : Brown, E.B. and Peticolas, W.L. 1975, Biopolymers 14, p. 1259.

(32) : Lu, K.C., Prohofsky, E.W. and Van Zandt, L.L. 1977, Biopolymers 16, p. 2491.

(33) : Tsuboi, M., Takahashi, S. and Harda, I. 1973, in
       Physicochemical Properties of Nucleic Acids, Ed.
       J. Duchesne.
(34) : Arnott, S., Chandrasekaran, R., Puigjaner, L.U.,
       Walker, J.K., Hall, I.H., Birdall, D.L. and Ratliff, R.L.
       1983, Nucleic Acids Res. 11, p. 1457.
(35) : Nishimura, Tsuboi, M., Nakano, T., Higushi, S., Sato, T.,
       Shida, T., Uesugi, S., Ohtsuka, E. and Ikehra, M. 1983,
       Nucleic Acids Res. 11, 1579.

RAMAN SPECTROSCOPY OF BIOMATERIALS ACTING AS BONE PROSTHESIS

A. Bertoluzza

Cattedra di Chimica e Propedeutica Biochimica, Centro
Studi Interfacoltà di Spettroscopia Raman,
Istituto Chimico "G.
Ciamician" University of Bologna, Italy.

1. INTRODUCTION

The purpose of this work is to illustrate the application of Raman
spectroscopy to the study of inorganic polymeric glasses which
can be utilized as biomaterials in bone prosthesis.
A biomaterial is defined as a systematically, pharmacologically
inert substance designed for implantation within or incorporation
with living systems. This definition was assigned during the 6th
Annual International Biomaterial Symposium (1974) and clearly
emphasized a biomaterial as an implant material.
As a rule, the term "biomaterial" implies the identification of im
plants replacing biological materials. As such, it must satisfy so
me specified conditions, of a physical and chemical nature. In par
ticular, it is expected to possess biocompatibility, namely the
capability of the prosthetic material to adapt to the host tissue.
Since the mineral component of bones is predominantly consitued
by phosphates, prosthetic materials originating from phosphates
and polyphosphates have recently attracted a rapidly expanding
research interest, since these materials seem to present a higher
degree of biocompatibility when compared to alternative materials.
In this work, the problem of the structure and property of sodium
and calcium metaphosphate and oligophosphate glasses will be
tackled with the aid of the Raman spectroscopy technique. These
vitreous biomaterials can be considered from two points of view,
namely they could constitute a bone-like prosthesis, themselves,
or alternatively they could provide a good coating for a metallic

191

C. Sandorfy and T. Theophanides (eds.), Spectroscopy of Biological Molecules, 191–211.
© 1984 by D. Reidel Publishing Company.

or ceramic prosthesis possessing superior mechanical properties but inferior biological tolerance. A final point that
deserves attention is the fact that their composition makes them
suitable in view of understanding the mechanism of bone growth.
Among the various structure-probing techniques available for the
study of this problem, Raman spectroscopy appears ideally suitable.
In fact, unlike other methods, it allows us not only to characterize the structure of the biomaterial with regard to application
requirements, but also in principle, to investigate the interactions
between biomaterial and tissue which affect the biocompatibility
of the material utilized as prosthesis.

2. SODIUM METAPHOSPHATE: GRAHAM SALT
A. Mechanism of formation.

Graham salt, i.e. vitreous sodium metaphosphate, forms a very interesting material from a technological point of view owing to its
linear polymeric structure and constitutes the building block of
polyphosphate biomaterials.
This product is obtained by heating $NaH_2PO_4 . 2 H_2O$ from room temperature up to 630°C (the melting point of $NaPO_3$). Alternatively,
as starting product $Na_3P_3O_9$, namely a product of the termal decomposition of $NaH_2PO_4 . 2 H_2O$, can be chosen.
Fig. 1 shows the thermogravimetric diagram of $NaH_2PO_4 . 2 H_2O$ when
heated from room temperature to 500°C.

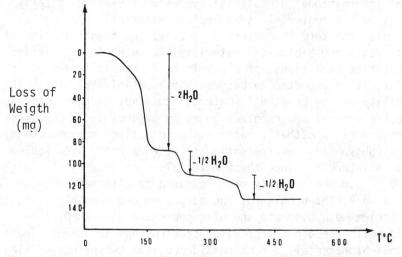

Fig. 1 - Thermogravimetric curve of $NaH_2PO_4 . 2 H_2O$ between room temperature and 500°C. (Weight of the sample 0.400 g).

Fig. 2 shows the I.R. spectra of the partial (at 175°C: - 2 H_2O; at 235°C: - 2,5 H_2O) and respectively exhaustive dehydratation products of the salt (at 370°C: - 3 H_2O).
The I.R. spectrum of the starting material (fig. 2a) shows a few optical absorption profiles which give clear structural indications. In fact, the component arising from v_{OH} asymmetric and symmetric stretching modes appear at 3515 and 3445 cm^{-1}, while the deformation band δH_2O appearing at 1650 cm^{-1} is partially covered by a wide band system, the principal components of which are situated at 2740, 2340, and 1700 cm^{-1}. This system is typical of compounds having O=X-OH groups capable of forming strong hydro gen bonds (1).

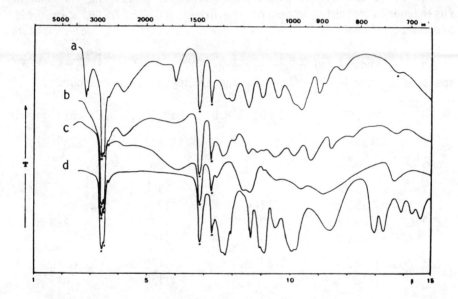

Fig. 2 - I.R. spectra of $NaH_2PO_4.2 H_2O$ (in Nujol) at different temperatures - a) room temperature; b) 175°C; c) 235°C; d) 370°C.

A comparison of the remaining absorption bands of $NaH_2PO_4.2 H_2O$ with those of NaH_2PO_4 and with those of the free $H_2PO_4^-$ ion (see Table I) makes it possible to identify in the spectrum of the hydrated salt the band arising from stretching vibrations of the phosphorus-oxygen bonds within PO_2 and $P(OH)_2$ groups, as well those due to POH deformation vibrations.
The appearance of more components in the crystal spectrum than would be expected for the single modes of the free $H_2PO_4^-$ ion is

due both to the lattice induced perturbation and to the vibrational coupling between ions contained within the unit cell.
The I.R. spectrum of the dehydration product of $NaH_2PO_4.2\ H_2O$ at 175°C (fig. 2b), which by the thermogravimetric study appears to correspond to the loss of two water molecules, by comparison with the spectra of a series of phosphates and polyphosphates of known structure, can be interpreted as the spectrum of anhydrous monosodium orthophosphoric acid, of formula NaH_2PO_4. Such spectrum is indeed characterized by the disappearance of the νOH stretching and of the δH_2O water bending bands: the A,B,C bands (situated at 2850, 2380 and~1680 cm^{-1} respectively) which arise from strong hydrogen bonds OHO are always present. Also, as shown in Table I, the in-plane and out-of-plane OH deformation modes and the phosphorus-oxygen stretching modes of the groups PO_2 and $P(OH)_2$ are still present.

Table I - Fundamental modes between 1300 and 800 cm^{-1} in the I.R. spectrum of the $H_2PO_4^-$ ion free and in crystalline compounds.

$H_2PO_4^-$ (2) cm^{-1}	NaH_2PO_4 cryst.(2) cm^{-1}	$NaH_2PO_4.2\ H_2O$ cryst. (3) cm^{-1}	$NaH_2PO_4.2\ H_2O$ at 175°C (3) cm^{-1}	Assignments (2)
			1300 s	
1230 s	1280 s	1281 s	1285 sh	δP-O-H
	1240 m	1242 s	1240 w	
	1098 m	1104 s	1100 w	
1150 vs	1166 s	1168 vs	1170 s	$\nu_{as}PO_2$
	1120 s		1130 m	
1072 vs	1053 vs	1044 s	1050 s	$\nu_s PO_2$
947 vs	989 s	987 sh	987 s	$\nu_{as}P(OH)_2$
	932 vs	962 vs	930 s	
878 s	875 m	906 s	872 m	$\nu_s P(OH)_2$
		890 m		
	820 w	846 w	820 w	γP-O-H

The second dehydration product of $NaH_2PO_4.2\ H_2O$ considered, formed at 235°C with a loss of 2.5 water molecules, and thus corresponding to the minimal formula $NaPO_3.0.5\ H_2O$ has also been studied, its I.R.

spectrum being compared with the spectra of phosphates and polyphosphates of known structure.
The comparison suggests that this material corresponds to an anhydrous disodium acid pyrophosphate $Na_2H_2P_2O_7$ (Table II).

Table II - Fundamental modes between 1300 and 700 cm^{-1} in the I.R.
spectrum of the $H_2P_2O_7^{--}$ ion free and in crystalline compounds.

$H_2P_2O_7^{--}$ (4) cm^{-1}	$Na_2H_2P_2O_7$ (5) cm^{-1}	$NaH_2PO_4 \cdot 2 H_2O$ at 235°C (3) cm^{-1}	Assignments (4)
1183 m	1191 s,sh	1190 vs	$\nu_{as}PO_2 + \delta P\text{-}O\text{-}H$
1110 s {1115 1108 1090	1155 s	1160 vs 1100 w	$\nu_{as}PO_3$, ν_sPO_2
1017	1027 w	1050 m,br	ν_sPO_3
954 m	957 m	965 s	ν P-OH
910 m	878 s,sh	895 vs,br	$\nu_{as}P\text{-}O\text{-}P$
	729 m	732 s	$\nu_sP\text{-}O\text{-}P$

In fact, its spectrum is essentially coincident with the spectrum
of $Na_2H_2P_2O_7$ reported in Table II, except for very minor drifts
in a few bands.
The anhydrous property of the species is confirmed by the missing
water bands in the spectrum (fig. 2c) and the acid character is
shown by the presence of the three bands A,B, and C (at 2850,
2380 and ~1650 cm^{-1}) which involve strong hydrogen bonds OHO.
Finally, the comparison summarized in Table II between the infra
red spectra of the free $H_2P_2O_7^{-2}$ ion, of crystalline $Na_2H_2P_2O_7$ and
of the 235°C dehydration product of $NaH_2PO_4 \cdot 2 H_2O$ enable us to iden
tify the phosphorus-oxygen stretching vibrations of groups PO_2, PO_3,
POH and POP in the material under investigation, along with the
deformation vibrations of POH. Thermogravimetric analysis suggests
the minimal formula $NaPO_3$ for the dehydration product of $NaH_2PO_4 \cdot$
2 H_2O obtained at 370°C. The infrared spectrum confirms this as-
sigment (fig. 2d), since the A,B,C and the POH deformation bands
typical of acid phosphates and polyphosphates are missing.

Table III - I.R. spectra of fully dehydrated $NaH_2PO_4.2\,H_2O$ at 370°C and of cyclic and linear sodium polyphosphate.

$NaH_2PO_4.2\,H_2O$ after heating to 370°C (3) cm^{-1}	$(NaPO_3)_3$ (3) cm^{-1}	$NaPO_3$ Maddrell II salt (5) cm^{-1}	$NaPO_3$ Maddrell III salt (3) cm^{-1}	$NaPO_3$ Kurrol IV salt (5) cm^{-1}
1650 vw,br				~1630 w,br
1315 sh	1314 vs	~1315 m,sh	1315 sh	
1298 vs	1295 vs	1298 vs	1295 vs,br	
				1268 vs,br
1262 m	1260 m			
1170 m	1168 m			
1163 s	1162 m	1160 m	1160 s	
				1150 m
1120 s	1120 s			
	1107 sh	1104 s	1110 sh	1110 s
1100 vs	1098 vs	1097 s	1096 s	
				1082 s
1055 m		1060 s	1060 s,br	
	1015 sh			
997 vs	996 vs			996 s,br
982 vs	984 vs			
878 m,br		878 vs,br		
				869 vs,br
			867 vs,br	778 w
773 s	772 s			
753 s	754 s			
742 sh		742 w	743 w	
				724 w
718 w		719 m	720 m	
700 w		702 m	702 m	
				698 m
686 m	686 m			

This spectrum has also been compared (Table III) with those of sodium polyphosphates of minimal formula $NaPO_3$ and of linear poly

meric structure: Maddrell salt in its various modifications
$NaPO_3$ II and $NaPO_3$ III, Kurrol salt $NaPO_3$ IV and cyclic sodium
trimetaphosphate $(NaPO_3)_3$. Since the spectrum of the 370°C dehydration product appears to be sum of the spectrum of the Maddrell
salt, containing the typical 878 cm^{-1} band, and of that of the
cyclic compound $(NaPO_3)_3$, the dehydration product examined is ap
parently a mixture of both these species.

Prolonged heating at 370°C of $NaH_2PO_4 \cdot 2 H_2O$ does not bring about
dramatic changes in the spectrum but shows only a slight intensity
increase of the bands typical of Maddrell salt II and the appearan
ce of a weak inflection point in the 860-870 cm^{-1} region where a
band of the other modification, Maddrell salt III, is expected.
A different behaviour is observed when the starting compound is
heated at 370°C for one hour. The completely dehydrated material
produced is a mixture of sodium trimetaphosphate $(NaPO_3)_3$ and
Maddrell salt $(NaPO_3)$III, as shown by the presence of a band at
868 cm^{-1}, while the one at 878 cm^{-1} is missing, all the other bands
remain essentially the same.

Under these conditions, changing the heating time does not produce
any further variation of the spectrum.

Finally, heating the starting material directly at 440°C, and
keeping it at this temperature for five hours, gives rise to the
complete transformation of the Maddrell salt into sodium trimetaphosphate, as shown by the resulting infrared spectrum.

The behaviour described above for the thermal dehydration of
$NaH_2PO_4 \cdot 2 H_2O$ is essentially in agreement with that observed by
other authors (6) who studied partial-and total- dehydration products, the shifts in formation temperature of the various compounds
being ascribable to the different experimental conditions, in par
ticular to the amount of different substances utilized, and to the
different speed of heating.

Our results give I.R. spectroscopic evidence for the structural
nature of the dehydration products of $NaH_2PO_4 \cdot 2 H_2O$ and contribute
a new piece of information regarding the formation mechanism of
Graham salt.

B. Raman spectra.

Fig. 3 shows the Raman spectrum of the glassy sodium metaphosphate
(Graham salt) and Figs. 4 and 5 show the Raman spectra of the
principal polymorphous crystalline modifications, i.e. Maddrell
salt $NaPO_3$ and sodium trimetaphosphate $(NaPO_3)_3$, which is   the

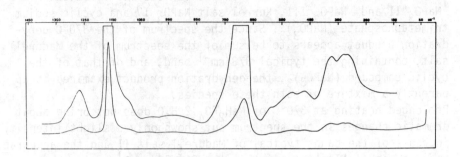

Fig. 3 - Raman spectrum of the glassy sodium metaphosphate
(Graham salt) $NaPO_3$.

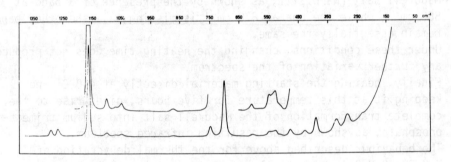

Fig. 4 - Raman spectrum of the crystalline sodium metaphosphate
(Maddrell salt) $NaPO_3$.

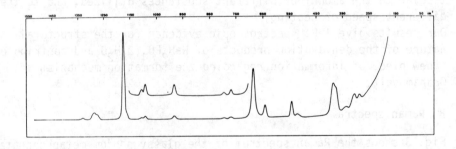

Fig. 5 - Raman spectrum of the crystalline sodium trimetaphosphate
$(NaPO_3)_3$.

final compound resulting from complete dehydration of $NaH_2PO_7.2\ H_2O$
and which give rise to Graham salt. Maddrell salt, i.e. a linear-
chain crystalline polyphosphate, is one of the typical forms iden
tified by means of X-ray analysis (7) within sodium metaphosphate,
the other being Kurrol salt. The structures of the two salts differ, for the different translation units, being the Maddrell salt
(light temperature form) belonging to the "Dreierkette" type and
the sodium Kurrol salt belonging to the "spiral" type (8). Sodium
trimetaphosphate on the other hand is a cyclic-chain polyphosphate
and contains six-atom units $P_3O_9{}^{3-}$.
The similarity of the properties of Maddrell salt and Graham salt
has been known in literature (9,10).
For the latter, it has also been established by means of diffusion
(11), ultracentrifuge (12) and X-ray (13) measurements that this
glassy metaphosphate is formed by linear polyphosphate chains of
high molecular weight, with the following structure

```
         0   0            0   0
         ||  ||           ||  ||
    HO-P-0-P-0-  .....-P-0-P-OH
         |   |            |   |
        ONa ONa          ONa ONa
```

In Table IV, the Raman spectra of Graham salt are reported and
compared with those of Maddrell salt and of sodium trimetaphosphate.
Table V collects the corresponding infrared spectra.
The spectral region within 1300 and 1050 cm^{-1} appears to contain
the bands due to valence vibrations of PO_2 groups both in linear
polyphosphate chains and in cyclic chains.
The region between 1050 and 600 cm^{-1} contains bands arising from
stretching vibrations of P-O-P bands of the linear chain (between
1050 and 700 cm^{-1}) and of the cyclic chains (between 1000 and 600
cm^{-1}). Finally, in the region between 600 and 200 cm^{-1} bands appear which are due to deformation vibrations of PO_2 groups (550-
350 cm^{-1}) and of P-0-P groups in linear and cyclic chains (350-200
cm^{-1}).
The differentiation between linear structure and cyclic structure
of the chain in crystalline polyphosphate must thus be sought
within the region 700-600 cm^{-1}. Two POP bands are observed at
∿685 and ∿640 cm^{-1} (stretching ring mode and breathing vibration
respectively) in the Raman and infrared spectra of sodium trimetaphosphate but not in the spectra of Maddrell salt. The absence of such
bands rules out the hypothesis that Graham salt possesses a polyphosphate structure formed by six-atom rings $P_3O_9{}^{3-}$, and supports

Table IV - Raman spectra of vitreous and crystalline sodium polyphosphates (14).

NaPO$_3$ cryst Maddrell salt cm^{-1}	NaPO$_3$ vitreous Graham salt cm^{-1}	NaPO$_3$ cryst trimetaphosphate cm^{-1}	Assignment
	1315 sh	1329 vw	
1288 w		1285 w	ν_{as} PO$_2$
1264 w	1272 w,br	1274 sh	
		1171 vs	
1159 vs	1167 vs	1156 sh	ν_s PO$_2$
	1101 sh	1104 vw	
1096 vw		1090 vw	
1065 vw			
1046 vw			
	1014 vw,br		
		980 vw	ν_{as} POP
	941 vw,br		
870 vw			
		793 vw	
		766 vw	
746 w			ν_s POP
	718 sh		
700 s	685 s		
		683 s	ν_s POP ring
		639 w	ν_s ring breathing
607 vw	591 vw,br		
549 v,br			
		539 w	
528 v,br			
512 sh	516 vw,br	518 vw	
477 w	471 vw,br		
428 w			PO$_2$ bend
391 m		383 m	POP bend
	377 m,br	373 sh	ring bend
	338 m	334 vw	
326 m	317 m		
		303 w	
284 m		284 vw	
235 vw			

Table V - Infrared spectra of vitreous and crystalline sodium
polyphosphates (14).

NaPO$_3$ cryst Maddrell salt cm^{-1}	NaPO$_3$ vitreous Graham salt cm^{-1}	NaPO$_3$ cryst trimetaphosphate cm^{-1}	Assignment
1315 sh		1314 vs	
1295 vs,br	1286 vs,br	1295 vs	$\nu_{as}PO_2$
		1260 m	
		1168 m	
1160 s	1153 sh	1162 m	
		1120 s	$\nu_s PO_2$
1110 sh		1107 sh	
1096 s	1095 s,br	1098 vs	
1060 s,br			
	1022 sh	1015 sh	
		996 vs	$\nu_{as}POP$
	974 s,br	984 vs	
867 vs,br	862 vs,br		
	770 m	772 s	
743 w		754 s	$\nu_s POP$
720 m	727 m		
702 m	690 sh		
		686 m	$\nu_s POP$ ring
		640 w	ν ring breathing
602 m	608 w,sh		
545 s			
		530 vs	
520 s	525 vs	520 vs	
514 s			PO$_2$ bend
465 vs,br	473 vs		POP bend
425 sh			ring bend
385 m		383 w	
		370 sh	
330 vw		333 w	
285 sh			
260 m			
		232 w	

the hypothesis that the glass has a linear-chain polyphosphate structure instead. This conclusion is buttressed by the appearance of intense infrared bands at $\sim$860 and $\sim$470 cm^{-1} in Graham salt and Maddrell salt, while these bands are missing in the trimetaphosphate spectrum.

A careful comparison of the vibrational spectra of glasses and of crystalline polyphosphates of known structure is necessary for the assignment of a given structure to the polyphosphate chain in the glass. In this connection, it must be remembered that within the glass a short-range order prevails, in the sense that there is or der around given fundamental structural units, in particular around PO_2 tetrahedral groups. The ordering of such groups along the chain occurs with a large distribution of $O_3P-O-PO_3$ angles among basic structural units, which bring about a certain loss of translation periodicity or a large distribution of the POP bands vibration of the chain skeleton. However, one can envisage in the glass a certain trend toward the disposition of structural units according to the ordering of typical linear chains in crystalline polyphosphates. To support this hypothesis one can mention the close analogy observed by Milberg e Daly (15) by means of X-ray data, between the structure of the polyphosphate chains of glassy $NaPO_3$ threaded in thin needles, and that of crystalline $RbPO_3$.

Table VI contains a comparison of the infrared spectra of $NaPO_3$ glassy Graham and crystalline Maddrell salts, obtained in this study, and those of $NaPO_3$ crystalline Kurrol salt (17) and crystal line $RbPO_3$ (form I at low temperature) (16).

A close analogy is observed between the infrared spectrum of the Graham salt and the I.R. spectra of $RbPO_3(I)$.

As far as Raman spectra are concerned, in the spectrum of glassy $NaPO_3$ there are two very intense bands at 1167 and 685 cm^{-1} corresponding to weak I.R. bands showing up as inflection points at 1153 and 690 cm^{-1}. A similar behaviour is observed for the related bands in the infrared and Raman spectra of $RbPO_3$, but not for $NaPO_3$ Maddrell salt.

In analogy with the suggestion of Poletaev and Urich, (16) the bands at 1167 and 685 cm^{-1} could be attributed respectively to symmetric PO_2 and POP stretching modes (probably, of symmetry A_g in the crystal), which are Raman active but I.R. inactive. This would confirm the analogy between the structure of glassy $NaPO_3$ and $RbPO_3$, for which the crystalline structure is known.

The results of this study and of our previous investigations (14) are in agreement with those other authors, who studied the problem

Table VI - Infrared spectra of vitreous and crystalline sodium phosphate with linear chain.

RbPO$_3$ cryst (16) cm^{-1}	NaPO$_3$ vitreous Graham salt (14) cm^{-1}	NaPO$_3$ cryst Maddrell salt (14) cm^{-1}	NaPO$_3$ cryst Kurrol salt (17) cm^{-1}	Assignment
		1315 sh		
1280 sh	1286 vs,br	1295 vs,br	1280 s,sh	ν_{as} PO$_2$
1260 vs			1270 vs	
1150 sh	1153 sh	1160 s	1140 m	ν_s PO$_2$
		1110 sh	1108 s	
1092 vs,br	1095 s,br	1096 s		
		1060 s,br		
1007 s	1022 sh			ν_{as} POP
	974 s,br		980 s	
870 vs	862 vs,br	867 vs,br	860 m	
800 s	770 m			
		743 w		ν_s POP
	727 m	720 m	716 m	
670 vw	690 sh	702 m	685 m	
590 sh	608 w,sh	602 m		
555 s		545 s		
	525 vs	520 s		
		514 s		PO$_2$ bend
480 s	473 vs	465 vs,br		POP bend
438 s		425 sh		
		385 m		
		330 vw		
		285 sh		
		260 m		

of the structure of Graham salt by infrared and Raman spectrosco py (4,17-22).

3. SODIUM OLIGOPHOSPHATE GLASSES.

Sodium oligophosphate glasses are characterized by a molar ratio
$R=Na_2O/P_2O_5$ comprised in the range 1.000 to 1.625, as opposed to
glassy sodium metaphosphate, for which R=1. These glasses   are
obtained by melting a mixture of a sodium phosphate for which R=1
(in particular sodium trimetaphosphate) and a sodium phosphate for
which R>1 (in particular sodium pyrophosphate, with R=2), both in
powder form (8).
The Raman spectra of sodium oligophosphates are reported in Fig. 6
(14). As observed also by Long et al. (22), increasing R value, a
weakening and a progressive shift toward lower wavenumbers of the
band from 1167 cm^{-1} (R=1)   to 1128 cm^{-1} (R=1.625) is shown.

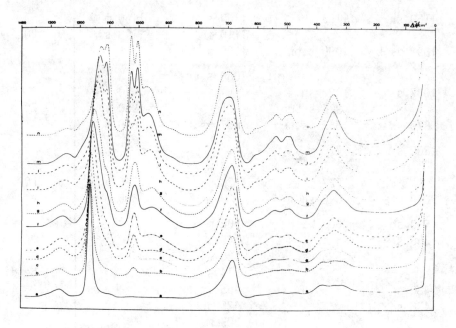

Fig. 6 - Raman spectra of sodium oligophosphate glasses with
composition $xNa_2O.P_2O_5$: a) x=1.000; b) x=1.140;c) x=1.249; d)
x=1.250; e) x=1.300; f) x=1.350; g) x=1.400; h) x=1.450; i)
x=1.500; 1) x=1.550; m) x=1.600; n) x=1.625.

At the same time, a band cleaning up as an inflection point at
∿1100 cm^{-1} in the R=1 glass becomes more intense and shifts to

$1108 \ cm^{-1}$ in the R=1.625 glass spectrum.
The region around $1000 \ cm^{-1}$ shows two main bands, the first of
which starts to split in the R=1.450 glass and becomes a well defined intense doublet (1025 and $1003 \ cm^{-1}$) in the R=1.625 glass
spectrum; the second, on the other hand, appears in the same glass
as a wide intense band at $972 \ cm^{-1}$.
The rest of the spectrum shows a broadening of the band close to
$700 \ cm^{-1}$, a weakening of bands at∿370 and∿320 cm^{-1} and the appearance of a medium intensity band at $341 \ cm^{-1}$ in the R=1.625 glass
spectrum.
The behaviour of Raman spectra of oligophosphate glasses are interpreted in terms of a breakdown of linear polymeric structure
as R increases, according to the hypothesis of other authors (8),
who studied these glasses by means of fractioned precipitation
measurements from organic liquid solutions.
In fact, the bands which appear around $1000 \ cm^{-1}$ absent in the
Graham salt spectrum and which are intensified in the oligophospha
te glass spectra with the increase of R may be attributed to symmetric stretching modes of PO_3 groups at the end of chains of dif
ferent lengths. These chain come from the breakdown of the indefi
nite polymeric structure of Graham salts caused by the interaction
of oxygen

```
     O   O                         O             O
     ||  ||                        ||            ||
    -P-O-P-O-   +  O⁻⁻    ─────→   -P-O⁻    +   ⁻O-P-O-
     |   |                         |             |
     O⁻  O⁻                        O⁻            O⁻
```

The other bands are, on the other hand, attributable to the vibra
tional modes of PO_2, P-O-P, etc. groups chains of different lengths.
A close relationship exists between the average chain length n̄
(calculated as phosphorous atoms) and the molar ratio R

$$R = \frac{\bar{n} + 2}{\bar{n}}$$

Therefore when R=1.6, n̄ is between 3 and 4.
The spectra of tri- and tetra sodium phosphate fused salts (18)
were chosen for comparison with the spectrum of sodium oligophospha
te glass with R=1.625. The similarity of Raman spectra is very clo
se, (Table VII) and provides spectroscopic evidence of the mechanism of degradation of Graham salt by oxygen ion in oligophosphate
glasses with R>1.

Table VII - Raman spectra of sodium oligophosphate glass
(R = 1.625) and of melted sodium polyphosphates with low
molecular weight.

Sodium oligophsophate glass $Na_2O/P_2O_5 = 1.625$ cm^{-1}	$Na_6P_4O_{13}$ cm^{-1}	$Na_5P_3O_{10}$ cm^{-1}
1248 w,br		
1108 vw,sh		
		1150 w,sh
1128 vs		
	1120 vs,br	1112 s,br
1108 vs		
1025 vs		
1003 vs	997 s	998 vs
972 s,br		
	951 m	
880 vw,sh	914 m	886 w,sh
	760 vw,sh	
702 s,br	680 s,br	692 s,br
610 vw	622 m	620 w,sh
558 vw,sh	560 vw	573 vw
536 w,br	534 m	537 w
508 vw,sh	502 w	
490 w,br		495 m
	421 vw	
		404 vw
374 vw,sh		
341 m,br	332 w,br	333 vw
	300 m,br	319 m,br
		284 vw,sh
	250 vw,sh	263 vw,sh
		225 vw,sh

4. CALCIUM METAPHOSPHATE AND CALCIUM OLIGOPHOSPHATE GLASSES.

These glasses containing calcium, are of importance and interest as
bone prosthesis with high biological tolerance, having a chemical
composition very similar to that of the bone.
On the basis of the previous study, it is possible to prepare
vitreous biomaterials of variable composition and therefore average

chain length is adjustable, and with correspondingly different
chemical, physical and mechanical properties.
Calcium metaphosphate glass was prepared by melting the crystal
line metaphosphate β-Ca(PO$_3$)$_2$ at 1300°C for 3 hours. The β-Ca(PO$_3$)$_2$
was obtained by dehydration of monocalcium phosphate monohydrate
Ca(H$_2$PO$_4$)$_2$.H$_2$O. Calcium oligophosphates glasses R CaO.P$_2$O$_5$ with a
molar ratio R=CaO/P$_2$O$_5$ ranging from 1.2 to 1.65 were obtained by
melting mixtures of β-Ca(PO$_3$)$_2$ and hydroxylapatite Ca$_{10}$(PO$_4$)$_6$(OH)$_2$
with a suitable composition at 1300°C for 3 hours (23).
In calcium oligophosphate glasses, hydroxylapatite, which undergoes
  thermal decomposition (24)

$$Ca_{10}(PO_4)_6(OH)_2 \longrightarrow 2\alpha\text{-}Ca_3(PO_4)_2 + Ca_4P_2O_9 + H_2O$$

is a calcium phosphate with R=3.3 and has been used here in place
of pyrophosphate salt (R=2) used for sodium oligophosphate glasses.
Is is known that fused calcium metaphosphate loses P$_2$O$_5$. Therefore,
it was necessary to verify the chemical composition of the samples,
and the values reported in Fig. 7 (Raman spectra) are experimental.
We also observed a similar loss when calcium metaphosphate glass
was prepared from crystalline metaphosphate. The chemical analysis
of the sample gave a molar ratio R=1.13 instead of 1.
The spectra show a trend which is almost identical to that observed
in the previous investigation on the sodium metaphosphate and sodium
oligophosphate glasses (24).
In particular, in the case of calcium metaphosphate glass (R=1.13
experimental value) a weak band appears at 1020 cm^{-1}. This band,
which is absent in the spectrum of sodium metaphosphate glass and
appears at the same frequency and intensity in the spectrum of
sodium oligophosphate glass with R=1.14, is to be attributed to
the symmetrical stretching modes of PO$_3$ terminal group in the linear polymeric chain with the average chain $\bar{n}$ about 15, according
to the previous formula

$$R = \frac{\bar{n} + 2}{\bar{n}}$$

This result is in agreement with previous chromatographic data of
other authors (25), who found that a calcium metaphosphate glass
sample, prepared under similar experimental conditions, gave
a molecular weight of about 20 phosphorus atom.

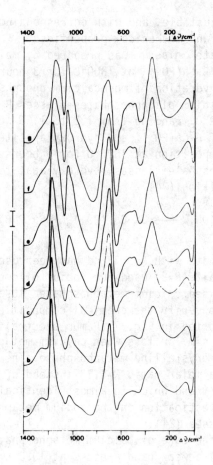

Fig. 7 - Raman spectra of vitreous biomaterials with molar ratio
R = CaO/P$_2$O$_5$ varying from 1.13 to 1.65. Sample a) R=1.13; b) R=1.20;
c) R=1.27; d) R=1.37; e) R=1.45; f) R=1.60; g) R=1.65.

As molar ratio R increases, significant changes are observed in
the Raman spectra and they are attributed to vibrational modes
of polyphosphate chains with length depending on the molar ratio,
according to the interpretation of the Raman spectra of oligophos
phate glasses.
The breakdown of the chains is shown by the appearance of the band
at 1020-1040 cm^{-1}, due to the symmetric stretching mode of terminal
PO$_3$ groups, which increases in intensity as R increases.
Other significant changes are observed in the spectra at about
1200-1100 cm^{-1} (symmetric stretching PO$_2$ modes), about 950-850 cm^{-1}
(asymmetric stretching P-O-P modes) and 750-650 cm^{-1} (symmetric

stretching P-O-P modes), and in the low frequency region of the
far infrared spectra (Fig. 8).
Here, a band located between 260 (R=1.13) and 280 (R=1.65) cm^{-1}
appears. This band, absent in the Raman spectra, can be attributed
to the cation vibration in the glass (26,27) and the variation of
its frequency indicates a different kind of electrostatic interaction due to the chains as a result of the breakdown.

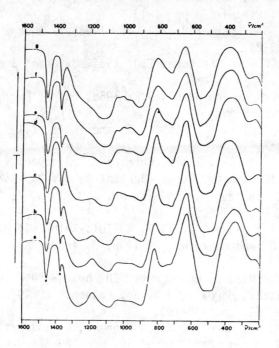

Fig. 8 - Infrared spectra of vitreous biomaterials with molar ratio
R = CaO/P$_2$O$_5$ varying from 1.13 to 1.65. Sample a) R=1.13; b) R=1.20;
c) R=1.27; d) R=1.37; d) R=1.45; f) R=1.60; g) R=1.65.

A programme of research is now in progress on the mechanism of
biocompatibility of these materials and on the correlation between
structure and biocompatibility. The molecular interaction between
the biomaterial and the biological environment in which it is im-
planted are also being studied.
The researches set out here have been partly carried out in collaboration with the School of Chemistry of Bradford University
(Prof. D.A. Long) in a bilateral project of research.

REFERENCES

1) Schuster, P., Zundel, G. and Sandorfy, C. "The hydrogen bond/II Structure and Spectroscpy",1976 (North-Holland Publishing Company).
2) Chapman, A.C. and Thirlwell, L.E. 1964, Spectroch. Acta 20, pp. 937-947.
3) Bertoluzza, A., Marinangeli, A. and Battaglia, M.A. 1974, Rend. Accad. Naz. Lincei 56, pp. 1-13.
4) Steger, E. and Fischer-Bartelk, C. 1965, Z. Anorg. Allgem. Chem. 338, pp. 15-21.
5) Corbridge, D.E.C. and Lowe, E.J. 1954, J. Chem. Soc. 76A, pp. 493-502.
6) Steinbrecher, L. and Hazel, J.F. 1968, Inorg. Nucl. Chem. Letters 4, pp. 559-562.
7) Doruberger-Schiff, K., Lieban, F. and Thilo, E.1955, Acta Cryst. 8, pp. 752-754.
8) Thilo, E. 1962, Adv. Inorg. Chem. and Radiochem. 4, pp. 1-75.
9) Thilo, E. 1952, Chem. Tech. (Berlin) 4, pp. 345-352.
10) Thilo, E., Schulze, G. and Wichman, E.M. 1953, Z. Anorg. Allgem. Chem. 272, pp. 182-200.
11) Korbe, K. and Jander, G. 1942, Kolloid-Beih. 54, pp. 1-146.
12) Von Lamm, O. and Malmgren, H. 1940, Z. Anorg. Allgem. Chem. 245, pp. 103-120.
13) Brady, G.W. 1958, J. Chem. Phys. 28, pp. 48-50.
14) Bertoluzza, A., Morelli, M.A. and Fagnano, C. 1973, Rend. Accad. Naz. Lincei 54, pp. 138-150.
    Bertoluzza, A., Morelli, M.A., Fagnano, C. and Battaglia, M.A. 1976, Rend. Accad. Naz. Lincei 59, pp. 269-273.
15) Milberg, M.E. and Daly, M.C. 1963, J. Chem. Phys. 39, pp. 2966-2973.
16) Poletaev, V. and Urikh, V.A. 1971, Izv. Akad. Nauk. Kaz. SSR, Ser. Khim. 21, pp. 16-22.
17) Bues, N. and Gehrke, H.W. 1956, Z. Anorg. Allgem. Chem. 288, pp. 307-323.
18) Bues, N. and Gehrke, H.W. 1956, Z. Anorg. Allgem. Chem. 288, pp. 291-306.
19) Kolesova, V.A. 1957, Opt.i.Spektrosk. U.S.S.R. 2, pp. 165-173.
20) Williams, D.J., Bradburg, B.T. and Maddocks, W.R. 1959, J. Soc. Glass Tech. 43, pp. 308-324.
21) Shih, C.K. and Su, G.J. 1965,"Proc. 7th. Int. Congr. Glass, Bruxelles"
22) Fawcett, V., Long, D.A. and Taylor, L.H. 1976, "Proceedings of

the Fifth International Conference on Raman Spectroscopy, edited by E.D. Schmid, J. Brandmüller, W. Kiefer, B. Schrader and H.W. Schrötter, - H.F. Schulz Verlag, Freiburg am Breisgan, pp. 112-113.

23) Bertoluzza, A., Battaglia, M.A., Simoni, R. and Long, D.A. 1983, J. Raman Spectrosc. 14, pp. 178-183.

24) Bertoluzza, A., Fagnano, C., Fawcett, W., Long, D.A. and Taylor, H. 1981, J. Raman Spectrosc. 11, pp. 10-13.

25) Ohashi, S. and Van Wazer, J.R. 1959, J. Am. Chem. Soc. 81, pp. 830-832.

26) Rouse, J.B., Miller, Ph.J. and Risen, W.M. 1978, J. Non-Cryst. Solids 28, pp. 193-207.

27) Nelson, B.N. and Exarhos, G.J. 1979, J. Chem. Phys. 71, pp. 2739-2747.

TEACHING THE NEW NMR: A COMPUTER-AIDED INTRODUCTION TO THE
DENSITY MATRIX FORMALISM OF MULTIPULSE SEQUENCES

G. D. Mateescu and A. Valeriu

Department of Chemistry
Case Western Reserve University
Cleveland, Ohio 44106

Abstract. This paper constitutes a guide for the use of density
matrix calculations in the description of multipulse nuclear magnetic resonance experiments. In keeping with its didactic nature,
the article follows in more detail than usual the evolution of the
density matrix elements from the thermal equilibrium stage to the
final magnetization behavior during the detection. Rotation
operators covering a wide variety of pulses are presented, together with the formalism necessary to predict the evolution of
the spin ensemble between and after pulses. The final calculation of the observables is then described. A two-dimensional
nmr sequence is used for "step-by-step" illustration.

1. INTRODUCTION

These notes are mainly intended as a theoretical background for
the computer generated movie presented at the Institute.
However, they have been written in such a way as to constitute an
independent study material. A "math reminder" is given in the
Appendix A for those who may need it. Our goal is to make it
easy for the reader, not only to understand in detail the density
matrix formalism but, also, to be able to use it for the
description of any other sequence not discussed here (existing
or to be invented).

2. THE DENSITY MATRIX

Before entering the formal treatment of the subject (see Appendix
B) let us build an intuitive picture of the density matrix. We

213

C. Sandorfy and T. Theophanides (eds.), Spectroscopy of Biological Molecules, 213–256.
© *1984 by D. Reidel Publishing Company.*

begin with the simplest system exhibiting both chemical shift and
scalar coupling: two spin ½ nuclei, A and X.   These coupled
species share four energy levels (Figure 1).

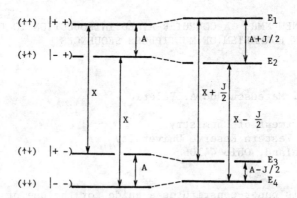

*Figure 1. Energy levels of an uncoupled (left) and
coupled (right) AX system. Transition frequencies A
and X and the coupling constant J are expressed in Hz.*

The possible relations between the states represented by "kets"
$|+ +)$, $|- +)$, $|+ -)$, and $|- -)$ are shown below (we assign the
first symbol in the ket to nucleus A and the second, to nucleus
X):

	$|+ +)$	$|- +)$	$|+ -)$	$|- -)$
$|+ +)$	P_1	$1Q_A$	$1Q_X$	$2Q_{AX}$
$|- +)$		P_2	$0Q$	$1Q_X$
$|+ -)$			P_3	$1Q_A$
$|- -)$				P_4

↑ ↑
$|A \ X)$

It can be seen that the off-diagonal elements connect pairs of different states. They are called "coherences", indicating possible transitions between the corresponding states. For instance in going from $|++)$ to $|-+)$ only the nucleus A is flipped. The corresponding matrix element will represent a single quantum coherence involving an A transition. We thus find two $1Q_A$ and two $1Q_X$ coherences (the matrix elements on the other side of the diagonal do not represent other coherences; they are mirror images of the ones indicated above the diagonal). There is also one double-quantum coherence, $2Q_{AX}$, in the box relating $|++)$ to $|--)$. The "zero-quantum" coherence can be considered as representing a flip-flop transition (E_2-E_3). The name of the coherence does not mean that the energy of the transition is necessarily zero. The diagonal elements represent populations. The density matrix description is more general than the spin temperature or vector magnetization descriptions. Populations, spin temperatures (when applicable), macroscopic magnetizations, can all be derived using the density matrix formalism even for sequences which defy the vector treatment

3. THE DENSITY MATRIX DESCRIPTION OF A PULSE SEQUENCE

The density matrix treatment of a pulse sequence includes the following calculation steps:

- thermal equilibrium populations (off diagonal elements are zero)
- effects of rf pulses (rotation operators)
- evolution between pulses
- evolution during acquisition
- determination of observable magnetization.

All these steps are presented here using the popular example of the two-dimensional heteronuclear correlation (2D HETCOR) sequence shown in Figure 2.

Equilibrium populations

At thermal equilibrium the four energy levels shown in Figure 1 are populated according to the Boltzmann distribution laws:

$$\frac{P_i}{P_j} = \frac{\exp(-E_i/kT)}{\exp(-E_j/kT)} = \exp\left(\frac{E_j - E_i}{kT}\right) . \qquad (1)$$

Taking the least populated level 1 as reference we have:

$$P_2/P_1 = \exp[(E_1 - E_2)/kT] = \exp[h(A + J/2)/kT] . \qquad (2)$$

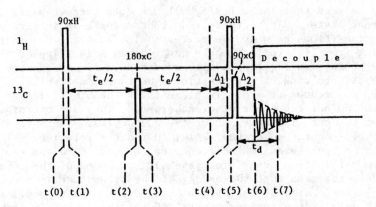

Figure 2. The two-dimensional heteronuclear correlation sequence:
$90xH - t_e/2 - 180xC - t_e/2 - \Delta_1 - 90xH - 90xC - \Delta_2 - AT.$

Since in our spectrometers transition frequencies are five orders of magnitude larger than coupling constants, we neglect the latter (only when we calculate relative populations; of course, they will not be neglected when calculating transition frequencies). Furthermore, the hA/kT and hX/kT ratios are much smaller than 1 (vide infra). This justifies a first order series expansion of the exponential [see (A11)]:

$$P_2/P_1 = \exp(hA/kT) = 1 + (hA/kT) = 1 + p, \tag{3}$$

$$P_3/P_1 = \exp(hX/kT) = 1 + (hX/kT) = 1 + q, \tag{4}$$

$$P_4/P_1 = 1 + [h(A + X)/kT] = 1 + p + q. \tag{5}$$

Let us take as an example the popular carbon-proton system. Their Larmor frequencies are in the ratio 1:4 (i.e., q = 4p). We now normalize the sum of populations:

$$P_1 = P_1$$

$$P_2 = (1 + p)P_1$$

$$P_3 = (1 + 4p)P_1$$

$$P_4 = (1 + 5p)P_1$$

$$\overline{}$$

$$1 = P_1(4 + 10p) = P_1 S. \tag{6}$$

Hence,

$$P_1 = 1/S$$

$$P_2 = (1 + p)/S$$

$$P_3 = (1 + 4p)/S$$

$$P_4 = (1 + 5p)/S,$$

and the density matrix at equilibrium is:

$$
D(0) = \begin{vmatrix} P_1 & 0 & 0 & 0 \\ 0 & P_2 & 0 & 0 \\ 0 & 0 & P_3 & 0 \\ 0 & 0 & 0 & P_4 \end{vmatrix} = \frac{1}{S} \begin{vmatrix} 1 & 0 & 0 & 0 \\ 0 & 1+p & 0 & 0 \\ 0 & 0 & 1+4p & 0 \\ 0 & 0 & 0 & 1+5p \end{vmatrix} =
$$

$$
= \frac{1}{S} \begin{vmatrix} 1 & 0 & 0 & 0 \\ 0 & 1 & 0 & 0 \\ 0 & 0 & 1 & 0 \\ 0 & 0 & 0 & 1 \end{vmatrix} + \frac{p}{S} \begin{vmatrix} 0 & 0 & 0 & 0 \\ 0 & 1 & 0 & 0 \\ 0 & 0 & 4 & 0 \\ 0 & 0 & 0 & 5 \end{vmatrix}
$$

The value of S is approximately 4 since p is very small [see (6)]. For instance, in a 4.7 Tesla magnet the C-13 Larmor frequency is $A = 50 \times 10^6$ Hz and

$$p = \frac{hA}{kT} = \frac{6.6 \times 10^{-34} Js \times 50 \times 10^6 s^{-1}}{1.4 \times 10^{-23} (J/K) \times 300K} = 0.785 \times 10^{-5} .$$

It is seen that the first term of the sum above is very large compared to the second term. However, this term is not important in NMR measurements because, containing the unit matrix [see (A20)-(A21)], it is not affected by any evolution operator (see Appendix B). Though much smaller, it is the second term which counts, because it contains the population differences (Vive la différence!). From now on we will work with this term only, ignoring the constant factor p/S and taking the licence to continue to call it D(0):

$$D(0) = \begin{vmatrix} 0 & 0 & 0 & 0 \\ 0 & 1 & 0 & 0 \\ 0 & 0 & 4 & 0 \\ 0 & 0 & 0 & 5 \end{vmatrix} \tag{7}$$

Equilibrium density matrices for other systems than C–H can be
built in exactly the same way.

First pulse: 90$^\circ_x$ proton

At time t(o) a selective 90° proton pulse is applied along the
x-axis. We now want to calculate D(1), the density matrix after
the pulse. The standard formula for this operation,

$$D(1) = R^{-1} D(0) R, \tag{8}$$

is explained in Appendix B. The rotation operator R for this
particular case is (see Appendix C):

$$R_{90xH} = \frac{1}{\sqrt{2}} \begin{vmatrix} 1 & 0 & i & 0 \\ 0 & 1 & 0 & i \\ i & 0 & 1 & 0 \\ 0 & i & 0 & 1 \end{vmatrix} . \tag{9}$$

Its inverse (reciprocal), R^{-1}, is readily calculated by transposition and conjugation [see (A22)-(A23)]:

$$R_{90xH}^{-1} = \frac{1}{\sqrt{2}} \begin{vmatrix} 1 & 0 & -i & 0 \\ 0 & 1 & 0 & -i \\ -i & 0 & 1 & 0 \\ 0 & -i & 0 & 1 \end{vmatrix} . \tag{10}$$

First we multiply D(0) by R (see Appendix A):

$$D(0)R = \begin{vmatrix} 0 & 0 & 0 & 0 \\ 0 & 1 & 0 & 0 \\ 0 & 0 & 4 & 0 \\ 0 & 0 & 0 & 5 \end{vmatrix} \times \frac{1}{\sqrt{2}} \begin{vmatrix} 1 & 0 & i & 0 \\ 0 & 1 & 0 & i \\ i & 0 & 1 & 0 \\ 0 & i & 0 & 1 \end{vmatrix} =$$

$$= \frac{1}{\sqrt{2}} \begin{vmatrix} 0 & 0 & 0 & 0 \\ 0 & 1 & 0 & i \\ 4i & 0 & 4 & 0 \\ 0 & 5i & 0 & 5 \end{vmatrix} . \tag{11}$$

Then we premultiply the result by R^{-1}:

$$D(1) = R^{-1}[D(0)R] = \frac{1}{\sqrt{2}} \begin{vmatrix} 1 & 0 & -i & 0 \\ 0 & 1 & 0 & -i \\ -i & 0 & 1 & 0 \\ 0 & -i & 0 & 1 \end{vmatrix} \times$$

$$\times \frac{1}{\sqrt{2}} \begin{vmatrix} 0 & 0 & 0 & 0 \\ 0 & 1 & 0 & i \\ 4i & 0 & 4 & 0 \\ 0 & 5i & 0 & 5 \end{vmatrix} =$$

$$= \frac{1}{2} \begin{vmatrix} 4 & 0 & -4i & 0 \\ 0 & 6 & 0 & -4i \\ 4i & 0 & 4 & 0 \\ 0 & 4i & 0 & 6 \end{vmatrix} = \begin{vmatrix} 2 & 0 & -2i & 0 \\ 0 & 3 & 0 & -2i \\ 2i & 0 & 2 & 0 \\ 0 . & 2i & 0 & 3 \end{vmatrix} \cdot (12)$$

A good check is to make sure the resulting matrix is Hermitian [see (A24)] (neither the rotation operators, nor the partial results need be Hermitian). *Comparing D(1) to D(0) we see that the 90° proton pulse created proton single quantum coherences, did not touch the carbon, and redistributed the populations.*

Evolution

The standard formula describing the time evolution of the density matrix elements in the absence of a pulse is:

$$d_{mn}(t) = d_{mn}(0) \exp(-i\omega_{mn}t). \qquad (13)$$

d_{mn} is the matrix element (row m, column n) and $\omega_{mn} = (E_m - E_n)/\hbar$ is the angular frequency of transition $m \rightarrow n$.

We observe that during the evolution the diagonal elements are invariant since $\exp[i(E_m - E_m)/\hbar] = 1$. The off diagonal elements experience a periodic evolution. Note that $d_{mn}(0)$ is the starting point of the evolution immediately after a given pulse. In our particular case the elements $d_{mn}(0)$ are those of D(1).

We now want to calculate D(2) at the time t(2) shown in Figure 2. The only elements that evolve are d_{13} and d_{24}. In a frame rotating with the proton transmitter frequency ω_x, after an evolution time $t_e/2$, their values are:

$$d_{13} = -2i \exp(-i\Omega_{13}t_e/2) = B \qquad (14)$$

$$d_{24} = -2i \exp(-i\Omega_{24}t_e/2) = C \qquad (15)$$

where $\Omega_{13} = \omega_{13} - \omega_x$ and $\Omega_{24} = \omega_{24} - \omega_x$.

Hence

$$
D(2) = \begin{vmatrix} 2 & 0 & B & 0 \\ 0 & 3 & 0 & C \\ B^* & 0 & 2 & 0 \\ 0 & C^* & 0 & 3 \end{vmatrix} . \tag{16}
$$

Second pulse: 180°x, carbon

The rotation operator for this pulse is:

$$
R_{180xC} = \begin{vmatrix} 0 & i & 0 & 0 \\ i & 0 & 0 & 0 \\ 0 & 0 & 0 & i \\ 0 & 0 & i & 0 \end{vmatrix} . \tag{17}
$$

Its reciprocal is:

$$
R_{180xC}^{-1} = \begin{vmatrix} 0 & -i & 0 & 0 \\ -i & 0 & 0 & 0 \\ 0 & 0 & 0 & -i \\ 0 & 0 & -i & 0 \end{vmatrix} . \tag{18}
$$

Postmultiplying D(2) by R gives:

$$
D(2)R_{180xC} = \begin{vmatrix} 0 & 2i & 0 & iB \\ 3i & 0 & iC & 0 \\ 0 & iB^* & 0 & 2i \\ iC^* & 0 & 3i & 0 \end{vmatrix} . \tag{19}
$$

Premultiplying (19) by R^{-1} gives:

$$D(3) = \begin{vmatrix} 3 & 0 & C & 0 \\ 0 & 2 & 0 & B \\ C^* & 0 & 3 & 0 \\ 0 & B^* & 0 & 2 \end{vmatrix} . \tag{20}$$

Comparing D(2) with D(3) we note that the 180° pulse on carbon interchanged the coherences B and C. This means that B, after having evolved with the frequency ω_{13} during the first half of the evolution time [see (14)] will now evolve with the frequency ω_{24}. The net result is that the evolution of B is characterized by the center frequency of the proton doublet, i.e., carbon de-coupling has taken place. The same can be said about C.

Evolution $t_e/2 + \Delta_1$

According to (13) the elements d_{13} and d_{24} become:

$$d_{13} = C \exp[-i\Omega_{13}(t_e/2 + \Delta_1)] , \tag{21}$$

$$d_{24} = B \exp[-i\Omega_{24}(t_e/2 + \Delta_1)] . \tag{22}$$

From Figure 1 we see that

$$\omega_{13} = 2\pi(X + J/2) = \omega_H + \pi J \tag{23}$$

$$\omega_{24} = 2\pi(X - J/2) = \omega_H - \pi J \tag{24}$$

Introducing the notations from (14) and (15)

$$d_{13} = -2i \exp[-i(\Omega_H - \pi J)t_e/2] \exp[-i(\Omega_H + \pi J)(t_e/2 + \Delta_1)] =$$

$$= -2i \exp[-i\Omega_H(t_e + \Delta_1) - i\pi J\Delta_1]$$

$$= -2i \exp[-i\Omega_H(t_e + \Delta_1)] \exp(-i\pi J\Delta_1), \tag{25}$$

$$d_{24} = -2i \exp[-i\Omega_H(t_e + \Delta_1)] \exp(i\pi J\Delta_1) \tag{26}$$

We can now discuss the role of the delay Δ_1. If we would further carry the calculations for $\Delta_1 = 0$ (i.e, $d_{13} = d_{24}$) we would find out that the signal is cancelled. To obtain maximum signal d_{13}

and d_{24} must be equal but of opposite sign. This is achieved by choosing $\Delta_1 = 1/2J$ since then (25) and (26) become:

$$d_{13} = -2 \exp[-i\Omega_H(t_e + \frac{1}{2J})] = -2(c - is), \qquad (27)$$

$$d_{24} = +2 \exp[-i\Omega_H(t_e + \frac{1}{2J})] = 2(c - is), \qquad (28)$$

where

$$c = \cos[\Omega_H(t_e + \frac{1}{2J})] \qquad (29)$$

$$s = \sin[\Omega_H(t_e + \frac{1}{2J})] . \qquad (30)$$

At this point the density matrix is:

$$D(4) = \begin{vmatrix} 3 & 0 & -2(c-is) & 0 \\ 0 & 2 & 0 & 2(c-is) \\ -2(c+is) & 0 & 3 & 0 \\ 0 & 2(c+is) & 0 & 2 \end{vmatrix} . \qquad (31)$$

Third and fourth pulses

Although physically these pulses are applied separately, we may save some calculation·effort treating them as a single nonselective pulse

$$R_{90xCH} = R_{90xC}R_{90xH} =$$

$$= \frac{1}{\sqrt{2}} \begin{vmatrix} 1 & i & 0 & 0 \\ i & 1 & 0 & 0 \\ 0 & 0 & 1 & i \\ 0 & 0 & i & 1 \end{vmatrix} \times \frac{1}{\sqrt{2}} \begin{vmatrix} 1 & 0 & i & 0 \\ 0 & 1 & 0 & i \\ i & 0 & 1 & 0 \\ 0 & i & 0 & 1 \end{vmatrix} =$$

$$= \frac{1}{2} \begin{vmatrix} 1 & i & i & -1 \\ i & 1 & -1 & i \\ i & -1 & 1 & i \\ -1 & i & i & 1 \end{vmatrix} \tag{32}$$

R_{90xCH}^{-1} is the reciprocal of (32):

$$\frac{1}{2} \begin{vmatrix} 1 & -i & -i & -1 \\ -i & 1 & -1 & -i \\ -i & -1 & 1 & -i \\ -1 & -i & i & 1 \end{vmatrix}$$

After the two pulses

$$D(5) = \frac{1}{2} \begin{vmatrix} 5 & i-4is & 0 & -4ic \\ -i+4is & 5 & 4ic & 0 \\ 0 & -4ic & 5 & i+4is \\ 4ic & 0 & -i-4is & 5 \end{vmatrix} \tag{33}$$

Detection

A second delay Δ_2 allows for a short evolution of the system in the absence of decoupling to preserve the proton information [see (29)-(30)]. Then, with the proton decoupler on, the carbon signal is detected. Since there is no additional pulse involved it is sufficient to follow from now on only the evolution of carbon co-herences d_{12} and d_{34}. At time $t(5)$

$$d_{12} = i/2 - 2is, \tag{34}$$

$$d_{34} = i/2 + 2is. \tag{35}$$

At time $t(6)$

$$d_{12} = i(1/2 - 2s)\exp(-i\Omega_{12}\Delta_2), \tag{36}$$

$$d_{34} = i(1/2 + 2s)\exp(-i\Omega_{34}\Delta_2) \quad . \tag{37}$$

where $\Omega_{12} = \omega_{12} - \omega_A$ and $\Omega_{34} = \omega_{34} - \omega_A$ indicate that now we are in the carbon rotating frame, which is needed to describe the carbon signal during the free induction decay.

From Figure 1 we deduce:

$$\Omega_{12} = \Omega_C + \pi J \quad , \tag{38}$$

$$\Omega_{34} = \Omega_C - \pi J \quad . \tag{39}$$

Hence,

$$d_{12} = i(1/2 - 2s)\exp(-i\Omega_C\Delta_2)\exp(-i\pi J\Delta_2) \tag{40}$$

$$d_{34} = i(1/2 + 2s)\exp(-i\Omega_C\Delta_2)\exp(i\pi J\Delta_2) \tag{41}$$

Analyzing the role of Δ_2 in (40) and (41) we see that for $\Delta_2 = 0$ the terms in s which contain the proton information are lost (the signal originates in the sum of d_{12} and d_{34}). As discussed for Δ_1, here also, the desired signal is best obtained for $\Delta_2 = 1/2J$, which makes $\exp(\pm i\pi J\Delta_2) = \pm i$ and

$$d_{12} = (1/2 - 2s)\exp(-i\Omega_C\Delta_2) \quad , \tag{42}$$

$$d_{34} = -(1/2 + 2s)\exp(-i\Omega_C\Delta_2) \quad . \tag{43}$$

From the time $t(6)$ on the system is proton decoupled, i.e., both d_{12} and d_{34} evolve with the frequency Ω_C:

$$d_{12} = (1/2 - 2s)\exp(-i\Omega_C\Delta_2)\exp[-i\Omega_C(t_d - \Delta_2)] =$$

$$= (1/2 - 2s)\exp(-i\Omega_C t_d) \quad , \tag{44}$$

$$d_{34} = -(1/2 + 2s)\exp(-i\Omega_C t_d) \quad . \tag{45}$$

As shown in Appendix B the transverse magnetization for C-13 is given by:

$$M = M_x + iM_y = -M_o(d_{12}^* + d_{34}^*) =$$

$$= 4M_o s \exp(i\Omega_C t_d) = 4M_o \sin[\Omega_H(t_e + \Delta_1)]\exp(i\Omega_C t_d) \tag{46}$$

where M_O is the z-magnetization at thermal equilibrium.

Equation (46) results from introducing in (B16) the angular momentum matrix for $I_x + iI_y$ shown in (C13) and keeping in mind that

$$M_o = - \gamma \hbar N_o \frac{P}{S} . \qquad (47)$$

We can verify (47) by equating M_o to $\dot{M}_z$ at thermal equilibrium.

We learn from (46) that the carbon magnetization rotates by $\Omega_c t_d$ while being amplitude modulated by the proton evolution $\Omega_H t_e$. Fourier transformation with respect to both time domains will yield the two-dimensional spectrum.

When transforming with respect to t_d, all factors other than $\exp(i\Omega_c t_d)$ are regarded as constant. A single peak frequency, Ω_c, is obtained. When transforming with respect to t_e, all factors other than $\sin[\Omega_H(t_e + \Delta_1)]$ are regarded as constant. Since

$$\sin \alpha = \frac{e^{i\alpha} - e^{-i\alpha}}{2i} \qquad (48)$$

both $+\Omega_H$ and $-\Omega_H$ are obtained (Figure 3a).

Imagine now that in the sequence shown in Figure 2 we did not apply the 180° pulse on carbon and suppressed Δ_1. During the evolution time t_e the proton is coupled to carbon. During the acquisition, the carbon is decoupled from proton. The result is that along the carbon axis we see a single peak, while along the proton axis we see a doublet due to the proton-carbon coupling. If we calculate the magnetization following the procedure shown in these notes, we find:

$$M_x + iM_y = 2M_o(\cos \Omega_{13} t_e - \cos \Omega_{24} t_e)\exp(+i\Omega_c t_d) \qquad (49)$$

Reasoning as for (46) we can explain the spectrum shown in Figure 3b.

Finally, if we also suppress the decoupling during the acquisition, we obtain

$$M_x + iM_y = iM_o(\frac{1}{2} - \cos \Omega_{13} t_e + \cos \Omega_{24} t_e)\exp(i\Omega_{12} t_d) +$$
$$+ iM_o(\frac{1}{2} + \cos \Omega_{13} t_e - \cos \Omega_{24} t_e)\exp(i\Omega_{34} t_d) , \qquad (50)$$

which yields the spectrum shown in Figure 3c.

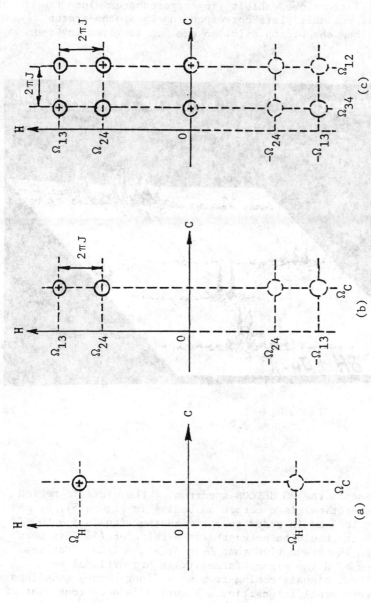

Figure 3. Schematic 2D heteronuclear correlation spectra (contour plot): (a) fully decoupled, (b) proton decoupled during the acquisition, and (c) fully coupled. The lower part of the spectrum is not displayed by the instrument, but proper care must be taken to place the proton transmitter beyond the proton spectrum. Such a requirement is not imposed to the carbon transmitter, provided quadrature phase detection is used.

So far we have treated the AX (CH) case. In reality, the proton
may be coupled to one or several other protons. In the sequence
shown in Figure 2 there is no proton-proton decoupling. The 2D
spectrum will therefore exhibit single resonances along the
carbon axis, but multiplets corresponding to proton-proton
coupling, along the proton axis. An example is given in Figure 4

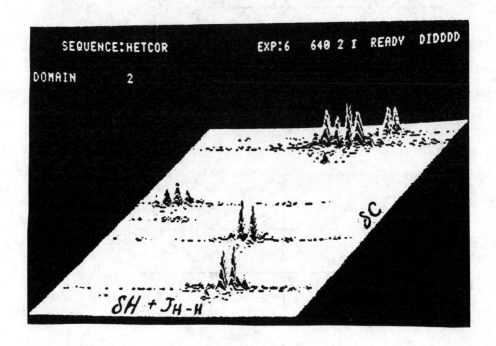

Figure 4.

which represents the 2D HETCOR spectrum of the olefinic region
of *all-trans*-retinal (see carbon numbering in Figure 5). Δ_1 and
Δ_2 were set to 3 ms in order to optimize the signals due to ^1J
(=160 Hz). It should be noted that the relation (46) has been
derived with the assumption that $\Delta_1 = \Delta_2 = \Delta = 1/2J$. For any
other values of J the signal intensity is proportional to
$\sin^2\pi J\Delta$. Thus, signals coming from weak (long range) couplings
will have very small intensities. Figure 5 is a contour plot of
the spectrum shown in Figure 4. It shows in a more dramatic
manner the advantage of 2D spectroscopy: the clear carbon-proton
correlation and the disentangling of the heavily overlapping
proton signals.

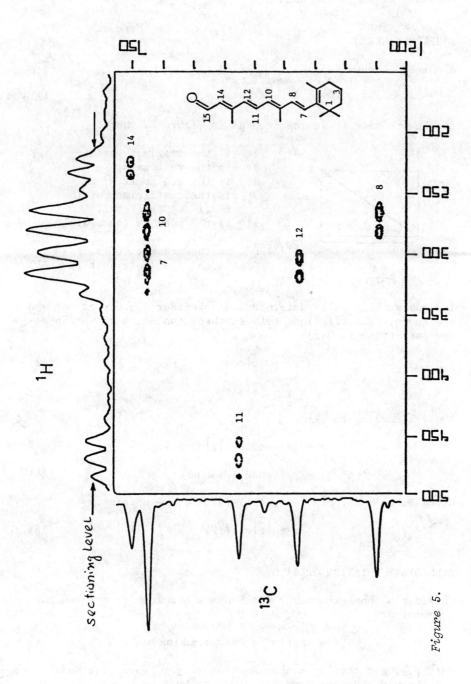

Figure 5.

APPENDIX A: MATH REMINDER

COMPLEX NUMBERS.

A complex number

$$z = x + iy \qquad\qquad (A1)$$

can be graphically represented as in Figure 1, where:

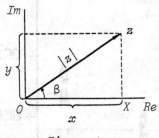

Im	is the imaginary axis,
Re	is the real axis,
x	is the real part of z,
y	is the coefficient of the imaginary part of z,
$\lvert z \rvert$	is the modulus (absolute value) of z,
β	is the argument of z,
i	is the imaginary unit ($i = \sqrt{-1}$).

Figure 1.

The number z is fully determined when either x and y or $\lvert z \rvert$ and β are known. The relations between these two pairs of variables are (see triangle OXz):

$$x = \lvert z \rvert \cos\beta \qquad\qquad (A2)$$

$$y = \lvert z \rvert \sin\beta \qquad\qquad (A3)$$

Thus, according to (A1),

$$z = \lvert z \rvert \cos\beta + i\lvert z \rvert \sin\beta =$$

$$= \lvert z \rvert (\cos\beta + i\sin\beta) \qquad\qquad (A4)$$

Using Euler's formula [see (A11)-(A16)], one obtains:

$$z = \lvert z \rvert \exp(i\beta) \qquad\qquad (A5)$$

ELEMENTARY ROTATION OPERATOR

Consider a complexnumber r which has a modulus $\lvert r \rvert$ = 1 and the argument α:

$$r = \exp(i\alpha) = \cos\alpha + i\sin\alpha \qquad\qquad (A6)$$

Multiplying a complex number such as z by r leaves the modulus of z unchanged and increases the argument by α:

$$zr = |z|e^{i\beta}e^{i\alpha} = |z|e^{i(\beta + \alpha)} \qquad (A7)$$

Equation (A7) describes the rotation of the vector Oz by an angle α (see Figure 2). We call r, the *elementary rotation operator*.

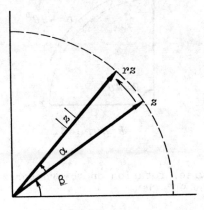

Figure 2.

Note: Although more complicated, the rotation operators in the density matrix treatment of multipulse NMR are of the same form as our elementary operator [cf(B25)].

Example 1. A -90° (clockwise) rotation (see Figure 3)

Let

$$\beta = 90°,$$

$$\alpha = -90°.$$

Then, from (A4) and A6),

$$z = |z|(\cos 90° + i \sin 90°) = i|z| \qquad (A8)$$

$$r = \cos(-90°) + i \sin (-90°) = -i \qquad (A9)$$

The product

$$zr = -i^2|z| = |z| \tag{A10}$$

is a real number (its argument is zero).

Equation (A10) tells us that the particular operator $-i$ effects

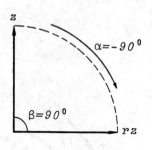

Figure 3.

a 90° CW (clockwise) rotation on the vector z. The operator $+ i$ would rotate z by 90° CCW.

Example 2. Powers of i (the "star of i")

Since i represents a 90° CCW rotation, successive powers of i are obtained by successive 90° CCW rotations (see Figure 4).

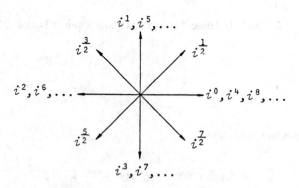

Figure 4. The Star of i.

Series expansion of e^x, sin x, cos x, and e^{ix}

The exponential function describes most everything happening in nature (including the +ve or -ve evolution of your investment accounts). The e^x series is

$$e^x = 1 + x + \frac{x^2}{2!} + \frac{x^3}{3!} + \ldots \qquad (A11)$$

The sine and cosine series are:

$$\sin x = x - \frac{x^3}{3!} + \frac{x^5}{5!} - \frac{x^7}{7!} + \ldots \qquad (A12)$$

$$\cos x = 1 - \frac{x^2}{2!} + \frac{x^4}{4!} - \frac{x^6}{6!} + \ldots \qquad (A13)$$

The e^{ix} series is:

$$e^{ix} = 1 + ix - \frac{x^2}{2!} - i\frac{x^3}{3!} + \frac{x^4}{4!} + i\frac{x^5}{5!} - \ldots \qquad (A14)$$

Separation of the real and imaginary terms gives

$$e^{ix} = 1 - \frac{x^2}{2!} + \frac{x^4}{4!} - \frac{x^6}{6!} + \ldots + \qquad (A15)$$

$$+ i(x - \frac{x^3}{3!} + \frac{x^5}{5!} - \ldots)$$

Recognizing the sine and cosine series one obtains the Euler formula

$$e^{ix} = \cos x + i \sin x . \qquad (A16)$$

MATRIX ALGEBRA

Matrices are arrays of elements disposed in rows and columns; they obey specific algebraic rules for addition, multiplication and inversion. The following formats of matrices are used in quantum mechanics:

$$A = \begin{vmatrix} a_{11}a_{12} - - - a_{1n} \\ \\ a_{n1}a_{n2} - - a_{nn} \end{vmatrix} \; ; A = [a_{11}a_{12} - - a_{1n}]; A = \begin{vmatrix} a_{11} \\ a_{21} \\ a_{n1} \end{vmatrix}$$

     Square matrix   *Row matrix*   *Column matrix*

Row and column matrices are also called row vectors or column vectors. Note that the first subscript of each element indicates the row number and the second, the column number.

Matrix addition

Two matrices are added element by element as follows:

$$\begin{vmatrix} a_{11} & a_{12} & a_{13} \\ a_{21} & a_{22} & a_{23} \\ a_{31} & a_{32} & a_{33} \end{vmatrix} + \begin{vmatrix} b_{11} & b_{12} & b_{13} \\ b_{21} & b_{22} & b23 \\ b_{31} & b_{32} & b_{33} \end{vmatrix} =$$

$$= \begin{vmatrix} a_{11}+b_{11} & a_{12}+b_{12} & a_{13}+b_{13} \\ a_{21}+b_{21} & a_{22}+b_{22} & a_{23}+b_{23} \\ a_{31}+b_{31} & a_{32}+b_{32} & a_{33}+b_{33} \end{vmatrix}$$

Only matrices with the same number of rows and columns can be added.

Matrix multiplication.

In the following are shown three typical matrix multiplications.

a) Square matrix times column matrix:

$$
\begin{vmatrix} a_{11} & a_{12} & a_{13} \\ a_{21} & a_{22} & a_{23} \\ a_{31} & a_{32} & a_{33} \end{vmatrix} \times \begin{vmatrix} b_{11} \\ b_{21} \\ b_{31} \end{vmatrix} = \begin{vmatrix} c_{11} \\ c_{21} \\ c_{31} \end{vmatrix}
$$

$$
\left. \begin{array}{l} c_{11} = a_{11}b_{11} + a_{12}b_{21} + a_{13}b_{31} \\ c_{21} = a_{21}b_{11} + a_{22}b_{21} + a_{23}b_{31} \\ c_{31} = a_{31}b_{11} + a_{32}b_{21} + a_{33}b_{31} \end{array} \right\} \quad \text{i.e.} \ c_{j1} = \sum_{k=1}^{3} a_{jk}b_{k1}
$$

$$(j = 1, 2, 3)$$

Example:

$$
\begin{vmatrix} 1 & 2 & 3 \\ 4 & 5 & 6 \\ 7 & 8 & 9 \end{vmatrix} \times \begin{vmatrix} 1 \\ 2 \\ 3 \end{vmatrix} = \begin{vmatrix} 14 \\ 32 \\ 50 \end{vmatrix}
$$

b) Row matrix times square matrix

$$
[a_{11} \ \ a_{12} \ \ a_{13}] \times \begin{vmatrix} b_{11} & b_{12} & b_{13} \\ b_{21} & b_{22} & b_{23} \\ b_{31} & b_{32} & b_{33} \end{vmatrix} = [c_{11} \ \ c_{12} \ \ c_{13}]
$$

$$
\left. \begin{array}{l} c_{11} = a_{11}b_{11} + a_{12}b_{21} + a_{13}b_{31} \\ c_{12} = a_{11}b_{12} + a_{12}b_{22} + a_{13}b_{32} \\ c_{13} = a_{11}b_{13} + a_{12}b_{23} + a_{13}b_{33} \end{array} \right\} \quad \text{i.e.} \ c_{1j} = \sum_{k=1}^{3} a_{1k}b_{kj}
$$

$$(j = 1, 2, 3)$$

c) Square matrix times square matrix

$$
\begin{vmatrix} a_{11} & a_{12} & a_{13} \\ a_{21} & a_{22} & a_{23} \\ a_{31} & a_{32} & a_{33} \end{vmatrix} \times \begin{vmatrix} b_{11} & b_{12} & b_{13} \\ b_{21} & b_{22} & b_{23} \\ b_{31} & b_{32} & b_{33} \end{vmatrix} = \begin{vmatrix} c_{11} & c_{12} & c_{13} \\ c_{21} & c_{22} & c_{23} \\ c_{31} & c_{32} & c_{33} \end{vmatrix}
$$

$$
c_{jl} = \sum_{k=1}^{3} a_{jk} b_{kl} \ .
$$

For instance, row 2 of the left hand matrix and column 3 of the right hand matrix are involved in the obtaining of the element c_{23} of the product. Example:

$$
\frac{1}{\sqrt{2}} \begin{vmatrix} 0 & 1 & 0 \\ 1 & 0 & 1 \\ 0 & 1 & 0 \end{vmatrix} \times \frac{1}{\sqrt{2}} \begin{vmatrix} 0 & -i & 0 \\ i & 0 & -i \\ 0 & i & 0 \end{vmatrix} =
$$

$$
= \frac{1}{2} \begin{vmatrix} i & 0 & -i \\ 0 & 0 & 0 \\ i & 0 & -i \end{vmatrix} = \begin{vmatrix} \frac{i}{2} & 0 & -\frac{i}{2} \\ 0 & 0 & 0 \\ \frac{i}{2} & 0 & -\frac{i}{2} \end{vmatrix} .
$$

We observe that the product inherits the number of rows from the first (left) matrix and the number of columns from the second (right) matrix. The number of columns of the left matrix must match the number of rows of the right matrix.

In general the matrix multiplication is not commutative:

$$
AB \neq BA \ . \tag{A17}
$$

It is associative:

$$
A(BC) = (AB)C = ABC \ , \tag{A18}
$$

and distributive:

$$A(B + C) = AB + AC \tag{A19}$$

The *unit matrix* is shown below:

$$\mathbb{1} = \begin{vmatrix} 1 & 0 & 0 & 0 \\ 0 & 1 & 0 & 0 \\ 0 & 0 & 1 & 0 \\ 0 & 0 & 0 & 1 \end{vmatrix} \tag{A20}$$

(it must be square and it can be any size)

$$\mathbb{1} \times A = A \times \mathbb{1} = A \tag{A21}$$

Matrix inversion

The inverse (A^{-1}) of a square matrix A is defined by the relation:

$$AA^{-1} = A^{-1}A = \mathbb{1} \tag{A22}$$

To find A^{-1}:

  1) replace each element of A by its algebraic minor determinant

  2) interchange the rows and the columns (this is called matrix transposition)

  3) divide all elements of the transposed matrix by the determinant of A.

Example:

Find the inverse of matrix A:

$$A = \frac{1}{2} \begin{vmatrix} 1 & \sqrt{2} & 1 \\ -\sqrt{2} & 0 & \sqrt{2} \\ 1 & -\sqrt{2} & 1 \end{vmatrix} = \begin{vmatrix} \frac{1}{2} & \frac{\sqrt{2}}{2} & \frac{1}{2} \\ -\frac{\sqrt{2}}{2} & 0 & \frac{\sqrt{2}}{2} \\ \frac{1}{2} & -\frac{\sqrt{2}}{2} & \frac{1}{2} \end{vmatrix}$$

1) We replace a_{11} by $\begin{vmatrix} a_{22} & a_{13} \\ a_{32} & a_{33} \end{vmatrix}$ i.e.,

$$\begin{vmatrix} 0 & \dfrac{\sqrt{2}}{2} \\ -\dfrac{\sqrt{2}}{2} & \dfrac{1}{2} \end{vmatrix} = (0 \times \dfrac{1}{2}) + (\dfrac{\sqrt{2}}{2} \times \dfrac{\sqrt{2}}{2}) = \dfrac{1}{2}$$

We replace a_{12} by $-\begin{vmatrix} a_{21} & a_{23} \\ a_{31} & a_{33} \end{vmatrix}$ i.e.,

$$-\begin{vmatrix} -\dfrac{\sqrt{2}}{2} & \dfrac{\sqrt{2}}{2} \\ \dfrac{1}{2} & \dfrac{1}{2} \end{vmatrix} = -[(-\dfrac{\sqrt{2}}{2} \times \dfrac{1}{2}) - (\dfrac{\sqrt{2}}{2} \times \dfrac{1}{2})] = \dfrac{\sqrt{2}}{2}$$

and so on, obtaining:

$$\begin{vmatrix} \dfrac{1}{2} & \dfrac{\sqrt{2}}{2} & \dfrac{1}{2} \\ -\dfrac{\sqrt{2}}{2} & 0 & \dfrac{\sqrt{2}}{2} \\ \dfrac{1}{2} & -\dfrac{\sqrt{2}}{2} & \dfrac{1}{2} \end{vmatrix}$$

2) The transposition yields:

$$\begin{vmatrix} \dfrac{1}{2} & -\dfrac{\sqrt{2}}{2} & \dfrac{1}{2} \\ \dfrac{\sqrt{2}}{2} & 0 & -\dfrac{\sqrt{2}}{2} \\ \dfrac{1}{2} & \dfrac{\sqrt{2}}{2} & \dfrac{1}{2} \end{vmatrix}$$

3) Calculate the determinant of A:

$$\det(A) = 0 + \frac{1}{4} + \frac{1}{4} - 0 + \frac{1}{4} + \frac{1}{4} = 1$$

then divide the transposed matrix by 1:

$$A^{-1} = \begin{vmatrix} \dfrac{1}{2} & -\dfrac{\sqrt{2}}{2} & \dfrac{1}{2} \\ \dfrac{\sqrt{2}}{2} & 0 & -\dfrac{\sqrt{2}}{2} \\ \dfrac{1}{2} & \dfrac{\sqrt{2}}{2} & \dfrac{1}{2} \end{vmatrix} = \frac{1}{2} \begin{vmatrix} 1 & -\sqrt{2} & 1 \\ \sqrt{2} & 0 & -\sqrt{2} \\ 1 & \sqrt{2} & 1 \end{vmatrix}$$

Check: $A^{-1}A = \mathbb{1}$

$$\frac{1}{2} \begin{vmatrix} 1 & -\sqrt{2} & 1 \\ \sqrt{2} & 0 & -\sqrt{2} \\ 1 & \sqrt{2} & 1 \end{vmatrix} \times \frac{1}{2} \begin{vmatrix} 1 & \sqrt{2} & 1 \\ \sqrt{2} & 0 & \sqrt{2} \\ 1 & -\sqrt{2} & 1 \end{vmatrix} =$$

$$= \frac{1}{4} \begin{vmatrix} 4 & 0 & 0 \\ 0 & 4 & 0 \\ 0 & 0 & 4 \end{vmatrix} = \begin{vmatrix} 1 & 0 & 0 \\ 0 & 1 & 0 \\ 0 & 0 & 1 \end{vmatrix}$$

Note: You will be pleased to learn that:
  a) There is a short cut for the inversion of
     rotation operator matrices $(R \to R^{-1})$ because
     they are of a special kind.

  b) In our calculations we will need to effect
     only $R \to R^{-1}$ inversions.

Here is the short cut:

  1) Transpose R (vide supra)

  2) Replace each element with its complex
     conjugate.

Example

$$R = \frac{1}{2} \begin{vmatrix} 1 & \sqrt{2} & 1 \\ -\sqrt{2} & 0 & \sqrt{2} \\ 1 & -\sqrt{2} & 1 \end{vmatrix}$$

1) Transpose: $\frac{1}{2} \begin{vmatrix} 1 & -\sqrt{2} & 1 \\ \sqrt{2} & 0 & -\sqrt{2} \\ 1 & \sqrt{2} & 1 \end{vmatrix}$

2) Conjugate: Because they are all real, all elements remain the same:

$$R^{-1} = \frac{1}{2} \begin{vmatrix} 1 & -\sqrt{2} & 1 \\ \sqrt{2} & 0 & -\sqrt{2} \\ 1 & \sqrt{2} & 1 \end{vmatrix}$$

Note: This is the same matrix as in the previous example; its being a rotation operator. matrix, allowed us to use the short cut procedure.

Another example:

$$R = \frac{1}{\sqrt{2}} \begin{vmatrix} 1 & 0 & i & 0 \\ 0 & 1 & 0 & i \\ i & 0 & 1 & 0 \\ 0 & i & 0 & 1 \end{vmatrix}$$

1) Transpose: you obtain the same matrix

2) Conjugate:

$$R^{-1} = \frac{1}{\sqrt{2}} \begin{vmatrix} 1 & 0 & -i & 0 \\ 0 & 1 & 0 & -i \\ -i & 0 & 1 & 0 \\ 0 & -i & 0 & 1 \end{vmatrix}$$

Note: *The matrix resulting from the transposition followed by complex conjugation of a given matrix A is called the* adjoint *matrix $A^{\dagger}$ of the original one. For all rotation operators,*

$$R^{-1} = R^{\dagger} .\qquad\qquad (A23)$$

In other words, our short cut for inversion (A23) is equivalent with finding the adjoint *of R.*

When

$$A = A^{\dagger} \qquad\qquad (A24)$$

we say that the matrix A is self adjoint or hermitian. The angular momentum, the hamiltonian and density matrix are all hermitian (the rotation operators never are).

APPENDIX B: DENSITY MATRIX FORMALISM

The density matrix is a tool used to describe the state of a
spin ensemble as well as its evolution in time. It allows the
passage from the probabilistic treatment of a *system* of a few
spins to the statistical treatment of a large *ensemble* of such
systems.

Since we are interested in the magnetization we want to describe
this observable in terms of the wave function Ψ of the system:

$$\langle \mu_x \rangle = (\Psi | \mu_x | \Psi) = \gamma \hbar (\Psi | I_x | \Psi) \quad . \tag{B1}$$

$\langle \mu_x \rangle$ is the expectation value of the x-component of the magnetic
moment and I_x is the x-component of the angular momentum operator
of the system. In order to calculate the macroscopic magnetization, the ensemble average is to be taken:

$$M_x = N_o \overline{\langle \mu_x \rangle} = N_o \gamma \hbar \overline{(\Psi | I_x | \Psi)} \tag{B2}$$

where N_o is the number of spins per unit volume.

Since I_x is an invariant operator, the time dependence is contained in Ψ which, in turn, may be expressed as a linear combin-
ation of the eigenstates $|n)$ of the system:

$$\Psi = \sum_{n=1}^{n} c_n |n) \quad . \tag{B3}$$

The Schrödinger equation

$$-\frac{\hbar}{i} \frac{\partial \Psi}{\partial t} = H\Psi \tag{B4}$$

becomes:

$$-\frac{\hbar}{i} \sum \frac{dc_n}{dt} |n) = \sum c_n E_n |n) \tag{B5}$$

where E_n is the eigenvalue corresponding to the state $|n)$ in the
time independent equation:

$$H|n) = E|n) \quad . \tag{B6}$$

Rearranging (B5) gives:

$$\sum (c_n E_n + \frac{\hbar}{i} \frac{dc_n}{dt}) |n) = 0 . \tag{B7}$$

Due to the orthogonality of the eigenfunctions, (B7) is satisfied
if and only if each term of the sum is null:

$$c_n E_n + \frac{\hbar}{i} \frac{dc_n}{dt} = 0 \quad . \tag{B8}$$

Hence

$$\frac{1}{c_n} \frac{dc_n}{dt} = - \frac{iE_n}{\hbar}$$

or

$$\frac{d}{dt}(\ell n c_n) = - \frac{iE_n}{\hbar} \quad . \tag{B9}$$

Integrating (B9) yields:

$$\ell n c_n = - \frac{iE_n}{\hbar} t + C \quad . \tag{B10}$$

$$c_n = \exp(-iE_n t/\hbar)\exp(C) \tag{B11}$$

The integration constant C may be related to the value of c_n at
time $t = 0$

$$c_n(o) = \exp(C)$$

and (B11) becomes

$$c_n = c_n(o)\exp(-iE_n t/\hbar) \quad . \tag{B12}$$

Knowing the evolution of all c_n's will allow us to predict the
time variation of Ψ in (B2), hence, the time variation of the
magnetization M_x.

We now calculate $(\Psi|I|\Psi)$ where I may be I_x, I_y or I_z:

$$(\Psi|I|\Psi) = \left| c_1^* \; c_2^* \ldots c_N^* \right| x \begin{vmatrix} I_{11} & \cdots & I_{1N} \\ \cdot & & \\ \cdot & & \\ \cdot & & \\ I_{N1} & \cdots & I_{NN} \end{vmatrix} x \begin{vmatrix} c_1 \\ c_2 \\ \cdot \\ \cdot \\ \cdot \\ c_N \end{vmatrix} = $$

$$= \left| c_1^* c_2^* \cdots c_N^* \right| \times \left| \begin{array}{c} \sum\limits_m I_{1m} c_m \\ \sum\limits_m I_{2m} c_m \\ \vdots \\ \sum\limits_m I_{Nm} c_m \end{array} \right| = \sum\sum\limits_{nm} I_{nm} c_m c_n^* . \tag{B13}$$

Since I_{nm} are constant throughout the ensemble, only $c_m c_n^*$ have to be averaged in (B2):

$$M = N_o \gamma \hbar \sum\sum\limits_{nm} I_{nm} \overline{c_m c_n^*} . \tag{B14}$$

Here M may be taken as M_x, M_y, or M_z using the corresponding matrix elements I_{nm} of the operators I_x, I_y, or I_z.

We notice that the only variable elements in (B14) are the averaged products $\overline{c_m c_n^*}$. There are N^2 such products which, arranged in a square table, form the density matrix:

$$D = \left| \begin{array}{cccc} d_{11} & d_{12} & \cdots & d_{1N} \\ d_{21} & & & \\ \vdots & & & \\ \vdots & & & \\ d_{N1} & \cdots & \cdots & d_{NN} \end{array} \right| \tag{B15}$$

with $d_{mn} = \overline{c_m c_n^*}$.

It follows that (B14) can be written

$$M = N_o \gamma \hbar \sum\sum\limits_{nm} I_{nm} d_{mn} = N_o \gamma \hbar \sum\sum\limits_{nm} I_{nm} d_{nm}^* , \tag{B16}$$

since D is Hermitian.

Therefore, the magnetization can be calculated by multiplying every element of I with the complex conjugate of the corresponding element of D and summing all products.

A more "elegant" but less practical avenue is calculating the trace (sum of diagonal elements) of the product I x D, e.g.:

$$M_x = Tr(I_x D) \tag{B17}$$

Note that in (B17) the constant factor $N_0\gamma\hbar$ has been omitted. This is general practice since one is interested only in relative intensities. Now, having established the relation between magnetization and density matrix, we discuss the time evolution of the matrix from the thermal equilibrium throughout the pulse sequence.

At equilibrium the nondiagonal elements are null because of the random phase distribution of the complex coefficients:

$$d_{mn} = \overline{c_m c_n^*} = \overline{|c_m||c_n|\exp[i(\alpha_m - \alpha_n)]} \quad . \tag{B18}$$

(A complex number, $\exp[i(\alpha_m - \alpha_n)]$, described as a vector in the complex plane, may be oriented in any direction defined by the phase difference $\alpha_m - \alpha_n$; hence, the average of a multitude of such vectors will be null). The diagonal elements represent populations since

$$d_{mm} = \overline{c_m c_m^*} = \overline{|c_m|^2} \tag{B19}$$

(In quantum mechanics $|c_m|^2$ is the probability of finding the system in the state m).

According to (B12) the evolution of the matrix elements is given by

$$d_{mn}(t) = d_{mn}(o)\exp[-i(E_m - E_n)t/\hbar] =$$

$$= d_{mn}(o)\exp(-i\omega_{mn} t) \quad . \tag{B20}$$

The populations are invariable because $E_m - E_m = 0$ (relaxation processes are neglected throughout this paper).

A more general form of the evolution equation is:

$$\frac{dD}{dt} = \frac{i}{\hbar}(DH - HD) \quad , \tag{B21}$$

The solution of which is:

$$D(t) = \exp(-iHt/\hbar) \; D(o) \; \exp(iHt/\hbar) \qquad (B22)$$

The advantage of (B22) is that it allows us to see the effect of a pulse, i.e., the effect of a field H_1, perpendicular to H_o. In the rotating frame the Hamiltonian

$$H = |\gamma|\hbar H_o I_z \qquad (B23)$$

may be replaced by $|\gamma|\hbar H_1 I_x$. (We established the following convention: the applied field H_o is oriented along the positive z-axis for $\gamma < 0$, and along the negative Z-axis for $\gamma > 0$). Then,

$$i\frac{H}{\hbar}t = i|\gamma|H_1 I_x t = iI_x\alpha . \qquad (B24)$$

Hence, for a pulse $\alpha = |\gamma|H_1 t$,

$$D(f) = \exp(-iI_x\alpha) \; D(i) \; \exp(iI_x\alpha) =$$
$$\qquad (B25)$$
$$= R^{-1}D(o)R ,$$

where R is the rotation operator [(cf (A6)-(A7)]; D(i) and D(f) denote the density matrix before and after the pulse.

Let us calculate $R_{x\alpha}$ for a spin ½. From angular momentum tables (see Appendix C)

$$I_x = \frac{1}{2}\begin{vmatrix} 0 & 1 \\ 1 & 0 \end{vmatrix} \qquad (B26)$$

$$R_{\alpha x} = e^{iI_x\alpha} = 1 + i\alpha I_x + \frac{(i\alpha)^2}{2!}I_x^2 + \frac{(i\alpha)^3}{3!}I_x^3 + \ldots \qquad (B27)$$

The powers of I_x may easily be calculated if one notices that

$$I_x^2 = \frac{1}{4}\begin{vmatrix} 1 & 0 \\ 0 & 1 \end{vmatrix}$$

$$I_x^3 = \frac{1}{4} I_x = \frac{1}{8} \begin{vmatrix} 0 & 1 \\ 1 & 0 \end{vmatrix}$$

$$I_x^4 = \frac{1}{16} \begin{vmatrix} 1 & 0 \\ 0 & 1 \end{vmatrix}$$

In general, for n = even we have:

$$I_x^n = \frac{1}{2^n} \begin{vmatrix} 1 & 0 \\ 0 & 1 \end{vmatrix} , \qquad \text{(B28)}$$

and for n = odd,

$$I_x^n = \frac{1}{2^n} \begin{vmatrix} 0 & 1 \\ 1 & 0 \end{vmatrix} . \qquad \text{(B29)}$$

Introducing (B28) and (B29) in (B27) and separating the even and odd terms gives:

$$R_{\alpha x} = [1 - (\frac{\alpha}{2})^2 \frac{1}{2!} + (\frac{\alpha}{2})^4 \frac{1}{4!} - \ldots] \times \begin{vmatrix} 1 & 0 \\ 0 & 1 \end{vmatrix} +$$

$$\text{(B30)}$$

$$+ i[\frac{\alpha}{2} - (\frac{\alpha}{2})^3 \frac{1}{3!} + (\frac{\alpha}{2})^5 \frac{1}{5!} - \ldots] \times \begin{vmatrix} 0 & 1 \\ 1 & 0 \end{vmatrix} .$$

According to (A12) and (A13)

$$R_{\alpha x} = \cos \frac{\alpha}{2} \times \begin{vmatrix} 1 & 0 \\ 0 & 1 \end{vmatrix} + i\sin \frac{\alpha}{2} \begin{vmatrix} 0 & 1 \\ 1 & 0 \end{vmatrix} =$$

$$
= \begin{vmatrix} \cos \frac{\alpha}{2} & 0 \\ \\ 0 & \cos \frac{\alpha}{2} \end{vmatrix} + \begin{vmatrix} 0 & i\sin \frac{\alpha}{2} \\ \\ i\sin \frac{\alpha}{2} & 0 \end{vmatrix} = \begin{vmatrix} \cos \frac{\alpha}{2} & i\sin \frac{\alpha}{2} \\ \\ i\sin \frac{\alpha}{2} & \cos \frac{\alpha}{2} \end{vmatrix}
$$

(B31)

For example

$$
R_{90x} = \frac{1}{\sqrt{2}} \begin{vmatrix} 1 & i \\ \\ i & 1 \end{vmatrix}
$$

(B32)

Appendix C contains angular momentum and rotation operators for a variety of spin systems and pulses.

APPENDIX C: ANGULAR MOMENTUM AND ROTATION OPERATORS

<u>System (spin):   A(1/2)</u>

$$I_x = \frac{1}{2} \begin{vmatrix} 0 & 1 \\ 1 & 0 \end{vmatrix} \quad ; \quad I_y = \frac{1}{2} \begin{vmatrix} 0 & -i \\ i & 0 \end{vmatrix} \tag{C1}$$

$$I_z = \frac{1}{2} \begin{vmatrix} 1 & 0 \\ 0 & -1 \end{vmatrix} \quad ; \quad I_x + iI_y \begin{vmatrix} 0 & 1 \\ 0 & 0 \end{vmatrix} \tag{C2}$$

$$R_{\alpha x} = \begin{vmatrix} \cos \frac{\alpha}{2} & i \sin \frac{\alpha}{2} \\ i \sin \frac{\alpha}{2} & \cos \frac{\alpha}{2} \end{vmatrix} \quad ; \quad R_{\alpha y} = \begin{vmatrix} \cos \frac{\alpha}{2} & \sin \frac{\alpha}{2} \\ - \sin \frac{\alpha}{2} & \cos \frac{\alpha}{2} \end{vmatrix} \tag{C3}$$

$$R_{90x} = \frac{1}{\sqrt{2}} \begin{vmatrix} 1 & i \\ i & 1 \end{vmatrix} \quad ; \quad R_{90y} = \frac{1}{\sqrt{2}} \begin{vmatrix} 1 & 1 \\ -1 & 1 \end{vmatrix} \tag{C4}$$

$$R_{180x} = \begin{vmatrix} 0 & i \\ i & 0 \end{vmatrix} \quad ; \quad R_{180y} = \begin{vmatrix} 0 & 1 \\ -1 & 0 \end{vmatrix} \tag{C5}$$

<u>System (spin):   A(1)</u>

$$I_x = \frac{1}{\sqrt{2}} \begin{vmatrix} 0 & 1 & 0 \\ 1 & 0 & 1 \\ 0 & 1 & 0 \end{vmatrix} \quad ; \quad I_y = \frac{1}{\sqrt{2}} \begin{vmatrix} 0 & -i & 0 \\ i & 0 & -i \\ 0 & i & 0 \end{vmatrix} \tag{C6}$$

$$I_z = \begin{vmatrix} 1 & 0 & 0 \\ 0 & 0 & 0 \\ 0 & 0 & -1 \end{vmatrix} ; \quad I_x + iI_y = \sqrt{2} \begin{vmatrix} 0 & 1 & 0 \\ 0 & 0 & 1 \\ 0 & 0 & 0 \end{vmatrix} \tag{C7}$$

$$R_{\alpha x} = \frac{1}{2} \begin{vmatrix} \cos\alpha + 1 & i\sqrt{2}\sin\alpha & \cos\alpha - 1 \\ i\sqrt{2}\sin\alpha & 2\cos\alpha & i\sqrt{2}\sin\alpha \\ \cos\alpha - 1 & i\sqrt{2}\sin\alpha & \cos\alpha + 1 \end{vmatrix} \qquad (C8)$$

$$R_{\alpha y} = \frac{1}{2} \begin{vmatrix} 1 + \cos\alpha & \sqrt{2}\sin\alpha & 1 - \cos\alpha \\ -\sqrt{2}\sin\alpha & 2\cos\alpha & \sqrt{2}\sin\alpha \\ 1 - \cos\alpha & -\sqrt{2}\sin\alpha & 1 + \cos\alpha \end{vmatrix} \qquad (C9)$$

$$R_{90x} = \frac{1}{2} \begin{vmatrix} 1 & i\sqrt{2} & -1 \\ i\sqrt{2} & 0 & i\sqrt{2} \\ -1 & i\sqrt{2} & 1 \end{vmatrix} ; R_{90y} = \frac{1}{2} \begin{vmatrix} 1 & \sqrt{2} & 1 \\ -\sqrt{2} & 0 & \sqrt{2} \\ 1 & -\sqrt{2} & 1 \end{vmatrix} \qquad (C10)$$

$$R_{180x} = \begin{vmatrix} 0 & 0 & -1 \\ 0 & -1 & 0 \\ -1 & 0 & 0 \end{vmatrix} ; R_{180y} = \begin{vmatrix} 0 & 0 & 1 \\ 0 & -1 & 0 \\ 1 & 0 & 0 \end{vmatrix} \qquad (C11)$$

<u>System (spin): A(1/2) X (1/2)</u>

$$I_{xA} = \frac{1}{2} \begin{vmatrix} 0 & 1 & 0 & 0 \\ 1 & 0 & 0 & 0 \\ 0 & 0 & 0 & 1 \\ 0 & 0 & 1 & 0 \end{vmatrix} ; \quad I_{yA} = \frac{1}{2} \begin{vmatrix} 0 & -i & 0 & 0 \\ i & 0 & 0 & 0 \\ 0 & 0 & 0 & -i \\ 0 & 0 & i & 0 \end{vmatrix} \qquad (C12)$$

$$I_{zA} = \frac{1}{2} \begin{vmatrix} 1 & 0 & 0 & 0 \\ 0 & -1 & 0 & 0 \\ 0 & 0 & 1 & 0 \\ 0 & 0 & 0 & -1 \end{vmatrix} ; (I_x + iI_y)_A = \begin{vmatrix} 0 & 1 & 0 & 0 \\ 0 & 0 & 0 & 0 \\ 0 & 0 & 0 & 1 \\ 0 & 0 & 0 & 0 \end{vmatrix} \qquad (C13)$$

$$I_{xX} = \frac{1}{2} \begin{vmatrix} 0 & 0 & 1 & 0 \\ 0 & 0 & 0 & 1 \\ 1 & 0 & 0 & 0 \\ 0 & 1 & 0 & 0 \end{vmatrix} \quad ; \quad I_{yX} = \frac{1}{2} \begin{vmatrix} 0 & 0 & -i & 0 \\ 0 & 0 & 0 & -i \\ i & 0 & 0 & 0 \\ 0 & i & 0 & 0 \end{vmatrix} \quad (C14)$$

$$I_{zX} = \frac{1}{2} \begin{vmatrix} 1 & 0 & 0 & 0 \\ 0 & 1 & 0 & 0 \\ 0 & 0 & -1 & 0 \\ 0 & 0 & 0 & -1 \end{vmatrix} \quad ; \quad (I_x + iI_y)_X = \begin{vmatrix} 0 & 0 & 1 & 0 \\ 0 & 0 & 0 & 1 \\ 0 & 0 & 0 & 0 \\ 0 & 0 & 0 & 0 \end{vmatrix} \quad (C15)$$

$$R_{90xA} = \frac{1}{\sqrt{2}} \begin{vmatrix} 1 & i & 0 & 0 \\ i & 1 & 0 & 0 \\ 0 & 0 & 1 & i \\ 0 & 0 & i & 1 \end{vmatrix} \quad ; \quad R_{90yA} = \frac{1}{\sqrt{2}} \begin{vmatrix} 1 & 1 & 0 & 0 \\ -1 & 1 & 0 & 0 \\ 0 & 0 & 1 & 1 \\ 0 & 0 & -1 & 1 \end{vmatrix} \quad (C16)$$

$$R_{180xA} = \begin{vmatrix} 0 & i & 0 & 0 \\ i & 0 & 0 & 0 \\ 0 & 0 & 0 & i \\ 0 & 0 & i & 0 \end{vmatrix} \quad ; \quad R_{180yA} = \begin{vmatrix} 0 & 1 & 0 & 0 \\ -1 & 0 & 0 & 0 \\ 0 & 0 & 0 & 1 \\ 0 & 0 & -1 & 0 \end{vmatrix} \quad (C17)$$

$$R_{90xX} = \frac{1}{\sqrt{2}} \begin{vmatrix} 1 & 0 & i & 0 \\ 0 & 1 & 0 & i \\ i & 0 & 1 & 0 \\ 0 & i & 0 & 1 \end{vmatrix} \quad ; \quad R_{90yX} = \frac{1}{\sqrt{2}} \begin{vmatrix} 1 & 0 & 1 & 0 \\ 0 & 1 & 0 & 1 \\ -1 & 0 & 1 & 0 \\ 0 & -1 & 0 & 1 \end{vmatrix} \quad (C18)$$

$$R_{180xX} \begin{vmatrix} 0 & 0 & i & 0 \\ 0 & 0 & 0 & i \\ i & 0 & 0 & 0 \\ 0 & i & 0 & 0 \end{vmatrix} ; R_{180yX} = \begin{vmatrix} 0 & 0 & 1 & 0 \\ 0 & 0 & 0 & 1 \\ -1 & 0 & 0 & 0 \\ 0 & -1 & 0 & 0 \end{vmatrix} \qquad (C19)$$

System (spin): A(1) X (1/2)

Spin labeling according to the figure below.

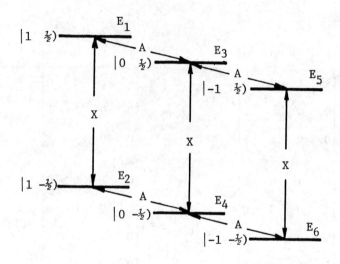

$$I_{xA} = \frac{1}{\sqrt{2}} \begin{vmatrix} 0 & 0 & 1 & 0 & 0 & 0 \\ 0 & 0 & 0 & 1 & 0 & 0 \\ 1 & 0 & 0 & 0 & 1 & 0 \\ 0 & 1 & 0 & 0 & 0 & 1 \\ 0 & 0 & 1 & 0 & 0 & 0 \\ 0 & 0 & 0 & 1 & 0 & 0 \end{vmatrix} \qquad (C20)$$

$$I_{xX} = \frac{1}{2} \begin{vmatrix} 0 & 1 & 0 & 0 & 0 & 0 \\ 1 & 0 & 0 & 0 & 0 & 0 \\ 0 & 0 & 0 & 1 & 0 & 0 \\ 0 & 0 & 1 & 0 & 0 & 0 \\ 0 & 0 & 0 & 0 & 0 & 1 \\ 0 & 0 & 0 & 0 & 1 & 0 \end{vmatrix} \qquad (C21)$$

Iy can be written in the same way, taking (C6) and (C1) as starting points. Examples of rotation operators for the AX(1, 1/2)
system:

$$R_{90yA} = \frac{1}{2} \begin{vmatrix} 1 & 0 & \sqrt{2} & 0 & 1 & 0 \\ 0 & 1 & 0 & \sqrt{2} & 0 & 1 \\ -\sqrt{2} & 0 & 0 & 0 & \sqrt{2} & 0 \\ 0 & -\sqrt{2} & 0 & 0 & 0 & \sqrt{2} \\ 1 & 0 & -\sqrt{2} & 0 & 1 & 0 \\ 0 & 1 & 0 & -\sqrt{2} & 0 & 1 \end{vmatrix} \qquad (C22)$$

$$R_{180xX} = \begin{vmatrix} 0 & i & 0 & 0 & 0 & 0 \\ i & 0 & 0 & 0 & 0 & 0 \\ 0 & 0 & 0 & i & 0 & 0 \\ 0 & 0 & i & 0 & 0 & 0 \\ 0 & 0 & 0 & 0 & 0 & i \\ 0 & 0 & 0 & 0 & i & 0 \end{vmatrix} \qquad (C23)$$

Reciprocals R^{-1} of all rotation operators can be found through
transposition and complex conjugation (see Appendix A).

APPENDIX D: FROZEN FRAMES FROM THE COMPUTER GENERATED MOVIE

The following figures represent static pictures of the density
matrix at different times of the pulse sequence (as indicated by
arrows). The filled areas represent either relative populations,
or the modulus of each coherence. Corresponding equation numbers
in the text are indicated under each matrix. One, two, three,
and four dots indicate angular frequencies 13, 24, 12, and 34,
respectively. The accordeon shape indicates amplitude modulation.

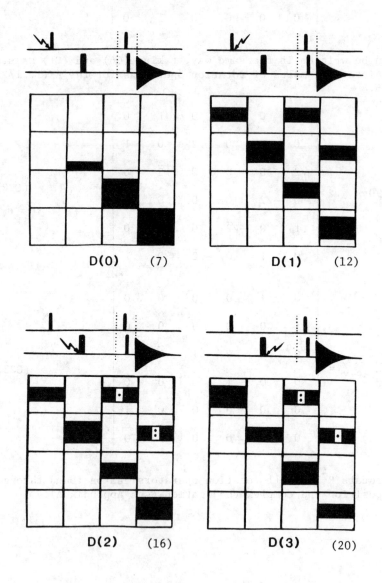

D(0)   (7)   D(1)   (12)

D(2)   (16)   D(3)   (20)

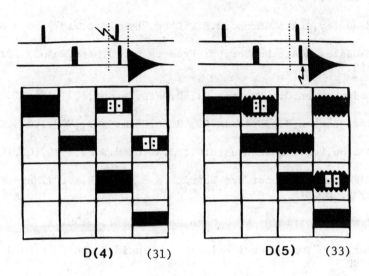

D(4)   (31)   D(5)   (33)

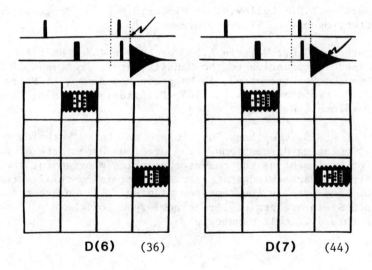

D(6)   (36)   D(7)   (44)

RECOMMENDED READING
1. Jeener, J., Ampère International Summer School, Basko Polje,
    Yugoslavia, 1971.

2. Maudsley,A.A.; Ernst,R.R., Chem.Phys.Lett.,1977,50,368.

3. Maudsley,A.A.; Müller,L.; Ernst,R.R., J.Magn.Reson.,1977,28,
    463.

4. Bodenhausen,G.; Freeman,R., J.Magn.Reson.,1977,28,471.

5. Bodenhausen,G.; Freeman,R., J.Am.Chem.Soc.,1978,100,320.

6. Bolton,P.H.; Bodenhausen,G., J.Am.Chem.Soc.,1979,101,1080.

7. Nagayama,K.; Kumar,A.; Wütrich,K.; Ernst,R.R., J.Magn.Reson.
    1980,40,321.

8. Bax,A.; Morris,G.A., J.Magn.Reson.,1981,42,501.

9. Bax,A., "Two-Dimensional NMR in Liquids", Reidel: Dordrecht,
    1982.

10. Slichter,C.P., "Principles of Magnetic Resonance", 2nd Ed.,
    Springer-Verlag: Berlin, 1980.

11. Buckley,P.D.; Lolley,K.W.; Pinder,D.N., in "Progress in NMR
    Spectroscopy", 1975,vol.10,part 1,1.

12. Mateescu,G.D.; Valeriu,A., "Teaching the New NMR: A Computer-
    Aided Introduction to the Density Matrix Treatment of Double-
    Quantum Spectrometry", in Magnetic Resonance, L. Petrakis and
    J. P. Fraissard (eds.), 1984, D. Reidel Publishing Company,
    Dordrecht, Holland.

ACKNOWLEDGMENTS. We thank Guy Pouzard and Larry Werbelow for
many stimulating discussions. The precious assistance of Randy
Harr for implementing the computer graphics program for the movie
is gratefully acknowledged. Astrid Ferszt is especially thanked
for her kind help in the computer filming. Eileen Green and
Florence Stein are gratefully acknowledged for their patience in
typing this difficult manuscript.

SOLUTION AND SOLID STATE C-13 and N-15 NMR
STUDIES OF VISUAL PIGMENTS AND RELATED SYSTEMS:
RHODOPSIN AND BACTERIORHODOPSIN[1a]

G. D. Mateescu,[1b] E. W. Abrahamson,[2a]
J. W. Shriver,[2b] W. Copan,[2c] D. Muccio,[2d]
M. Iqbal, and V. Waterhous

Department of Chemistry, Case Western Reserve
University, Cleveland, Ohio 44106

ABSTRACT.   This paper constitutes a chronological
review of C-13 and N-15 solution and solid state NMR
investigations of bovine rhodopsin and the purple
membrane of <u>Halobacterium halobium.</u>   Model retinyli-
deneimines and their protonated species yielded rather
informative variation in C-13 chemical shifts, as a
function of structure and conformation.   Retinals,
specifically enriched with C-13 at key atom positions,
were incorporated in rhodopsin and bacteriorhodopsin
during the bleaching-regeneration process.   The labeled pigments exhibited C-13 signals which were interpreted in terms of the state of protonation and
conformation of the Schiff base linkage between the
chromophore and opsin (15-anti in rhodopsin and light
adapted bacteriorhodopsin) and chromophore-protein
interactions.   Peaks arising from both all-<u>trans</u> and
13-<u>cis</u>-retinylidene moieties were observed in dark
adapted bacteriorhodopsin.   Upon protonation, the
Schiff base nitrogen of model compounds displayed
dramatic (up to 146 ppm) high field N-15 shifts (an
order of magnitude larger than C-13).   Hydrogen bonding also led to important (up to 50 ppm) nitrogen
shielding. <u>H. halobium</u> grown on a medium containing
ε^{15}N-lysine incorporated, with high yield, the labeled
amino acid in all of the seven lysine sites of bacteriorhodopsin.   Of these, one forms the Schiff base
linkage with retinal; its N-15 chemical shift could be
observed only in lyophilized preparations.   It appears
that the degree of protonation of the Schiff base
linkage depends on the degree of hydration of the

C. Sandorfy and T. Theophanides (eds.), Spectroscopy of Biological Molecules, 257–290.
© 1984 by D. Reidel Publishing Company.

specimen. Double labeling ($^{13}C=^{15}N$) is expected to provide valuable information concerning the structure and conformation of this moiety. The six lysine residues not involved in retinal binding yielded, in a micellar (octyl-β-glucoside) solution, N-15 signals of various linewidths, spread over a narrow (2-4 ppm) range. This could be interpreted in terms of lysine residue interactions within the protein. Preliminary results indicate the feasibility of in vivo measurements which avoid alterations inherent to extraction, purification, and/or lyophilization procedures.

1.   INTRODUCTION.

The **duplicity theory of vision**[3] describes the **rods** (Fig. 1) as photoreceptors responding to very low light intensities. These cells are associated with the **scotopic**[4], black and white, vision. In contrast, the cones are associated with the **photopic**[5] vision, in which colors can be distinguished. In the **mesopic range** (full moonlight) both the scotopic and photopic mechanisms are involved. The great majority of spectroscopic research is conducted with rods, apparently because of the ease of separation of their outer segments (**ROS**). During the initial stage of visual transduction the photic (absorbed light) signal is converted, within milliseconds, into a highly amplified electric potential (the ultimate step is an alteration of the polarization of the plasma membrane **PM**, with a rapid change in the polarization at the rod inner segment, **RIS**, membrane and the release of neurotransmitters at the synapse).[6] This intricate process involves four domains vastly different in dimension and dynamic activity:

a) a twenty-carbon polyene chromophore, **retinal**, derived from vitamin A (**retinol**),

b) a glycolipoprotein, **opsin** (M.W. ca. 39,000D), which binds retinal via the ε-NH$_2$ group of a lysine residue to form **rhodopsin** (the **visual pigment**),

c) a disc **organelle** (2-5 micrometer diameter) which contains up to 500,000 rhodopsin molecules in a bilayer phospholipid matrix,

d) the **plasma membrane** (**PM**) enclosing 500-1000 discs in the rod outer segment (**ROS**).

Knowing the structure and conformation of the native retinylidene chromophore and its interactions with the surrounding opsin (e.g., negatively and positively charged groups of appropriately situated amino acid

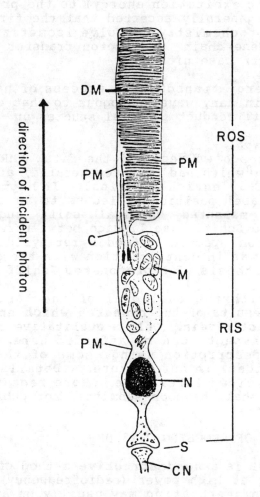

*Figure 1.   The structure of the bovine rod cell showing the
rod outer segment (ROS), inner segment (RIS), disc
membranes (DM), plasma membrane (PM), connecting
cilium (C), mitochondria (M), cell nucleus (N),
synapse (S), and a portion of the connecting neuron
(CN).*

units) will contribute to the understanding of the
mechanisms which regulate the wavelength of the absorbed light and determine the transfer of light energy (electronic excitation energy) to the protein domain. It is generally accepted that the first events
in vision photochemistry involve isomerization of
the retinylidene chain and proton transfer to and/or
from the Schiff base nitrogen.

Since the energy transduction process of halophilic
bacteria is in many ways analogous to that of vision,
it is useful to conduct parallel studies on bacteriorhodopsin.

Several years ago we reported the first NMR spectrum
of rhodopsin[7,8]which had been regenerated, after bleaching, with C-13 enriched retinal. It happened that
choosing to label position 14 led us to an unexpected
result: the measured chemical shift suggested a
nonprotonated Schiff base linkage between retinal and
opsin (based on spectra of model retinylidenealkyli-
mines). This was in contradiction with the generally
accepted hypothesis of a **protonated** Schiff base.[9,10]

In keeping with the dual goal of the Institute we
present the results of the **research which ensued** in a
somewhat **didactic** vein. Only a qualitative discussion
of selected aspects can be afforded here. We will
conduct our description using some of the figures
shown, as slides, in our lecture. Details can be
found in the cited literature. More recent results
have been or shall be soon submitted for publication.

2.   RATIONALE OF CHOOSING C-13 NMR

Generally, NMR is a **nondestructive** method of investigation. Even at high power (radiofrequency) irradiation specimen overheating may usually be avoided by
choosing appropriate transmitter and decoupler pulse
sequences and adequately cooling the probe.

The low excitation energy of NMR ensures examination
of molecules in their **ground state**. This is essential
for studies of visual pigments whose inherent **photochemistry** is characterized by very **high quantum
yields**.

C-13 NMR is particularly suited for **structural** investigations because it probes the backbone of molecules,

unlike proton NMR where chemical shifts may easily
vary with the medium and concentration. However, C-13
shifts are sensitive to **pH variations** and to secondary
internal fields created by the unpaired electron of
lanthanide shift reagents.   They are also sensitive to
fields induced by charged groups (e.g., carboxylate
ions) found at an appropriate distance from the measured carbon nucleus. Finally, **steric interactions**
may often be characterized by C-13 NMR.

The chemical shift variations found in the polyenic
chromophores of rhodopsin and bacteriorhodopsin can be
qualitatively discussed in terms of the equation given
by Karplus and Pople[11] for the paramagnetic screening
constant σ_{δ}^{p}:

$$\sigma_{\delta}^{p} = \frac{e^2\hbar^2}{2m^2c^2(\Delta E)} \left\langle r^{-3}\right\rangle_{2p} \Sigma Q_{ij} \tag{1}$$

where: e, $\hbar$, m, and c are the usual constants
         ΔE is the estimated mean electronic excitation
         energy
         r is the 2p electron orbital size term
         ΣQ_{ij} is the sum over all atoms of the elements
         of the bond order and charge density matrix.

Of course, **all** terms of this equation are interdependent but it shall prove useful to invoke them separately when we will want to estimate the importance of
various contributions of inductive, steric, electric
field, mesomeric, or other effects to chemical shift
changes.

The fact that carbon nuclei experience quite different
paramagnetic shieldings in different classes of compounds (e.g., saturated vs. unsaturated molecules)
leads to a chemical shift (δ) scale which is at least
an order of magnitude larger for C-13 (>200 ppm) than
for proton (<20 ppm).   As the C-13 linewidths are
narrow, it is not uncommon to find separate resonance
lines for each carbon in relatively large molecules
(see, for instance, Figure 2).

Homo- and heteronuclear scalar coupling can be interpreted in terms of **electronic structure** and **molecular
configuration.**

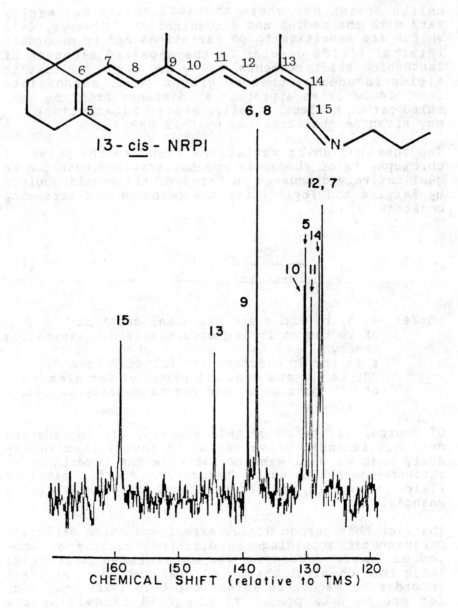

Figure 2.   Alkene region of the 25.16 MHz C-13 NMR spectrum of
            13-cis-NRPI (0.03M) in CDCl$_3$ at -65°C (from the Ph.D.
            thesis of J.W. Shriver[8]). With the higher field
            superconducting magnet spectrometers which are now
            increasingly available, dispersion up to five times
            better is possible. Chemical shifts are listed in
            Table 1.

TABLE 1. ^{13}C NMR Chemical Shifts of N-13-cis-retinylidenepropylimine, and N-13-cis-retinyledenepropyliminium Chloride[a]

Carbon	NRPI	NRPI-HCL	Predicted NRPI-HCl[b]
1	34.06	34.10	34.17
2	38.91	38.92	39.05
3	18.91	18.86	18.88
4	32.88	33.04	33.14
5	129.85	131.61	131.95
6	137.36	137.18	137.03
7	127.31	131.45	131.95
8	137.36	137.18	136.92
9	138.74	145.05	145.63
10	130.09	129.13	129.62
11	128.99	136.77	138.58
12	127.31	124.41	125.22
13	143.86	161.32	162.19
14	127.76	118.25	118.45
15	158.40	162.01	161.79
1,1CH$_3$	28.83	28.87	28.92
5CH$_3$	21.95	22.00	22.15
9CH$_3$	13.04	13.36	13.40
13CH$_3$	21.23	22.70	22.33
1'	63.66	54.60	
2'	24.00	22.79	
3'	12.00	11.23	

[a]Spectra taken on approximately 0.03M solutions in CDCl$_3$ at -65°C at 25.16 MHz. Chemical shifts are reported as ppm downfield from internal TMS.
[b]Predicted by adding difference between all-trans-NRPI-HCL (-65°C) and all-trans-retinal (RT) chemical shifts to 13-cis-retinal chemical shifts (see References 8, 17 and 18).

Determination of nuclear relaxation times and nuclear Overhauser enhancements of these carbons will give an estimate of the **dynamic properties** of molecules.

Finally, the low natural abundance (1.1%) of C-13 allows **selective enrichment** at precisely determined sites of the chromophore and detection of the corresponding signal in the pigment regenerated with the labeled chromophore (the rotational correlation time of the rhodopsin/octylglucoside micelle with micellar weight of less that 60,000 will allow linewidths smaller than 100 Hz). Characterization of the pigment in the **solid (lyophilized) state** is also possible.[12,13] This eliminates, in part, possible specimen alterations due to extraction and purification procedures.

3.   STRATEGY

Our strategy (Figure 3) for the NMR study of visual pigments and related systems is described below:

     a)   Use readily available apoprotein sources: retina from bovine eyes (commercially available) and purple membrane from **H. halobium** cultures.
     b)   Synthesize and study appropriate chromophore models in order to establish which carbon nuclei exhibit chemical shifts most sensitive to structural and conformational changes (a carbon-13 **"fingerprint"**, i.e., enrichment of all positions, turned out to be necessary).
     c)   Optimize the synthesis of C-13 enriched retinals.
     d)   Optimize the regeneration of pigments with labeled retinal and natural opsin or bacterioopsin.
     e)   Establish best conditions (purification, solubilization, lyophilization) for obtaining satisfactory spectra.
     f)   Interpret the results based on model systems and theoretical calculations.

4.   STUDIES OF MODEL CHROMOPHORES

It has been established that rhodopsin contains 11-cis-retinal.[14]   Bacteriorhodopsin contains all-trans- and 13-cis-retinal; the all-trans isomer is predominant in the light-adapted membrane, while approximately equal populations of the two isomers are found in the dark-adapted membrane.[15] The prosthetic chromo-

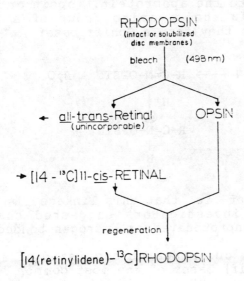

Figure 3a. *Strategy used for C-13 labeling of Rhodopsin.*

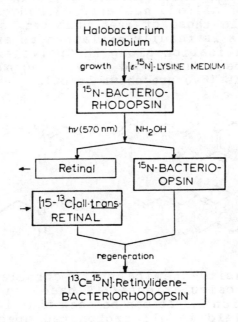

Figure 3b. *Strategy used for C-13 and N-15 labeling of Bacteriorhodopsin.*

phores are bound to the apoprotein, opsin or bacteriorhodopsin, via the ε-amino group of a lysine
residue with which they form a Schiff base:

$$R-C=O + H_2N-OPSIN \longrightarrow R-C=N-OPSIN + H_2O \qquad (2)$$

$$H^+ \downarrow$$

$$\overset{\oplus}{R-C=N-OPSIN}$$

$$|$$

$$H$$

The general belief is that this linkage is protonated.[9] Several investigators suggested that the
ground state is nonprotonated or hydrogen bonded.[10]

The first stage of our research consisted of synthesizing retinal Schiff bases of the most common isomers
and measuring their NMR spectra in natural abundance.[17,18] The results are shown in Figure 4. It is
seen that upon protonation dramatic changes occur in
the C-13 chemical shift pattern of retinylidenimines.
The signals of all odd numbered carbons are shifted
downfield while those of even numbered carbons shift
upfield. This is readily explained as an effect of
charge delocalization along the polyolefinic chain
which could be illustrated by writing all possible
canonical forms (two are shown below).

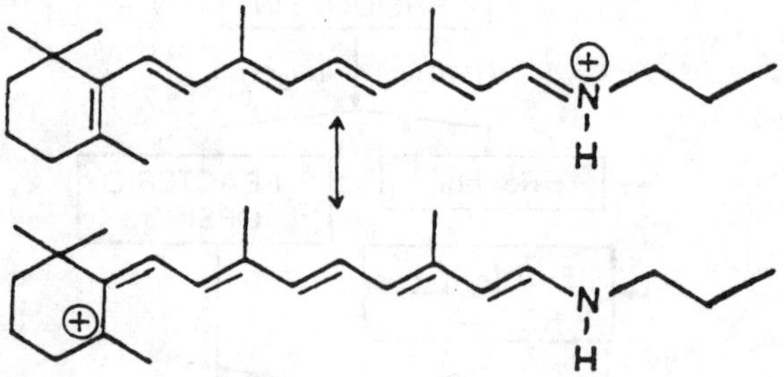

The chargedensity-chemicalshift correlation[16] was
confirmed by calculations[17,18] based on the Pople-
Karplus equation 1. It can be seen that 14-C absorbs
at highest field in all protonated species. This
position was therefore chosen for the first C-13 labeling experiment.[7] As it shall be seen later, the
proper interpretation of C-13 spectra in terms of

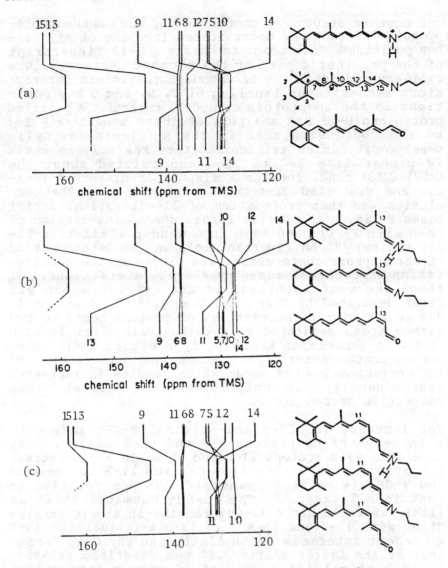

Figure 4. Schematic presentation of the observed changes in the alkene region of the C-13 NMR spectra of:

(a) all-trans retinal, all-trans-NRPI and all-trans-NRPI·HCl

(b) 13-cis-retinal, 13-cis-NRPI and 13-cis-NRPI·HCl

(c) 11-cis-retinal, 11-cis-NRPI and 11-cis-NRPI·HCl

in CDCl₃ at room temperature.

chromophore structure, conformation, and chromophore-
protein interactions necessitates labeling of all carbon positions. This means to obtain a **C-13 fingerprint**
of the prosthetic part of the pigment. Moreover, the
complete understanding of chromophore-protein interactions necessitates **labeling of C, N, and O key positions in the apoprotein** or models thereof. A detailed
proton and C-13 NMR analysis of these models[17],[18] led
to the conclusion that in the N-(11-cis-retinyli-
dene)propyliminium trifluoroacetate the polyene chain
is planar from 7-C to 12-C and twisted about the
C(12)-C(13) bond, forming a mixture of distorted 12-s-
cis and distorted 12-s-trans conformations. The conclusion was that protonation of 11-cis-retinal Schiff
bases relieves the steric strain due to interaction of
10-H with 13-CH_3 and 14-H in a manner similar to 11-
cis retinal.[19] An important role in the determination
of chromophore conformation was played by the consideration of effects caused by steric interactions. Of
these the most important is the effect which was
first described by Grant and Cheney.[20] This effect is
due to the steric interactions of protons bound to two
carbon atoms separated by a carbon-carbon double bond
(i.e., in γ-position starting from either end). The
proton-proton interaction causes the polarization of
the corresponding C-H bonds which results in increased
charge density (i.e. upfield chemical shifts) of the
respective carbon atoms.

For instance in all-trans- retinal 9-CH_3 interacts
with neighboring protons 7-H and 11-H while 13-CH_3
interacts with protons 11-H and 15-H. In 9-cis-retinal
the steric interaction of 9-CH_3 and 11-H is removed
and 9-CH_3 is shifted downfield 7.95 ppm relative to
that in all trans.[19] The 13-CH_3 chemical shift is
little affected since 13-CH_3 remains in steric interaction with 11-H and 15-H. In 13-cis-retinal the 11-H
no longer interacts with 13-CH_3 and the C-13 resonance of the latter shifts 7.94 ppm downfield relative
to that of all-trans. Relief of steric interaction
between 13-CH_3 and 11-H shifts the methyl resonance
downfield in 11-cis-retinal but not as much as in 13-
cis-retinal because in the 11-cis isomer 13-CH_3 now
interacts with 10-H. Obviously the steric interaction
hypothesis can be successfully applied to visual pigment chromophores because of the ability of the retinylidene chain to twist about formal single bonds
and remove steric interactions without large perturbations in orbital hybridization.

Assuming that the spectrum of 11-cis retinal may be
interpreted on the basis of that of all-trans retinal
with altered steric interactions, differences between
shifts of corresponding carbon atoms in retinal, reti-
nylideneimine and retinylideniminium ions for all-
trans and 11-cis can be considered to be due to different conformations of 11-cis isomer in the three model
chromophores. On this basis and dipole moment considerations,[21] as well as nuclear Overhauser effects,
we concluded that protonation of 11-cis,12-s-cis-N-
retinylidenepropylimine in CDCl$_3$ at -65°C results in a
change of the more stable conformation from distorted
12-s-cis to distorted 12-s-trans. We, therefore,
proposed that the conformation of the chromophore of
rhodopsin is distorted 11-cis, 12-s-trans. In studying the series of all-trans retinal, N-all-trans-
retinylidenepropylimine and N-all-trans-retinylidene-
propyliminium chloride we observed only one resonance
for 1'-C and 15-C which meant that only one isomer
with respect to the C=N bond was present. Using

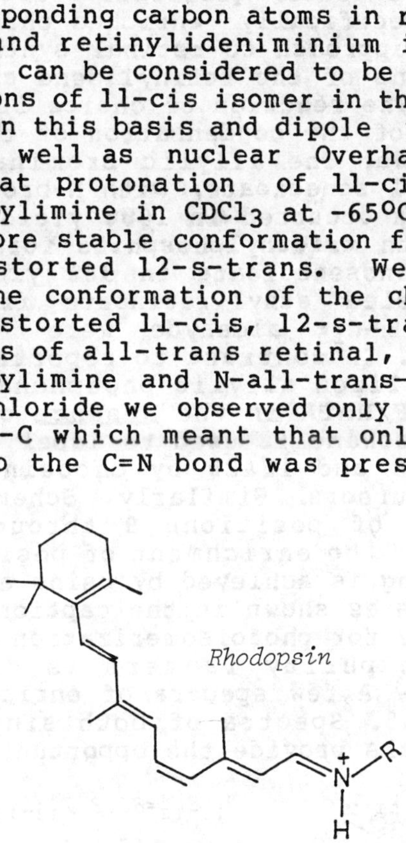

Rhodopsin

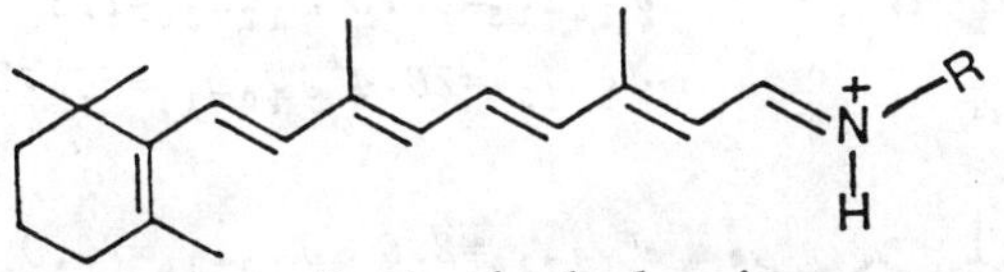

Bacteriorhodopsin

Eu(fod)$_3$ we found a contact shift which demonstrates
that the protonated all-trans-retinylideneimine has a
15-anti configuration[18]. This should also be true in
the case of **bacteriorhodopsin.** Since in rhodopsin the
conformation of the chromophore in the vicinity of the
Schiff base linkage is similar to that found in bacteriorhodopsin we suggest that the preferred conformation of the retinylidene chain in **rhodopsin** is also
15-anti, as shown above.

5. SYNTHESIS OF C-13 LABELED RETINALS

We give here a brief, qualitative account on the synthesis of specifically enriched chromophores.[7,12,22]
An efficient approach to retinal synthesis is to join
larger portions of the retinylidene chain as shown in
Scheme 1.   The reaction of choice is the phosphonate
modification of the condensation of the acrylate with
β-C[15] aldehyde. The allylic brominationof 5, also
known as ethyl senecioate, with N-bromosuccinimide in
CCl[4] at reflux occurred in 100% yield.   Arbusov reaction of 6 with triethylphosphite formed the ethyl-γ-
diethylphosphonosenecioate in 98% yields.   Very high
yields of labeled ethyl retinoate from condensation
of 7 with β-C[15] aldehyde were obtained using
NaH/THF/HMPA. In contrast to reports in the literature, a stabilized allylic phosphonate carbanion is
formed in THF/HMPA in the <u>absence</u> of aldehyde substrate. This scheme is used to label carbon positions
13 through 15 and 12&20 by choosing appropriately
enriched precursors. Similarly, Scheme 2 illustrates
the labeling of positions 9 through 11 and 8&19.
Scheme3 is for the enrichment of positions 6 and/or 7.
Double labeling is achieved by using appropriate starting materials as shown in the caption of each scheme.
The techniques for photoisomerization and HPLC separation of high purity isomers is described elsewhere.[7,8,13,23] A few spectra of enriched retinals are
shown in Fig.5. Spectra of both single and double
labeled retinals provide the opportunity to obtain C-C

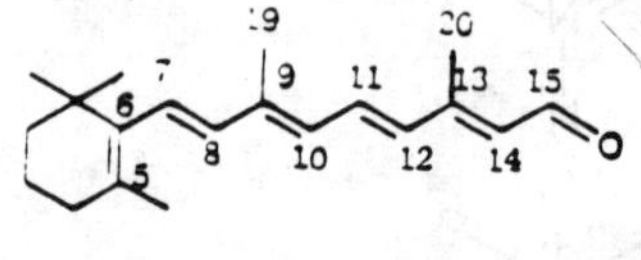

$J_{11-12}=69.8$ $J_{12-13}=54.8$ $J_{13-14}=66.7$

$J_{14-15}=57.5$ $J_{12-20}=1.8$ $J_{13-20}=40.3$

$J_{9-10}=70.4^*$ $J_{10-11}=58.7^*$ $J_{9-19}=43.4^*$

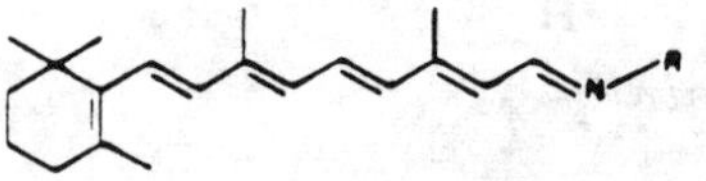

$J_{12-20}=2.0$ $J_{13-14}=70.0$ $J_{14-15}=64.9$

$J_{15-N}=6.5$ $J_{14-N}=7.3$ $J_{13-N}=5.7$

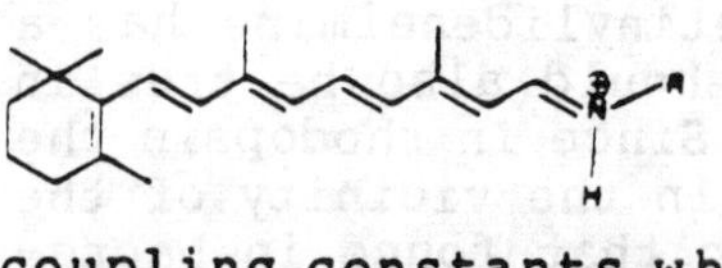

$J_{12-20}=1.8$ $J_{15-N}=17.6$

R =hexyl   * See reference 22

coupling constants which may yield important structural and conformational indications. It is thus seen
that there is more double bond character (larger
coupling) wherever we formally draw a double bond.

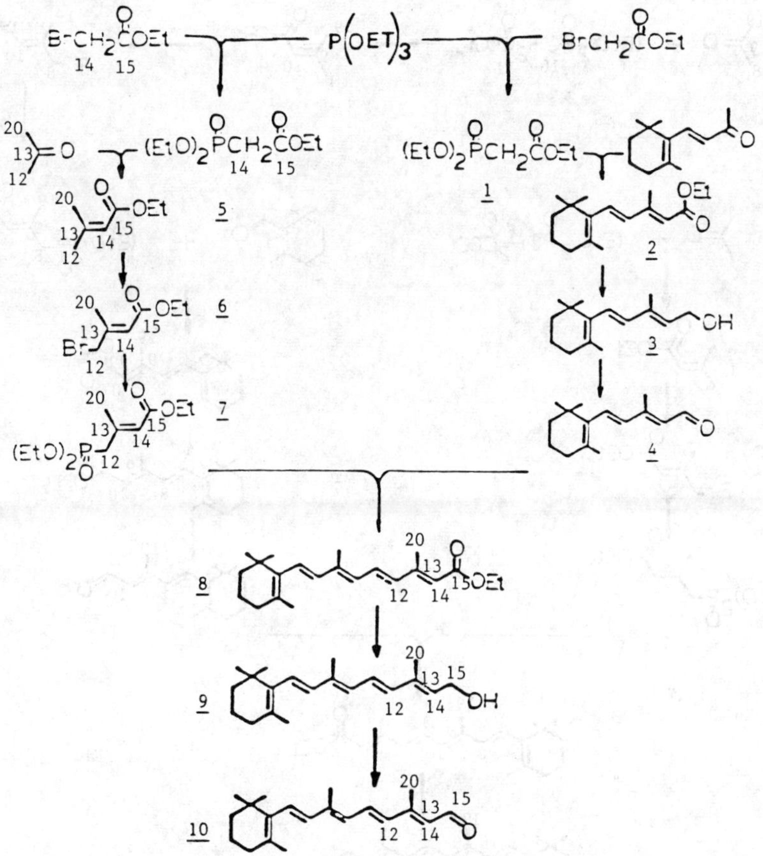

Synthesis of retinals enriched at positions 13 through 15, and 12,20. The numbers indicate the origin of the labels in the appropriate precursors.

PRECURSOR	POSITIONS ENRICHED
1-*C ethylbromoacetate	15
2-*C ethylbromoacetate	14
1,2-*C ethylbromoacetate	14,15
2-*C acetone	13
2-*C acetone and 2-*C ethyl-bromoacetate	13,14
1,3-*C acetone	12,20

Scheme 1.

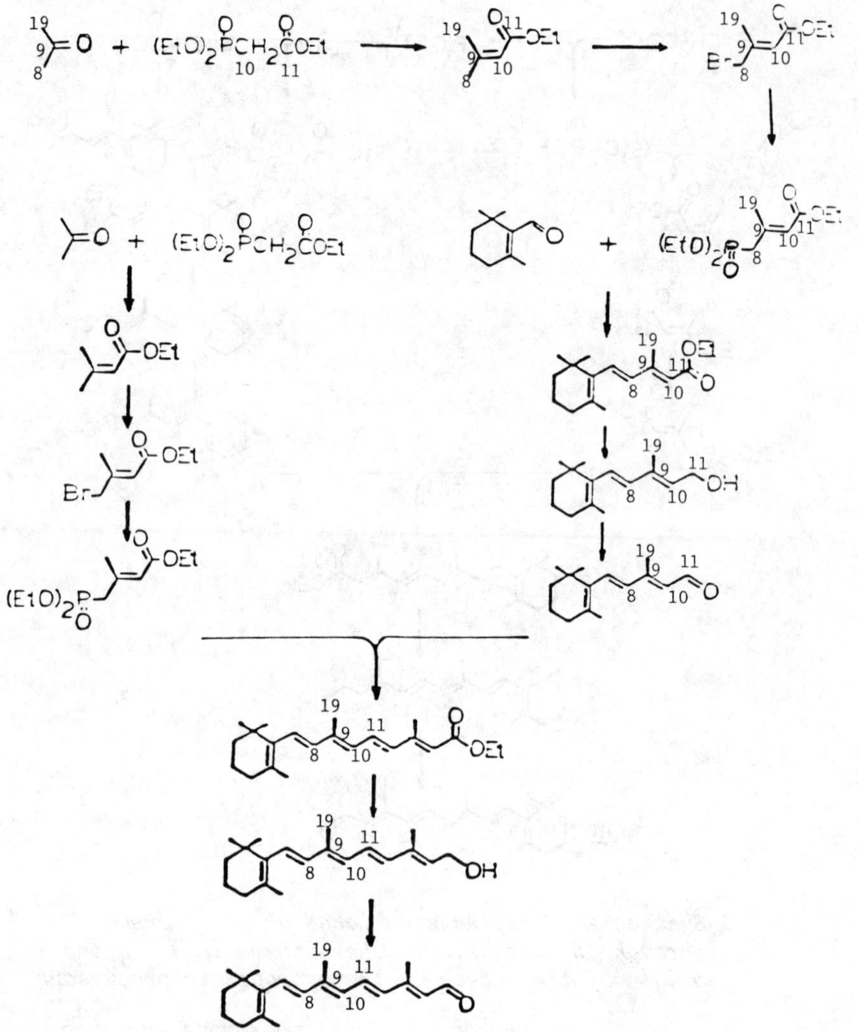

Scheme 2. Synthesis of retinals enriched at positions 9
          through 11, and 8,19.

PRECURSOR	POSITIONS ENRICHED
1-*C ethylbromoacetate	11
2-*C ethylbromoacetate	10
1,2-*C ethylbromoacetate	10,11
2-*C acetone	9
2-*C acetone and 2-*C ethyl-	
bromoacetate	9,10
1,3-*C acetone	8,19

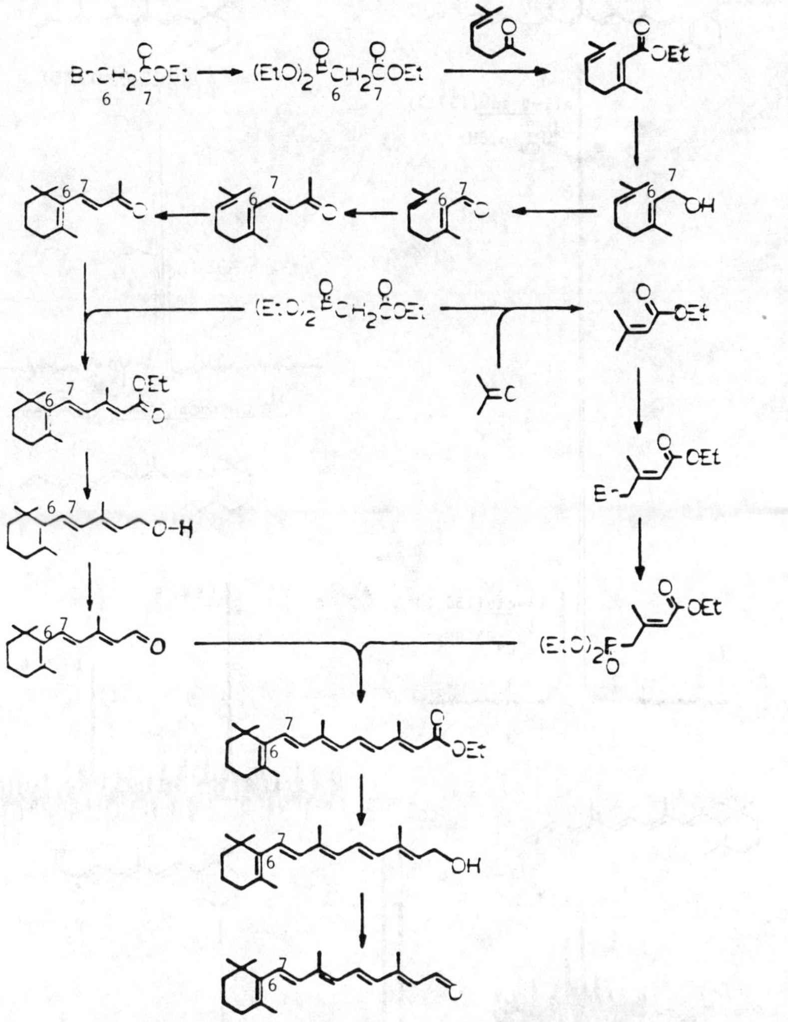

Synthesis of retinals enriched at positions 6 and 7.
The numbers indicate the origin of the labels in the
appropriate precursors.

PRECURSOR	POSITIONS ENRICHED
1-*C ethylbromoacetate	7
2-*C ethylbromoacetate	6
1,2-*C ethylbromoacetate	6,7

Scheme 3.

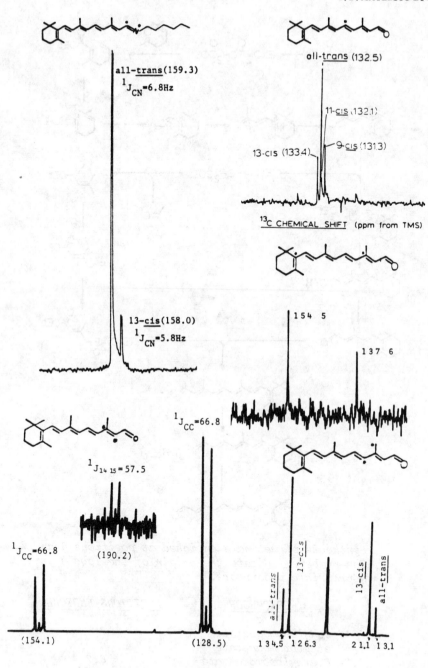

Figure 5

6. SOLUTION AND SOLID STATE C-13 SPECTRA OF LABELED PIGMENTS

Labeled retinals were incorporated in rhodopsin according to the scheme shown in Figure 3. Each experiment required 250 retinas from bovine eyes. Gentle homogenization, buoyant density flotation, and centrifugation (sucrose gradients) yielded rod outer segments which were bleached with visible light and regenerated in dark with the desired C-13 labeled retinal. The protein was solubilized with octyl-β-glucoside(OBG). Affinity chromatography (concanavalin A/Sepharose 4B) yielded pure labeled rhodopsin ($A_{278}/A_{498}=1.6$). Ultra filtration and dialysis led to a final rhodopsin concentration of 0.5 to 1.0 mM in OBG. Solution NMR spectra were taken at temperatures between 2 and 4°C.

Optimal growth of H. halobium, mutant strain R1 (gift of Dr. W. Stoeckenius)[15] was achieved at 39°C in a 10 liter homebuilt Virtis-type fermenter on a complex medium with basal salts and aeration at 17 SCFH under mechanical stirring.[24] The growth reached stationary phase, with little purple membrane formation; thereafter the aeration was reduced to 3 SCFH and the suspension illuminated by 18 cool white 15 watt flourescent tubes. The formation of bacteriorhodopsin was monitored spectrophotometrically. After six or seven days the cells were harvested by 19 minutes of centrifugation at 12,000 rpm. Typical bacteriorhodopsin yields were 15-20 micromoles per 10 liters based on an ε = 55,000[25]. After lysis in double distilled water and dilution to approximately 800 ml, an overnight stirring with a few mg deoxyribonuclease followed by differential centrifugation yielded cell fragments which were sedimented at 12K rpm for 60 minutes. After several differential centrifugations the purple membrane fragments were separated from cell debris. Final purification was accomplished by discrete sucrose density gradient centrifugation (90,000xg for 60 minutes) with repeated removal of sucrose by dialysis vs. 50mM phosphate buffer, pH 7. Bacteriorhodopsin was bleached in a stirred supension in water at 4°C in the presence of 100 mM hydroxylamine, pH 7, with intense illumination (500 watt projector light filtered through Kodak Wratten 2B filter and 5 cm water). The apomembrane (5 uM) with associated retinaloxime was washed several times by resuspension and repeated centrifugation. Labeled bacteriorhodopsin was prepared by regeneration with an EtOH solution (5mM) of C-13 enriched retinals in water suspension at 4°C.

The process was monitored spectrophotometrically.
Retinaloxime and retinal were removed by hexane extraction. The solubilization and concentration procedure was similar to that described for rhodopsin.

Figure 6 illustrates the results obtained with rhodopsin enriched at position 13 of the retinylidene chain.
The spectrum on the top of the figure contains a
resonance at 168.1 ppm which is not observed in the
spectrum of unlabeled preparations (Figure 7). The
intense natural abundance protein resonances centered
at 128 and 174 ppm originate from the ring carbon
atoms of aromatic residues and amide carbonyls, respectively. Less intense protein reso-nances observed
at 157 and 182 ppm are attributed to tyrosine and
arginine carbon atoms and carboxylate residues.
Therefore, the 168.1 ppm resonance is assigned to the
labeled carbon atom of retinal bound to rhodopsin.
The linewidth of 20 Hz is consistent with the expected
rotational correlation time forspherical micelles
containing molecules of rhodopsin size. The UV-visible
spectrum sampled during the course of the NMR experiment remained identical to that of purified rhodopsin.
The assignment was further checked by the fact that
heat denaturation or bleaching with light led to the
diminishing of the intensity of the 168.1 ppm signal
and concomitant increase of the narrow line at 156.5
ppm assigned to solubilized (nonbound) 11-cis-[13-^{13}C]
retinal (Figure 6b). This is accompanied by a corresponding decrease in the intensity of the 498 nm absorption band of the intact pigment and an increase in
the intensity of the 370 nm (free retinal) band.
Furthermore, the reduction with NaBH$_4$ led to the total
disappearance of the retinal signal and appearance of
the vitamin A (retinol) resonance at 135.5 ppm (Figure
6c). The change was also confirmed by the 320 nm band
in the UV-visible spectrum.

Similar measurements with labeled bacteriorhodopsin
yielded, for the enriched position, a signal at 168.9
ppm. The linewidth is narrower due to the lower molecular weight of the pigment.

These results suggest that both rhodopsin and bacteriorhodopsin experience similar electron distribution
at the site of the carbon atom in position 13 of their
retinylidene chain. Moreover, the position of these
signals suggest a significant protein-chromophore interaction, since the resonances appear at even lower
field than in models measured in strongly proton-donor

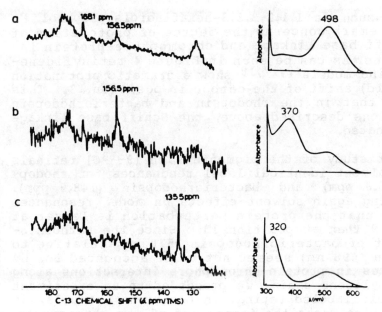

Figure 6. The 45.6 MHz ^{13}C NMR spectra of (13-^{13}C)retinylidene rhodopsin (0.6 mM in OβG/80 mM phosphate buffer, pH 7.0) obtained at 4°C using 90° pulse angles; 0.4 sec delay time; (a) in its "native" solubilized form (10,600 transients); (b) partly photobleached with released all-trans(13-^{13}C)-retinal (4,760 transients); and (c) reduced with NaBH$_4$ to the (13-^{13}C)-retinol (4,000 transients).

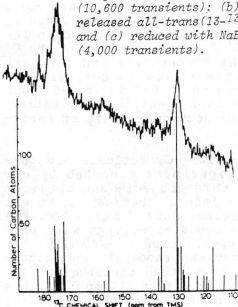

Figure 7.
Low field region in the 45.6 MHz ^{13}C NMR spectrum of unlabelled, solubilized rhodopsin at 4°C in 150 mM OβG and 80 mM phosphate buffer, pH 7.0 using 90° pulses and 5 sec recycle times, 10,000 transients; rhodopsin concentration is 1 mM. The bar diagram refers to the expected chemical shift distribution of the protein carbons based on the amino acid analysis of rhodopsin.

solvents such as 1,1,1,3,3,3-hexafluoroisopropanol.
A final remark concerns the degree of protonation of
the Schiff base linkage and chromophore-protein interactions. As can be seen in Figure 4 retinylidenepropylimine models[12,17,18] show a dramatic protonation
(low field) shift of the carbon in position 13. This
suggests that in the rhodopsin and bacteriorhodopsin
preparations described above, the Schiff base linkage
is protonated.

A similar study of the pigments with [9-^{13}C] retinals
yielded almost identical label resonances for rhodopsin   (148.8 ppm)   and   bacteriorhodopsin   (148.5 ppm).
Considering again solvent effects on model resonances
suggests that the protein perturbation is lesser at
position 9 than at position 13.   Since the bathochromic shift of bacteriorhodopsin (570 nm) relative to
rhodopsin (498 nm) seemed not to be accounted for by
differences in protein-chromophore interactions along
the retinylidene chain, we proceeded to an experiment
with double labeled retinal in positions 6 and 13, in
order to also probe the ionone end of the chromophore.

The chemical shift of carbon 6 in isomeric retinylide-
neimines and iminium ions is very insensitive to protonation (Figure 4) and hydrogen bonding.[8] Variations
due to solvent effects are 0.5 ppm or less. Therefore,
this resonance is an excellent probe of the effects of
protein folding, hydrophobic binding, or protein induced perturbation.   At 137.0 ppm in rhodopsin, the C-
6 resonance indicates that the protein induces very
little perturbation at this site.   In contrast, in
bacteriorhodopsin the resonance of C-6 is at 135.6 ppm
indicating the possibility of a specific perturbation.
This may account for the bathochromic shift of bacteriorhodopsin.

The results obtained with [14-^{13}C] retinals are puz-
zling. We repeated the experiment executed by our
firstgraduate student on this project[8] and obtained
the same results, i.e., the labels' chemical shift was
close to that of nonprotonated Schiff bases (cf. Figure 4). Moreover, the analogous experiment with bacteriorhodopsin yielded a similar result.[13] While this
confirms the similarity between rhodopsin and bacteriorhodopsin with respect to protein perturbation, it
seems difficult to explain the extent (10 ppm) of this
effect. Another perturbation, in addition to a strong
external anionic charge,[26] appears to operate at the
binding site of retinal. It is also conceivable that

in solution measurements a peak is masked by the strong detergent absorption at 106 ppm. Should this be true, the hypothesis of the existence of a negatively charged group[26] in the proximity of carbons 12 and 14 would be invalidated.

Important results were obtained with lyophilized specimens of bacteriorhodopsin which were measured in the solid state using a cross polarization/magic angle spinning (CP/MAS) probe. The signal of the labeled carbon appeared at 143.7 ppm indicating the presence of a nonprotonated Schiff base (see Figure 8). Experiments conducted with bacteriorhodopsin enriched at position 15 may also be interpreted in terms of a nonprotonated Schiff base (Figure 9).

Interestingly, a brief account[27] of a CP/MAS measurement of [13-^{13}C]bacteriorhodopsin indicated the detection of a resonance close to that of a protonated Schiff base; this result supports the suggested existence of a protonation-deprotonation process which depends on the degree of hydration.

The C-13 results are summarized in Table 2.

TABLE 2. Solution and Solid State ^{13}C NMR
         Chemical Shifts of Labeled Rhodopsin
         and Bacteriorhodopsin Preparations

| C | RHODOPSIN | | | | BACTERIORHODOPSIN | | | |
	δ	$\Delta\delta$[a]	PSB[b]	SB[c]	δ	$\Delta\delta$[a]	PSB[b]	SB[c]
15	165.9	+2.0	163.9	159.6	166.0	+3.1	162.9	159.7
					*150	−10.9		
14	130.0	+8.7	121.3	130.0	130.0	+10.3	119.7	129.0
13	168.1	+5.5	162.6	145.5	168.9	+6.5	162.4	144.4
					*143.7	−18.7		
9	148.8	+2.2	146.6	139.3	148.5	+3.0	145.5	138.0
6	137.0	−0.2	137.2	137.5	135.6	−1.4	137.0	137.4

a) Difference from the NRPI.HCl; b) Protonated Schiff base = NRPI.HCl; c) Nonprotonated Schiff base = NRPI (see Figure 4). *CP/MAS results.

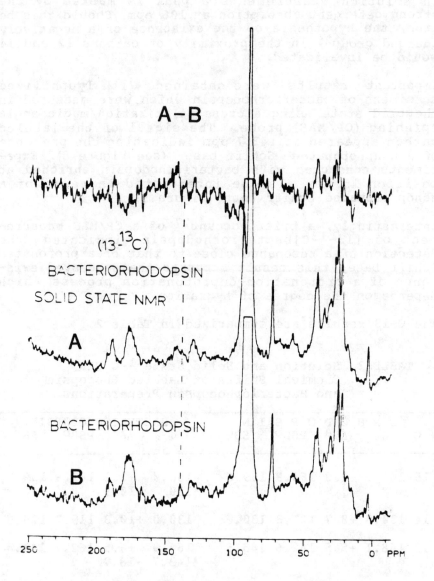

A–B

(13-^{13}C)

BACTERIORHODOPSIN

SOLID STATE NMR

A

BACTERIORHODOPSIN

B

250   200   150   100   50   0   PPM

.Figure 8.   Solid state ^{13}C NMR spectra of (13-^{13}C)-bacteriorho-
dopsin (A), unlabelled bacteriorhodopsin (B), and
the computer-subtracted difference spectrum of A
minus B.   Acquisition parameters are spinning rate =
3.75 KHz; 90° pulse angle; acquisition time = 205
msec; delay = 16 sec; transients = 1,400.   Spectra
taken at 37.75 MHz.

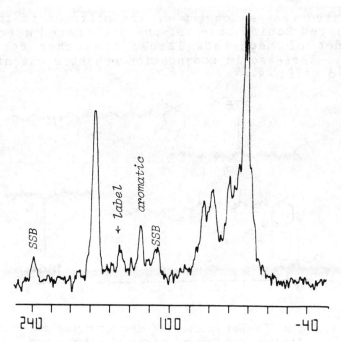

Figure. 9   C-13 CP/MAS(50 MHz) of lyophilized purple membrane enriched with {15-^{13}C}retinal.   Spinning rate = 3kHz;   contact time=0.8ms; 14744 pulses at 1 second intervals.   The label peak centered at 150ppm appears to be split by approximately 1.5ppm due to the presence of all-trans and 13-cis isomers coexisting in the dark adapted bacteriorhodopsin.   The unusually high field shift seems to indicate not only the presence of a non-protonated Schiff base but also a very strong shielding interaction yet to be explained and substantiated.   A remarkable aspect of this spectrum is the very low intensity and wide separation of the spinning side bands (SSB) which allow observation of important details in this region.

7.   RATIONALE OF CHOOSING N-15 NMR

The question addressing the Schiff base protonation-deprotonation process can best be answered by N-15 NMR studies. We have indeed demonstrated that protonation of N-retinylidenealkylimines results in very large (150 ppm) upfield shifts of the N-15 resonance.[28,29] (See Figure 10) This is primarily due to drastic changes in the E and charge density terms of Equation 1.   The former (.4 eV) is due to the loss, upon protonation, of the n-π* transition; the latter, to (positive) charge localization. The difference in the calculated charge density of the nitrogen in the non-protonated Schiff base vs. the protonated species is an order of magnitude larger than that found for carbon. Increase in conjugation enhances the nitrogen upfield shift.[28,29]

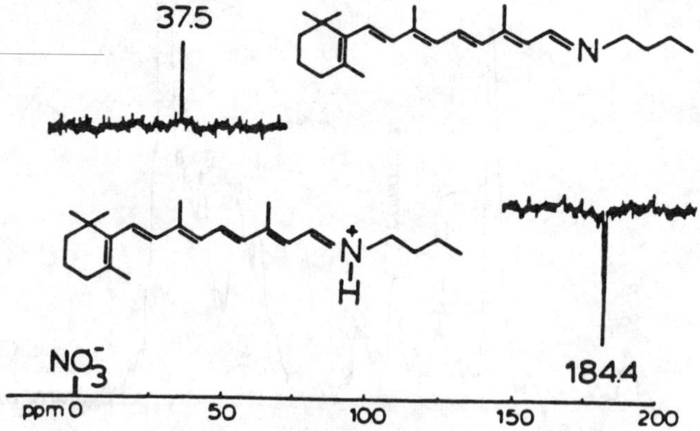

Figure 10. The ^{15}N NMR spectra of nonprotonated and protonated
          N-retinylidenebutylimine: $\Delta\delta = 146.9$ ppm

8. SOLUTION AND SOLID STATE N-15 SPECTRA OF LABELED BACTERIORHODOPSIN AND H. HALOBIUM CELLS

Halobacterium halobium grown on a medium containing [ε-^{15}N]lysine incorporates, with high yields, the labeled amino acid in all of the seven lysine sites of bacteriorhodopsin. Of these, one forms the Schiff base linkage with retinal. If protonated, the corresponding nitrogen should exhibit a high field shift comparable to that seen in Figure 10. CP/MAS measurements of severallyophilized preparations of biosynthetically enriched purple membrane yielded spectra represented in Figure 11. All of these are characterized by a strong resonance centered at 12.8 ppm from $^{15}NH_4NO_3$ assigned to the six labeled lysine residues incorporatedinto bacterioopsin,and a smaller peak at100ppm assigned to approximately 250 natural abundance peptide nitrogens of the apoprotein. In all preparations the relative positions of these two resonances were remarkably constant: δ(amide)- δ(Lys) = 87 ppm. In contrast to this feature, the nitrogen atom involved in the formation of the Schiff base linkage between retinal and bacterioopsin yielded, in prepara-

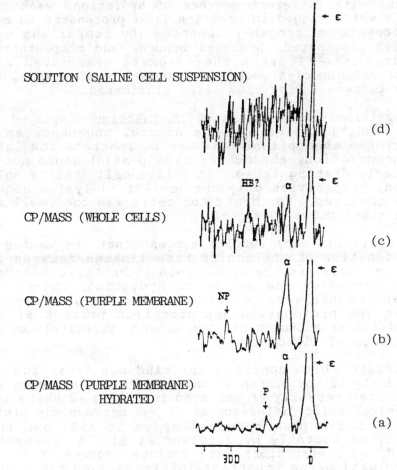

SOLUTION (SALINE CELL SUSPENSION)

(d)

HB?   α   ← ε

CP/MASS (WHOLE CELLS)

(c)

α   ← ε

CP/MASS (PURPLE MEMBRANE)

NP
↓

(b)

α   ← ε

CP/MASS (PURPLE MEMBRANE)
HYDRATED

P
↓

(a)

300   0

*Figure 11. 20.28 MHz ^{15}N NMR spectra of biosynthetically en-
riched H. halobium preparations. (a) Cross polar-
ization/magic angle sample spinning (CP/MASS) of ~4
μM purified purple membrane (PM) after lyophilization
and storage in a water saturated atmosphere (26,000
90° pulses, 0.4s acquisition time (AT), 3.6s pulse
delay (PD), 1.5 ms contact time, and linebroadening,
LB = 40 Hz). (b) The CP/MASS of a preparation immediately after lyophilization (~2 μM PM, 7920 90°
pulses, 0.064s AT, 15s PD, 0.4ms contact time, LB =
60 Hz). (c) The CP/MASS of lyophilized whole cells
of H. halobium (27,000 90° pulses, 0.4s AT, 2.0s PD,
1.5ms contact time, LB = 53 Hz). (d) Solution spectra of packed volume of whole cells in 4.0M salts
(42,000 90° pulses, 0.4s AT, 0.0s PD, 16mm sample
tube, $4^{O}C$, and LB = 60 Hz).*

tions with different degrees of hydration, weak signals which ranged in position from protonated to nonprotonated nitrogen resonances (by comparison with
model protonated, hydrogen bonded, and nonprotonated
imines).[28,29] It seems that signals associated with
protonated species occur in hydrated preparations, but
the pattern is far from being elucidated.

Lyophilized whole cells of H. halobium displayed the
$[\epsilon\text{-}^{15}NH_3^+]$ lysine and the natural abundance amide
nitrogen absorptions but, due to inherent low label
concentration, the Schiff base peak(s) could not be
clearly distinguished. In a live cell (saline solu-
tion) preparation only the $[\epsilon\text{-}^{15}NH_3^+]$ lysine signal
was observed. The number of cells was too small for
the other peaks to appear.

These preliminary results suggest that the degree of
protonation of the Schiff base linkage between the
chromophore and the apoprotein in bacteriorhodopsin
may depend on the degree of hydration. Since the
measurements were performed in the dark, it appears
that the protonation-deprotonation process at the
Schiff base nitrogen may not depend exclusively on the
presence of light.

A result which confirms our findings regarding the
mobility of the proton at the Schiff base nitrogen was
reported recently.[30] As seen in Figure 12 there is a
chemical shift difference of 19 ppm between the protonated imine peak reported by us in 1982 and that
reported recently by Harbison et al. A systematic
study on preparations with various degrees of hydration will bring important information on the proton
dynamics at the imine linkage. Our preliminary results stress the importance of the work reported by
Rafferty and Shichi on the involvement of water at the
retinal binding site in rhodopsin and early light-
induced intramolecular proton transfer.[31]

Careful analysis of the spectrum in the epsilon-
amino region of the 6 lysine residues not involved in
retinal binding reveals a number of peaks of various
line widths (Figure 13a). There is little doubt that
these peaks owe their position and shape to
interactions with the surrounding protein moiety.
With the spectrometer set in identical conditions
(Figure 13b) $\epsilon\text{-}^{15}NH_2$-lysine·HCl gives a sharp peak
aligned with the central absorption of the pigment.
Would these spectra have been taken in isolated

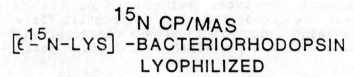

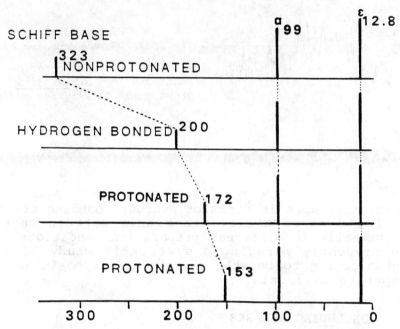

Figure 12.   Schematic summary of ^{15}N chemical shifts measured in
              lyophilized ^{15}N enriched bacteriorhodopsin prepara-
              tions. The top 3 spectra recorded in our laboratory
              in 1981 and 1982 show imine nitrogen absorptions at
              positions varying from nonprotonated to "protonated"
              Schiff bases. In the recently reported spectrum by
              Harbison, et al. the resonance reaches a higher
              field; the peak has been interpreted as being due to
              a fully protonated species, but the trend clearly
              seen in this figure may suggest that an even higher
              field resonance may be detected.

experiments, one would be tempted to align the lysine reference with the sharp highest field resonance of the pigment. However, being sure of the relative positions one can affirm that while still "free", that epsilon amino group experiences a characteristic local shielding effect. The width of the other peaks should

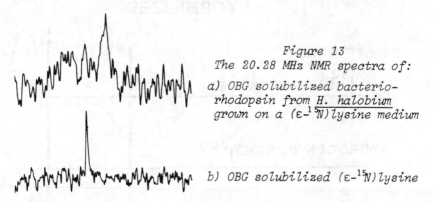

Figure 13
The 20.28 MHz NMR spectra of:

a) OBG solubilized bacteriorhodopsin from H. halobium
   grown on a (ε-^{15}N)lysine medium

b) OBG solubilized (ε-^{15}N)lysine

be interpretable in terms of hydrogen bonding effects. It should be emphasized that this pattern changes occasionally in different preparation conditions. We are presently pursuing a systematic study of this region hoping to be able to relate line positions and shapes to C-13 data.

10. CONCLUDING REMARKS

Due to its particular sensitivity to electron distri-bution in conjugated systems, C-13 NMR spectroscopy is excellently suited for studies of chromophore struc-ture and conformation in visual pigments and related systems. Thus, charge delocalization, steric, and ex-ternal charge effects are characterized by changes in chemical shifts and coupling constants which can be related to well studied models. It should be empha-sized, however, that although we can obtain clear, separate signals for a given carbon nucleus in a labeled pigment, it is mandatory to obtain a **cmr fingerprint**, i.e. the shifts for **all** carbon atoms of the retinylidine moiety. With a single exception (14-C) solubilized rhodopsin and bacteriorhodopsin yield signals close to those of model protonated N-retinyl-idinealkylimines. Differences from these models could be interpreted in terms of chromophore-protein inter-actions.

CP/MAS spectra of lyophilized specimens display, in different preparations, signals corresponding to either protonated or nonprotonated Schiff base linkages. This is tentatively attributed to variations in the degree of sample hydration which obviously alters protein conformation with consequent protonation or deprotonation.

Nitrogen-15 NMR is by far the best method of investigation of the degree of protonation. Hydrogen bonding and protonation induce N-15 shifts which are an order of magnitude larger than those of C-13.

There is a strong need for a concerted spectroscopic (NMR-UV-vis-IR-Raman,...) study of models mimicking the combined wavelength regulation function of the quaternary (Schiff base) nitrogen and the functional (carboxylic) group capable of placing a negative charge in the proximity of the retinylidine chain. Charge delocalization, counterion, and external charge effects could be evaluated starting with molecules such as retinylproliniminium perchlorate and continuing with Schiff bases synthesized with olgopeptides having the same structures as found in the two pigments.[33]

With the instrumentation improving at the present pace, the ideal in vivo experiments (already demonstrated by us to be feasible) will become a reality. We believe and hope that the subtleties of NMR (and the other methods) will lead us far enough to have a meaningful insight into the intricacy of phototransduction.

ACKNOWLEDGMENTS. The support of this research by the National Eye Institute of NIH (Grant EY 02998) is gratefully acknowledged. Instrumentation grants were awarded by NIH and NSF. We thank Dr. Edward Kean and his colleagues from the Department of Ophthalmology for their generous assistance.

REFERENCES AND NOTES

1. a) Part 8 of the series: Spectral Studies of Visual Pigments and Related Systems.
   b) Address correspondence to this author.
2. Present address: a) Department of Chemistry, University of Guelph, Ontario; b) Department of Chemistry, University of Southern Illinois,

Carbondale; c) The Lubrizol Corporation, Wickliffe, Ohio; d) Department of Chemistry, University of Alabama, Birmingham, Alabama.

3. H. Davson, Physiology of the Eye, Fourth Edition, Academic Press: New York & San Francisco, 1980.

4. The noun **scotopia** came into the new Latin from the Greek **skotos** which means **darkness**. Scotopia is defined by the American Heritage Dictionary as the "ability to see in dim light". **Scotoma** is an area of pathologically diminished vision within the visual field.

5. **Photopia** is the adaptation of the eye to light: daylight vision. **Photophobia** is the abnormal intolerance of light.

6. For recent reviews see: (a) R. Uhl & E.W. Abrahamson, Chem.Rev. 981, $\underline{81}$, 291; (b) R.R. Birge, Ann. Rev. Biophys. Bioeng. 1981, $\underline{10}$, 315.

7. J. W. Shriver, G. D. Mateescu, R. Fager, D. Torchia, E.W. Abrahamson, Nature(London), 1977, $\underline{270}$, 271, and references therein.

8. J. W. Shriver, Ph.D. Thesis, 1977, Case Western Reserve University, Cleveland, Ohio.

9. (a) R.A. Marton & G.A.J. Pitt, Biochem J.1955, $\underline{59}$, 128; (b) A. Kropf & R. Hubbard, Ann.N.Y.Acad. Sci. 1958, $\underline{74}$,266; (c) D. Bownds,Nature (London),1967, $\underline{216}$,1178;(d) R.Hubbard,Nature(London),1969,$\underline{221}$, 432; (e) L.Rinini, R.G. Kilpon & D. Gill, Biochem. Biophys. Res. Commun. 1970, $\underline{41}$, 492; (f) A.Lewis, R.S. Fager & E.W.Abrahamson, J. Raman Spectrosc. 1973,$\underline{1}$,465;(g) A.R.Oseroff & R.H. Callender,Biochemistry, 1974,$\underline{13}$,4243; (h)R.Mathies & L.Stryer, Proc.Natl.Acad.Sci. USA, 1976,$\underline{73}$, 2169;(i) R.H. Callender, A. Doukas, R. Crouch & K.Nakanishi, Biochemistry, 1976,$\underline{15}$, 1621; (j) F. Siebert and W. Mantele, Eur. J. Biochem. 1983, $\underline{130}$, 565; M. Stockburger, W. Klussmann, H.Gatermann, G. Massig and R. Peters, Biochemistry, 1980,$\underline{8}$, 4886.

10. The existence of a nonprotonated or hydrogen bonded Schiff base was suggested by: (a) K. Van der Meer,J.J.C.Mulder & J.Lugtemburg, Photochem. Photobiol. 1976, $\underline{24}$, 363; (b) K. Peters, M.L. Applebury & P. Rentzepis, Proc. Natl. Acad. Sci. USA 1977,$\underline{74}$, 3119; (c)F. I. Harosi, J. Favrot, J. M.Leclercq, D.Vocelle & C.Sandorfy, Rev. Can. Biol.1978,$\underline{37}$,257; (d)J.Favrot, J.M. Leclercq, R. Roberge, C. Sandorfy & D. Vocelle, Chem. Phys. Letters,1978,$\underline{53}$,433; (e) F. Siebert & W. Mantele, Biophys. Struct. Mech. 1980, $\underline{6}$, 147; (f) J. Leclercq, J.M. Leclercq & C. Sandorfy, Chem. Phys. 1982,$\underline{91}$,246,and references therein; (g) P. Dupuis,

F. I. Harosi, C. Sandorfy, J. M. Leclercq & D. Vocelle, Rev. Can. Biol. 1980, 39,247.

11. M. Karplus & J.A.Pople, J.Chem.Phys. 1963,38,2803.

12. G. D. Mateescu, W.G. Copan, D.D. Muccio, D.V. Water-
    hous & E.W. Abrahamson, Proc.Int.Symp.Synt. and
    Appl. of Isotop. Label. Compds., W.P. Duncan
    and A.B. Susan (Eds.), Elsevier: Amsterdam, 1983.

13. W.G. Copan, Ph.D. Thesis, 1982, Case Western
    Reserve University, Cleveland, Ohio.

14. (a) G. Wald & P.K. Brown, Proc. Natl. Acad. Sci.
    1950, 36, 84;(b) R. Hubbard & G. Wald, J. Gen.
    Physiol. 1952, 36, 269.

15. (a) D. Oesterhelt & W. Stoeckenius, Nature New
    Biol. 1971,233,149; (b) J.J. Pettei, A.P. Yudd,
    K. Nakanishi, R.Henselman & W. Stoeckenius, Bio-
    chemistry, 1977,16, 1955.

16.(a) G.A. Olah and G.D. Mateescu,J. Am. Chem. Soc.
    1970,92,1430; (b) H. Spiesecke and W.G. Schneider,
    Tetrahedron Lett. 1961,14,468.

17. (a)J.W. Shriver, E.W.Abrahamson, and G.D. Mateescu,
    J. Am. Chem. Soc. 1976, 98, 2407; (b) Y. Inoue,
    Y. Tokito, R. Chojo and Y.Miyoshi,ibid.1977,99,5592.

18. J.W. Shriver, G.D. Mateescu, E.W. Abrahamson,
    Biochemistry, 1979, 18, 4785.

19. (a) R. Rowan,III & B.D. Sykes, J. Am .Chem. Soc.
    1974,96, 7000; (b)R.Rowan, III, A. Warshel, B.D.
    Sykes & M. Karplus,Biochemistry 1974, 13, 970.
    (c) Y. Inoue, A. Takahashi, Y. Tokito, R. Chujo,and
    Y. Miyoshi, Org. Magn. Reson.1974,6,487; ibid. 1975
    7,485.

20. D.M. Grant & B.V. Cheney, J. Am. Chem. Soc. 1967,
    89, 5315.

21. R. Maties & L. Stryer, Proc.Natl.Acad.Sci.U.S.A.
    1976, 73, 2169.

22. See also J. Lugtenburg and coworkers, this volume.

23. D. D. Muccio, W. G. Copan, E. W. Abrahamson, and
    G. D. Mateescu, Biochemistry, submitted for publi-
    cation.

24. B.Becher & J.Y.Cassim, Biophys.J., 1976, 16,1183.

25. D. Oesterhelt & W. Stoeckenius, Proc. Natl Acad.
    Sci. USA 1973, 70, 2853.

26. (a) M. Arnaboldi, M. Motto, K. Tsujmoto, N.
    Balogh-Nair, K. Nakanishi, J. Am. Chem. Soc. 1979,
    101, 7082. (b) B. Honig, V. Dinur, K. Nakanishi,
    V. Balog-Nair, M. Gawinowicz, M. Arnaboldi, M.
    Moto, ibid. 1979, 101, 7084; (c) T. Kakitani and
    H. Kakitani, J.Phys.Soc.Japan,1977,44,1287;
    H. Kakitani, T. Kakitani, B. Honig and R.H. Cal-
    lender,Biophys.J.1982a,37,228a.

27. E.A. Dratz,W. Gartner, D. Oesterhelt, W.S. Veeman,

Biophys. J. 1983, 42, 12a; see also, solution results by A. Yamaguchi, T. Unemoto, and I. Ikegami,
Photochem.Photobiol.1981,33,511.
28. D. D. Muccio, W. G. Copan, E. W. Abrahamson and G.
    D.Mateescu, Fed. Proc. Fed. Am. Soc. Exp. Biol.
    1980,39, 2096.
29. D. D. Muccio, W. G. Copan, W. W. Abrahamson and G.
    D. Mateescu, Org.Magn.Resn., to be published.
30. G. S. Harbison, J. Herzfeld, R. G. Griffin,
    Biochemistry 1983, 22, 1.
31. C. N. Rafferty, H.Shichi, Photochem. Photobiol.,
    1981, 33, 229.
32. (a) Y.A. Ovchinnikov, N.G. Abdulaev, M.Y. Feigina,
    I.D. Artamnov, A.S. Zolotarev, M.B. Costina, A.S.
    Bogachuk, A.I. Moroshinkov, V.I. Martinov, & A.B.
    Kudelin, Bioorg. Khim. 1982, 8, 1011;
    (b) P.A. Hargrave, J.H. McDowell, D.R. Curtis, J.
    K. Wang, E. Juszczczak, S.L. Fong, J.K. Mohana Rao
    and P. Argos, Biophys. Struct. Mech. 1983,9,235.
    (c) E.A. Dratz and P.A. Hargrave, TIBS, 1983, 128.
    (d) Y.A. Ovchinnikov, N.G. Abdulaev, M.Y. Feigina,
    A.V. Kiselev, N.A. Lovanov, FEBS Lett. 1979, 100,
    219; (e) H.G. Khorana, G.E. Gerber, W.C. Herlihy,
    C.P. Gray, R.J. Anderegg, K. Nihei, and K. Biemann
    Proc. Natl. Acad. Sci. USA, 1979,76, 5046.

HIGH RESOLUTION ^{1}H NMR STUDIES OF MONONUCLEOTIDES WITH METALS

T. THEOPHANIDES and M. POLISSIOU[†]

Département de Chimie, Université de Montréal
C.P. 6210, Succ. A, Montréal, Québec, H3C 3V1   CANADA

ABSTRACT

Proton nuclear magnetic resonance spectra (^{1}H NMR) of deoxy-guanosine-5'-monophosphate 5'-dGMP and guanosine-5'-monophosphate (5'-GMP) in D_2O solutions with $MgCl_2$ and with cis-Pt(NH$_3$)$_2$Cl$_2$ or trans-Pt(NH$_3$)$_2$Cl$_2$ salts (cis-DDP, trans-DDP) have been obtained and will be discussed. The ^{1}H NMR spectra show perturbations of the H8 proton which is shifted to lower fields in the presence of the metal ions employed, indicating that the metal is coordinated to the N7 site of guanine. The coordination of the magnesium atom to these mononucleotides forming hydrated complexes, [Mg(H$_2$O)$_5$5'-(dGMP)]$^{++}$ and [Mg(H$_2$O)$_5$(5'-GMP)]$^{++}$ is a dynamic and reversible fixation of magnesium to the N7 site of the base, since it depends on the concentration of magnesium chloride, while the cis- and trans-DDP formed inert coordination compounds of the formulae, cis-[Pt(NH$_3$)$_2$(5'-GMP)$_2$]Cl$_2$ and trans-[Pt(NH$_3$)$_2$-(5'-GMP)$_2$]Cl$_2$ in which the platinum moiety is blocking the N7 site by forming an inert covalent bond with the nitrogen atom (similarly with 5'-dGMP).

The coordination of the metal to the N7 site also caused shifts of the sugar protons which in conjunction with the coupling constant data suggest that the purine ring has been reoriented when the metal ion is fixed at the N7 site of the guanine base. The ^{3}E$\leftrightarrow$E$_3$ sugar equilibrium is slightly shifted to the left, favoring the ^{3}E (3'-endo) conformers and the gg$\leftrightarrow$gt/tg and g'g'$\leftrightarrow$g't'/t'g' equilibria are also perturbed favoring the gg and g'g' rotamers.

C. Sandorfy and T. Theophanides (eds.), Spectroscopy of Biological Molecules, 291–301.
© 1984 by D. Reidel Publishing Company.

INTRODUCTION

One of the most important features of nucleic acids is the sugar-phosphate backbone flexibility in the double helix (1). This is dramatically shown (2) by the discovery of Z-DNA with its left helix conformation and from the other structures of DNA, i.e., A-DNA and B-DNA (3). In these DNA structures the parameters defining the diameter of the helix, the base-base distance in the sequence of bases and the half turn distance are different in the three conformations (4) (see Fig. 1). In aqueous solutions these macro-molecules are very flexible and they may exist in a whole family of conformations giving equilibria with different populations of conformers with disordered and dynamic structures which may be altered by the presence of the counterions or the ionic environment, in general, and other factors such as temperature, concentration and pH. Divalent counterions, for instance, Mg^{++}, Ca^{++} and Ba^{++} might be used to bind negatively charged phosphate groups in the helix. These metal ions according to their concentration and nature can stabilize particular conformations. Some of these conformations could be related to the different pucking possibilities of the sugar ring resulting in a variety of helical structures, which are stabilized by hydrogen bonding with free water molecules or metal-coordinated water molecules, i.e., $Na(H_2O)_6^+$, $K(H_2O)_6^+$ and $Mg(H_2O)_6^{++}$ (1). From X-ray structural analysis it can be shown that the counterions are always hydrated and coordinated or hydrogen bonded through coordinated water molecules to specific sites of the bases or bound to the exocyclic functions of the bases such as carbonyl, amine, and may also be linked directly or indirectly to the phosphate oxygens of the backbone. From all these interaction forces the electric charges of the counterions are predominent compared to H-bonding, coordination of water molecules to the metals, stacking and dispersion forces.

The dynamic properties of nucleic acids stem from the unique geometrical characteristics of the five-membered sugar ring (5) and the exocyclic chain $-O-CH_2-O-PO_2^--O-$ which links the two ends of a sugar and is attached to the two bases of a strand (6). The presence or absence of a hydroxyl group in the position C'2 of the sugar molecule determines the two main classes of nucleic acids, namely, DNA and RNA, which have different structural properties and biological functions. The DNA molecule does not have a hydroxyl group on its sugars at C'2, whereas in RNA there is a hydroxyl group at the C'2 position. In the polymer backbone and in the presence of counterions this may play an important role. The ring mobility and the influence of the metal ions as well as the puckering of the sugar molecules will determine to a great extent the structural aspects of DNA and RNA.

Flexibility in these molecules results also from the possibility of restricted rotation around the C'4-C'5 (ψ) and C'5-O5 (ϕ) bonds (see Fig. 2). It is known that flexibility in

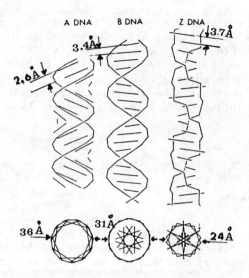

Fig. 1.   Model Conformational Isomers of DNA.   Both views are
shown, i.e., perpenticular and along the helical axes (adapted
from Ann. Rev. Biochem., 1982, 51: 395-427, consult for more
details)

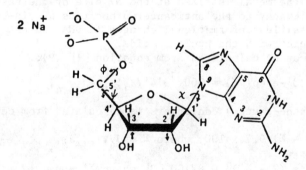

Fig. 2.   The Numbering and Molecular Structure of G⁵'pNa₂ Salt
in the anti, gg Conformation

bio-macromolecular structures is very important for their biological activity. The results obtained by NMR in D_2O solutions
with the nucleic acid constituents 5'-dGMP, 5'-GMP and their
interaction with magnesium chloride or with cis-dichloroammine-
platinum(II) which is an antitumor drug will be discussed here.
The trans-isomer does not have any antitumor activity (7,8).
Both isomers have planar structures shown first by Alfred Werner

in 1893. In this report the interactions of these two isomers
and magnesium chloride with the mononucleotides 5'-dGMP and
5'-GMP has been examined and studied by ^{1}H NMR spectroscopy.

RESULTS AND DISCUSSION

N7 Base Binding
        The results obtained from the above study have been used
to investigate the following conformational equilibria:
        syn $\Longleftrightarrow$ anti   (χ, rotation around the N9-C'1 bond)
        $^3E \Longleftrightarrow {}^2E$ (flip-flap of the sugar)
        gg $\Longleftrightarrow$ gt, tg   (ψ, rotation around the C'4-C'5 bond)
        g'g'$\Longleftrightarrow$ g't', t'g'   (ϕ, rotation around the C'5-O'5 bond)
        The metal attack can change the above equilibria. We have
set out to investigate these conformational populations in water
solutions when the metal is fixed at the N7 site of the base or
to find any relevancy to the anticancer properties of cis-DDP
and suggest possible conformational changes in DNA and RNA when
these metals react with the nucleic acids.
        The % ^3E can be calculated from equation [1] (9);

$$\% \ 3'\text{-endo}(^3E) = 100 \ {}^J3'4'/J_{1',2'}+J_{3',4'} \qquad [1]$$

The % of gg(gauche-gauche) conformers is calculated from equation
[2]

$$\% \ gg = (13.7 - \Sigma)/9.7 \times 100, \text{ where } \Sigma = J_{4',5'}+J_{4',5''} \qquad [2]$$

and the % g'g' = $(25- \Sigma')/20.8 \times 100$ [3] where $\Sigma' = J_{5'p}+J_{5''p}$.

        From the above empirical equations one can calculate the
populations of the rotamers present by measuring the [1]H-[1]H
coupling constants (J). These populations are compared in Table
1 for magnesium (10) and for cis- and trans-DDP (11). Results of
metallation at N7 have been included and are compared with
protonation (12) and methylation (13). An interesting decrease
in gg population has been observed when the negative platinum

moiety $PtCl_3^-$ was fixed at N7 (14). This has not been observed
with cis- and trans-DDP. Interaction of magnesium at N7 is
milder than that of cis-DDP (see Table).

The interaction of cis-DDP with DNA in vitro and in vivo
was demonstrated by many laboratories. The antitumor drug,
cis-DDP is believed to attack DNA in order to kill the tumor
cells (15). The mechanism of the antitumor activity of cis-DDP,
and of this attack on DNA is not yet clear. There are several
NMR studies on the interaction of platinum compounds, in general,
with mononucleotides and oligonucleotides(L) (11,16) which point
out that in all these complexes the main target of platinum is
the guanine base to which it is coordinated the cis-DDP through
the N7 nitrogen atom to form a complex with the two adjacent
guanine bases giving adducts of the formula $[cis\text{-}Pt(NH_3)_2\text{-}(N7\text{-}L)_2]^{++}$, as it was first shown by us in the reaction of
guanosine with cis-DDP resulting in a defined complex of the
formula, $[cis\text{-}Pt(NH_3)_2(N7\text{-}Guo)_2]Cl_2$ (17).

Cis-DDP reacts with DNA and denatures the double helix by
disrupting the Watson-Crick base pairs (18). What is the inti-
mate mechanism of this reaction is not yet known. Early spectroscopic studies (19) suggested that cis-DDP reacts with the
guanine bases in DNA, in particular with the N7 position and with
the nearby equally reactive site O6 position forming a specific
chelate. The platinum binding to O6 through chelation or bridging could be direct or indirect through hydrogen bonding by a
water (or ammine) molecule coordinated to platinum (19,20).

COMPLEX CHELATE O₆N₇   Complex Bridging O₆N₇   (NH₃)₂Pt-G at N-7

The binding of cis-DDP to N7 of DNA will undoubtedly disrupt
the hydrogen-bonding of the G-C pairs which could lead to the
observed denaturation of the double helix. Disruption or interaction of cis-DDP with the sugar-phosphate of DNA has also been
observed (21). The N7-N7 "chelation" of the cis-DDP with two
adjacent guanines of oligonucleotides or DNAs' is not surprising,
since the N7 of guanine is the most reactive site of the nucleotide for cis-DDP shown from early studies (8,17). The N7-N7
"chelation" has been further demonstrated by recent ^{1}H NMR
studies (22,23). We shall be concerned here with the study of
metal-molecule interactions, i.e., the fixation of the metal at a
particular site of the base, and with the study of the structures

of the mononucleotides 5'-dGMP and 5'-GMP and their metal adducts.

Fig. 3 shows ^{1}H NMR spectra of cis-DDP with guanosine
taken previously (17) in our laboratory. The effect of the
metal on the chemical shift of H8 is clearly shown, in the case
of inosine/xanthosine and guanosine. Previous studies (10) also
showed that the H8 resonance is shifted towards higher fields
by the addition of a 10 fold excess of $MgCl_2$ or by reaction with
$PtCl_4$ (14) or cis- and trans-DDP (11). This indicates that
binding of the metal occurs at N7. Platinum coordination to the
N7 atom of the guanine base is expected and it has been confirmed
by X-ray structural data (20). The NMR spectra were obtained
from approximately neutral solutions, however small variations
of the pH value may give slight changes in chemical shifts, due
to the protonation of the phosphate group whose pKa value is
about 6.0 (24). Reaction of cis-DDP with two equivalents of
5'-GMP or 5'-dGMP leads to the formation of the corresponding
substituted analogs of cis or [trans-Pt(NH$_3$)$_2$L$_2$]$^{++}$, where L =
5'-GMP and 5'-dGMP. Small amounts of intermediates have been
detected, probably [Pt(NH$_3$)$_2$LCl]$^+$ and [Pt(NH$_3$)$_2$L(H$_2$O]$^{++}$. The
overall reaction, however, is:

$$\text{cis- or trans-Pt(NH}_3)_2\text{Cl}_2 + 2L \rightarrow \text{cis or trans-[Pt(NH}_3)_2\text{L}_2]^{++}\text{Cl}_2$$

The intermediates [Pt(NH$_3$)$_2$ClL]$^+$ and [Pt(NH$_3$)$_2$Cl(H$_2$O)]$^{++}$ have
also been observed in the spectra. Greater amounts of these
intermediates were shown in the spectra of the trans-isomer
than in those of the cis-compound. The ^{1}H NMR spectra of both
cis- and trans-DDP complexes with 5'-GMP and 5'-dGMP show
considerable changes in the H5', H5" resonances which indicates
an interaction of Pt(NH$_3$)$_2$$^{++}$ with the phosphate group. The
above reaction was complete with cis-DDP, but not with trans-DDP.

The studies (10) with magnesium chloride showed that magnesium coordinates reversibly at the N7 atom of the guanine base.
This was also shown by X-ray crystal analysis (2). The conformation of the hexanucleotide molecule in Z-DNA crystals (2)
is interesting. In the crystals the two strands are stabilized
by spermine and magnesium cations. The two chains in the crystal
are similar except for the position G4pG5, where there is a
linkage with pentahydrated magnesium complexed at N7 of guanine
G6. The phosphate group which links G4 to G5 has been rotated
in such a manner as to bring its oxygens closer to the hydrated
magnesium atom and forms a hydrogen bond with one of the five
coordinated water molecules. The magnesium interaction with N7
is a dynamic and reversible linkage in solution (10). The
magnesium is bound indirectly to the phosphate, through a coordinated water molecule which is hydrogen bonded to the negatively charged phosphate oxygens (25). The coordination compounds of magnesium with 5'-dGMP and 5'-GMP molecules synthesized

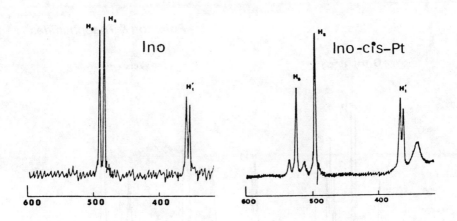

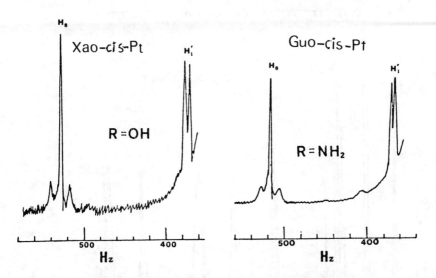

Fig. 3a. Proton Nuclear Magnetic Resonance (^{1}H NMR) Spectra of
the Nucleosides Inosine (Ino), Xanthosine (Xao) and Guanosine
(Guo) and their 1:2 Complexes with cis-Pt(NH$_3$)$_2$Cl$_2$ (cis-Pt).
The satellite bands are the ^{195}Pt couplings (adapted partly
from ref. 17).

Polissiou & Theophanides

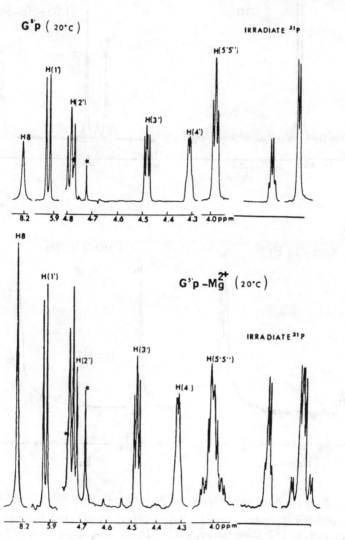

Fig. 3.b. Proton Nuclear Magnetic Resonance (^{1}H NMR) Spectra of of 5'-GMPNa$_2$ (0.02M) and Interaction with Magnesium Chloride (0.2M) in D$_2$O Solution (adapted from ref. 10).

at pH values of 7-9 indicate from infrared spectral analysis
that the magnesium atom is coordinated to N7 and that the
coordinated water molecules are hydrogen bonded to the phosphate
group, while at pH = 4-6 the coordination compound isolated and
characterized indicate direct coordination of magnesium to the
phosphate group evidenced from the splitting of the phosphate
infrared vibrations (26). This shows that magnesium reacts with
nucleotides, polynucleotides and nucleic acids differently in
acidic media or in near neutral conditions. At physiological pH
values magnesium interacts preferably with the N7 site and is
hydrogen bonded through a coordinated water molecule to the
phosphate group.

REFERENCES
† Permanent address Agricultural School of Athens, Votanikos, Athens, Greece.
1. E. Clementi in "Structure and Dynamics: Nucleic Acids and Proteins" Proceedings of the International Symposium on Structure and Dynamics of Nucleic Acids and Proteins. Eds. E. Clementi and R.H. Sarma, Adenine Press, P.O. Box 355, Guilderland, New York 12084, 321 (1982); ibid. M.U. Palma, 125.
2. A. Rich, Science, Vol. 211, 9, 172 (1981).
3. R.E. Dickerson, M.C. Kopka, and H.R. Drew, "Structure and Dynamics: Nucleic Acids and Proteins", Eds. E. Clementi and R.H. Sarma, Adenine Press, NY p. 149 (1982).
4. H.M. Wu, N. Dattagupta, and D.M. Crothers, Proc. Natl. Acad. Sci. USA, 78, 6806 (1981).
5. J.D. Dunitz, X-Ray Analysis and the Structure of Organic Molecules, Cornell University Press, Ithaca (1980).
6. M. Sandaralingam and E. Westhof, Biomolecular Stereo-dynamics I (R.H. Sarma, Ed.) Adenine Press, NY 301 (1981).
7. B. Rosenberg in "Nucleic Acid-Metal Ion Interactions", T.G. Spiro (Ed.) Weily, New York, 1, 1980.
8. A.W. Prestayko, S.T. Crooke, and S.K. Carter (Eds.) "Cis-Platin Status and New Developments", Academic Press, New York, 1980; T. Theophanides, Chemistry in Canada, 32, 30 (1980).
9. C.H. Lee, F.S. Erza, N.S. Kondo, R.H. Sarma, and S.S. Danyluk, Biochemistry, 15, 3627 (1976).
10. T. Theophanides and M. Polissiou, Inorg. Chim. Acta, 56, L1 (1981); M. Polissiou and T. Theophanides, Biomolecular Stereodynamics II (R.H. Sarma, Ed.) Adenine Press, NY 497 (1981).
11. M. Polissiou, M.T. Phan Viet, M. St-Jacques, and T. Theophanides, to be published.
12. Tran-Dinh Son and W. Gushlbauer, Nucl. Acids Research, 2, 873 (1975).
13. C.H. Kim and R.H. Sarma, J. Amer. Chem. Soc., 100, 1571 (1978).

14. M. Polissiou, M.T. Phan Viet, M. St-Jacques, and T. Theophanides, Can. J. Chem., 59, 3297 (1981).
15. J.J. Roberts and A.J. Thomson, Progr. Nucleic Acid Res. Molec. Biol., 22, 79 (1979).
16. J.P. Girault, G. Chottard, J.-Y. Lallemand, J.-C. Chottard, Biochemistry, 21, 1352 (1982); A.P. Marcelis, J.H.J. den Hartog, J. Reedijk, J. Amer. Chem. Soc., 104, 2664 (1982).
17. P.C. Kong and T. Theophanides, Inorg. Chem., 13, 1167 (1974).
18. J.K. Barton and S.J. Lippard, in "Nucleic Acid — Metal Ion Interactions", T.G. Spiro, Ed. (Wiley, New York, 1980) pp. 31-113.
19. M.M. Millard, J.P. Macquet, and T. Theophanides, Biochim. Biophys. Acta, 402, 166 (1975); Idem, Bioinorganic Chem., 5, 59 (1975).
20. R.G. Gellert and R. Bau, in "Metal Ions in Biological Systems", H. Siegel, Ed., Marcel Dekker, New York, Vol. 8, p. 1 (1979).
21. T. Theophanides, Appl. Spectrosc., 35, 461 (1981).
22. J.-C. Chottard, J.-P. Girault, G. Chottard, J.Y. Lallemand, and D. Mansey, J. Amer. Chem. Soc., 102, 3565 (1980).
23. J.P. Caradonna, S.J. Lippard, M.J. Gait, and M. Singh, J. Amer. Chem. Soc., 104, 5793 (1982).
24. K.H. Scheller, V. Scheller-Krattinger, and R.B. Martin, J. Amer. Chem. Soc., 103, 6833 (1981).
25. M. Manfait and T. Theophanides, Magnesium, In Press.
26. A. Tajmir-Riahi and T. Theophanides, Can. J. Chem. 61, 1813 (1983).

TABLE

Conformational Parameters of Guanosine-5'-Monophosphate with
Protonation, Methylation and Metallation at the N7 Site of the Guanine Base

Compound[a]	pD	T(°C)	Ribose ^3E	% Conformational Isomers		ref.
				C(4')-C(5') gg	C(5')-O(5') g'g'	
5'-GMP	8.3	20	36	65	76	10
N7-H$^+$-5'-GMP	0.9	20	57	87	70	9,13
N7-CH$_3^+$-5'-GMP	7.0	20	54	97	77	9
N7-PtCl$_3^-$-5'-GMP	7.8	20	51	36	72	14
N7-Mg^{++}-5'-GMP	7.6	20	44	75	72	10
N7-cis-Pt^{++}-5'-GMP[b]	7.4	43	50	69	76	11
N7-trans-Pt^{++}-5'-GMP[b]	6.9	35	39	86	80	11

a: N7- indicates fixation at the N7 position of guanine of the various cations;
b: similar results have been obtained with 5'-dGMP.

RESONANCE RAMAN DETERMINATION OF RETINAL CHROMOPHORE STRUCTURE
IN BACTERIORHODOPSIN

Richard A. Mathies

Chemistry Department
University of California
Berkeley, CA   94720

ABSTRACT: Resonance Raman scattering is a versatile technique
for studying the mechanism of proton-pumping in bacteriorhodopsin.   This method is particularly valuable when spectra of isotopic derivatives of retinal are used to assign the vibrational
features.   Isotopic substitution provides a basis for the *interpretation* of the spectra and the *in situ* determination of chromophore structure.   To accomplish this it is important to obtain
reliable spectra and to understand the bases behind the spectral
interpretations.   First, current techniques for obtaining high-
quality spectra of bacteriorhodopsin and its intermediates are
summarized, and their relative advantages and disadvantages are
discussed.   Then the vibrational structure of all-*trans* and 13-
cis retinal chromophores is outlined in light of the recent normal coordinate analyses on these isomers.   With this foundation,
the spectra of light-adapted bacteriorhodopsin, dark-adapted
bacteriorhodopsin, K_{610}, M_{412}, and O_{640} are presented.   The
"isotopic fingerprint" method is used to examine the geometric
configuration of retinal in each intermediate.   Finally, a model
summarizing current knowledge about retinal structural changes
during proton pumping in bacteriorhodopsin is presented.

1.   INTRODUCTION

      Proton-pumping in bacteriorhodopsin (BR) is driven by the
photochemical 13-*trans*→13-*cis* isomerization of its retinal prosthetic group.   The all-*trans* retinal chromophore in BR is bound
to lysine 216 by a protonated Schiff base linkage.   Light absorption causes isomerization about the $C_{13}=C_{14}$ bond and the deprotonation of the Schiff base nitrogen (see Figure 1).   This results

303

C. Sandorfy and T. Theophanides (eds.), Spectroscopy of Biological Molecules, 303–328.
© *1984 by D. Reidel Publishing Company.*

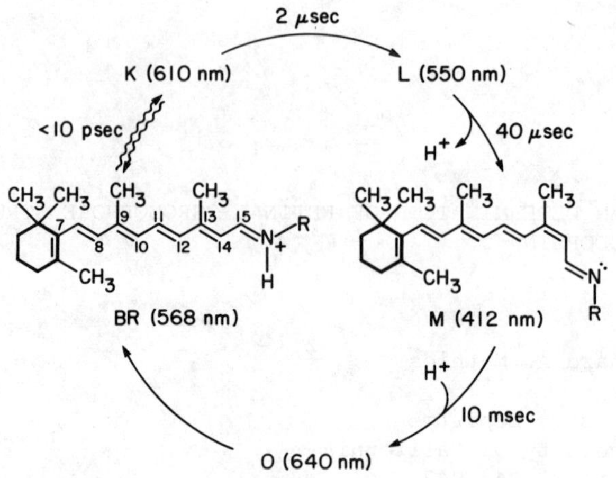

Figure 1. Proton-pumping photocycle of bacteriorhodopsin. Absorption maxima and room temperature decay times are indicated. Light-adapted BR$_{568}$ contains an all-*trans* protonated Schiff base chromophore, while M$_{412}$ contains an unprotonated 13-*cis* Schiff base. In the dark, BR$_{568}$ slowly converts to dark-adapted bacteriorhodopsin which contains an approximately equal mixture of all-*trans* and 13-*cis* protonated Schiff base chromophores (denoted BR$_{568}$ and BR$_{548}$, respectively).

in the vectorial transport of protons across the cell membrane. The resulting electrochemical gradient is used by *Halobacterium halobium* to synthesize ATP. BR has the additional unique feature of operating in a photocycle--the pigment returns to light-adapted BR$_{568}$ *in the dark* in ∿10 msec. This makes BR very convenient for biophysical studies. It also suggests that when the molecular mechanism of this pump is determined, it may be possible to harness this efficient and photochemically stable protein as a useful light-energy converter. Toward this end we are developing the resonance Raman technique as a probe of the molecular mechanism of this proton pump.

An unusually wide variety of techniques have been used to study the structure of the retinal chromophore in BR (for reviews see 1,2). One of these, chemical extraction, has shown that BR$_{568}$ contains an all-*trans* chromophore while L$_{550}$ and M$_{412}$ contain 13-*cis* chromophores (3,4). However, this approach can provide only the crudest information on chromophore structure because the chromophore must be removed from the protein, thereby destroying the protein-chromophore interactions that are vital for the

pigment's function. Also, it is very difficult to apply chemical
extraction to transient intermediates. Raman spectroscopy, however, provides an *in situ* probe of chromophore structure that does
not perturb the pigment and that is applicable to a wide range of
transient species.

Raman spectroscopy does have its challenges. First, we are
using an optical technique to study a photochemical system. This
requires that proper care be exercised to ensure that the act of
measurement does not perturb the composition of the sample. Solutions to this problem are discussed in METHODS. Second, the vibrational spectra are so rich in information that their interpretation is non-trivial. To address this challenge we have been working with Prof. Johan Lugtenburg and co-workers at Leiden University.
They have synthesized an extensive series of isotopic retinal derivatives that have been used to assign and interpret the vibrational
spectra of the retinal isomers and of the retinal-containing pigments. The assignments that have been developed for the all-*trans*
and 13-*cis* isomers are discussed in detail in the third section.
The general vibrational features and trends that are useful in
the interpretation of pigment spectra will be emphasized. In the
final section the spectra of bacteriorhodopsin and its intermediates are presented. The utility of isotopic substitution is illustrated by employing the "isotopic fingerprint" method to determine the chromophore configuration in the K_{610} and O_{640}
intermediates. In addition, features of the spectra that report
on chromophore-protein interactions are identified. The experimental and interpretive methods presented here should be directly
useful in studies of visual pigments and may also be valuable in
studies of heme and chlorophyll-based protein systems.

2.   METHODS

It is the selective resonance enhancement of scattering from
the chromophore that makes the Raman technique so useful in biophysical studies of BR. For example, the cross-section for the
ethylenic line of BR_{568} with 568 nm excitation is more than 10^5
times larger than that for the strong 992 cm^{-1} line in benzene (5).
This allows us to obtain high signal-to-noise spectra with relatively small amounts of sample and short experimental times.
Also, the resonance effect allows us to select the scattering
species by our choice of excitation wavelength. By exciting in
the main $\pi \rightarrow \pi^*$ absorption band of the retinal chromophore we selectively enhance its scattering over that of the protein side chains
and peptide backbone. In addition, by exciting in a spectral region where just one intermediate absorbs, the scattering from that
BR intermediate can be *selectively* enhanced. However, the cross-
section for Raman scattering into all modes is only $\sim 1 \times 10^{-6}$ Å^2/
molecule, which is much smaller than the cross-section for

absorption (2.4 $\overset{o}{A}{}^{2}$/molecule at 568 nm [5]). Therefore, for photochemical systems such as BR, experiments must be designed so that
the Raman probe beam does not photochemically perturb the initially prepared pigment sample. A wide variety of such techniques
have been described in a recent volume of *Methods in Enzymology*
(6). Three of the most useful experimental configurations are
discussed below.

2.1 Rapid-Flow Experiments

The rapid-flow technique was first developed for Raman studies
of visual pigments (7,8). The photolabile sample is recirculated
through a jet stream nozzle as depicted in Figure 2. If the velocity (ν, cm/sec) is sufficiently high, the photochemical products
formed by the probe laser beam will be swept out of the irradiated
volume before they can accumulate to a significant level. The
rate of photoproduct production depends on the laser wavelength
(λ), the laser power (P, photon/sec), the focused beam size (ω,
the $1/e^2$ radius in cm), and the photochemical quantum yield (Φ).
With these definitions the fraction of molecules F that are photochemically converted during one pass of the sample through the
laser beam is given by (7):

$$F = 3.824 \times 10^{-21} \cdot \frac{P \; \varepsilon(\lambda) \; \Phi}{\sqrt{\pi} \; \omega \; \nu}$$

With this equation the optimum conditions for a rapid-flow experiment can be easily selected. Clearly it is advantageous to have
as high a flow rate as possible. Velocities from 4 to 20 m/sec
are practical. Typically we work with $F \overset{\Delta}{=} 0.1$ which corresponds
to an *effective* photolysis level of 5%. For a bacteriorhodopsin
sample flowing at 500 cm/sec and a $1/e^2$ laser beam radius of 20 µm
(λ=568 nm, ε=62,700 $\underline{M}^{-1}cm^{-1}$, Φ=0.3), the laser power must be
2.5×10^{15} photon/sec or $\sim$1 mW. This low laser power clearly illustrates the difficulty of obtaining a low-photoalteration spectrum.

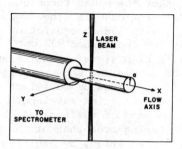

Figure 2. Jet stream apparatus for obtaining rapid-flow resonance
Raman spectra of photolabile molecules [from Mathies *et al.* (7)].

Note that for some rapid-flow experiments, a significant S/N advantage can be obtained by cylindrically rather than spherically focusing the probe laser beam because higher laser powers can be employed (9).

Because the sample is being recirculated one must also worry about the rate of photolysis or bleaching of the entire pool. This is described by the bulk photoalteration parameter (10):

$$F_{bulk} = 3.824\text{x}10^{-21} \cdot (2\ a\ P\ \Phi\ \varepsilon(\lambda)T)/V$$

Here a is the jet stream radius, T is the overall irradiation time and V is the sample volume. In addition to the obvious linear dependence on P, ε and T, it is evident that a large sample volume should be used if bulk photoalteration is a problem.

2.2 Time-Resolved Pump-Probe Experiments

The rapid-flow technique is limited to species that can be prepared or isolated in bulk quantities. To study transient intermediates, time-resolved techniques must be employed. The most extensive time-resolved methods have been developed by El-Sayed and co-workers (11,12 and references cited therein). The best design is to *prepare* a state with an actinic or *pump* beam which initiates the photochemistry. Then, after a suitable delay (see Figure 1), a *probe* beam is used to excite the Raman scattering from the desired intermediate. The primary means for assigning Raman lines to an intermediate is their temporal dependence. They should appear in the correct time-window. A second important method for assigning the scattering from an intermediate is to use a probe wavelength that selectively enhances the scattering from that intermediate. This requires that one perform a "two-color" experiment that is simple in concept but often difficult in practice. The difficulty arises because the photoalteration of *both* beams must be considered. The pump beam must be strong enough to get a significant fraction of the molecules photocycling ($\hat{F}=1$) but not so strong as to produce deleterious photochemistry. The probe beam must, of course, be low-photoalteration. Such two-color experiments have been used to obtain spectra of M_{412} and O_{640} (9,13). Also, a nsec time-resolved, two-color experiment has recently been performed on the K intermediate (14).

Figure 3 describes the apparatus that we have used to obtain spectra of BR intermediates on the 20 μsec-20 msec time scale. A 514-nm pump beam from an Ar^+ laser is cylindrically or spherically focused on a flowing stream of BR so that the entire sample is irradiated. The flowing sample is then allowed to evolve as it flows downstream to a 412-nm probe beam from a Kr^+ laser for the M_{412} experiments (or 752 nm from a Kr^+ laser for the O_{640} experiments). The time-window is selected by translating the pump beam along the capillary with a micrometer.

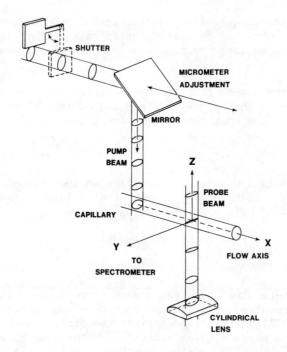

Figure 3. Apparatus for obtaining two-color, time-resolved Raman spectra of BR intermediates. The useful time range is 20 μsec- 20 msec [from Smith *et al.* (13)].

2.3 Photostationary Steady-State Experiments

    Time-resolved techniques can be extended to study the very fast early intermediates of BR. Preliminary experiments on the nsec and psec (10^{-12} sec) time scale have been reported (14,15). These experiments are technically difficult because a pulsed laser must be built and because the duty cycle of the experiment is often quite low so the S/N is poor. An alternative approach involving the production of low-temperature photostationary steady-states permits structural studies of the primary photoproduct with technology available to almost all laboratories. The decay of K_{610} to L_{550} can be blocked by cooling to 77°K (see Figure 1). Under these conditions a steady-state mixture of K_{610} and BR_{568} can be produced because the BR→K transition is photoreversible. Based on the earlier discussion, the obvious two-color experiment is to pump the low-temperature BR sample at 514 nm to produce K_{610}, and then to probe this sample with a red laser (676 or 752 nm) that will selectively enhance the K scattering. This experiment fails when two coaxial beams are used because the pump beam excites fluorescence from BR_{568} in the steady-state mixture that totally obscures the Raman scattering from the probe beam. This problem has forced most workers to use a less than optimal probe wavelength (16,17). The solution to this problem is to devise an apparatus that allows the spatial separation of the pump and probe events so that the pump beam-induced fluorescence is not "seen" by the Raman spectrometer.

Figure 4 presents a schematic of apparatus designed to spin BR samples at $77^{\circ}K$ (18). The light-adapted purple membrane sample is spread into a circular groove cut into a conical copper tip. This tip is mounted in a liquid-N_2 dewar on a shaft that is spun at 1700 rpm. The pump beam enters on one side of the dewar and produces K_{610} along with the BR_{568} fluorescence. This sample is then rotated to the other side of the dewar where a red probe beam is used to excite the scattering from K_{610}. Note that this method allows the effective use of much more sample than would otherwise be possible. In this way thermal and photochemical damage to the sample can be minimized. For example, we have used this apparatus to obtain much better spectra of rhodopsin and its primary photoproduct than would otherwise have been possible.

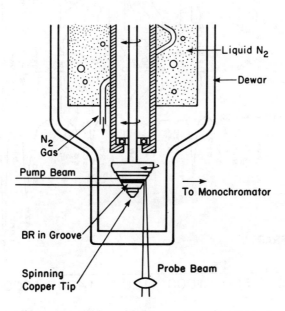

Figure 4. Dewar for obtaining spectra of spinning BR samples at $77^{\circ}K$ [from Braiman and Mathies (18)].

3.   ALL-TRANS AND 13-CIS VIBRATIONAL SPECTRA

      The interpretation of the bacteriorhodopsin resonance Raman
spectra has awaited the vibrational analysis of the all-*trans*
and 13-*cis* retinal isomers.   Recently Curry *et al.* have published
a detailed analysis of these isomers based on the isotopic shifts
observed in an extensive series of ^{13}C and ^{2}H derivatives (19,20).
The Raman and infrared spectra for all-*trans* retinal are given in
Figure 5.   The empirically derived assignment for each vibrational
line is listed in the figure.   The body of isotopic data used in

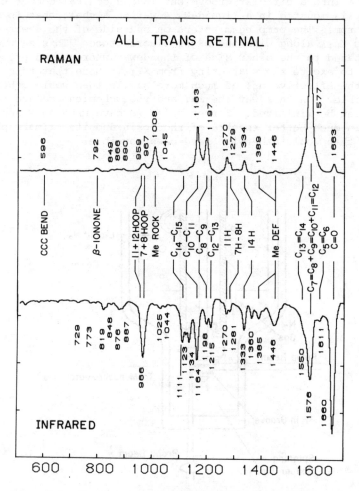

Figure 5.   Vibrational assignments of Raman and infrared spectra
of all-*trans* retinal [adapted from Curry *et al.* (19)].   The dominant internal coordinate in each normal mode is indicated.   The
atomic numbering of retinal is given in Figure 1.

this analysis is much too extensive to reproduce here. However, it is instructive to give a few examples so that the reader can understand *how* to use isotopes to assign retinal and pigment vibrational spectra.

3.1 All-*Trans* Retinal

In Figure 6 we compare unmodified all-*trans* retinal with its $10,11-^{13}[C]$ derivative. The 1163 cm^{-1} line shifts down to 1147 cm^{-1}, confirming that the 1163 cm^{-1} line is predominantly due to the localized $C_{10}-C_{11}$ stretch. This substitution would also be expected to cause the $C_9=^{13}C_{10}$ and $^{13}C_{11}=C_{12}$ stretches to shift down. The drop of the 1577 cm^{-1} line to 1556 cm^{-1} argues that the 1577 cm^{-1} line involves significant motion of C_{10} and C_{11}. Finally, the 2 cm^{-1} downshift of the 1270 cm^{-1} line supports its assignment to the $C_{11}-H$ in-plane hydrogen rock. The latter assignment is confirmed by the 11-D all-*trans* spectrum presented in Figure 7. Here the $C_{11}-H$ rock shifts from 1270 cm^{-1} to 966 cm^{-1} as evidenced by the dramatic increase of intensity at 966 cm^{-1}. The vinyl hydrogen motion is also mixed with the carbon skeletal modes. For

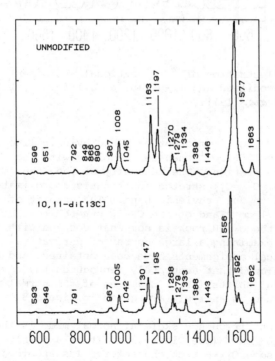

Figure 6. Illustration of $^{13}[C]$ vibrational assignments with unmodified and $10,11-^{13}[C]$ labeled all-*trans* retinal [adapted from Curry *et al.* (19)].

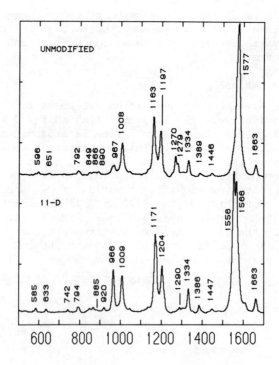

Figure 7.   Illustration of ^{2}H vibrational assignments with Raman spectra of unmodified and 11-D labeled all-*trans* retinal [adapted from Curry *et al*. (19)].

example, the $C_{11}=C_{12}$ component of the ethylenic line shifts down by $\sim$20 cm^{-1} to 1556 cm^{-1} when its coupling with the C_{11}-H rock is removed by deuteration. The intensity of the 1556 cm^{-1} line suggests that the $C_{11}=C_{12}$ stretch contributes approximately half of the intensity to the ethylenic band. The upshift of the $C_{10}-C_{11}$ stretch to 1171 cm^{-1} and of the C_8-C_9 stretch to 1204 cm^{-1} demonstrates that the C_{11}-H rock is somewhat coupled with these modes as well. By examining a large number of derivatives a complete set of empirical assignments have been obtained, and a force field has been developed that accurately reproduces these vibrational frequencies and shifts (19). Examples of the normal modes that result from such an analysis are given in Figure 8.

The highest frequency skeletal normal mode is the carbonyl stretch at 1663 cm^{-1} (see Figure 5). It is a fairly pure C=O stretch with some C_{15}-H rock character. Its atomic displacements are sufficiently simple that it has not been drawn in Figure 8. The ethylenic stretching region contains five normal modes, one for each possible combination of the five C=C internal stretching

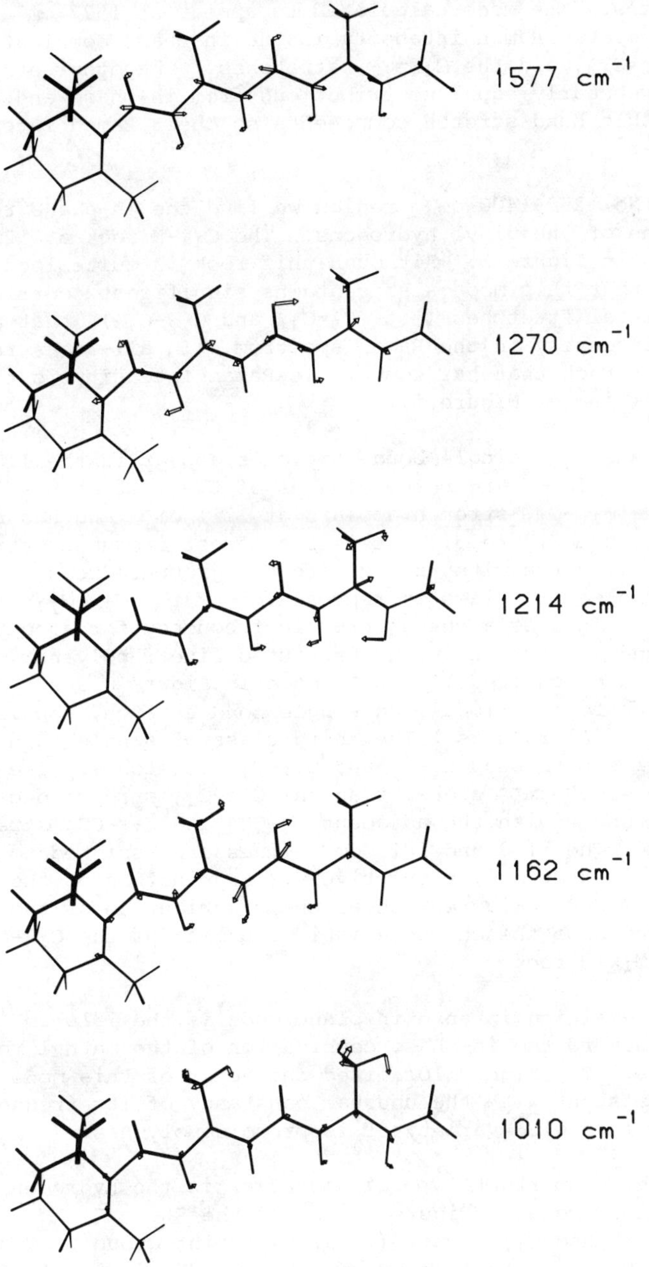

Figure 8. Examples of the mass weighted atomic displacements in the prominent in-plane normal modes of all-*trans* retinal. The mass weighting magnifies the carbon motion by $\sqrt{12}$. The hydrogen amplitudes are ~8 times their rms zero point values.

coordinates. The mode calculated to appear at 1577 cm^{-1} with the majority of the Raman intensity is the in-phase combination of the $C_7=C_8$, $C_9=C_{10}$, and the $C_{11}=C_{12}$ stretches. The other ethylenic modes have nearly equal contributions from in-phase and out-of-phase double bond stretch components so their Raman intensity is much lower.

In the 1250-1400 cm^{-1} region we find the in-plane rocking vibrations of the vinyl hydrogens. The C_{11}-H rock at 1270 cm^{-1} is presented in Figure 8. Although this rock is quite localized, it is clear that this mode also contains significant contributions from skeletal stretches (esp. $C_9=C_{10}$ and $C_{12}-C_{13}$), that account for its intensity in the Raman spectrum. In all-*trans* retinal the only other rock that has comparable Raman intensity is the C_{14}-H at 1334 cm^{-1} (see Figure 5).

The skeletal single bond stretches appear in the 1100-1250 cm^{-1} range. There are three classes of these stretches in retinal. First, the $C_{14}-C_{15}$ stretch appears at 1111 cm^{-1} and has no Raman intensity (see Figure 5). It is the lowest frequency stretch because the electron-withdrawing carbonyl group reduces its bond order. The second class is represented by the $C_{10}-C_{11}$ stretch at 1163 cm^{-1}. This is a characteristic frequency for a conjugated single bond stretch in an unsubstituted linear polyene. Examination of the calculated 1162 cm^{-1} mode in Figure 8 shows the $C_{10}-C_{11}$ stretch to be a localized single bond stretch, consistent with the 10,11-[13][C] results. The third class of single bonds includes the methyl substituted C_8-C_9 and $C_{12}-C_{13}$ stretches. These vibrations are 40-50 cm^{-1} higher than the $C_{10}-C_{11}$ stretch because they couple strongly with the adjacent C_9-CH_3 and C_{13}-CH_3 stretches at $\sim$850 cm^{-1}. The 1198 and 1215 cm^{-1} lines are assigned to the C_8-C_9 and $C_{12}-C_{13}$ stretches, respectively. Examination of the calculated 1214 cm^{-1} mode in Figure 8 shows the contribution of the C_{13}-CH_3 stretch and also the characteristic coupling of the $C_{12}-C_{13}$ stretch with the C_{14}-H rock.

The remaining intense in-plane mode is the 1010 cm^{-1} line which is due to the in-phase combination of the methyl rocking vibrations. The highly localized character of this mode (Figure 8) is consistent with the unusual constancy of its frequency and intensity in a wide variety of isoprenoid polyenes.

Another important type of vibration is the hydrogen out-of-plane or HOOP mode. Figure 9 depicts the 960 cm^{-1} A_u combination of the C_{11}-H and C_{12}-H wags (local C_{2h} point group). Two such 1,2-disubstituted ethylene groups are found in retinal ($C_{11}H=C_{12}H$ and $C_7H=C_8H$). They give rise to the 959 and 967 cm^{-1} lines which are characteristically strong in the IR. A second class of HOOP vibrations arise from chain double bonds that have one hydrogen and one methyl substituent. Characteristic HOOP frequencies for

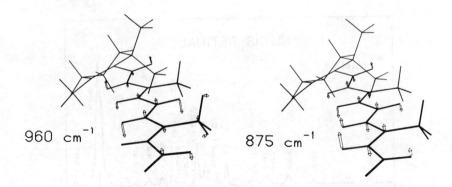

Figure 9.   Mass weighted atomic displacements in representative hydrogen out-of-plane (HOOP) normal modes of all-*trans* retinal.

these 1,1,2-trisubstituted ethylene groups are somewhat lower than the A_u modes.   For example, the C_{14}-H wag in all-*trans* is found at 875 cm^{-1}.

3.2   13-*Cis* Retinal

The Raman and IR spectra of 13-*cis* retinal are presented in Figure 10 along with the empirical assignments (20).   As might be expected, the assignments are not greatly different from those of all-*trans* retinal except near the C_{13}=C_{14} bond where the configurational change has occurred.   Several more lines are observed in the Raman spectrum in the vinyl hydrogen rock region.   The new lines at 1311 and 1399 cm^{-1} are assigned to the C_{12}-H and C_{15}-H rocks, respectively, which pick up intensity because of non-bonding effects unique to the 13-*cis* configuration.   The C_{12} and C_{15} hydrogens are in van der Waals contact in 13-*cis* retinal.   This causes these rocks to couple slightly with the skeletal stretches, accounting for their Raman intensity.

In the single bond stretch region, lines are observed at nearly the same frequencies as in all-*trans* retinal.   However, the C_{14}-C_{15} stretch has much more Raman and IR intensity and the C_{12}-C_{13} stretch has much more Raman intensity than is observed in the all-*trans* isomer.   The $\sim$1110 cm^{-1} normal mode in the two isomers has very similar atomic displacements and is accurately described as a localized C_{14}-C_{15} skeletal stretch.   Thus its dramatic increase in IR intensity in 13-*cis* must be due to an increase in the polarization of this bond due to the increased contribution from ground state resonance structures like $^+C_{13}$-C_{14}=C_{15}-O$^-$.   The increased Raman intensity argues that 13-*cis* retinal experiences a much larger change in the C_{14}-C_{15} bond order upon electronic excitation than does all-*trans*.

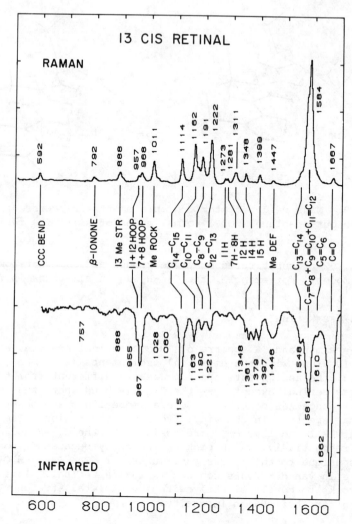

Figure 10. Vibrational assignments of the Raman and infrared spectra of 13-*cis* retinal [adapted from Curry *et al*. (20)]. The dominant internal coordinate in each normal mode is indicated.

The $C_{12}-C_{13}$ stretch at 1222 cm^{-1} also undergoes a dramatic increase in Raman intensity in the 13-*cis* isomer. This increase can be easily understood by comparing the calculated 1226 cm^{-1} normal mode for 13-*cis* in Figure 11 with the corresponding 1214 cm^{-1} mode of all-*trans*. In 13-*cis* the "$C_{12}-C_{13}$" stretch is more accurately described as an in-phase combination of the C_8-C_9, $C_{10}-C_{11}$ and $C_{12}-C_{13}$ stretches. As we observed for the double bond

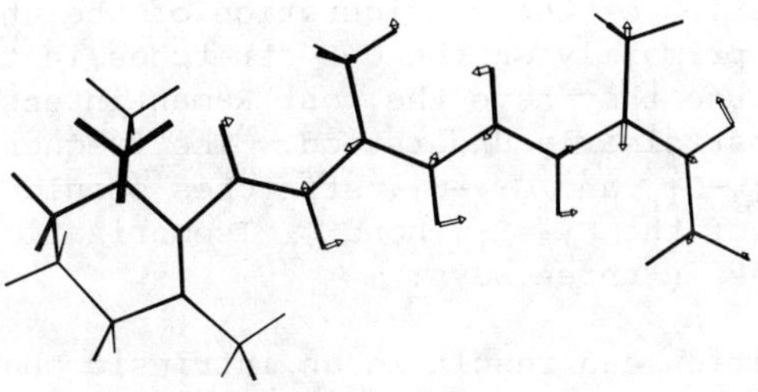

$$1226 \ cm^{-1}$$

Figure 11. Mass weighted atomic displacements in the "C_{12}-C_{13}" stretch of 13-*cis* retinal.

stretches, the in-phase combination of the single bond stretches will be strong in the Raman spectrum. Thus, in 13-*cis* retinal the C_{12}-C_{13} stretch mixes with the other nearby skeletal stretches and "borrows" intensity from them. The C_{12}-C_{13} stretch itself would be expected to have at least some intrinsic Raman intensity in both isomers. In all-*trans* the weak intensity of the 1214 cm^{-1} mode must result from out-of-phase mixing of the C_{12}-C_{13} stretch with the C_8-C_9 and C_{10}-C_{11} stretches (see Figure 8).

4.   CHROMOPHORE STRUCTURE IN BACTERIORHODOPSIN

Now that a basic understanding of the vibrational structure of all-*trans* and 13-*cis* retinal chromophores has been developed, it is possible to learn much more about the protein-bound chromophore structure from the pigment spectra. The initial problem is the determination of the chromophore configuration. This question was first addressed by simply comparing the pigment spectra with spectra of appropriate model compounds. This approach was quite successful for the visual pigments rhodopsin and isorhodopsin (10). However, analogous studies of BR and its intermediates have resulted in ambiguous conclusions (21-24). The ambiguity arises because the all-*trans* and 13-*cis* protonated Schiff base spectra are not sufficiently different *when compared with typical protein-induced frequency and intensity changes*. We have resolved this problem by using isotopic substitution of the retinal chromophore to assign and characterize the individual normal modes. This is the "isotopic fingerprint" method (9).

4.1 BR$_{568}$

The rationale for the isotopic fingerprint method is as follows. The intensities and frequencies of the in-plane C-C stretches

and C-C-H bends between 1150 and 1400 cm^{-1} empirically provide the
clearest characterization of the configuration of the chromophore
(10). We will focus primarily on the C-C stretches in the 1150 to
1300 cm^{-1} region because they have the most Raman intensity and
are, at this time, most clearly understood. The frequencies and
intensities of the C_{14}-C_{15} and C_{12}-C_{13} stretches should be sensitive to the geometry of the C_{13}=C_{14} bond. Isomerization can
affect these stretches in three ways:

First, isomerization can result in an intrinsic change in the
C-C stretching frequency. For example, it has been convincingly
argued that the C_{14}-C_{15} stretch in Schiff bases and protonated
Schiff bases should be significantly lower in 13-*cis* isomers than
in all-*trans* isomers because of altered coupling with the skeletal
bends (20).

Second, the mixing of especially the C_{14}-C_{15} and C_{12}-C_{13} internal or basis coordinates is expected to be different because
the kinetic (geometric) and potential interactions between these
two atomic motions will be significantly altered by isomerization.
This will result in normal modes that have significantly different
character. Compare, for example, the "C_{12}-C_{13}" stretches in all-
trans and 13-*cis* retinal (Figures 8 and 11). Altered normal
mode character may lead to a change in Raman intensity that is
not a result of altered intrinsic Raman intensity for a particular
internal coordinate.

Finally, there may be an intrinsic change in the Raman intensity of these stretches similar to that seen for the C_{14}-C_{15}
stretch in 13-*cis* retinal. For a homologous series of molecules
this may also be diagnostic for a particular isomer.

With these ideas in mind we can examine the spectra of native
and isotopically substituted BR_{568} and use the isotopic shifts to
assign the C_{14}-C_{15} and C_{12}-C_{13} stretches. Their frequencies, intensities and isotopic shifts should be different in the all-*trans*
and 13-*cis* isomers, thereby providing our isotopic fingerprint.
The methods for regenerating and purifying isotopic bacteriorhodopsin derivatives have already been given in detail (9,13,25).
We first employed 15-deuterio substitution which is synthetically
easy and should allow us to assign at least the C_{15}-H rock and the
C_{14}-C_{15} stretch (9). In 15D BR_{568} (Figure 12) the only frequency
change in the hydrogen rock region is the disappearance of the
1252 cm^{-1} line which we assign as the C_{15}-H rock in agreement with
the conclusions of Stockburger (26). The C-C stretches in native
BR are very similar in frequency to those observed in all-*trans*
retinal. The obvious assignments of the 1169, 1200 and 1211 cm^{-1}
lines are to the C_{10}-C_{11}, C_{12}-C_{13}, and C_8-C_9 stretches. The location of the C_{14}-C_{15} stretch cannot be determined by analogy with
retinal because protonated Schiff base formation is expected to

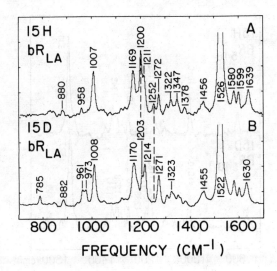

Figure 12. Comparison of native (15H) light-adapted BR$_{568}$ with its 15-deuterio (15D) derivative [adapted from Braiman and Mathies (9)].

result in a significant increase in its frequency from the 1111 cm^{-1} value in all-*trans* retinal. The dramatic shift of intensity from 1200 to 1214 cm^{-1} in 15D BR$_{568}$ argues that the C$_{14}$-C$_{15}$ stretch is at 1200 cm^{-1}, degenerate with the C$_{12}$-C$_{13}$ stretch. When its coupling with the C$_{15}$-H rock is removed by deuteration, the C$_{14}$-C$_{15}$ stretch shifts up by $\sim$14 cm^{-1}. These assignments have now been confirmed by further isotopic studies. Since BR$_{568}$ is known to contain an all-*trans* chromophore, this pattern of 15D-induced changes will be our all-*trans* "isotopic fingerprint."

4.2 BR$_{548}$

In Figure 13 spectra of native and 15-deuterio BR$_{548}$ are presented (13). Pure spectra of the "13-cis" component of dark-adapted bacteriorhodopsin have been isolated by subtraction of the all-trans component following published procedures (24,27). Upon 15-deuterio substitution, a weak line at 1233 cm^{-1} increases in intensity, the 1168 cm^{-1} shoulder moves up in frequency resulting in a broadening and increased intensity of the 1182 cm^{-1} band, and the 1201 cm^{-1} line loses some intensity. As anticipated, the C$_{14}$-C$_{15}$ stretch at 1168 cm^{-1} is $\sim$30 cm^{-1} lower than that in BR$_{568}$. This assignment is confirmed by its upshift upon 15D substitution. The C$_{12}$-C$_{13}$ stretch is now observed at 1233 cm^{-1}. It is very weak in native BR$_{548}$, presumably because of vibrational mixing with the C$_{14}$-C$_{15}$ or C$_8$-C$_9$ stretches which cause a cancellation of intensity.

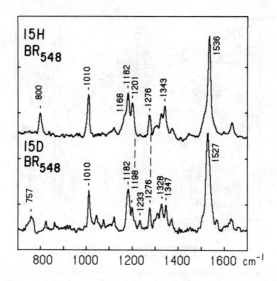

Figure 13.   Comparison of native and 15-deuterio BR_{548}. The spectrum of BR_{548} is generated by first obtaining a spectrum of dark-
adapted BR which is an equal mixture of BR_{548} and BR_{568}. The pure
spectrum of BR_{548} is then obtained by subtracting a light-adapted
spectrum [from Smith *et al*. (13)].

15-deuterio substitution alters this mixing so some of the intensity in the 1201 cm^{-1} line shifts up to 1233 cm^{-1}. This pattern
of intensity and frequency changes is characteristic of 13-*cis*
chromophores in BR.

4.3 M_{412}

     The configuration of the chromophore in the M_{412} intermediate
has been well characterized by both chemical extraction and resonance Raman (3,4,9). The chromophore is known to be a 13-*cis* unprotonated Schiff base. The pattern of changes observed in the
native and 15D M_{412} spectra in Figure 14 conform to the pattern
expected for 13-*cis* chromophores. In M_{412} three strong fingerprint
lines are observed at 1180, 1200 and 1228 cm^{-1}, assignable to the
$C_{10}-C_{11}$, C_8-C_9 and $C_{12}-C_{13}$ stretches in direct analogy with 13-
cis retinal (see Figure 10). Upon 15D substitution the 1228 cm^{-1}
line undergoes a significant increase in intensity very similar to
the intensity increase at 1233 cm^{-1} in 15D BR_{548}. As discussed
earlier, this intensity change is attributable to altered mixing
between the $C_{12}-C_{13}$ and the $C_{14}-C_{15}$ or C_8-C_9 stretches upon 15D
substitution. The location of the $C_{14}-C_{15}$ stretch in M_{412} is unclear because it evidently has little intrinsic Raman intensity.

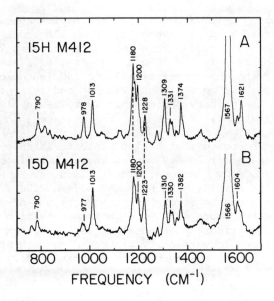

Figure 14. Comparison of native and 15-deuterio M_{412} spectra obtained with the two-color, time-resolved apparatus using a 412 nm
probe wavelength [adapted from Braiman and Mathies (9)].

4.4 K_{610}

Now that the 15-deuterio isotopic fingerprint pattern is well
established, we can turn to more transient species in the photocycle. Figure 15 presents spectra of native and 15D K_{610} obtained

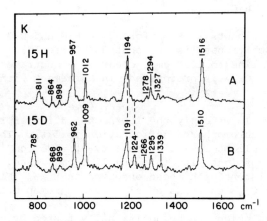

Figure 15. Comparison of native and 15-deuterio substituted K_{610}
spectra obtained at 77°K with the spinning sample cell [adapted
from Braiman and Mathies (18)].

with the spinning cell method (18). The native spectrum consists
of a single broad band at 1194 cm^{-1}. Upon 15D substitution a new
line appears at 1224 cm^{-1}. This intensity increase is very similar to that observed in the 13-*cis* pigments, BR_{548} and M_{412}.
Observation of the characteristic 13-*cis* 15D isotopic fingerprint
in K_{610} provides the most direct evidence and the only *in situ*
evidence that the primary photochemistry in BR involves a 13-*trans* →
13-*cis* isomerization.

It should be noted that K also exhibits intense lines at 811
and 957 cm^{-1} in the frequency region characteristic of hydrogen
out-of-plane (HOOP) vibrations. These lines have been assigned
to the C_{14}-H and $C_{11}H=C_{12}H$ A_u HOOP modes (see Figure 9). Out-of-
plane HOOP modes can be strongly enhanced by out-of-plane distortions of the chromophore (28,29). Their observation in K_{610} indicates that the primary photochemical isomerization produces a
product that is conformationally distorted through steric interaction with the protein residues in the binding site.

4.5 O_{640}

The O_{640} intermediate has only recently been studied in detail
(13). It is the key intermediate in resetting the cyclic proton
pump. For these reasons its resonance Raman spectra will be discussed in more detail. Figure 16 presents an example of how the
two-color, time-resolved O_{640} experiments were performed. In A
we have obtained a 752-nm probe spectrum of BR_{568}. In B the pump
beam is turned on, resulting in a spectrum containing both BR_{568}
and O_{640}. Because a red-probe beam is used to selectively enhance
the O_{640} scattering, new features unambiguously attributable to
O_{640} can be used to guide the spectral subtraction to isolate the
O_{640} spectrum in C.

O_{640} experiments were performed in H_2O and D_2O to determine
whether the Schiff base is protonated (Figure 17). The shift of
the 1628 cm^{-1} line to 1589 cm^{-1} clearly demonstrates that the
Schiff base in O_{640} is protonated, in agreement with earlier work
(12).

In Figure 18 we apply the 15D isotopic fingerprint method to
determine the chromophore configuration in O_{640}. The major
changes are a slight loss of intensity at ~1250 cm^{-1} and a more
obvious shift of intensity into the 1215 cm^{-1} line. These changes
are very consistent with those observed in BR_{568} (Figure 12). This
suggests that O_{640} contains an all-*trans* chromophore.

This conclusion can be additionally tested by employing the
12,14-dideuterio derivative of retinal. Smith *et al*. (13) have
obtained spectra of native and 12,14-D_2 O_{640} and BR_{568}. Figure
19 shows that this substitution causes much more dramatic shifts

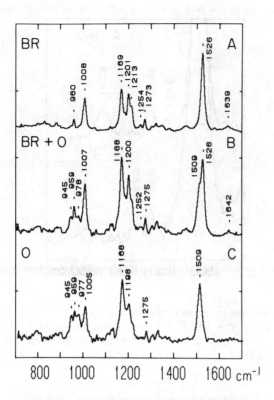

Figure 16.   Time-resolved resonance Raman spectra of BR_{568} and O_{640}.   In A, a 752-nm probe beam is used to obtain an unphotolyzed BR_{568} spectrum.   In B, a 514-nm pump beam is turned on 4 msec upstream from the probe beam, thereby producing a significant O_{640} concentration.   In C, the spectrum of O_{640} is obtained by subtracting 60% of spectrum A from spectrum B [from Smith *et al.* (13)].

of the $C_{12}-C_{13}$ and $C_{14}-C_{15}$ stretches. As in the 15D derivative, 12,14-D_2 substitution is expected to shift the $C_{14}-C_{15}$ stretch up by ~30 cm^{-1}. The $C_{12}-C_{13}$ stretch should shift up much more than this because it is strongly coupled with *both* the $C_{12}-H$ rock and the $C_{14}-H$ rock (see Figure 8). Thus, in 12,14-D_2 BR_{568} the intensity increase at 1315 cm^{-1} is due to the upshifted $C_{12}-C_{13}$ stretch character and the 1240 cm^{-1} line is due to the upshifted $C_{14}-C_{15}$ stretch, while the 1191 and 1224 cm^{-1} lines are due to the relatively unaffected $C_{10}-C_{11}$ and C_8-C_9 stretches, respectively. Virtually the same pattern of isotopic shifts is seen in 12,14-D_2 O_{640}, where new lines appear at 1319 and 1240 cm^{-1}. This unambiguously confirms the all-*trans* assignment for the chromophore in O_{640}.

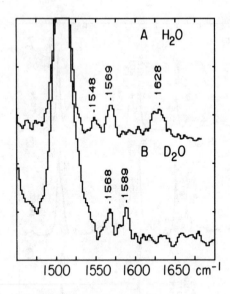

Figure 17. Resonance Raman spectra of O_{640} in H_2O and D_2O obtained as in Figure 16. The shift of the C=NH stretch from 1628 to 1589 cm^{-1} demonstrates that the Schiff base is protonated in O_{640} [from Smith *et al.* (13)].

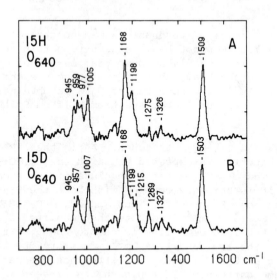

Figure 18. Comparison of native and 15-deuterio regenerated O_{640} spectra [adapted from Smith *et al.* (13)].

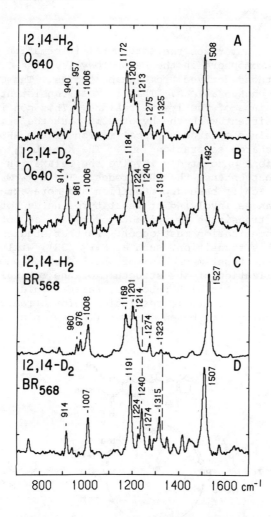

Figure 19.   Comparison of native and 12,14-dideuterio regenerated
O_{640} and BR_{568} spectra in ^2H_2O [adapted from Smith *et al.* (13)].

    Finally, it should be noted that O_{640} exhibits significant
HOOP intensity at 945, 959 and 977 cm^{-1} (see Figure 18).   This
demonstrates that the chromophore in O_{640} is conformationally distorted. O_{640} is the direct product of a protein-catalyzed dark
isomerization. K_{610}, which is also the direct product of a
protein-bound isomerization, also exhibits intense HOOP modes.   It
is evident that when retinal is in a restrictive protein environment, conformational relaxation to a planar structure does
not occur immediately after *cis-trans* isomerization.

5.   SUMMARY

    Figure 20 summarizes the structural information known about
the retinal chromophore in the BR photocycle. The $C_{13}=C_{14}$ configurations are those derived from Raman studies. The C=N configuration has been chosen as *cis*. This is most consistent with the
steric constraints of the lysine residue. This configuration is
also most consistent with the hypothesis that the primary photochemistry involves a large physical motion of the NH⊕ moiety,
resulting in charge separation (30). The $K_{610} \rightarrow L_{550}$ and the $O_{640} \rightarrow$
BR_{568} transitions undoubtedly involve the relaxation of the protein
constraints that enhance the HOOP modes. The $L_{550} \rightarrow M_{412}$ transition
involves both Schiff base deprotonation and more extensive protein
structural changes involving at least tyrosine deprotonation (31).
The $M_{412} \rightarrow O_{640}$ transition is evidently complex, as both reprotona-
tion and reisomerization must occur. Further studies on intermediates in this transition, such as N_{520} (32), will be needed to
characterize the reset mechanism of the BR proton pump. The further characterization of the structure of the retinal chromophore
in these intermediates with resonance Raman should now move rapidly as a result of the atomic resolution information provided by
vibrational assignments based on specific isotopic substitution.

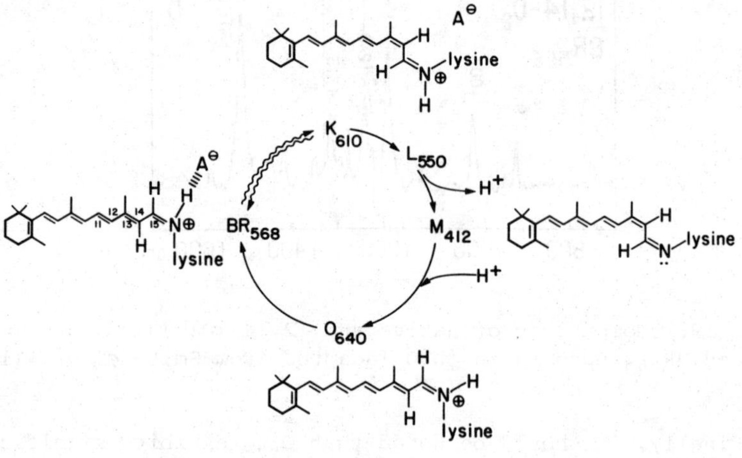

Figure 20.   Schematic of the retinal structural changes known to
occur during the proton-pumping photocycle of BR. The chromophore
is assumed to be C=N *cis* because the C=N *trans* configuration would
require large lysine distortions in the primary $C_{13}=C_{14}$ isomerization and would result in a mechanistically less useful motion and
orientation of the C=NH group in K (see text).

ACKNOWLEDGMENT

This work was performed in collaboration with Prof. Johan
Lugtenburg and his co-workers Albert Broek, Hans Pardoen and
Patrick Mulder at Leiden University, The Netherlands. At
Berkeley, Mark Braiman, Bo Curry, Steve Smith, Anne Myers and
Ilona Palings carried out this research and helped to prepare
this manuscript. This research was supported by grants from
the NSF (CHE 8116402) and the NIH (EY-02051). RM is an NIH
Research Career Development Awardee (EY-00219).

REFERENCES

(1) Stoeckenius, W. and Bogomolni, R.A. 1982, Annu. Rev. Bio-
     chemistry 51, 587.
(2) Birge, R.R. 1981, Annu. Rev. Biophys. Bioeng. 10, 315.
(3) Tsuda, M., Glaccum, M., Nelson, B. and Ebrey, T.G. 1980,
     Nature (London) 287, 351.
(4) Pettei, M.J., Yudd, A.P., Nakanishi, K., Henselman, R. and
     Stoeckenius, W. 1977, Biochemistry 16, 1955.
(5) Myers, A.B., Harris, R.A. and Mathies, R.A. 1983, J. Chem.
     Phys. 79, 000.
(6) Packer, L., ed. 1982, Methods in Enzymology 88, pp. 561-667.
(7) Mathies, R.A., Oseroff, A.R. and Stryer, L. 1976, Proc.
     Natl. Acad. Sci. USA 73, 1.
(8) Callender, R.H., Doukas, A., Crouch, R. and Nakanishi, K.
     1976, Biochemistry 15, 1621.
(9) Braiman, M. and Mathies, R. 1980, Biochemistry 19, 5421.
(10) Mathies, R., Freedman, T.B. and Stryer, L. 1977, J. Mol.
     Biol. 109, 367.
(11) El-Sayed, M.A. 1982, Methods in Enzymology 88, 617.
(12) Terner, J., Hsieh, C.-L., Burns, A.R. and El-Sayed, M.A.
     1979, Biochemistry 18, 3629.
(13) Smith, S.O., Pardoen, J.A., Mulder, P.P.J, Curry, B., Lugten-
     burg, J. and Mathies, R. 1983, Biochemistry, submitted.
(14) Smith, S.O., Braiman, M. and Mathies, R. 1983, in Proceed-
     ings of the First International Conference on Time-Resolved
     Vibrational Spectroscopy (G. Atkinson, ed.) Academic Press.
(15) Hsieh, C.-L., Nagumo, M., Nicol, M. and El-Sayed, M.A. 1981,
     J. Phys. Chem. 85, 2714.
(16) Pande, J., Callender, R.H. and Ebrey, T.G. 1981, Proc. Natl.
     Acad. Sci. USA 78, 7379.
(17) Terner, J., Hsieh, C.-L., Burns, A.R. and El-Sayed, M.A.
     1979, Proc. Natl. Acad. Sci. USA 76, 3046.
(18) Braiman, M. and Mathies, R. 1982, Proc. Natl. Acad. Sci.
     USA 79, 403.
(19) Curry, B., Broek, A., Lugtenburg, J. and Mathies, R. 1982,
     J. Am. Chem. Soc. 104, 5274.

(20) Curry, B., Palings, I., Broek, A., Pardoen, J.A., Mulder,
      P.P.J., Lugtenburg, J. and Mathies, R.   J. Phys. Chem., sub-
      mitted.
(21) Marcus, M.A. and Lewis, A.   1978, Biochemistry 17, 4722.
(22) Terner, J., Campion, A. and El-Sayed, M.A.   1977, Proc.
      Natl. Acad. Sci. USA 74, 5212.
(23) Aton, B., Doukas, A.G., Callender, R.H., Becher, B. and
      Ebrey, T.G.   1977, Biochemistry 16, 2995.
(24) Stockburger, M., Klusmann, W., Gattermann, H., Massig, G. and
      Peters, R.   1979, Biochemistry 18, 4886.
(25) Mathies, R.   1982, Methods in Enzymology 88, 633.
(26) Massig, G., Stockburger, M., Gärtner, W., Oesterhelt, D. and
      Towner, P.   1982, J. Raman Spectrosc. 12, 287.
(27) Aton, B., Doukas, A.G., Callender, R.H., Becher, B. and
      Ebrey, T.G.   1979, Biochim. Biophys. Acta 576, 424.
(28) Eyring, G., Curry, B., Mathies, R., Fransen, R., Palings, I.
      and Lugtenburg, J.   1980, Biochemistry 19, 2410.
(29) Warshel, A. and Barboy, N.   1982, J. Am. Chem. Soc. 104,
      1469.
(30) Honig, B., Ebrey, T., Callender, R.H., Dinur, U. and Otto-
      lenghi, M.   1979, Proc. Natl. Acad. Sci. USA 76, 2503.
(31) Kalisky, O., Ottolenghi, M., Honig, B. and Korenstein, R.
      1981, Biochemistry 20, 649.
(32) Nagle, J.F., Parodi, L.A. and Lozier, R.H.   1982, Biophys.
      J. 38, 161.

STRUCTURAL AND KINETIC STUDIES OF BACTERIORHODOPSIN BY RESONANCE
RAMAN SPECTROSCOPY

Thomas Alshuth, Peter Hildebrandt, Manfred Stockburger

Max-Planck-Institut für biophysikalische Chemie
Göttingen, Germany (FRG)

RESONANCE RAMAN EXPERIMENT

This lecture is devoted to resonance Raman (RR) spectroscopic studies of Bacteriorhodopsin (BR) which is a protein
in the so-called purple membrane of *Halobacteria*. It is so
famous because it has like the visual pigment, rhodopsin, a
retinal molecule at its chromophoric site and, on the other
hand, constitutes a light-driven proton pump (1). This latter
function is of great importance for bioenergetics and has
attracted the interest of scientists from various disciplines.
The subject is documented by several review articles (2-4).
At BR's chromophoric site the retinal molecule is attached
to the protein via a Schiff base (SB) group (Fig. 1). Based
on biochemical extraction- and reconstitution experiments it
is generally accepted that in the dark a 1:1 equilibrium
exists between two components in which the retinal moiety is
in the *all-trans* and *13-cis* configuration, respectively. The
trans chromophore, BR-570, has its maximum absorption at 570 nm,
while the *13-cis* chromophore, BR-548, is slightly blue-shifted
to 548 nm.

(1) D.Oesterhelt and W.Stoeckenius, Proc.Natl.Acad.Sci. USA
    70, 2853 (1973).
(2) W.Stoeckenius, R.H.Lozier, and R.A.Bogomolni, Biochim.
    Biophys. Acta 505,215 (1979).
(3) M.Ottolenghi, Adv. Photochem. 12, 97 (1980).
(4) Methods in Enzymology, Vol.88 (1982), I.

C. Sandorfy and T. Theophanides (eds.), Spectroscopy of Biological Molecules, 329–346.
© *1984 by D. Reidel Publishing Company.*

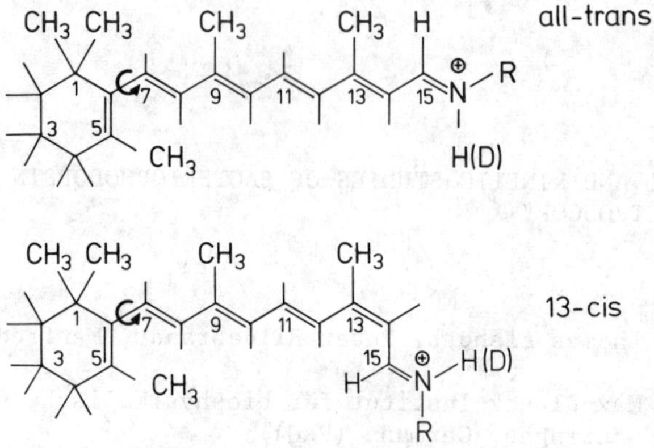

Fig. 1. *Retinal-chromophores of bacteriorhodopsin in all-trans and 13-cis configuration.*

BR-chromophores exhibit a strong absorption band in the visible which corresponds to a $\pi\pi^*$ electronic transition. In the RR experiment one irradiates with a sharp laser line into the chromophoric absorption band and detects the dispersed scattered light. Here we are only interested in the Stokes-shifted vibrational RR spectrum. One obtains three pieces of information: the frequency of vibrational modes in the electronic groundstate, the intensity and width of RR bands. All three pieces obtain information on various structural components of the chromophore. Evidently, the vibrational frequency of the C=N stretching vibration indicates the state of protonation of the nitrogen and C-C frequencies are characteristic of the configuration of the retinal chain. Both structural components play an important role in the proton pump mechanism.

In the present case where we are dealing with a strong electronic transition the intensity of an RR band is determined by Franck-Condon overlap integrals. This implies that a certain normal mode is only RR active if it involves a molecular coordinate, bond length or -angle, which changes its magnitude on electronic excitation. As a consequence mainly C=C, C-C and C=N stretching modes will occur in the RR spectra.

But on the other hand, hydrogen out-of-plane bending modes may also occur if the molecule in the ground state is non-planar and
changes its geometry on electronic excitation. The appearance of
such bands in the RR spectra indicates a distorsion of the retinal chromophore in the protein environment (5).

The width of a band mainly depends on relaxation mechanisms
of the final vibrational state which is reached in the RR transition. Later we shall give an example where the width of the C=N
stretching vibrational band is used to derive structural information.

If one excites within the visible absorption band resonance
enhancement is confined to the chromophoric site which therefore
is selectively probed in the RR experiment. In this way the RR
spectroscopist is able to follow structural changes of the chromophore, which controls the light-induced proton pump mechanism, in
great detail. This finally should contribute to an understanding
of this important process. Valuable contributions were already
made by various groups (6-14).

PHOTOCHEMICAL CYCLE

Figure 2 shows the famous photochemical cycle of bacteriorhodopsin as was obtained in its essential parts by flash photolysis experiments. Even at very low light levels the system is
completely in the all-*trans* chromophoric state, BR-570, since the
thermal backreaction to BR-548 is slow. BR-570, therefore, is the
parent chromophore of the photochemical cycle. On illumination it
runs through several intermediates and is reconstituted after a
few milliseconds. Note, that only the first step from BR-570 to
K-590 is photoinduced while the consecutive steps are due to thermal relaxation.

The important parameter of the cycle is the rate constant l_0
of the photochemical step. This is given by the excitation probability times the quantum efficiency of the photoprocess. In each
experiment one has a free choice of this parameter by setting the
irradiance of the laser to an appropriate level. For studying the
unphotolyzed species it is convenient to use a flow system in combination with CW irradiation.

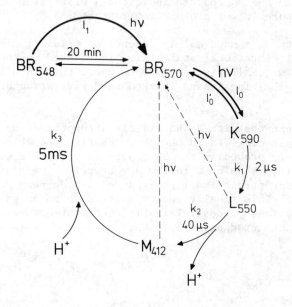

Fig. 2. Photochemical cycle of bacteriorhodopsin

This means that the sample, an aqueous suspension of the purple
membrane, flows across the laser beam, being illuminated during
its transit time Δt. If the product $l_0 \Delta t$ is small compared to
unity photolysis in the laser beam is neglegible and one obtains
the spectrum of BR-570. Since the photocycle is completely reversible one can use a cyclic flow system under the "fresh sample
condition" which means that the period of the flow cycle is at
least five times larger than the 5 ms period of the photocycle. A
spectrum of a 1:1 mixture of BR-570 and BR-548 can be also recorded with a properly designed flow-system (11). The pure spectrum
of BR-548 then is obtained by computer subtraction of the BR-570
contribution.

SPECTRA OF THE PARENT CHROMOPHORE BR-570 AND THE DARK ADAPTED
COMPONENT BR-548

The spectra of BR-570 in H_2O and D_2O suspension are depicted
in Figure 3. In D_2O the proton at the SB nitrogen is replaced by
a deuteron which induces several important spectral changes.

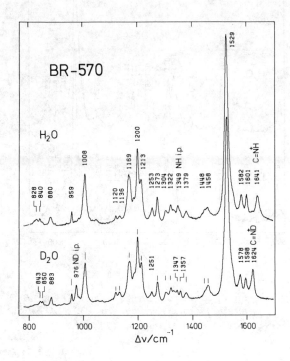

Fig. 3. RR spectra of BR-570 in H_2O and D_2O
suspension. Excitation at 514 nm, $I_0\Delta t \approx 0.1$.

Thus the C=N stretching vibration at 1641 cm^{-1} in H_2O shifts down
to 1624 cm^{-1} in D_2O. One reason for this effect obviously is due
to the increased reduced mass of the SB group. This shift, therefore, is often used as evidence for a protonated Schiff base. The
band which appears at 1349 cm^{-1} in H_2O and shifts down to 976 cm^{-1}
in D_2O can be assigned to the N-H in-plane bending mode of the
SB group. This isotope effect is the most convincing evidence for
the existence of a protonated SB group in BR-570.

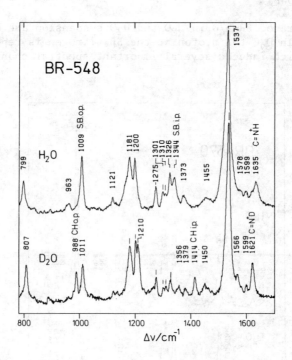

Fig. 4. RR spectra of BR-548 in H_2O and D_2O
suspension. Excitation at 514 nm,
see also reference (11).

The normal modes of the retinal moiety cannot be characterized
in a simple way, since they are more or less delocalized. In this
respect, recent work on the vibrational analysis of retinal based
on deuterio-derivatives and normal mode calculations was of great
help for a better understanding of such modes (15).

In this lecture we can only give a brief and qualitative description of the most characteristic vibrational bands. The
strongest band at 1529 cm^{-1} in the BR-570 spectra is due to a C=C
stretching mode which obtains its main contributions from the
$C_{11}=C_{12}$ and $C_9=C_{10}$ central bonds but on the other hand, is extended over the whole π-electron system of the chromophore. It is
well known that its frequency depends sensitively on the π-elec-
tron delocalization and therefore is inversely proportional to the
wavelength of the chromophore's absorption maximum. The two weak

bands at 1601 and 1582 cm^{-1} must be ascribed to C=C stretches which probably are more localized in the terminal region.

The prominent bands between 1150 and 1230 cm^{-1} mainly involve a C-C stretching motion which is more or less coupled to an in-plane bending vibration of adjacent hydrogen atoms. For instance, the band at 1200 cm^{-1} shifts to 1215 cm^{-1} when the C_{15} carbon is deuterated (12). This spectral region is highly sensitive to structural changes of the retinal chain and is a "fingerprint" of its configurational state.

The bands between 1300 and 1400 cm^{-1} predominantly incorporate a hydrogen in-plane bending motion while the hydrogen out-of-plane modes are reflected by the weak bands between 800 and 1000 cm^{-1} in the spectra of BR-570 (H_2O). The rather strong band at 1008 cm^{-1} could be assigned to the in-plane rocking mode of the CH_3-groups of the retinal chain (15).

The spectra of the dark-adapted *13-cis* component, BR-548, are depicted in Fig. 4 and shall be compared with those of the *all-trans* component, BR-570. The shift of the absorption maximum from 570 to 548 nm correlates with an increase in frequency of the C=C stretch from 1529 to 1537 cm^{-1}. The characteristic intensity and frequency changes in the fingerprint region are characteristic of the *all-trans* to *13-cis* isomerization. The analysis of the characteristic modes of the SB group reveals that its structure is different from that in BR-570. It is interesting to note that rather strong hydrogen out-of-plane modes are observed in the spectra of BR-548. The band at 988 cm^{-1} in D_2O could be assigned to the C_{15}-H o.p. bending mode, while the two bands at 799 cm^{-1} and 807 cm^{-1} were ascribed to the C_{14}-H mode. This implies that in the BR-548 chromophore the terminal region is distorted by interaction with the protein.

RR SPECTRA OF PHOTO-INTERMEDIATES

The RR spectra of intermediates of the photochemical cycle
can be obtained in time-resolved double-beam experiments. RR spectroscopists prefer CW lasers for excitation, since they allow
continuous sampling of the weak Raman signal. Thus it is desirable
to design a time-resolved RR experiment with CW lasers. This can
be done by using a flow-system. The essential features of such an
experiment are sketched in Figure 5.

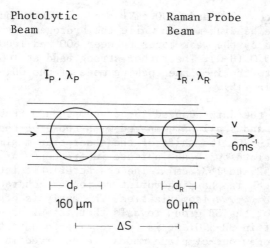

Fig. 5. Double beam flow experiment.

The time which a molecule needs to flow from the photolytic
to the probe beam provides a scale on which the temporal evolution of intermediates can be measured. This delay time is given
by

$$\delta = \Delta s \, \upsilon$$

where Δs is the lateral distance between the foci of the two
beams in the cell and υ was kept constant (6 ms^{-1}) and δ was
varied by changing Δs. The focal diameters of the two beams were
160 μm and 60 μm for the photolytic and probe beam, respectively.
A lower limit δ_{min} is given by Δs_{min} for which the two beams no
longer overlap.

In the present case this is about 20 μs. This time-resolution is sufficient to separate the intermediates L-550 and M-412. Studies of the first photo-intermediate K-590 would require a much higher resolution which only can be achieved by the application of pulsed laser systems.

For the two beams the following conditions have to be considered. Firstly, the Raman probe beam should not induce additional photoreactions. This can be easily achieved by an appropriate choice of the laser irradiance. Secondly, the photolytic beam should induce sufficient photoreactions of the parent species while photoalteration of intermediates should be avoided. The first part of this condition can again be easily fulfilled. The second part, however, requires careful examination as will be demonstrated below.

It was the objective of our experiment to detect Raman signals of the intermediate L-550. As photolysis beam the 647 nm line of a krypton laser was chosen. At this wavelength the absorption coefficient of L-550 is about ten times smaller than that of the parent BR-570 (see Fig. 6). This allows to select a laser irradiance at which the photoalteration of BR-570 is sufficiently high but can be neglected for L-550. On the other hand, photoreactions of the first intermediate K-590 are possible. But it is well known that the most efficient reaction reproduces the parent chromophore BR-570. It can be easily seen that under the same conditions the intermediate M-412 can be accumulated.

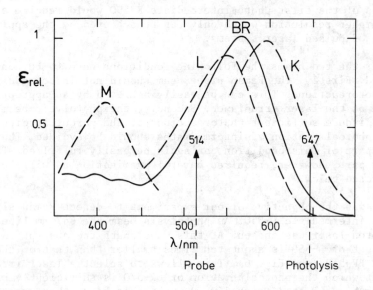

Fig. 6. Absorption spectra of the parent chromo-
phore BR-570 and the intermediates K-590,
L-550 and M-412 on a relative scale.

A typical design of a double-beam flow experiment is depicted
in Fig. 7. Two CW lasers are combined with a rotating cell. The
photolysis beam can be interrupted periodically by a chopper which
allows to accumulate simultaneously two spectra and to avoid
mechanical errors of the spectrometer. Spectrum I reflects the
parent species while spectrum II involves the photo-intermediate.

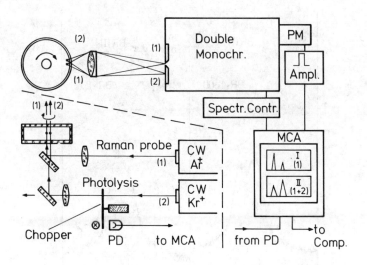

Fig. 7. Experimental design of a double-beam
flow experiment.

In Fig. 8 the procedure we use is demonstrated for the C=C stretching region. The intensity of the C=C stretch of BR-570 at 1529 cm^{-1} decreases on illumination with the photolytic beam by about 30 %. When the contribution of BR-570 to the composed spectrum is subtracted one obtains the pure spectrum of L-550. In this region this consists of a characteristic doublett which can be used for kinetic studies. Note that the intensity distribution of the doublett is changed in D$_2$O when the SB proton is replaced by a deuteron.

RR SPECTRA OF L-550

The RR spectra of L-550 between 800 and 1700 cm^{-1} are depicted in Figure 9 for H$_2$O and D$_2$O suspension. A comparison with the spectra of the two native chromophores BR-570 (*all-trans*) and BR-548 (*13-cis*) suggests that in L-550 the retinal chain is in a "*13-cis*-like" configuration as in the dark-adapted chromophore BR-548. A careful analysis of the characteristic vibrations of the Schiff-base group, however, reveals that the two chromophores are entirely different in the terminal region between the C_{13}-atom and the SB nitrogen.

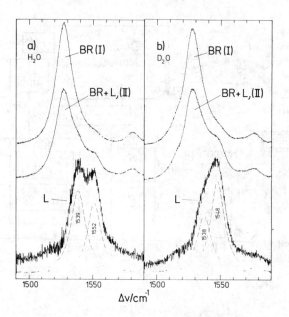

Fig. 8. *RR Spectra of bacteriorhodopsin and its photointermediate L-550 in the C=C stretching region;* λ_P=647 nm, λ_R=514 nm, δ=25 µs

Indeed, from the physical point of view the two *13-cis* chromophores must have different structural components. Thus L-550 must have a higher free energy than the parent BR-570 to which it reacts back completely. BR-548 on the other hand exists in thermal equilibrium with the parent. Since both chromophores have a *13-cis* like structure it is suggested that at least part of the free energy in L-550 is stored in the terminal part which differs from that of BR-548.

KINETIC MEASUREMENTS. Once the characteristic parameters, frequency and bandwidth, of the C=C bands of the various chromophores have been derived from the individual spectra, these bands can be used for studying the kinetic behaviour of the various species in the course of the photochemical cycle. This is demonstrated by the spectra in Figure 10. The upper spectrum (a) was obtained 0.2 ms after photolysis. The spectrum was fitted using the parameters of the C=C bands of BR-570, L-550 and M-412.

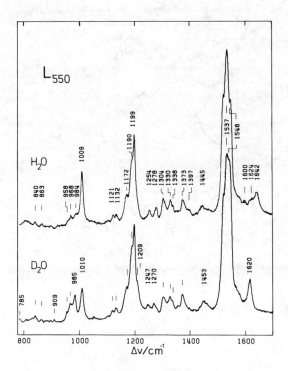

Fig. 9. RR spectra of the photointermediate L-550.

In this time domain a mixture of the parent species and the intermediates L-550 and M-412 is observed. Very surprisingly it turned out that the dark reaction from L-550 to M-412 is not complete. Extension of the measurements into the ms time-domain revealed that L-550 and M-412 do not decay with the same time-constant (Figure 10b). We conclude that only part of the L-intermediates react to M-412 with a time-constant of about 50 μs. The residual part of L(≈30-50%) is completely decoupled and has a lifetime which is even longer than that of M-412. - This behaviour has not been found previously with other techniques.

DEHYDRATION OF THE MEMBRANE

When a highly concentrated purple membrane suspension is spread over a silica substrate one obtains a membrane film after drying the sample. As long as the film still entails sufficient water the optical spectrum is unchanged with respect to the solution spectrum. However, when the film is dehydrated rigourously in the vacuum a blueshift is observed (16).

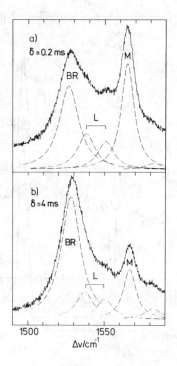

Fig. 10. Time-behaviour of the intermediates L-550
and M-412 in the photochemical cycle.
Photolysis: λ_P=647 nm, RR probe: λ_R=476 nm.
Delay time: (a), δ=0,2 ms; δ=4 ms.

The similarity in the fingerprint region of the spectra of
Figure 11 with those of BR-548 (Fig. 4) suggests that the retinal
moiety in the dehydrated chromophores is predominantly in the
13-cis configuration. - On the other hand, no spectral changes
are observed if the films were prepared from H_2O or D_2O suspension.
Since, as is seen from Figure 4, the protonation of the SB group
is manifested by various isotope effects, the complete lack of
any such effect in the spectra of the dehydrated species (Fig. 11)
is a clear evidence that in the dehydrated chromophore the SB
group is not protonated.

From the experiments with dehydrated films we conclude that
water molecules are an essential part of the intact BR-chromophore.
When water is rigourously extracted the SB group also looses its
proton. Probably water molecules are forming a bridge between the
proton donor and the accepting SB group.

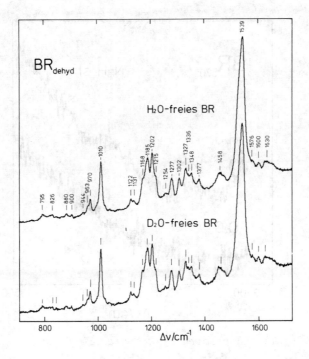

*Fig. 11. RR spectra of dehydrated purple membrane
films, obtained from H_2O and D_2O suspensions.
$T \approx 115$ K, $\lambda_R = 514$ nm.*

When the chromophore looses its well-defined structure it
also looses its functional capability as a proton pump. It is interesting to note that in spite of this on illumination the dehydrated chromophore also runs through a cyclic process but with
different intermediates and kinetics.

In order to obtain further evidence for the state of protonation of the SB group we tried to record spectra in the N-H stretching region. There is no chance to detect such a band in aqueous PM-
suspensions since the strong and very broad Raman band of water in
this region obscures any vibrational features of the membrane. –
In hydrated membrane films, however, the contribution of the water
band can be neglected. In Fig. 12 a weak feature is observed at
3379 cm^{-1} in the H_2O spectrum which disappears in D_2O. – This
band which is a few hundred times weaker than the C=C stretch is
attributed to the N-H stretch of the chromophore's Schiff base
group.

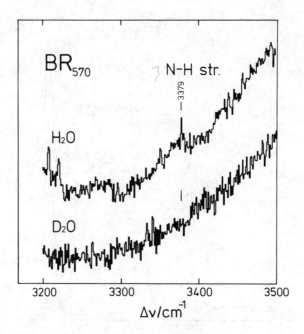

Fig.12. RR spectra of hydrated PM films.

VIBRATIONAL BAND WIDTH

     Structural information can also be involved in the width of a
vibrational band. This will be demonstrated by the spectra in Figure 13, where the two weak C=C stretches and the C=N stretch were
recorded in PM suspensions of H_2O, D_2O and a 1:1 mixture of both.
It nicely comes out that the C=N stretch is narrowed from 16.3 to
12.5 cm^{-1} when the solvent is changed from H_2O to D_2O. This effect
is either due to the exchange of the proton by a deuteron at the
SB group or to the exchange of the solvent. If it were the solvent
then the width of the two C=N stretches should be the same if we
choose a 1:1 mixture of H_2O and D_2O. This is indeed the case as
is seen from the spectrum at the bottom. But there is still additional information in this spectrum. Note that the width of 13cm^{-1},
respective 13.5 cm^{-1}, is only slightly above the 12.5 cm^{-1} in D_2O.
This suggests that only the H_2O isomer which has a statistical
weight of 25% affects the linewidth, but not the three other isomers. Thus a broadening mechanism must exist which operates for
H_2O but not for the deuterated isomers. In this respect it is of
great importance that the bending vibration of H_2O has its maximum at 1645 cm^{-1} which is nearly in resonance with the C=N
stretch.

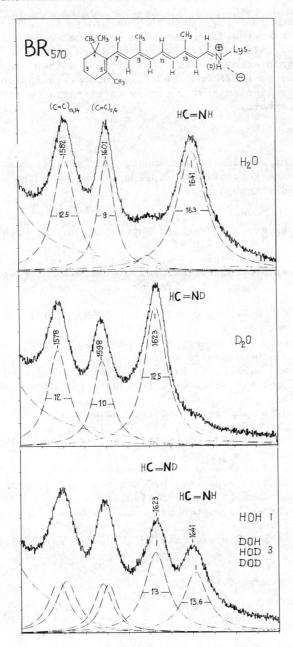

Fig. 13. RR spectra of BR-570 in aqueous suspension.

This suggests that the broadening mechanism is resonance energy exchange between the SB group and a water molecule. If this is true a water molecule must be closely attached to the SB group since the exchange mechanism is certainly not a long range effect but rather requires close contact. This interpretation would be in line with the finding that water is an essential part of the intact BR-chromophore.

ACKNOWLEDGEMENTS

We thank Prof. D. Oesterhelt and Dr. W. Gaertner for leaving us purple membrane suspension and preparations with the C(15)-D modified chromophore.

(5) G.Eyring, B.Curry, A.Broek, J.Lugtenburg and R. Mathies, Biochemistry 21, 384 (1982).
(6) A.Lewis, J.Spoonhower, A.Bogomolni, R.H.Lozier, and W. Stoeckenius, Proc.Natl.Acad.Sci. USA 71, 4462 (1974).
(7) B.Aton, A.G.Doukas, R.H.Callender, R.H.Becher, and T.G. Ebrey, Biochemistry 16, 2995 (1977).
(8) J.Terner, C.L.Hsieh, and M.A.El-Sayed, Biophys.J. 26, 527 (1979).
(9) M.Stockburger, W.Klusmann, H.Gattermann, G.Massig, and R.Peters, Biochemistry 18, 4886 (1979).
(10) M.Braiman and R.Mathies, Biochemistry 19, 5421 (1980).
(11) Th.Alshuth and M.Stockburger, Ber.Bunsenges.Phys.Chem. 85 484 (1981).
(12) G.Massig, M.Stockburger, W.Gärtner, D.Oesterhelt and D. Towner, J.Raman Spectrosc., 12, 287 (1982).
(13) M.Braiman and R.Mathies, Proc.Natl.Acad.Sci. USA, 79, 403 (1982).
(14) K.J.Rothschild, P.V.Argade, Th.N.Earnest, K.S.Huang, E. London, M.J.Liao, H.Bayley, H.G.Khorana, and J.Herzfeld, J.Biological Chem., 257, 8592 (1982).
(15) B.Curry, A.Broek, J.Lugtenburg and R.Mathies, J.Am.Chem.Soc., 104, 5274 (1982).
(16) R.Korenstein and B.Hess, FEBS LETTERS, 82, 7 (1977).

STATIC AND TIME-RESOLVED INFRARED DIFFERENCE
SPECTROSCOPY APPLIED TO RHODOPSIN AND BACTERIORHODOPSIN

F. Siebert

Institut für Biophysik und Strahlenbiologie
Freiburg im Br., FRG

The principles of static and time-resolved infrared difference spectroscopy and its application to
biochemical systems is described. Important experimental aspects of both methods are discussed. The application of the methods to the investigation of the
light-induced reactions of bacteriorhodopsin and rhodopsin is presented. The results demonstrate, that the
method allows the detection of molecular changes of
the chromophores as well as of the protein, and that
it is especially sensitive for the retinylidene Schiff
base binding site of these retinal-proteins. Regarding
the chromophore similar results, in some cases also
complementary, are obtained as in resonance Raman
spectra. Completely new information is obtained with
respect to molecular events in the protein.

INTRODUCTION

In the case of many biochemical systems, molecular changes in a small part of a complex functional
unit, e.g. protein, protein complex or membrane, cata-
lyse chemical reactions. The catalytic center may be
formed by amino acid side chains as in the case of
chymotrypsin, where a proton transfer is thought to
play a major role in the mechanism of peptide cleavage.
In other systems, the active center is part of a redox
chain. In photobiological units the active center is
formed by chromophores which catch a light quantum and
subsequently undergo molecular transformations. In
order to understand the mechanism of all these re-

347

C. Sandorfy and T. Theophanides (eds.), Spectroscopy of Biological Molecules, 347–372.
© 1984 by D. Reidel Publishing Company.

actions, it is necessary to gain information on the
molecular events occurring in the catalytic centers.
The task requires a method, which allows to monitor
the molecular changes of such small units against the
large complex background system. Vibronic spectroscopy
as Raman and infrared spectroscopy allow in principle
to detect these molecular changes. However, since the
total system, catalytic center and background system,
contribute to the vibronic spectra, a normal infrared
or Raman spectra of such complex systems is generally
not detailed enough to detect molecular events in the
active center, which constitutes a small part of the
total system only. In recent years resonance Raman
spectroscopy has proved to be a powerful tool for the
investigation of such systems which contain a chromophore absorbing in the visible or near ultraviolet
spectral range: the resonance enhancement effect ampli-
fies the Raman scattering caused by certain vibrations
of the chromophore by several order of magnitudes.
Therefore, the chromophore is selectively mirrored in
the resonance Raman spectra of such systems. This
method will be dealt with by others during this course.
In infrared spectroscopy there is no comparable mechanism, which would pick out certain vibrations of the
chromophore or other catalytic centers. This nonselec-
tivity, however, can be overcome by employing infrared
difference spectroscopy: if it is possible to record
spectra of the same sample before and after the reaction has taken place, the formation of the difference
spectrum between the two spectra will unveil the catalytic center and the molecular changes which have
taken place. The method does not depend on the presence
of a chromophore. Since the difference spectrum originates from a relatively small part of the total system,
it can be hoped, that it can be interpreted in molecular terms. Several methods, which can help the interpretation and whose applicability depends on the system to be investigated, will be discussed later on. In
many cases it is possible to stop the reaction path at
well defined steps by manipulating external parameters
such as temperature and pH. In this way also intermediates of the reaction can be investigated. Examples,
which will be discussed at greater detail below, are
rhodopsin and bacteriorhodopsin: by variation of the
temperature between 77 K and room temperature the complete light-induced reaction sequence can be sampled.

So far only the formation of difference spectra
between stable or artificially stabilized states has
been discussed. This method will be called static in-

frared difference spectroscopy. Many systems, however,
do not allow to take static difference spectra. In
photosynthesis, for example, the photo-oxidized reaction centers are fast reduced by substrates or cyclic reactions. Therefore, the initial state is re-
formed within milliseconds. In addition, it is difficult to stabilize intermediates in these systems.
There are also cases, where the procedures to fix the
intermediates may introduce artefacts. If one still
wants to measure difference spectra, one is forced
to employ time-resolved spectroscopy. In addition to
the advantage, that intermediates can be investigated
directly and under natural conditions, time-resolved
infrared difference spectroscopy offers the possibility to measure the time course of the reaction. In
this way kinetic parameters which are important for
the understanding of the molecular mechnism of the
reaction, can be obtained. Also, the measurement of
the time course may help to discriminate between reactions taking place directly in the catalytic center,
e.g. in the chromophore, and reactions in the neighborhood, which are triggered by the catalytic center.
Generally, they will proceed with different kinetics.
Examples for this will be presented later on. Time-
resolved spectroscopy in the visible or ultraviolet
spectral range has been applied for a long time to
the investigation of chemical or biochemical reactions.
The spectral range requires the presence of a chromophore. Among others, the following methods to start
the reactions have been employed: flash photolysis (1),
pulse radiolysis (2), temperature jump (3) and rapid
mixing (4). In principle all these methods can also
be applied to time-resolved infrared spectroscopy.
In this lecture, however, we will only deal with reactions of photobiological systems and which are triggered by a light flash. This method is also easiest
to incorporate into an apparatus for time-resolved
infrared spectroscopy.

EXPERIMENTAL CONSIDERATIONS

      Since, as has been pointed out in the introduction,
in the difference spectra only a small part of the
total system is reflected, the total system, however
contributes to the background absorption, the apparatus
for this kind of difference spectroscopy has to be very
sensitive. The visual pigment rhodopsin, which will be
discussed at greater detail below, may serve as an
example. Its active center, the chromophore 11-cis

retinal, has a molecular weight of 290 daltons. The
protein, however, to which the chromophore is bound,
has a molecular weight of 37000 daltons. In spectral
regions, where all the constituents contribute to the
absorption the part caused by the chromophore will be
at most in the region of 1%. For the application of
static infrared difference spectroscopy to problems
described above, FTIR spectroscopy offers several
advantages as compared to conventional infrared spectroscopy:
1. The multiplex advantage and the higher spectral
energy of the infrared beam shorten the measuring
time considerably.
2. Since the scan time for one interferogram is of
the order of one second or shorter, slow fluctuations
of the infrared beam will only produce a base-line
shift in the difference spectrum, in the course of
which several hundreds of scans have been accumulated
for a single beam spectrum.
3. Slow unspecific changes of the biological sample,
e.g. light scattering changes, will only shift the
base-line of the difference spectrum in addition to
a weak bending.

     In spectral regions where high background ab-
sorption is observed phase errors in the interferogram may present some problems. They can cause base-
line distortions in the difference spectrum, almost
reproducing the background absorption bands.In Fig.1 a
typical infrared spectrum of a biological sample,
here bacteriorhodopsin is shown in the region between 2000 cm^{-1} and 1000 cm^{-1}. It is dominated by the
two strong absorption bands at 1650 cm^{-1} and at
1550 cm^{-1}. The former band is composed of the amide
I band of the protein and of the water absorption
band, the latter band is caused by the amide II band
of the protein. These two bands will be most susceptible for such base-line distortions. However, the use
of the computer in FTIR spectroscopy offers a convenient way of controlling such errors: the total
number of a single beam spectrum may be subdivided
into several blocks and the difference spectra may
be computed between the different blocks. If a flat
base-line is obtained for the various difference
spectra, base-line distortions may be neglected.

     A typical experiment for recording a static in-
frared differences spectrum by means of FTIR spe-
troscopy consists of the following steps: Recording
the subdivided single beam spectrum of the sample in

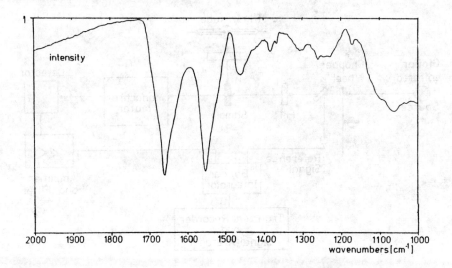

Fig. 1 FTIR single beam spectrum of a hydrated film of purple membranes

the initial state, initiating the reaction in situ and stabilizing the intermediate or the end-product; recording the subdivided single-beam spectrum of this state; forming the difference spectrum between the initial state and the state produced by the reaction.

The same arguments concerning the sensitivity of the apparatus hold also for time-resolved infrared spectroscopy. A schematic diagram of an apparatus with flash excitation developed by us (5) is shown in Fig. 2. It resembles a conventional flash photolysis instrument. The globar infrared source, sample chamber and monochromator are part of a Perkin-Elmer model 180 infrared spectrophotometer. The thermopile infrared detector of the instrument cannot be used for time-resolved spectroscopy. It is too slow and not sensitive enough. HgCdTe-detectors offer both high sensitivity and high time resolution (better than 500 ns). Since the 1/f noise of the detector dominates below ca. 2 kHz, chopping of the infrared beam and boxcar integration of the signal proved to be

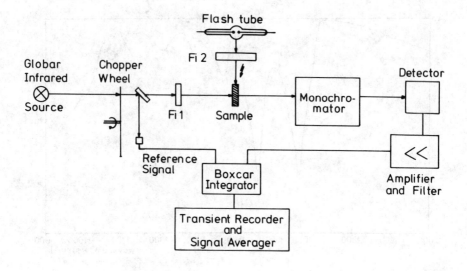

Fig. 2 Schematic diagram of an apparatus for time-resolved IR spectroscopy with flash excitation

necessary for measurements with time constants longer than one ms. For shorter time constants the 1/f noise can be filtered off by a high-pass filter, thus eliminating the need of a chopper. In this figure the sample is excited by a xenon flash. But now we use an excimer laser pumped dye laser whose light pulse has an energy of several millijoules. The effective electronic time constant determining the noise in the static difference spectra is in the order of several minutes, since several hundreds of interferograms are co-added for one single beam spectrum. To achieve a comparable sensitivity for time-resolved spectroscopy, several hundreds of flash-induced signals have to be averaged using a digital signal averager.

Since the transition moments for infrared absorption is much smaller than that for absorption in the visible spectral range of typical chromophores, it is important to excite as much as possible of the system. Therefore, systems which undergo cyclic reaction and which thus can be excited repeatedly are most suitable for time-resolved spectroscopy. For systems which bleach, after each flash the sample has to be replaced.

After having outlined the general ideas which prompted us to apply infrared difference spectroscopy to biochemical systems and after having discussed the experimental prerequisites, the application of both the static and time-resolved method to the photo-biological systems bacteriorhodopsin and rhodopsin will be described below. Most of the data have been published recently (5, 6, 7, 8).

BACTERIORHODOPSIN

Bacteriorhodopsin represents a light-driven pro-ton pump of the bacterium Halobacterium halobium. An extensive review-article on this simple photosynthetic system has been published by Stoeckenius et al. (9). The pump is located in special regions of the plasma membrane, the so-called purple membrane, due to its purple color. Upon absorption of light it transfers protons across the plasma membrane, thus transforming light into electrochemical energy by creating a pro-ton gradient. The proton pump is a chromoprotein, con-sisting of a single protein with a molecular weight of about 27.000 daltons and the chromophore, which is all-trans retinal (vitamin A aldehyde). The amino acid sequence has been determined recently (10). The chro-mophore is bound to the protein via a protonated Schiff base to the ε-amino group of lysine 216. The chromophore together with its binding site is shown in Fig. 3. Bacteriorhodopsin has its absorption maxi-mum at 570 nm, whereas model compounds of protonated retinylidene Schiff bases exhibit an absorption maxi-mum between 460 and 500 nm. Since it is very unlikely, that an isolated positive charge exists in a protein, the presence of a negative counterion has to be assumed. The importance of the counterion in regulating the absorption maximum has been stressed by several authors (11) Evidence will be presented below, that an aspartatic acid nearby the binding site is the counterion. However, other charged groups have to be assumed to explain the absorption maximum of bacterio-rhodopsin (11). Upon absorption of a proton, it under-goes a cyclic reaction via several intermediates as shown in Fig. 4. The numbers behind the intermediates indicate their absorption maximum. At room temperature the formation of K610 takes several picoseconds, the formation of L550 about one μsec, the formation of M412 60 μsec, and the reformation of BR570 10 msec. For the understanding of the proton pumping mecha-nism it is important to obtain information on the

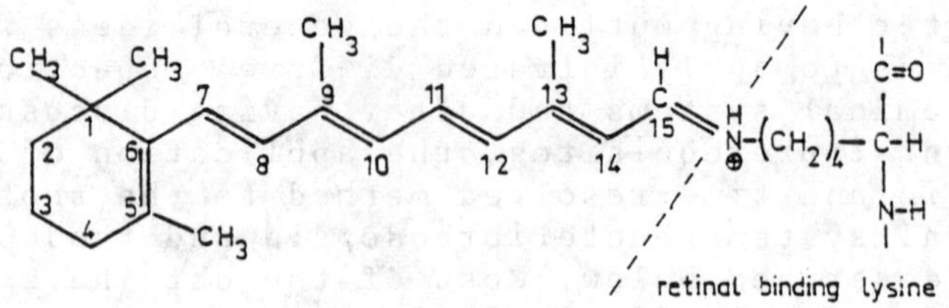

Fig. 3 All-trans retinal and its binding site in
bacteriorhodopsin

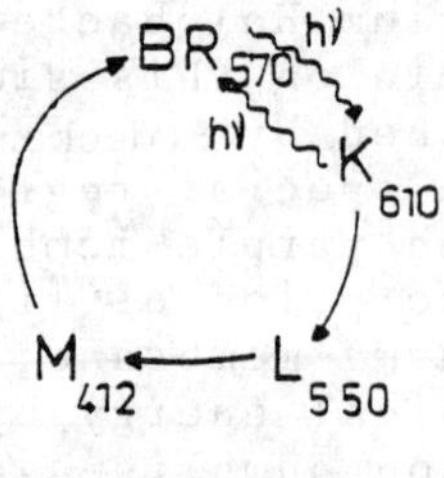

Fig. 4 Photocycle of bacteriorhodopsin

molecular processes in the chromophore and in the
protein, how these processes are linked together and
how they are coupled to the proton transport processes,
which occur at the approximate time scale of the formation of M412 and reformation of BR570.

In the following, I will present the infrared
difference spectra between BR570 and the various intermediates, partly static and partly time-resolved.
It will be shown, that the spectra contain information
on the chromophore as well as on the protein. Regarding the chromophore the spectra will often be compared with resonance Raman spectra of the intermediates,
some of which will be shown in the paper by Stockburger.

BR-570-K610 DIFFERENCE SPECTRUM

In Fig. 5 a static BR570-K610 difference spectrum
is presented. It was measured at 77 K to stabilize the
intermediate. A detailed discussion of the difference
spectrum has been published by us recently (7), and
similar data have also been obtained by two other
groups (12, 13). Here I will only emphasize some

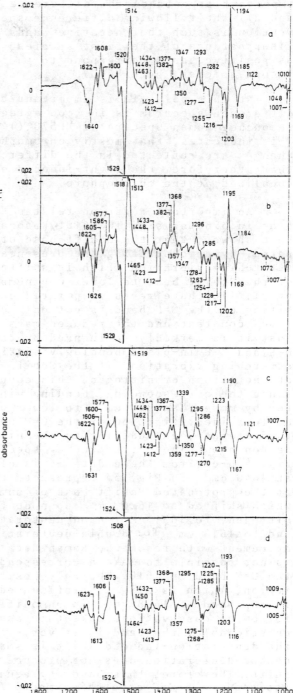

Fig. 5 Static
BR570-K610 dif-
ference spectra
a: unmodified BR;
b: BR with deuter-
ation of the Schiff
base nitrogen;
c: deuteration of
the Schiff base
carbon;
d: double deuter-
ation

important aspects of the difference spectrum. In this
and all the following difference spectra the convention is such that negative bands correspond to the
disappearance of the first species, here BR570, whereas positive bands correspond to the appearance of the
second species, here K610.

At this point it is important to note that the
difference spectrum is in good agreement with the
resonance Raman spectra of BR570 (14) and K610 (15).
This indicates, that mostly chromophore molecular
changes are reflected in the difference spectra, and
that, therefore, the light-induced excitation is still
localized at the chromophore.

An important question relates to the retinal-
protein binding site in K610. Resonance Raman spectra
have shown , that the Schiff base of BR570 is protonated, by demonstrating the presence of the C=N
stretching vibration which is shifted by deuteration
of the Schiff base to lower wavenumbers (14). Aton
et al. (16) have recently pointed out, that the position of the C=N band is not only determined by the
force constant and the reduced mass of the system but
that it is markedly influenced by the coupling of the
N-H and C-H in-plane bending vibrations to the C=N
stretching vibration. If the Schiff base is deuterated
at the carbon or nitrogen, this coupling is removed,
since the corresponding bending vibrations are shifted by more than 300 cm^{-1} to lower wavenumbers. For
a clear identification of the C=N band it is important
to deuterate, in addition to the nitrogen, also the
carbon atom. The corresponding spectra are shown in
Fig. 5 b-c. From these data it is clear, that the band
at 1640 cm^{-1} in Fig. 5a represents the C=N vibration
of the protonated Schiff base of unmodified BR570,
which shifted to 1626 cm^{-1} upon deuteration of the
nitrogen, to 1631 cm^{-1} upon deuteration of the carbon,
and to 1613 cm^{-1} for double-deuteration. This is in
agreement with resonance Raman data (14). However, it
is not possible to make a corresponding assignment for
the C=N band in K610. The only positive band in this
region, which is slightly influenced by deuteration
is the band at 1608 cm^{-1} in unmodified BR570. The
following observations make this assignment highly
questionable: the band is at very low wavenumbers;
the deuteration-induced shift is small (2-3 cm^{-1});
double-deuteration does not produce an additional
shift. Therefore, this band may equally well be
assigned to a C=C stretching mode. Additional

labelling methods are required to elucidate the structure of the retinal-binding site in K610. Using 13-C
and 14-N labels would be very useful, since they would
produce only reduced mass effects, avoiding the complexity, which is brought about by shifting the N-H
and C-H in-plane bending vibrations. In any case, one
must conclude, that major changes occur in the Schiff
base region. This may be a hint, that the Schiff base
has a special role in the proton pumping mechanism.
Further evidence for this will be presented below.

The difference spectra are dominated by the
strong C=C stretching bands of BR570 (at 1529 cm^{-1})
and of K610 (at 1514 cm^{-1}). The fact that the bands
are partially influenced by deuteration indicates,
that mostly the terminal part of the retinal is reflected in these bands.

The region between 1500 cm^{-1} and 1300 cm^{-1}
appears very complex, and an interpretation is beyond the scope of this presentation. Several bands
are candidates for protein molecular changes. We have
measured spectra of bacteriorhodopsin, which was
labelled with completely deuterated tyrosines. The
labelling was made by the group of D. Engelman at
Yale University by growing bacteria in a synthetic
culture medium containing deuterated tyrosine in the
amino acid mixture. Clear spectral changes around
1450 cm^{-1} were produced by this labelling, indicating,
that even at this early stage of the photocycle protein molecular changes are involved. Another interesting aspect of this spectral region are the bands at
1347 cm^{-1} in unmodified K610 and at 1339 cm^{-1} in
K610, for which the Schiff base carbon was deuterated.
Both bands are removed by deuteration of the Schiff
base nitrogen. These results indicate, that the bands
can be assigned to the N-H bending vibration, and
that in K610 the nitrogen has still its proton attached.

The fingerprint region (1300 cm^{-1}-1100 cm^{-1})
deserves special interest. For free retinals and
model compounds it is indicative for the isomeric
state of the retinals. Chemical extraction methods
present evidence, that the retinal isomerizes during
the photocycle from all-trans to 13-cis and back to
all trans (17). Many models for the early photochemistry of bacteriorhodopsin assume, that this
isomerization takes place already during the BR570-
K610 transition. Our difference spectra neither do
support this hypothesis nor do they disprove it.

The bands of K610 are dominated by the strong band
at 1194 cm^{-1}, whose intensity equals that of the
C=C stretching band. No comparable results have been
found in the infrared spectra of model compounds
(retinylidine Schiff bases) of any isomeric composition. Therefore, no conclusions on the isomeric
state can be drawn. The strong band is splitted by
deuteration of the terminal carbon of the retinal.
Again this demonstrates that mostly this terminal
part of the retinal is reflected in the difference
spectrum.

BR570-L550 and BR570-M412 difference spectra

In Figs.6 and 7 the static and time-resolved BR570-
L550 difference spectra are shown, respectively. The
agreement, besides minor deviations, is obvious. This
demonstrates in this case the equivalence of the two
methods. It is striking, that the difference spectra
are dominated by negative bands, i.e. the bands of
BR570 are much stronger than those of L550. For
comparison, in Fig.8 the time-resolved BR570-M412
difference spectrum is presented. The spectra are
very much alike. It is generally accepted, that in
M412 the retinylidene Schiff base is deprotonated.
Evidence for this is obtained from resonance Raman
spectra and from the infrared difference spectrum.
The positive band at 1620 cm^{-1}, which is shifted to
1615 cm^{-1} by deuteration of the terminal carbon of
the retinal, indicates the presence of a deprotonated
Schiff base. There is a plausible argument, that for
a deprotonated retinylidene Schiff base the absorption strength of most of the retinal bands is
reduced as compared to the absorption strength of
the protonated Schiff base: protonation causes a reduced bond alternation in the conjugated retinal
system, and therefore an increased charge alternation,
rendering the molecule more polar and thereby increasing the absorption strength of most of the retinal bands. Infrared spectra of model compounds
support this supposition (18).

     The similarity of the BR570-L550 and BR570-M412
difference spectra could indicated, that already at
L550 the Schiff base is deprotonated. Support for
this is obtained from the negative band at 1640 cm^{-1},
representing the C=N stretching vibration of the
protonated Schiff base. The band is equally present
in both difference spectra. The absorption maximum

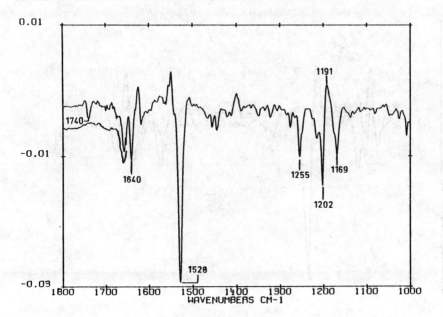

Fig. 6 Static BR570-L550 difference spectrum;
lower trace between 1800 cm^{-1} and 1650 cm^{-1}:
measurement in ^{2}H$_2$0

of L550 at 550 nm, however, is not conceivable with
L550 representing a deprotonated Schiff base. Due to
the weak absorption bands of L550 it is difficult to
make definitive statements on the structure of L550.
Two observations appear remarkable: there is still a
positive band around 1195 cm^{-1} in L550, which disappears during the L550-M412 transition (compare
Figs 6, 7 and 8); the positive band at 1608 cm^{-1} in
the BR570-K610 difference spectrum, for which it could
not completely be excluded that it represents the
C=N stretching band of the protonated Schiff base in
K610, disappears during the K610-L550 transition. To
explain all the results concerning the L550 intermediate, appreciable charge movements have to be

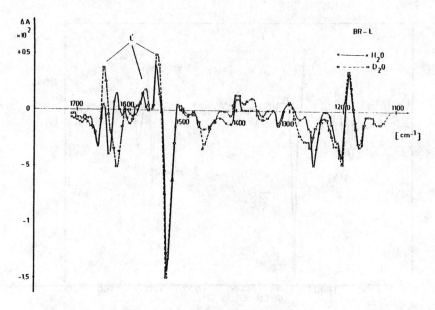

Fig. 7 Time-resolved BR570-L550 difference spectrum in H_2O and ^2H_2O

assumed, comparable to those occuring during the de-protonation step. These charge movements could render the retinal less polar. On the other hand such charge movements could play a crucial role in the proton pumping mechanism, by influencing the protein. Evidence for such processes will be presented below.

In Fig. 6 there is a peculiar negative band at 1740 cm^{-1}. The position could indicate the presence of a carbonyl group. The band is abolished by ^2H_2O. The only C=O stretching vibration, which is influenced by ^2H_2O is that of protonated carboxylic groups. However, in this case, a shift of 10 cm^{-1} to lower frequencies is usually observed. From the static

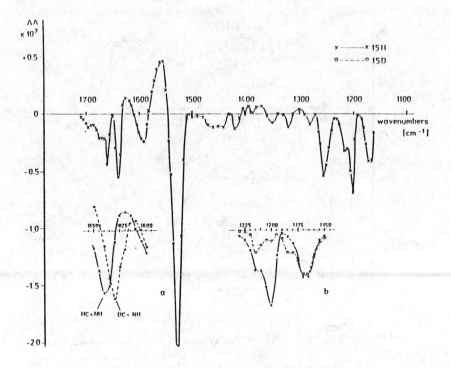

Fig. 8 Time-resolved BR570-M412 difference spectrum.
Inserts show special spectral regions and the effect
of deuteration of C-15 of the retinal

difference spectra, therefore, the band cannot be
identified with a carboxylic group. To obtain also
kinetic information, we investigated this spectral
region with time-resolved infrared spectroscopy. The
results are shown in Fig. 9, for H_2O and ^2H_2O
The insert shows signals representing the flash-induced transmission changes at 1740 cm^{-1} and at
1730 cm^{-1}. The fast phase of the signals exhibits
a half-time characteristic for the rise of L550,
whereas the slower phase of the signals proceeds with
a half-time characteristic for the rise of M412. It
is noteworthy, that the slower phase is slowed down
by a factor of about 5 by ^2H_2O. This kinetic isotope
effect is in agreement with that observed for the

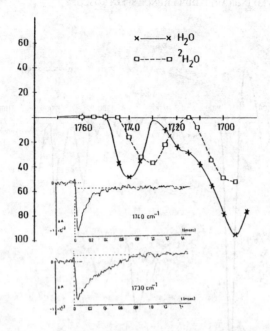

Fig. 9 Time-resolved BR570-L550 difference spectrum
in H_2O and ^2H_2O. The lower traces represent the
corresponding time-resolved transmission changes at
1740 cm^{-1} (H_2O) and 1730 cm^{-1} (^2H_2O)

rise-time of M412. The time-resolved BR570-L550 difference spectrum also exhibits the negative band at
1740 cm^{-1} for H_2O. Contrary to the static difference
spectra, this band is clearly shifted by ^2H_2O to
1730 cm^{-1}, indicating, that it represents a carboxylic
group. The discrepancy between the static and time-
resolved difference spectra is not resolved as yet. We
feel, however, that the time-resolved spectra are more
reliable in this case, since the measurement of the
kinetics of the processes provides additional information, which clearly helps to identify the L550 intermediate. The difference spectra can now be interpreted as follows: with the formation of L550 a carboxylic group is deprotonated and reprotonated with the

formation of M412. The charge movement   infered from
the BR570-L550 difference spectra (Fig. 6 and 7) could
initiate this deprotonation step.

Two other proton movement steps are shown in
Fig. 10 (6). It shows the BR570-M412 difference spectrum for H_2O (a) and ^2H_2O (b) in the spectral region
of carboxylic groups. The corresponding signals are
also presented. The spectral feature is shifted 10 cm^{-1}
to lower frequencies by ^2H_2O. It is evident, therefore,
that it represents the protonation of carboxylic
groups. Two kinetics can be discerned for the protonation steps: a faster process with the half-time of
that of the M412 formation, the corresponding band
being located at 1765 cm^{-1} and a slower process
having no correlation with the intermediates of the
photocycle. The corresponding band is located at
1755 cm^{-1}. A detailed discussion of these processes
has been given in (6). These two carboxylic groups
have also been detected by static difference spectroscopy (13, 19). The faster group is a plausible
candidate for a proton acceptor for the deprotonation
of the Schiff base during the M412 formation. The
role of the slower carboxylic group is more difficult
to visualize.

For the understanding of the role of these carboxylic groups it would be important to know, which is
part of a glutamic acid and which is part of an aspartic acid residue. If this question is settled, a
tentative assignment of these amino acids to specific
amino acids of the known primary structure can be
made, taking into consideration the retinal binding-
site and the suggested fit of the primary structure
into the seven helical columns (20). In collaboration
with M. Engelhardt at the Max Planck-Institute in Dortmund we could show, that both bands at 1765 cm^{-1} and
at 1755 cm^{-1} are causes by aspartic acids. The group
in Dortmund succeeded in incorporating aspartic acid
into bacteriorhodopsin, whose β-carboxylic group is
labelled with ^{13}C by growing bacteria in a synthetic
medium containing this labelled amino acid in the
amino acid mixture. Control experiments with radioactive labelled aspartic acid indicate an incorporation of about 70%. Approximately 5% of the radio-
activity migrated also to the glutamic acid. However,
if the general biosynthetic pathway is assumed to be
valid also for halobacteria, only the α-carboxylic
group should be labelled. Fig.11 shows the time-resolved difference spectrum of bacteriorhodopsin containing labelled aspartic acid. It corresponds to the
difference spectrum of Fig.10. It is evident, that

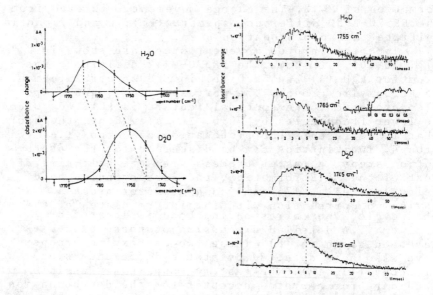

Fig. 10 Time-resolved BR570-M412 difference spectrum.
The traces to the right represent the time-resolved
transmission changes in H_2O and H_2O, at wavenumbers
as indicated

both bands at 1765 cm^{-1} and 1755 cm^{-1} are reduced to
about 30% and that a new band arises around 1720 cm^{-1}.
A kinetic analysis shows, that again the two protonation steps as in Fig.10 are reflected in this band.
Therefore, we have to conclude, that both carboxylic
groups infered from Fig.10 are aspartic acids. Quite in
the neighborhood of the retinal binding-site (lysine
216) is the aspartic acid 212. Due to the proximity to
the retinal Schiff base it probably represents the
proton acceptor for the deprotonation step of the
Schiff base during the M412 formation.

The labelling experiments indicate, that the band
at 1740 cm^{-1} in Fig.9 is not influenced. Therefore, the

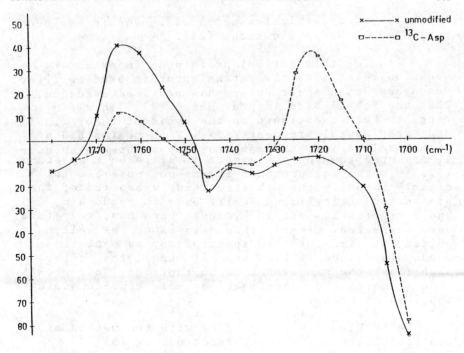

Fig. 11 Time-resolved BR570-M412 difference spectrum pf unmodified BR and BR containing ^{13}C-aspartic acid

corresponding carboxylic group is a glutamic acid. Further experiments are required to identify this glutamic acid with one of the primary structure of bacteriorhodopsin.

RHODOPSIN

The visual pigment rhodopsin is, as bacterio-rhodopsin, a chromoprotein containing retinal as the chromophore, which is bound to a lysine via a Schiff base. Here, the chromophore is 11-cis retinal, i.e. as compared to Fig. 3 a rotation around the 11, 12 double bond is present. An extensive review article

on the physico-chemical aspects of rhodopsin has been
published by Honig and Ebrey (21).

Also for this retinal protein an important
question relates to the retinal-protein binding site.
Resonance Raman spectroscopy has provided evidence,
that the Schiff base is protonated (22). This finding has been questioned on the basis of theoretical
and experimental arguments (23). With the method of
time-resolved infrared spectroscopy we have previously measured the rhodopsin-meta II difference spectrum.
(5). In these measurements it was not possible to
clearly identify the C=N stretching vibration of the
protonated retinylidene Schiff base in rhodopsin,
and a protonation via an hydrogen bond has, therefore,
been suggested. Clearly rhodopsin is not as well
suited for time-resolved spectroscopy as bacteriorhodopsin, since it bleaches; it means, that after
each flash the sample has to be replaced. This procedure makes the investigations very time-consuming
and less accurate.

We tackled now the problem with the method of
static infrared difference spectroscopy (8). In the
following some of the results will be presented
which appear to be important with respect to the retinal-protein link. As for bacteriorhodopsin, the
first photoproduct of rhodopsin, i.e. bathorhodopsin
can be stabilized at liquid nitrogen temperature.
However, if bathorhodopsin absorbs light, it converts
not only back to rhodopsin, but also, with lower
quantum yield, to isorhodopsin. This last species is
stable at room temperature. It contains 9-cis retinal as the chromophore. If bathorhodopsin is warmed
up it converts to lumirhodopsin, then to metarhodpsin
I and finally to metarhodopsin II. Whereas all the
former products exhibit an absorption maximum shifted
far to the red as compared to an unprotonated Schiff
base, metarhodopsin II (meta II) shows an absorption
maximum almost coincident with that of an unprotonated Schiff base. Meta II remains stable at 0°C for
at least 30 minutes, before it decomposes into free
retinal and the apoprotein opsin. With the formation
of meta I at the latest, the retinal has isomerized
from 11-cis to all-trans.

In Figs.12, 13 and 14 the rhodopsin-bathorhodopsin, rhodopsin-isorhodopsin and rhodopsin-meta II
difference spectra are shown, respectively. a) refers
to measurements in H_2O, b) to measurements in ^2H_2O.

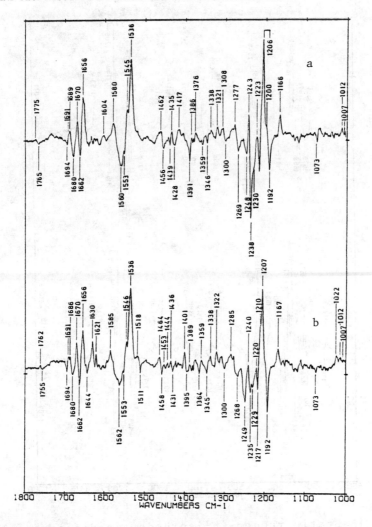

Fig. 12 Static rhodopsin-bathorhodopsin difference
spectra at 77 K; a: H_2O; b: ^2H_2O
Contrary to what one would expect for difference
spectra between protonated retinylidene Schiff bases
many bands are present in the spectral region between
1600 cm^{-1} and 1700 cm^{-1} (C=N stretching band region)
in Figs. 12 and 13. The same holds also for Fig. 14,
the difference spectrum beween a protonated Schiff
base and an unprotonated Schiff base. ^2H_2O has little
influence on the rhodopsin-bathorhodopsin difference
spectrum (Fig. 12). It merely causes a splitted band
at 1621 cm^{-1} and 1630 cm^{-1} for bathorhodopsin, without
influencing the bands of rhodopsin. However, in the

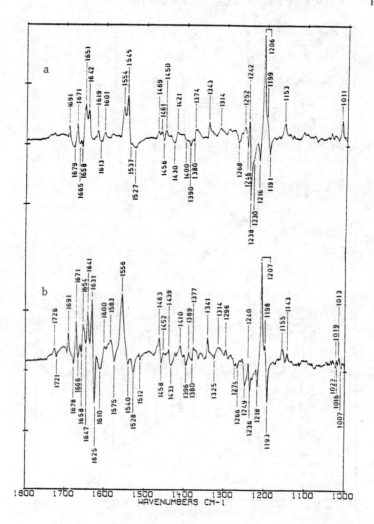

Fig. 13 Static rhodopsin-isorhodopsin difference
spectra at 77 K; a: H_2O; b: ^2H_2O

rhodopsin-isorhodopsin difference spectrum (Fig.13)
^2H_2O causes two new strong and sharp bands, one for
rhodopsin at 1625 cm^{-1} and one for isorhodopsin at
1631 cm^{-1}. From their position one can conclude that
they represent the stretching vibration of the HC=ND
group. To explain the lack of corresponding bands
in Fig. 12b one has to assume, that the bands for
rhodopsin and bathorhodopsin coincide, causing in
the difference spectrum this splitted structure. This
has been demonstrated by measuring the isorhodopsin-
bathorhodopsin difference spectrum (data not shown here),

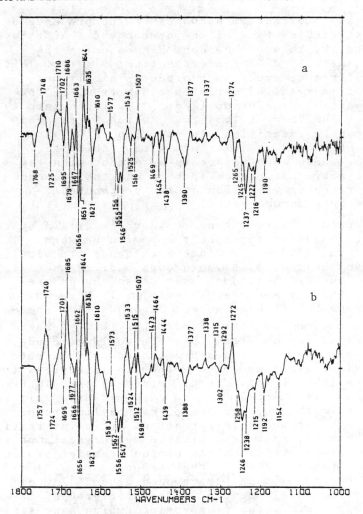

Fig. 14 Static rhodopsin-meta II difference spectra at 0°C, pH 6; a: H_2O; b: ^2H_2O

in which the band at 1631 cm^{-1} for isorhodopsin and the band at 1625 cm^{-1} are present (8).

From these results one would expect the C=N stretching vibration of the protonated Schiff base around 1650 cm^{-1}. However ^2H_2O does not provide evidence for the presence of such bands in causing their removal. To explain the discrepancy one has to assume, that the C=N stretching bands of rhodopsin, isorhodopsin and bathodhodopsin all coincide and, therefore cancel each other in the difference spectra.

It still is important to demonstrate the presence of
the C=N stretching band of the protonated Schiff base
more directly. This can be done by measuring the
rhodopsin-meta II difference spectrum, Fig. 14. This
static difference spectrum reproduces nicely our
previous time-resolved difference spectrum, if the
higher spectral resolution (2 cm^{-1}) is taken into
account. A shoulder at 1651 cm^{-1}, which is removed
by H_2O can be identified. In addition H_2O causes and
increase of a band at 1621 cm^{-1} with a shift to
1623 cm^{-1}. These results are in agreement with the
low-temperature difference spectra and the presence
of the stretching vibration of the protonated C=N
group has been demonstrated.

It would be interesting ro know the origin of the
many other bands in the spectral region between
1600 cm^{-1} and 1700 cm^{-1}. Since such bands are also
present in the rhodopsin-isorhodopsin difference
spectrum, one could assume, that these bands are
caused by the chromophore. However, in the IR spectra of
retinylidene Schiff base model compounds (18) there
is no indication for other bands than the C=N band in
this spectral region. From this one would conclude,
that the absorbance changes are caused by the protein.
A clear discrimination between the two interpretations
is not possible as yet. The differential band in Fig.12
at 1770 cm^{-1} is shifted to 1760 cm^{-1} by H_2O. This indicates again the involvement of carboxylic groups
during the bathorhodopsin formation. It is interesting
in this respect, that from kinetic measurements at very
low temperature eveidence for proton transfer processes
has been obtained (24). In the rhodopsin-meta II difference spectrum a much larger band around 1750 cm^{-1}
indicates the involvement of carboxylic groups. It is
surprising, that in the fingerprint region many similarities can be observed between the rhodopsin-bathorhodopsin and rhodopsin-isorhodopsin difference spectra. In
addition, the rhodopsin-bathorhodopsin difference spectrum resembles the BR570-K610 difference spectrum. This
shows, that it is difficult to draw conclusions, without
additional information, from the fingerprint region on
the isomeric state of the retinal.

I would like to thank my coworkers W. Mäntele and
K. Gerwert for their contributions to this work. The
financial support from the Deutsche Forschungsgemeinschaft (Ac 16/14, SFB 60-D10) is gratefully acknowledged.

REFERENCES

1 Porter, G. in Techniques of Organic Chemistry, 2nd
   ed. Vol.VIII Part 2, 1055-1106, Interscience Publishers, New York, 1963
2 Matheson, M.S., Dorfmann, L.M. Pulse Radiolysis,
   MIT Press, Cambridge 1969
3 Czerlinski, G., Eigen, M., 1959, Z. Elektrochem.63,
   652-661
4 Chance, B., in The Enzymes, Vol.II, 428, Academic
   Press, New York, 1951
5 Siebert, F., Mäntele, W. Kreutz, W., 1981 Can. J.
   Spectrosc.26, 119-125
6 Siebert, F., Mäntele, W., Kreutz, W. 1982 FEBS Lett.
   14, 82-87
7 Siebert, F., Mäntele, W. 1983 Eur. J. Biochem. 130,
   565-573
8 Siebert, F., Mäntele, W., Gerwert, K. 1983 Eur. J.
   Biochem. in press
9 Stoeckenius, W., Lozier, R.H., Bogomolni, R.A. 1979
   Biochim. Biophys. Acta 505, 215-278
10 Ovchinnikov, Yu. A., Abdulaev, N.G., Feigina, M. Yu.,
   Kiselev, A. V., Lobanov, N. A. 1979 FEBS Lett. 100,
   219-224
11 Favrot, J., Leclercq, J.M., Roberge, R., Sandorfy,
   C., Vocelle, D. 1979 Photochem. Photobiol. 29,
   99-108
   Nakanishi, K., Balogh-Nair, V., Arnaboldi, M. Tsaji-
   moto, K., Honig, B. 1980 J. Am. Chem. Soc. 102,
   7945-7947
12 Rothschild, K.J., Marrero, H. 1982 Proc. Natl. Acad.
   Sci. USA 79, 4045-4049
13 Bagley, K., Dollinger, G., Eisenstein, L., Singh,
   A.K., Zimanyi, L. 1982 Proc. Natl. Acad. Sci. USA
   79, 4972-4976
14 Stockburger, M., Klusmann, Gattermann, H., Massig,
   G., Peters, R. 1980 Biochemistry 19, 4886-4900
15 Braimann, M., Mathies, R. 1982 Proc. Natl. Acad.
   Sci. USA 79, 403-407
16 Aton, B., Doukas, A.G., Narva, D., Callender, R.H.,
   Dinur, U., Honig, B. 1980 Biophys. J. 29, 79-94
17 Tsuda, M., Glaccum, M., Nelson, B., Ebrey, T.G.
   1980 Nature (Lond.) 287, 351-353
   Mowery, P.C., Stoeckenius, W. 1981 Biochemistry, 20
   2302-2306
18 Siebert, F., Mäntele, W. 1980 Biophys. Struct. Mech.
   6, 147-164
19 Rothschild, K.J., Zagaeski, M., Cantore, W. 1981
   Biochem. Biophys. Res. Commun. 103, 483-389

20 Engelman, D.M., Henderson, R., McLachland, A.D.,
   Wallace, B.A. 1980 Proc. Natl. Acad. Sci. USA 77,
   2023-2027
21 Honig, B., Ebrey, Th.G. 1974 Ann. Rev. Biophys.
   Bioeng. 3, 151-177
22 Mathies, R., Oseroff, A.R., Stryer, L. 1976 Proc.
   Natl. Acad. Sci. USA 73, 1-5
   Callender, R.H., Doukas, A., Crouch, R., Nakanishi,
   K. 1976 Biochemistry 15, 1621-1629
23 Shriver, J., Mateescu, G., Fager, R., Torchia, D.,
   Abrahamson, E.W. 1977 Nature 270, 272-275
   van der Meer, K., Mulder, J.J.C. Lugtenburg, J. 1976
   Photochem. Photobiol.24, 363-367
   and see references 11 and 5
24 Peters, K., Applebury, M.L., Rentzepis, P.M. 1977
   Proc. Natl. Acad. Sci. USA 74, 3119-3123

Intermediate States in Vision

P. M. Rentzepis
Bell Laboratories
Murray Hill, New Jersey 07974

Picosecond spectroscopy is more than fifteen years old and since its first experimental data on the relaxation of Azulene it has evolved into a rather mature field which may be utilized for probing picosecond events in the picosecond time scale. Absorption, emission and Raman picosecond spectroscopy have been utilized to probe primary events in chemistry, biology and physics. The events leading to the visual transduction processes after the initial absorption of a photon by the rhodopsin chromophore have been extensively investigated and varying aspects quite adequately presented in this conference.

The investigation of the primary event in vision was first performed by Busch et al.[1] These authors were able to measure directly, by means of a picosecond spectroscopic system similar to the one depicted schematically in Fig. 1, the rate of conversion of rhodopsin to bathorhodopsin at room temperature. These data are shown in Fig. 2. Also the conversion of bathorhodopsin to lumirhodopsin was determined by the same picosecond techniques. The fact that the rate of this first step is extremely fast, i.e., less than 6 ps was interesting since it was one of the fastest isomerization rates measured for large molecules. These original results have been confirmed by other investigations and catalyzed a large amount of research including Raman, infrared and model calculations aimed at mechanisms of the formation of the first intermediate. Our effort was directed

C. Sandorfy and T. Theophanides (eds.), Spectroscopy of Biological Molecules, 373–384.
© 1984 by D. Reidel Publishing Company.

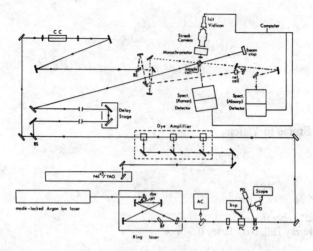

Fig. 1   Schematic diagram of a synchronously pumped dye laser system. AC: autocorrelator, P: polarizer, PC: Pockels cell, CP: crossed polarizer, hvp: high-voltage pulser, PD: photodiode, SHG: Second harmonic generating crystal, BS: beam splitter, CC: continuum cell.

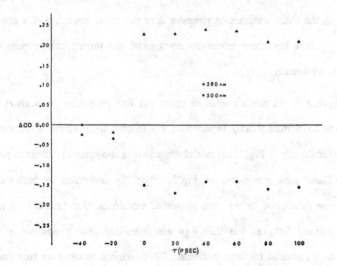

Fig. 2   Formation of the prelumirhodopsin band with a maximum at 580 nm induced by a 530 nm 6 ps pulse Δ 580 nm; ● 480 nm.

toward the measurement and understanding of the electronic spectra changes after irradiation with a very low energy picosecond duration 530 nm picosecond pulse.

The spectral changes are shown in Fig. 3. It is evident from these data that the rhodopsin band bleaches within the duration of the pulse and a new band, assigned to bathorhodopsin, appears within this period of time. In view of the remarkably fast formation of rhodopsin we decided to lower the temperature in order to slow down the isomerization as is the case with practically all other isomerization processes.

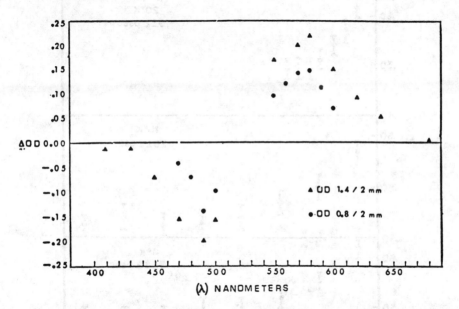

Fig. 3   Difference spectra of rhodopsin (bleaching) and bathorhodopsin (absorption). These spectra were taken 100 ps after excitation of bovine rhodopsin with a 530 nm 6 ps pulse.

The formation of bathorhodopsin was studied in optically clear glasses of the rhodopsin preparation in ethylene glycol over temperatures ranging from 300 K to 4 K (Fig. 4). Even at as low temperature as 4 K, the risetime of the appearance of bathorhodopsin was found to be extremely fast, 36 ps.[2]

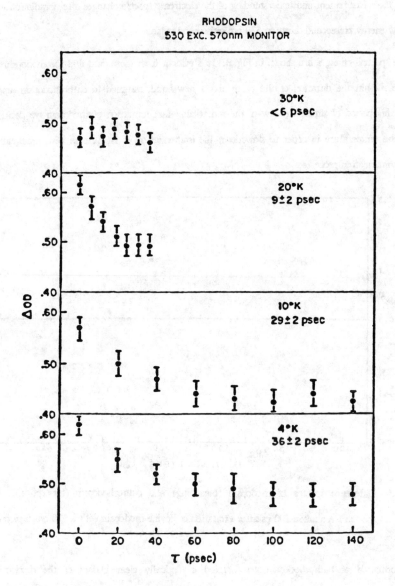

Fig. 4   Formation kinetics of bathorhodopsin monitored at 570 nm and at various temperatures.

A 5-mJ 530-nm, 6-ps excitation pulse was used. The risetime is given for each

temperature.

Originally the primary event which results in the formation of bathorhodopsin has been described as the isomerization of the 11-*cis*-retinal moiety of rhodopsin to the all-*trans* form.[3] Prior to our low temperature studies[1,2] the possibility of the occurrence of complete isomerization of the retinal chromophore within <6 psec has been questioned. In view of the very fast formation of bathorhodopsin at 4 K, we considered the possibility of proton translocation either as an alternate mechanism or in conjunction with partial isomerization for the primary event of visual transduction, at least at low temperature.

Samples of deuterated rhodopsin (D-rhodopsin) by shaking the rhodopsin preparation with D_2O. Under these conditions, only readily exchanged protons would be substituted with deuterium, and hydrogen atoms, for example, directly bonded to the hydrocarbon skeleton of the retinal moiety would be unaffected. A pronounced deuterium isotope effect was observed for the rate of formation of bathorhodopsin, $k_H/k_D = 7$ (Fig. 5). The temperature dependence of the rate of formation of bathorhodopsin also exhibited a non-Arrhenius behavior and at very low temperatures the rate was practically independent of temperature. These data (Fig. 6) revealed another interesting and as yet undetected experimentally process, namely proton translocation, which occurs via quantum mechanical tunneling at low temperatures. Analysis of the data allows for the calculation of the energy barrier through which the proton tunnels and, therefore, the distance required for the proton to translocate. The calculated distance of ~0.5 Å and the fact that the Schiff base proton was the only deuterium-exchangeable proton within the retinal moiety provided the basis for proposing that translocation of the Schiff base proton is at least a prominent component of the first event in vision.

A statement which has been used to support the hypothesis that *cis-trans* isomerization is the primary event in the visual transduction process is the commonly occurring statement that excitation of both 11-*cis*-rhodopsin and 9-*cis*-rhodopsin lead to a common bathorhodopsin. In a recent review, after a consideration of the various proposed mechanisms, it was concluded that the first event in the visual process is *cis-trans* isomerization at the 11-*cis* double bond of the retinal skeleton. This conclusion was based on the observation that photoexcitation of both 11-*cis*-rhodopsin and 9-*cis*-

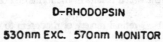

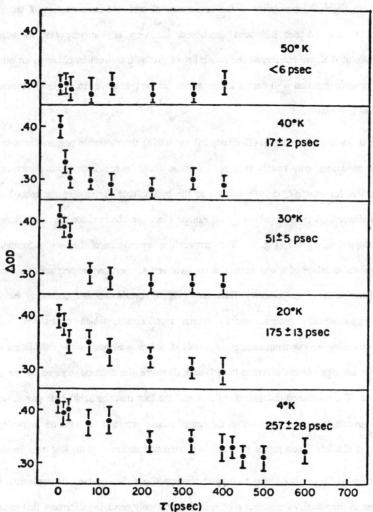

Fig. 5   Formation kinetics of deuterium-exchanged bathorhodopsin at 570-nm and at various temperatures. Excitation was performed with a 5-mJ, 530-nm, 6-ps pulse. The risetime is given for each temperature.

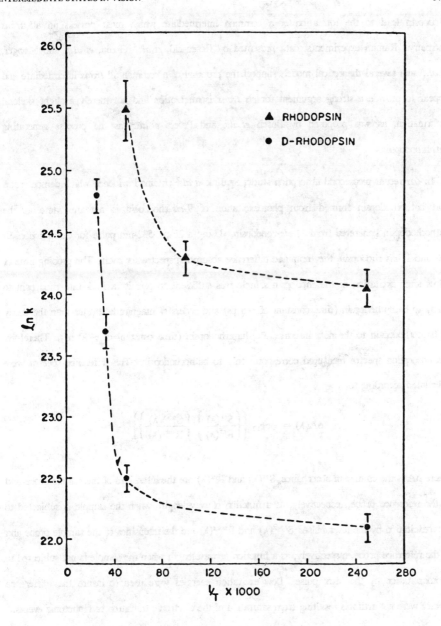

Fig. 6   An Arrhenius plot of the kinetic data in Figs. 4 and 5, ln k for formation of
prelumirhodopsin versus 1/T (K) × 10³.

rhodopsin lead to the formation of a common intermediate which must possess an all *trans* geometry. Raman experiments, data presented by Rosenfeld, *et al.*,[4] Green, *et al.*,[5] and Monger, *et al.*,[6] and several theoretical models support the proposal of a common all-*trans* intermediate and appear to provide a strong argument for *cis-trans* isomerization and not merely a slight skeletal deformation, as was proposed by Busch *et al.*[1] and Peters *et al.*,[2] as the process generating bathorhodopsin.

In our recent picosecond absorption study, Spalink *et al.*[7] obtained evidence which demonstrates that bathorhodopsin formed from photoexcitation of 9-*cis*-rhodopsin is not the same as the bathorhodopsin generated from 11-*cis*-rhodopsin. Using a 25-ps, 532-nm pulse for excitation of 9-*cis*- and 11-*cis*-rhodopsin, they obtained difference absorption spectra for each. The probing time of 85 ps after excitation of the rhodopsin sample was sufficient to permit excited-state rhodopsin to decay to bathorhodopsin (time constant of ∼6 ps) and orders of magnitude shorter than the decay of bathorhodopsin to the next intermediate, lumirhodopsin (time constant of ∼30 ns). Therefore, the absorption spectra measured correspond only to bathorhodopsin. The difference spectra were calculated according to

$$\Delta A(\lambda) = -\log_{10}\left\{\left\{\frac{S^{ex}(\lambda)}{R^{ex}(\lambda)}\right\} \Big/ \left\{\frac{S^{noex}(\lambda)}{R^{noex}(\lambda)}\right\}\right\}$$

where ΔA is the change in absorbance, $S^{ex}(\lambda)$ and $R^{ex}(\lambda)$ are the intensities of the sample probe and of the reference probe, respectively, as a function of wavelength, when the sample is subjected to photoexcitation by the laser pulse. $S^{noex}(\lambda)$ and $R^{noex}(\lambda)$ are the intensities of the sample probe and of the reference probe, respectively, as a function of wavelength when the sample is not subjected to photoexcitation by the laser pulse. Low excitation energies were used to insure that difference spectra were not artifacts resulting from saturation of the electronic transition or biphotonic events.

Figure 7 illustrates the difference spectra obtained 85 ps after excitation of 9-*cis*-rhodopsin and 11-*cis*-rhodopsin at 290 K. Two important observations not previously noted are quite evident upon

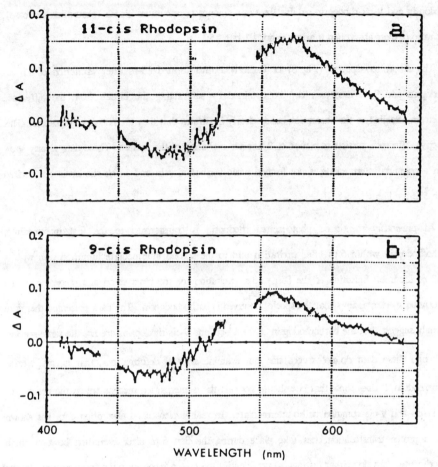

Fig. 7   Difference spectra of (a) 11-*cis*-rhodopsin (average of 58 laser shots) and (b) 9-*cis*-rhodopsin (average of 55 laser shots) obtained 85 ps after excitation with a 532 nm pulse.

inspection of both spectra. First, the absorption and bleached bands of the difference spectrum resulting from 9-*cis* excitation and its isosbestic point are shifted to longer wavelengths by ~10 nm relative to the corresponding positions in the 11-*cis* difference spectrum. This shift is similar to that observed for the ground-state absorptions of 11-*cis*- and 9-*cis*-rhodopsin. Second, while both 11-*cis*-

rhodopsin and 9-*cis*-rhodopsin exhibit the same amount of maximum bleach (absorption decrease), the ratio of their absorption maxima is not 1:1 but 1.5:1.

The absorption spectra (Fig. 8) of bathorhodopsins formed 85-ps after excitation of 11-*cis*-rhodopsin and 9-*cis*-rhodopsin can be obtained by subtracting the bleach from the difference spectrum. Scaling of the absorption bands clearly illustrates that they are not superimposable. This spectral difference persists to a time of 8 ns after excitation of the sample. These data clearly show that a common intermediate is not formed when 9-*cis*-rhodopsin and 11-*cis*-rhodopsin are excited with 532-nm light!

As previously mentioned photoexcited rhodopsin is transformed to the first intermediate, bathorhodopsin within 6 ps after excitation and to the second intermediate lumirhodopsin with a ~30 ns lifetime. Selection of the 85 ps after excitation for the observation and recording of the absorption spectra assures that the species observed is bathorhodopsin. The data presented therefore strongly suggest that the bathorhodopsin of the 11-cis and 9-cis rhodopsins are not the same species. Although these data do not record nor can examine skeletal deformations, that is, cis $\rightleftharpoons$ trans isomerization, it does show that one should not rest the isomerization mechanism on the premise of an 11-cis and 9-cis common batho-intermediate. In reality picosecond absorption data has shown that a proton translocation does take place during the first 6 ps after excitation however these experimental data do neither distinguish nor identify the proton involved in the translocation. It was proposed that the proton attached to the Schiff base would translocate and further that it would move towards the Schiff base nitrogen. Studies by Sandorfy and co-workers[8] suggest that this proposal is correct. Lately Raman studies presented in this symposium suggest that the Schiff base is fully protonated and conclude that the Schiff base proton does not translocate,[9] however the resonance Raman (nanosecond) spectra do not probe directly the N-H although the evidence indirect as it may be is quite convincing. Which proton translocates? No experimental proofs exists yet. All that is agreed upon is that a configurational change occurs, shown by Raman and depicted by the proton translocation mechanism.

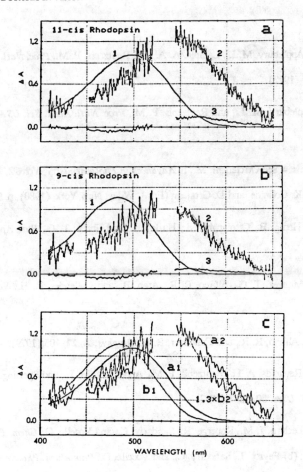

Fig. 8   (a) (1) 11-cis-rhodopsin spectrum; (2) calculated spectrum of bathorhodopsin from 11-

cis-rhodopsin (resolved from difference spectrum (curve 3) by assuming a 15% bleach);

(3) difference spectrum; (b) (1) 9-cis-rhodopsin spectrum (2) calculated spectrum of

bathorhodopsin from 9-cis-rhodopsin (resolved from difference spectrum (curve 3) by

assuming a 13% bleach; (3) difference spectra; (c) comparison of the two calculated

bathorhodopsin spectra (a_2 and b_2). Both are scaled to the same maximum absorption to

illustrate the spectral shift. a_1 and b_1 are the respective 11-cis- and 9-cis-rhodopsin

spectra.

REFERENCES

[1] Busch, G. E., Applebury, M. L., Lamola, A. A., and Rentzepis, P. M., *Proc. Natl. Acad. Sci. USA*, **69**, 2802 (1972).

[2] Peters, K., Applebury, M. L., and Rentzepis, P. M., *Proc. Natl. Acad. Sci. USA*, **74**, 3119 (1977).

[3] For a recent review see Ottolenghi, M., in *Advances in Photochemistry*, Vol. 12, J. N. Pitts, G. Hammond, K. Gollnide, and D. Grosjean (Eds.), Wiley, New York, (1980), p. 97.

[4] Rosenfeld, T., Honig, B., Ottolenghi, M., Hurley, J., and Ebrey, T. G., *Pure Appl. Chem.*, **49**, 341 (1977).

[5] Green, B. K., Monger, T. G., Alfano, R. R., Aton, B., and Callender, R. H., *Nature*, **269**, 179 (1977).

[6] Monger, T. G., Alfano, R. R., and Callender, R. H., *Biophys. J.*, **27**, 105 (1979).

[7] Spalink, J. D., Reynolds, A. H., Rentzepis, P. M., Applebury, M. L., and Sperling, W., *Proc. Natl. Acad. Sci. USA*, **80**, 1887 (1983).

[8] (a) Favrot, J., Leclerq, J. M., Roberge, R., Sandorfy, C., and Vocelle, D., *Chem. Phys. Lett.*, *53*, 433 (1978); (b) Favrot, J., Sandorfy, C., and Vocelle, D., *Photochem. Photobiol.*, **28**, 271 (1978).

[9] Mathies, R., Oseroff, A. R., and Stryer, L., *Proc. Natl. Acad. Sci. USA*, **73**, 1 (1976).

CORRELATION OF CELLULAR AND MOLECULAR CHANGES IN VISUAL PHOTORECEPTORS BY LIGHT SCATTERING RELAXATION PHOTOMETRY

By E.W. ABRAHAMSON, T.J. BORYS, B.D. GUPTA,
R. JONES, R. UHL, A. GEISTEFER AND
S. DESHPANDE
The Guelph - Waterloo Centre for Graduate
in Chemistry
The University of Guelph
Guelph, Ontario N1G 2W1

ABSTRACT:
    The correlation of molecular events with structural changes within a cell requires a non-destructive relaxation technique that can be adapted to measure such cellular changes in a time range of milliseconds to minutes. Light scattering relaxation techniques have proved useful for such studies as they can often be measured simultaneously or in parallel with spectral changes characterizing molecular or macromolecular processes. Such techniques are proving useful in the study of photobiological processes particularly in cases where specific cytological changes produced photochemically can be effected by alternate controlled perturbations such as osmotic shrinking or swelling of cell organelles and/or whole cells.

INTRODUCTION

    Since the early 50's biophysical and biochemical processes have been studied in terms of their molecular and macromolecular changes using a variety of relaxation methods such as flash photolysis, stopped flow, temperature and pressure jump techniques. Generally such processes have been monitored by rapid absorption and/or emission spectrometry. In intact cells many of these rapid molecular changes are coupled to equally rapid or somewhat slower cytological changes, e.g. electrical polarization and/or shrinking and swelling of organelle substructures as well as the entire cell itself. Electrophysiological techniques have been widely used to monitor electrical

385

C. Sandorfy and T. Theophanides (eds.), Spectroscopy of Biological Molecules, 385–407.
© 1984 by D. Reidel Publishing Company.

changes in cyto and in some cases a limited correlation of such changes with molecular changes has been
achieved.

    For the study of the broad range of intra-
cellular changes light scattering techniques are
coming into prominent use. The Biophysics Laboratories of the University of Freiburg have been pio-
neers in the development of light scattering relaxation techniques. In the early 70's they developed a
flash photolysis unit capable of measuring light
scattering changes in the millisecond range and have
initiated studies of the very rapid cellular dynamics
associated with the transduction process in visual
photoreceptors (1-3). In the late 70's Dr. Ranier Uhl
of the Freiburg laboratories visited our laboratory at
the University of Guelph for several years and introduced us to these techniques. Since that time our
laboratory and several others have followed the Freiburg lead and have joined them in the study of visual
photoreceptors.

    In this paper we report a selected summary of our
own studies on visual photoreceptors to illustrate the
utility of the method in correlating molecular and
cellular changes in the outer segments (ROS) of
vertebrate (bovine) rod cells.

THEORY OF LIGHT SCATTERING

    The light scattering properties of molecules and
particles in condensed media are theoretically in
three size domains (shown in figure 1): (i) Rayleigh
Scattering when the molecule or particle is less than
$\lambda/20$ where λ is the wavelength of the scattered light,
(ii) Rayleigh-Gans-Debye Scattering when the size of
the particle is in the range $\lambda/20$ to approximately
$\lambda/2$, and (iii) Mie-van de Hulst Scattering for particles with at least one dimension larger than $\lambda/2$.
Einstein (4) and Debye (5) have developed equation (1)
for application to particles of size less than $\lambda/20$
where $I\theta$ is the intensity of light of wavelength
scattered at angle θ and I_o is the incident intensity
of the unpolarized light. The parameters, r, M, c
and No

$$I_{\theta} = I_{o} \ \frac{2\Pi^2 (1+\cos^2\theta)(dn/dc)^2 Mc}{N_o\lambda^4 r^4} \tag{1}$$

are respectively the distance between the scattering
particle and the detector, the molecular weight of

Light Scattering Theory

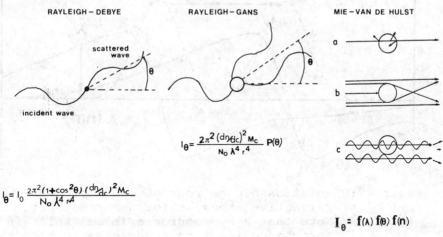

Figure 1.

the particle, the concentration of particles, and Avogadro's number. The derivative dn/dc is the differential increase in refractive index, n, with particle (solute) concentration. It is clear from equation (1) that Io will depend on the size of the particle as measured by M, the concentration of particles, C and the refractive index, n (real part).

If the particle absorbs radiation in addition to scattering this will have a pronounced effect on the scattering properties; the closer an absorption band is to the wavelength of scattered light the larger will be the refractive index n. Thus if a change in the light absorption of a particle occurs during a light scattering measurement there is a shift in the profile of the anomalous dispersion spectrum which, of course, will markedly change n and dn/dc (figure 2).

Most solutions of macromolecules and small particles have at least one dimension which is larger than $\lambda/20$ and so fall in the Rayleigh-Gans region. This situation gives rise to the phenomenon of destructive interference of the scattered light arising from different regions of the particle. Complete destructive interference can set in when the particles approach $\lambda/2$ in size. The angular dependence $P(\theta)$ in this case is much more complex than the simple $(1 + \cos^2\theta)$ of Rayleigh scattering and is given by equation (2).

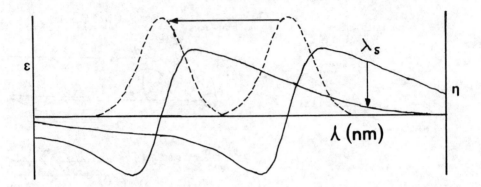

Figure 2:   Relationship between absorbance, ε and
            refractive index, n for a chromophore.
            Note that a hypsochromic (blue) shift in
            absorbance leads to a decrease in n at the
            wavelength λ_s for scattered light.

$$P(\theta) = \frac{2}{w^2} \ (e^{-w} + w-1) \tag{2}$$

where $w = (4\Pi/\lambda) \ (\sin^2\theta \ /2)R_G$ where R_G is the radius
of gyration of the particle which varies markedly
with shape. For a particle of $M = 5 \times 10^5 g$. R_G is
calculated to be 45 Å for a sphere and 2500 Å for a rod
10 Å in diameter (6).

     Of particular interest to our study and the
general study of biological cells and organelles is
the Mie scattering region where the particles are
larger than $\lambda/2$ in at least one dimension. For particle suspensions in this size range the scattered
light intensity is a complicated function of λ and θ
dependent on size, shape and internal structure of
the particle. Mie (7) early in the century calculated the scattering in this region for a sphere and
more recently van de Hulst (8) has extended the
treatment to oriented cylindrical particles. But as
biological particles have quite varied shapes such
simple, idealized treatments as these do not adequately take into account the shape factor as well as
scattering within the particle itself, i.e. internal
scattering.

     Although the scattering and transmission of
light by solutions or suspensions of biological particles are sensitive and convenient indicators of a

variety of physiological processes, the complications described above have made it difficult to relate, theoretically, changes in light scattering intensity as a function of λ and θ to a particular physical or conformational change of (or within) the particle, e.g. a volume change (9). To identify the conformational change giving rise to any given change in the scattered light intensity therefore requires alternate experimental methods that clearly produces the same scattered light λ and θ intensity patterns.

Bateman (10) osmotically induced volume changes in rod-like bacteria, measuring scattered light intensity changes $\Delta I\theta$ at 90° but his results when fitted to Mie theory gave only partial agreement as the optical changes proved to be very sensitive to particle size and suspension concentration. The latter parameter is particularly important as "multiple particle" scattering can occur and the onset of this complicating situation is a function not only of particle concentration but also of the shape and volume of the containing vessel. Reliable, interpretable data requires that "single particle" scattering conditions be maintained. A further complication arises if the particles are not "mono-dispersed", i.e. if they are not homogeneous in size, shape and internal structure. Thus meaningful light scattering measurements in terms of interpretations of qualitative and quantitative structural changes in a particle depend on how carefully the conditions described above are met.

LIGHT SCATTERING MEASUREMENTS

The apparatus used for measuring light scattering transients is similar to a conventional flash photolysis unit (figure 3). As an actinic flash source a commercial PRL xenon flash unit was used which could deliver pulsed flashes in the wavelength range 450-550 nm. of controlled intensity and time duration (50 micro sec.). For the scattered light source a tungsten lamp directed through a Bausch and Lomb monochromator with a grating blazed at 750 nm. was used. The emergent parallel light was directed through a water thermostated cuvette containing an aqueous ROS suspension. By means of a protractor mounting the scattered light detector, an EG&G photodiode (UV 44B), could be set at the desired angle θ.

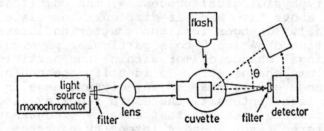

fig.3 light scattering apparatus.

The light scattering cell employed was a 1 cm.
square cuvette with 4 transparent optically clear
faces mounted in a spherical transparent water thermostat. An air thermostat was occasionally employed
using a Cambion model 801-2001-01 thermoelectric
module for temperature regulation. The square cuvette
limited the practical angular range from 0 to 30° and
60-90° over which $I\theta$ could be studied. In the range
30-60° a cylindrical cuvette can be employed.

Routine measurements were generally made at θ =
10° and at λ = 810 nm. Occasionally measurements at
0° were made along the axis of a 1 cm. cylindrical
cell.

Scattering signals from the photodiode detector
were preamplified and compensated to zero voltage by
means of the DC offset mode of a Tektronix differential amplifier. These were collected and displayed
in a Nicolet Explorer III digital storage oscillo-
scope which also stored the signals on floppy disks.

In this way signals could be retrieved from the floppy
disks and compared visually or printed out on an
Omnigraph X-Y recorder interfaced to the Explorer III
scope.

More elaborate versions of the apparatus shown
in figure 3 now exist in the University of Freiburg
laboratory of K. P. Hofmann and in Rainer Uhl's laboratory at the Max Planck Institute at Gottingen.
Hofmann's unit is equipped to measure absorption
spectral changes as well as light scattering changes
(11). He also has a facility for making studies in
magnetic fields which orient the ROS axially in the
field (12).

THE STRUCTURE OF THE VERTEBRATE PHOTORECEPTOR CELL

The vertebrate rod and photoreceptor cell has
two morphologically and functionally distinct regions;
(figure 4) the inner segment (RIS) containing organelles concerned with the metabolic functions of the
cell such as the nucleus and mitochondria which are
mediated by a oubain sensitive Na^+/K^+ ATPase, and the
outer segment (ROS), containing the visual pigment
rhodopsin, joined to the RIS by a tubular ciliary
process. The vertebrate ROS consists of a stack of
500-2000 disks (depending on the species) which is
enveloped by a plasma membrane (13). In the bovine
case the ROS is quite small, about 10μ long and 1μ in
diameter. The disks are flat saccules with a phospholipid bilayer membrane about 60-70 Å thick, topologi-
cally enclosing an inner aqueous space (lumen) between
10 and 30 Å in depth (14). Separating the disks is a
cytoplasmic region about 150 Å thick. A substantial

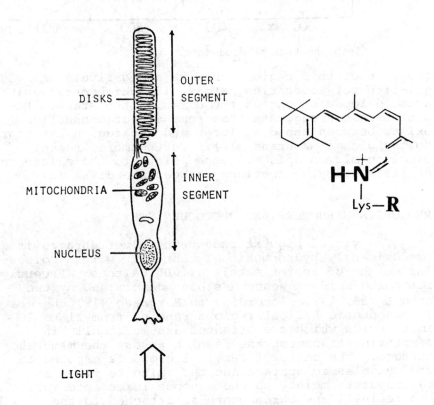

fig.4 Rod cell and 11-cis-retinal Schiff base.

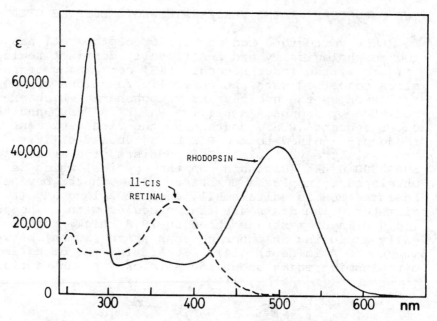

fig.5 Spectrum of Rhodopsin.

fraction of this region contains a negatively charged,
gel-like polysaccharide matrix (15) which presumably
acts to loosely connect the disks in the stack. Recent evidence suggests that some direct connection
exists between the disk lumen and the region exterior
to the plasma membrane (16). Furthermore, recent
electron microscopic evidence points to the existence
of filamentary connections between the disks (17).

RHODOPSIN STRUCTURE AND PHOTOCHEMISTRY

        The visual pigment rhodopsin in vertebrates is a
predominantly hydrophobic glycoprotein with a molecular weight of approximately 39,000 daltons. Recently
its amino acid sequence has been determined by two
groups (18, 19). According to Hargrave (19) there are
7 hydrophobic helical regions ranging from 21 to 28
amino acids which are oriented assymetrically in
serpentine transmembrane fashion across the membrane
surface. The carboxyl terminal group is exposed to
the cytoplasmic surface and the amino terminus and
carbohydrate moiety to the aqueous lumen interior.
The retinylidene chromophore is attached to the
epsilon-amino group of a lysine residue situated in
the interior of the membrane on helix number 7.

The 11-cis retinylidene chromophore derived from
the aldehyde 11-cis retinal through formation of a
Schiff's base linkage with lysine is shown in figure
4. If one considers that the 7 helices are arranged
in a cylindrical tertiary structure an estimated diameter of the rhodopsin molecule is about 900 $Å^2$.

The absorption spectrum of rhodopsin (figure 5)
has three characteristic regions: an intense long
wavelength band in the visible generally with a maximum at 500 nm. in most vertebrates, a weaker band at
about 350 nm. arising from the cis configuration of
the polyene chromophore and a third maximum at about
280 nm. ascribed to aromatic amino acids of the protein moiety (13).

Absorption of light by the rhodopsin molecule
results in a complex sequence of chromophoric spectral changes covering a time range from picoseconds to
minutes. In figure 6 this sequence is shown. The
times shown are the approximate transition times at
physiological temperatures (37°C) and the temperatures
are approximately those at which the intermediate can
be stabilized (13). The very rapid primary photochemical process appears to be a photoisomerization of the
11-cis retinylidene chromophore to a somewhat torsion-
ally distorted all-trans form (bathorhodopsin). But
this process may also involve a displacement in a
hydrogen bond that could conceivably link the nitrogen
atom of the chromophoric Schiff's base linkage to
some electronegative group on the apoprotein opsin
(13).

The visual photobiologist is primarily interested
in the process of visual transduction by which the ROS
converts an absorbed photon of light into highly amplified hyperpolarizing receptor potential in times of
the order of a few milliseconds. The amplitude of
this potential is controlled over more than 5 log
units of incident light intensity. Obviously one or
more processes in the sequence shown in figure 6 must
couple with cellular activity in the ROS and it is
the purpose of studies such as will be described here
to develop a viable model for this coupling. Since
visual transduction occurs in the millisecond time
range the process Metarhodopsin I → Metarhodopsin II
is the latest process in the photolytic sequence that
can couple to cellular activity leading to transduction. Furthermore this process is particularly
attractive in this regard since it involves reaction
with protons in its disk membrane environment.

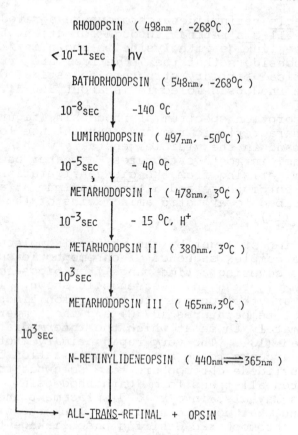

fig.6 sequence of Rhodopsin bleaching.

METARHODOPSIN I_{478} → METARHODOPSIN II_{380} (MI/MII)

This process appears to have a number of complications. Early studies of the kinetics in digitonin
micelles (20,21) indicated several isochromic forms
of MI each decaying by a first order process to MII.
Studies of an excised eye of an albino rabbit (22)
and bovine ROS suspensions (23,24) were interpreted
as decaying in a single first order process. For the
ROS suspensions at pH=7 and 37°C the half life of the
process was about 0.5 msec. The enthalpy of activa-
tion, ΔH^{*} was on the average about 32 kcal mole^{-1} and
the entropy of activation, ΔS^{*}, 55 cal/deg.$^{-1}$ mole^{-1}.
Recent studies of this process in ROS (25,26) favour
two separate first order decay processes for MI.

At temperatures near 4°C in digitonin suspension
MI and MII appear to exist in a quasi stable equili-

brium (27,28) according to equation (3) where X has
a value near unity. In ROS suspensions X varies

$$XH^+ + MI \rightleftharpoons MII \tag{3}$$

between 1 and 2 (29). The appearance of more than
one form of metarhodopsin I may account for this vari-
ability in X. In uncharged detergent micelle preparations Plantner and Kean (30) have isolated three
different forms of micellar rhodopsin by isoelectric
focusing. This would suggest that the MI/MII conversion rate in cyto could be controlled by the degree
of protonation of the rhodopsin molecule.

CELLULAR CONFORMATION STUDIES IN ROS

     Most of our studies were performed on bovine rod
outer segments from young heifers excised within
several minutes of the animal's death. ROS were prepared from excised eyes by sucrose density gradient
technique (31) and were then passed through a syringe
needle packed with glass wool. Electron micrographs
showed such preparations were essentially whole rod
outer segments with perforated plasma membranes.
Such preparations were stored in plastic vials under
liquid nitrogen until ready for use. This treatment
had only a minimal effect on the scattering properties
as compared to fresh preparations.

     For most measurements a particle density equiva-
lent to 1µM rhodopsin was used to insure single particle scattering. Flash intensities for the bulk of
the studies was less than 15% for a single flash.

     Several distinct types of light scattering sig-
nals can be observed in ROS suspensions depending on
conditions. These are discussed below.

The A Signal:

     The A signal is unique in that it shows both a
dark (A_D) and a light-stimulated component (A_L) (33
35).

     The dark A_D signal - This is produced when to an
ROS suspension, buffered to pH=7.3 with Tris HCl, varying amounts of Mg-ATP are added. A relative decrease
in scattered light intensity ($-\Delta I/I$) is observed which
appears to reach near saturation at room temperature
in approximately 10 minutes (figure 8). This decrease,
plotted as an increase in ($-\Delta I/I$), is accompanied by

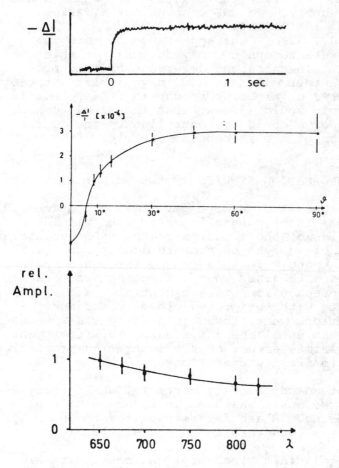

Figure 7.

At the top is an N signal obtained from an ROS suspension by bleaching 20% of the rhodopsin with 1 flash of actinic light. T = 21°C, pH 7.3, $\Theta = 10°$, $\lambda = 800$ nm.

Below are the dependences of N on measuring angle Θ, and wavelength of the measuring light λ.[38]

an increase in the appearance of inorganic phosphate (35) indicating that ATP has been hydrolyzed enzymatically via a Mg-ATPase in the disk membranes (33,34). Since the A_D signal is insensitive to ouabain and succinate dehydrogenase activity is negligible the ATPase cannot be a Na^+/K^+ ATPase arising from the inner segment or mitochondrial contamination.

The properties of the dark A_D light scattering signal can be summarized as follows:

(i) A_D is substantially inhibited by the non-hydrolyzable ATP analog, AMP-PNP in concentrations comparable to ATP (31,34). GMP-PNP has no effect on A_D.

(ii) Both vanadate and calcium at concentrations of 1/10 that of magnesium showed substantial inhibiting effects on A_D (31,32)).

(iii) The nucleotide cGMP when present at 1/10 the concentration of ATP shows a substantial enhancement of A_D.

(iv) The A_D signal is independent of the presence of the so-called 'G' protein (34) as a low ionic strength extraction (4mM NaCl at 4°C) leaves A_D unaffected.

(v) The kinetics of A_D is essentially first order in the presence of excess Mg-ATP with an activation energy, $E_a = 24.0$ kcal mole^{-1} (31)

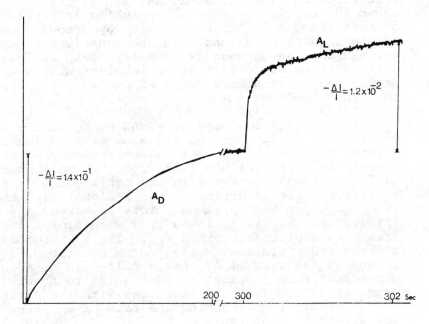

Figure 8.

An ROS suspension in Tris HCl pH 7.3 is incubated in the dark for 300 sec. to give A_D. At 300 sec. it is flashed with actinic light and monitored for an additional 2 sec. to detect A_L. $\Theta = 10°$. $\lambda = 800$ nm.

In addition to the above points the A_D signal is sensitive to the ROS particle concentration. In our apparatus the range "single particle" scattering was between zero and about 1mM rhodopsin. All our experiments were done in this range. Above this concentration range the amplitude ($-\Delta I/I$) of A at $\theta = 10° \pm 1°$ decreases with increasing particle concentration and actually is inverted above 8 mM rhodopsin concentration. We stress here that angular resolution is very important particularly in the range 0-10° as the variation of $-\Delta I/I$ with θ varies sharply and shows biphasic behaviour in this region.

In order to identify the particular cellular conformation change responsible for A_D perforated ROS suspensions were subjected to an osmotic perturbation. If one assumes that in the perforated ROS the individual disks within the stack are collectively the sole osmotic compartments and that the osmotic perturbations produce only volume changes, then the volume changes presumably reflect changes in the volumes of the individual disks. In figure 9 is shown the results of osmotic shrinking and swelling of perforated ROS suspensions with hypertonic and hypotonic NaCl solutions. The reasonable agreement between angular dependency profiles of A_D and osmotic disk swelling suggests that A_D involves disk swelling occasioned by an ATPase mediated pumping of hydrated ions into the disk lumen.

The Light Stimulated A_L signal. - Following the development of the dark A_D signal over a period of several minutes, if the ROS suspension is then flash-illuminated with actinic light (500-540 nm.) there is an instantaneous increase in the amplitude of ($-\Delta I/I$) similar to that occurring in the N signal save that it is much larger in amplitude. This is termed A_{Lf} (figure 8). A_{Lf} is followed by a much slower increase in amplitude, termed A_{Ls}. The A_{Ls} process is a first order process. At a photobleaching of 15% rhodopsin the rate of A_{Ls} has a half-life of about 30 seconds at room temperature decreasing to about 300 msec. at 37°C. Its rate increases with the extent of photobleaching, approaching the rate of A_D at very low levels of photolysis.

A_{Ls} in all its properties save rate duplicates those of A_D (31). Angle θ and wavelength λ dependencies of A_{Ls} are the same as A_D. Enhanced inorganic phosphate appearance is observed in the initial

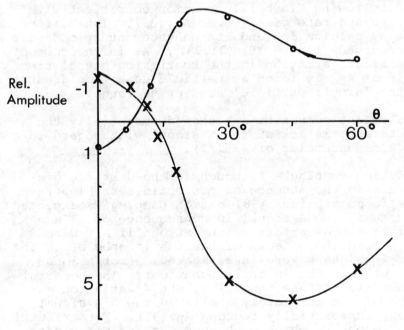

Figure 9.

The angular dependence of osmotically induced swelling
(o) and shrinking (x) of ROS as measured at λ = 800 nm.

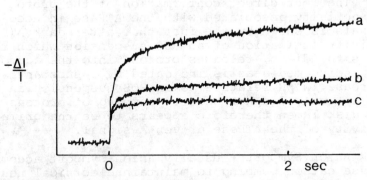

Figure 10.

a) A_L signal obtained from ROS in 60 mM Tris HCl, pH 7.3,
   1 mM $MgCl_2$ and 1 mM ATP incubated for 3 minutes in the
   dark and then flash illuminated at 0 sec.

b) as in a) except 0.5 mM $CaCl_2$ is also present.

c) is as a) but in the presence of 1 mM $CaCl_2$. Note that
   when Ca^{+2} = Mg^{+2} there is essentially 100% inhibition of
   both A_D and A_L since curve c) is the same as an N
   signal obtained without any added ATP. θ = 10°

period following flash illumination consistent with
the increased rate of A_{Ls} over A_D (35). Inhibition
of A_{Ls} by calcium ion and its enhancement by cGMP are
also observed (figure 10) (31,35). As in the case of
A_D a similar variation in the scattering signal time
profile of A_{Ls} is found at particle densities greater
than the 'single particle' scattering range.

The fast component A_{Lf} showed essentially the
same characteristics of the N signal with regard to
θ and λ dependencies of $(-\Delta I/I)$.

Cation Pumping. - Although in the dark the disk
membranes in the absence of nucleotides are imper-
meable to most anions (38) cation pumping is expected
to accompany the A signal in consequence of ATPase
activity. To ascertain which cations if any were
being pumped the effects of various ionophores on the
A_D and A_L signals were investigated. Proton ionophores were unique in their behaviour. The powerful
mitochondrial ATPase uncoupler, SF6847 at a concentration of 9µM completely abolished the A_{Ls} signal
and very substantially reduced A_D (31). For tributyl
tin an ionophore which exchanges OH^- and Cl^- across
membranes both the A and N signals were identical (31).
Ionophores specific for cations other than protons
did not show this behaviour.

The most direct confirmation of the proton pumping activity associated with the ATPase induced A
signals is shown by the fact that after dark ATPase
activity sonication of an ROS suspension which rup-
tures the disks, releases protons into the aqueous
suspending media as is indicated by a substantial
decrease in pH relative to an ROS suspension in which
ATP activity did not occur. Pumping of protons into
the disk lumen therefore appears to be the primary
activity of the ATPase driven A signal.

Anion Symport - Usually anion symport accompanies
ATPase cation pumping to maintain electrical neutrality. The anion symported is usually a normally im-
permeable one which requires a protein carrier closely coupled to the cation ATPase pump. Diisothiocyano-
stilbene sulfonic acid (DIDS) generally is a blocking
reagent for such anion carriers. In our case we
found that DIDS very substantially reduced the normal
A_L signal. For the case of the normally permeant
anion SCN^- which requires no protein carrier the rates
of both A_D and A_L are enhanced over their normal behaviour. Chloride ion which is normally present in

Tris buffer is the most effective anion of all the impermeant anions studied. Bicarbonate anion can replace Cl^- but in this case the A_{Lf} signal is considerably reduced suggesting a buffering action of bicarbonate within the lumen that changes the MI/MII ratio.

The A Signal in Whole Bovine ROS.- The A signal can be observed in ROS preparations in which the plasma membrane is left intact (31). No exogenous Mg-ATP need be added to such preparations. Apparently endogenous Mg-ATP present in the ROS begins to act as soon as the frozen ROS preparation is brought to room temperature by the addition of Tris buffer solution. The A_D and A_L signals seen under these conditions are comparable to those observed in the presence of Cl^- rather than HCO_3-.

The G Signal:

If instead of Mg-ATP one adds GTP (250 μM.) along with a divalent anion (Mg^{2+}, Ca^{2+} or Mn^{2+}) at a concentration of 1.2 mM to a suspension of ROS in 60 mM Tris HCl buffer one observes, on flash illumination, a scattering signal at room temperature in which $(-\Delta I/I)$ increases with time, approaching near saturation in 5 to 10 msec. This is termed the G signal. Unlike the A signal there is no dark G signal comparable to A_D and since G is studied at levels of rhodopsin bleaching $\leq$ 10%, compared to 25% usually used for A_L, one does not observe a pronounced initial sharp amplitude increase in the G signal comparable to A_{Lf} associated with the A signal (figure 11).

Kühn et al. (36) have studied the G signal under conditions of high, 'multiple particle' scattering concentrations and very low levels of (inhomogeneous) photobleaching of rhodopsin. Their scattering profiles are very much different than our G signals taken under conditions of 'single particle' scattering (< 1 μM rhodopsin) (35,37).

The G signal differs from the A_{Ls} signal in the following respects:
(i)   G is inhibited by the non hydrolyzable analog GMP-PNP but not AMP-PNP. Since the A signal is inhibited by AMP-PNP but not GMP-PNP the two signals A and G are completely independent, the A signal arising from an ATPase activity and the G signal arising from a GPTase activity.
(ii) The G signal requires the presence of a G protein as shown by Kühn et al. (35) while the A signal

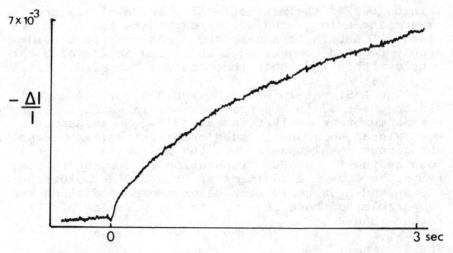

Figure 11.

The G signal obtained upon flash illumination of ROS
in the presence of GTP instead of ATP. $\Theta = 10°$. The
flash bleached about 6% of the rhodopsin.

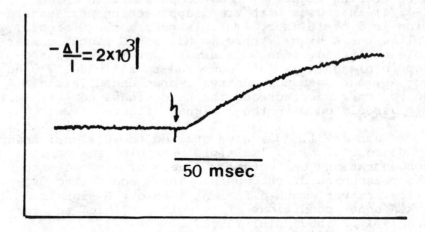

Figure 12.

The P signal obtained from an ROS suspension at pH 7, 18°C
$\Theta = 0°$. The flash bleached about 3% of the rhodospin.[39]

       is independent of the G protein.
(iii) The G signal does not show Ca^{2+} inhibition as
       does A.
(iv) cGMP present in concentrations < 0.1 that of GTP
       inhibits G while it enhances A.
(v)   The G signal increases in rate with the extent of
       photobleaching, reaching a maximum at 10% bleach-
       ing after which it decreases.
(vi) The kinetics of G show 2 distinct processes (29).
       The first of these is pH independent and first
       order in MII and G protein-GTP complex
               $$R = k_1 \, [G\text{-}GTP] \, [MII] \tag{4}$$
       with activation parameters $\Delta H^{\ddagger} = 15.74$ kcal $mole^{-1}$
       and $\Delta S^{\ddagger} = -4.8$ cal $deg.^{-1}$ $mole^{-1}$. The second pro-
       cess is pH dependent and requires G protein (37).
       Its rate equation is given by equation (5).
               $$R = k_2 \, [MII\text{-}G\text{-}GTP] \, [H^+] \tag{5}$$
       with activation parameters $\Delta H^{\ddagger} = 22.2$ kcal $mole^{-1}$
       and $\Delta S^{\ddagger} = +17.5$ cal $deg.^{-1}$ $mole^{-1}$. GTP hydrolysis
       occurs in this second step as the analog GMP-PNP
       inhibits this portion of the signal.

The P Signal:

     This signal has been studied in the Freiburg
laboratories (3,40) on ROS suspension with perforated
plasma membranes in Pipes buffer. Figure 12 shows
the time course of a typical P signal which saturates
at about 15% photobleaching of rhodopsin. It sould
be stressed that the P signal is measured on ROS preparations free of nucleotides. Uhl's measurements
show the P signal to have a θ and λ dependence consistent with disk shrinkage. Furthermore the kinetics
of P are similar to the MI/MII process although somewhat slower, The P signal, therefore, appears to be
the cellular conformational change that directly
communicates with the macromolecular MI/MII conformational change (1).

     It has proved rather difficult to establish the
cause of the disk shrinkage giving rise to the P signal. Our recent studies (4) together with the proton
pumping action of the A signal stand to support the
notion that protons together with a large amount of
H_2O are rapidly ejected from the disks accompanying
the P signal. Further experiments, however, are required to confirm this.

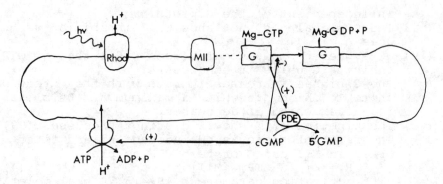

Figure 13. Overview of the processes described in this paper

A PROVISIONAL MECHANISM

Based on the scattering studies reported here as
well as those of the Julich and Freiburg laboratories
referred to herein a tentative partial mechanism ex-
plaining the various scattering signals is summarized
in figure 13.

The function of the A signal under physiological
conditions appears to be one of regulating the MI/MII
process and the possible ejection of protons by the
disk.   Temporally this latter process could be involved in transduction but at present it appears to be
merely the first step and perhaps the rate controlling one in the involved sequence terminating in the
generation of the receptor potential.

Neither of the two processes associated with the
G signal appears fast enough to be involved in visual
transduction.   This would suggest that the stimulation
of phosphodiesterase (PDE) activity, i.e. the hydrolysis of cGMP is not the key step in receptor potential generation.   It is interesting that cGMP, so far,
appears to be the only link between the A and the G
signals, accelerating the former and inhibiting the
latter.

ACKNOWLEDGEMENTS

The work at our laboratory reported here has been
supported by an NSERG grant of the National Research
Council of Canada to E. W. Abrahamson.

REFERENCES

1. K.P. Hofmann, R. Uhl, W. Hoffmann and W. Kreutz, (1976) Biophys. Struct. Mech. $\underline{2}$, 61.

2. R. Uhl, K.P. Hofmann and W. Kreutz, (1977) Biochim. Biophys. Acta, $\underline{469}$, 113.

3. R. Uhl, K.P. Hofmann and W. Kreutz, (1978) Biochemistry $\underline{17}$, 5347.

4. A. Einstein, (1908) Ann. Physik, $\underline{25}$, 205.

5. P. Debye, (1944) J. Appl. Phys., $\underline{15}$, 338.

6. C. Tanford 'Physical Chemistry of Macromolecules', Chap. 5, John Wiley and Sons, New York, London, 1963.

7. G. Mie (1908) Ann. Physik, $\underline{25}$, 377.

8. H.C. van de Hulst, 'Light Scattering by Small Molecules', John Wiley and Sons, New York, London, 1957.

9. P. Latimer, D. Moore, F. Bryant, (1968) J. Theor. Biol., $\underline{21}$, 348.

10. J. Bateman, (1968) J. Coll. Interf. Sci., $\underline{27}$, 458.

11. K.P. Hofmann and D. Emeis, (1981) Biophys. Struct. Mech., $\underline{8}$, 23.

12. K.P. Hofmann, A. Schleicher, D. Emeis and J. Reichert, Biophys. Struct. Mech., (1981), $\underline{8}$, 67.

13. R. Uhl and E.W. Abrahamson, Chem. Rev. (1981) $\underline{81}$, 291.

14. C.R. Worthington, (1974) Ann. Rev. Biophys. Bioeng., $\underline{3}$, 53.

15. D. Corliss, (1976) Ph.D. Thesis, University of Alabama at Birmingham.

16. P. Schnetkamp, (1980) Biochim. Biophys. Acta, $\underline{598}$, 66.

17. M.L. Applebury - private communication

18. Y.A. Ovchinnikov, N.G. Abdulaev, M.Y. Feigina,
    I.D. Artamov, A.S. Zolatarev, A.S. Bogachuk, A.I.
    Miroshnikov, V.I. Martinov, A.B. Kudelin, (1982)
    Bioorg. Khim.,$\underline{8}$, 1011

19. P.A. Hargrave, J.H. McDowell, D.R. Curtis, J. Wang,
    E. Juszczak, S.L.Fong, J.K. Mohanna-Rav, and
    P. Argos,(1983) Biochemistry

20. H. Linschitz, V. Wulff, R. Adams and E.W. Abraham-
    son,(1957) Arch. Biochem.,$\underline{68}$, 233

21. E.W. Abrahamson, J. Marquisee, P. Gavuzzi and
    J. Roubie, (1960) Z. Elektrochem., $\underline{64}$, 177

22. W.A.Hagins, Nature(London), (1956), $\underline{177}$, 989

23. J. Rapp, J.R. Wiesenfeld and E.W. Abrahamson, (1970)
    Biochim. Biophys. Acta, $\underline{701}$, 119

24. G. von Sengbusch and H. Stieve, (1971)
    Zeit.Naturforsch. $\underline{26b}$, 488

25. J. Stewart, B. Baker, T.P. Williams, (1977)
    Biophys. Struct. Mech., $\underline{3}$, 19

26. W. Hoffmann, F. Siebert, K.P. Hofmann and W.
    Kreutz, (1978) Biochim. Biophys. Acta, $\underline{503}$, 450

27. R. Matthews, R. Hubbard, P. Brown and G. Wald,
    (1963) J. Gen. Physiol., $\underline{47}$, 215

28. S. Ostroy, F. Erhardt, and E.W. Abrahamson (1966)
    Biochim. Biophys. Acta, $\underline{112}$, 265

29. H.M. Emrich, (1971) Zeit. Naturforsch., $\underline{26b}$, 352

30. J. Plantner and E. Kean - unpublished data

31. T.J. Borys, (1981) Ph.D. Thesis, University of Guelph

32. B.D. Gupta, T.J. Borys and E.W. Abrahamson -
    unpublished ms.

33. R. Uhl, T.J. Borys and E.W. Abrahamson, (1979)
    Photochem. Photobiol., $\underline{29}$, 703

34. R. Uhl, T.J. Borys and E.W. Abrahamson, (1979)
    FEBS Letters, $\underline{107}$, 317

35. T.J. Borys and R. Uhl and E.W. Abrahamson, Nature (London), in press.

36. H. Kuhn, N. Bennett, M. Michel-Villaz and M. Chabre (1981) Proc. Nat. Acad. Sci., U.S.A., 78, 6873.

37. B.D. Gupta, T.J. Borys and E.W. Abrahamson, unpublished MS.

38. R. Uhl, P. Kuras, K. Anderson and E.W. Abrahamson, (1980) Biochim. Biophys. Acta, 601, 462.

39. J. Reichert and K.P. Hofmann, (1981) Biophys. Struct. Mech., 8, 95.

40. R. Uhl, (1976) Ph.D. Thesis, University of Freiburg.

41. R. Uhl, (1981) Can. J. of Spectr., 26, (3), 72.

42. T.J. Borys and E.W. Abrahamson, unpublished data.

SPECTROSCOPY OF PLANT TETRA-PYRROLES IN VITRO AND IN VIVO

Hugo Scheer

Botanisches Institut der Universität
Menzinger Straße 67
8ooo München 19   F.R. Germany

ABSTRACT

The spectroscopy of chlorophylls and biliproteins is discussed with examples selected from recent investigations.
These include the structure elucidation of the phytochrome Pfr chromophore and the current status of the work
on chlorophyll-RC I, work on the biosynthesis of chlorophylls by non-invasive techniques, studies on the in situ
structure of chlorophylls and biliprotein chromophores,
energy transfer studies in photosynthetic antenna systems.

Abbreviations:

Chl = Chlorophyll, Bchl = Bacteriochlorophyll, Phe = Pheophytin, Bphe = Bacteriopheophytin, Mephe = Methylpheophorbide, Bmephe = Bacteriomethylpheophorbide, PC = Phycocyanin
Pr and Pfr = Phytochrome in the red and far-red absorbing
form, respectively, uv-vis-nir = absorption in the ultraviolet, visible and near infrared spectral range, CD =
circular dichroism, FDMR = fluorescence detected magnetic
resonance, NMR = nuclear magnetic resonance, ESR = electron
spin resonance, ENDOR = electron nuclear double resonance,
HPLC = high performance liquid chromatography.

Acknowledgements: The cited work of the author was supported by the Deutsche Forschungsgemeinschaft, Bonn and by the
Universitätsgesellschaft, München.

C. Sandorfy and T. Theophanides (eds.), Spectroscopy of Biological Molecules, 409–445.

1a: R = vinyl

2a: R = ethyl

1b, 2b

1c, 2c

3a   3b   4b   4a

A : INTRODUCTION

Spectroscopic techniques have become increasingly important
in biological studies. This is mainly due to great improvements in their sensitivity and selectivity, but also to
advancements in the underlying theories which made less
ambigous interpretations possible. The technical details of
these improvements are beyond the scope of this talk and
have in part been dealt with by other authors of this
summer school. In view of the vast amount of material on
the spectroscopy of plant tetrapyrroles, I have also not
tried to give a complete review, but have rather focused
on comparatively few examples which are meant to illustrate
the current approaches, potentials and limitations of some
selected methods. I apologize for any undue bias in this
selection.

Plant tetrapyrroles comprise the chlorophylls and the biliproteins serving as photoreceptor pigments, the cytochromes
functioning in the electron transport of photosynthesis and
respiration, the sirohemes and cobalamine(s) acting as co-
factors in several important enzymes, and also the biosynthetic precursors and degradation products of all these
pigments. I shall only deal with recent investigations on
chlorophylls, biliproteins and metabolically related
structures. The chosen examples focus on two   pects.
The first is the structure elucidation of isolated or at
least enriched compounds (section B). The second is the
spectroscopy of increasingly complex systems up to whole
organelles or organisms (section C). It has been used
study the in situ structure of the pigments, but also to
study their biosynthesis and to investigate the inter-
relations among the different pigment molecules.

B: SPECTROSCOPY OF ISOLATED PIGMENTS

Improved separation techniques have recently allowed the
isolation of many new plant tetrapyrroles and provided
evidence for an even larger number of pigments present
only in small amounts besides the major chlorophylls, e.g.
Chl a and Chl b in plants and Bchl a, b, c and d in bacteria. Optical spectroscopy, in particular uv-vis-nir
absorption, is the most sensitive and rapid technique to
classify among the different pigments. Useful structural
hints can be obtained from comparison with pigments of
known structure, but the method is unable to provide more
details. The two major methods for structure elucidation

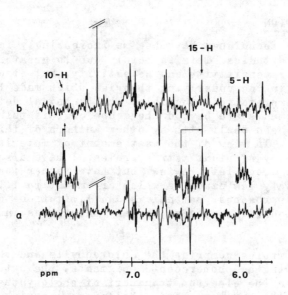

Fig. 1: ^{1}H-NMR-spectra (500MHz) of phytochrome chromopeptides in the methine regions. (a) 1:1-
mixture of <u>Pfr</u> (<u>1b</u>) and <u>Pr</u> (<u>1a</u>) peptide obtained from the photoequilibrium mixture of
<u>Pfr</u>. (b) The same sample after irradiation
with white light. The pairs of lines associated with the two forms collapse to a
single line of the Z,Z,Z-peptide (<u>Pr</u>=<u>1a</u>),
the largest shift is observed at the site
of isomerisation (C-15).
From Thümmler <u>et al</u>. (7).

5b   5a   5c   6

are NMR and mass spectroscopy, which are both well established, and have been considerably improved recently. The limiting factor in NMR spectroscopy is its sensitivity. It is now sufficiently increased to allow work in the μg range, which puts stringent requirements on the sample preparation and purity. The limiting factor of mass spectroscopy has been the ionisation without alteration. It has also been considerably improved recently by new ionisation techniques, which now allow the direct analysis of e.g. underivatized chlorophylls. Both methods are most useful in conjunction with suitable chemical correlation methods. Two structures are chosen to illustrate their potentials.

B.a: CHROMOPHORE OF PHYTOCHROME IN THE Pfr FORM

The biliprotein, phytochrome is the only photomorphogenetic pigment in plants of which some structural details are known. It is a photoreversibly photochromic pigment, with one form, Pr, absorbing maximally around 660 nm, and the other, Pfr, around 730 nm in the long-wavelength spectral region. Phytochrome occurs only in small amounts and concentrations in plants, and its photochromic properties have been used as analytical tool during the isolation. The similarities in the optical absorption spectra of Pr and the readily accessible phycobiliprotein, phycocyanin (PC), were substantial in assigning a bile pigment structure, too, to the chromophore of the former (see RÜDIGER and SCHEER (1) for leading references). The detailed structure analysis of phytochrome was complicated by the rather stable thioether bond between the chromophore and the apoprotein, which precluded a chromophore cleavage without alterations of the chromophore, and also by the poor accessibility of phytochrome. The same chemical degradation techniques, developed with PC and then applied to phytochrome (2) provided together with studies on model pigments like 3, 4 (3) and with the total synthesis of bile pigments related to the native pigments (e.g. 5, refs. 4) the structures 1 a and 2 a for the chromophores of Pr and PC, respectively. The evidence for the 18-vinyl group in 1 was, however, only circumstantial (5).

The same structures and a proof of the latter substituent have been arrived at by proton NMR of bilipeptides from the two pigments (6). The NMR spectrum of linear tetrapyrroles covers the spectral range between $\delta = 1$ and 8 ppm. There is a considerable overlap with the NMR lines of the peptide moiety of these chromopeptides. The structure analysis was nonetheless possible by the use of high-field NMR spectrometers (360 MHz in the particular case). This

414   H. SCHEER

Table 1: Incremental chemical shifts (Δ) of the methine
proton NMR signals in Z,E-isomeric bilindiones.
Δ (in ppm) is given as $\delta_E - \delta_Z$.

Compound	5-H	15-H	10-H	solvent	ref.
phytochrome chromophore (**1b** minus **1a**)	-0,057	+0,195	+0,012	acetone-d$_6$/ H$_2$O/CF$_3$COOH	(7)
phycocyanin chromophore (**2b** minus **2a**)	-0,056	+0,195	+0,018	acetone-d$_6$/ H$_2$O/CF$_3$COOH	(7)
octaethyl-bilindione (**3b** minus **3a**)	0	+0,27	0	CDCl$_3$/CF$_3$COOH	(8)
15E A-dihydro-octaethyl-bilindione (**4b** minus **4a**)	0	+0,23	0	CDCl$_3$	(8)
15E A-dihydro-bilindione (unsubstituted Ring A)	-0,07	+0,23	-0,01	CDCl$_3$	(4b)
15E 3-dihydro-7,8,12,13,18-methyl-2-di-methyl-17-ethyl-bilindione (**6b** minus **6a**)	-0,09	+0,20	-0,03	CDCl$_3$	(9)
4E 3-dihydro-7,8,12,13,18-methyl-2-di-methyl-17-ethyl-bilindione (**6c** minus **6a**)	+0,43	+0,08	+0,26	CDCl$_3$	(9)

did not only allow for sufficient resolution to separate
the overlapping lines from the peptide and the chromophore,
but gave also the necessary increased sensitivity required
for the work with phytochrome. An assignment of all chromophore proton signals in the PC-peptide was achieved by
comparison with a synthetic peptide of the identical sequence, but lacking the chromophore. This was in turn the
basis for the assignment of the chromophore derived signals
(including those of the 18-vinyl group), in the Pr peptide,
leading independently to structure 1 a.

The structure elucidation of the Pfr chromophore posed
additional problems. Pfr peptides revert spontaneously to
the Pr form. This reversion is pH-dependent and sufficiently retarded only at moderately low pH. The preparation of
Pfr peptides required therefore peptic digestion and subsequent work-up under conditions which maintained a pH $\approx$ 3
throughout. This work-up can be done in protonated solvents,
but an exchange with deuterated solvent at the last
chromatographic step is necessary for NMR (7). Unfortunately, at least one (5-H) of the three methine protons is
readily exchangeable, too, with deuterium under acidic
conditions. Such an exchange was untolerable, because the
methine signals were essential to differentiate between
the discussed mechanisms in the Pr - Pfr interconversion,
e.g. a redox mediated substitution and a Z,E-isomerisation.

The extensive tests to circumvent these obstacles made
again use of the readily accessible PC and of synthetic
bilidiones like 3 and 4 as models for phytochrome, which
were used to optimize the measuring conditions. It could
be shown, that a mixture of deuterated acetone with protic
water and protic trifluoroacetic acid (85:10:5, v/v) was
appropriate. It led to only partial exchange of the 5-H,
which at the same time facilitated its assignment, and it
also fulfilled another spectroscopic requirement. The
presence of a considerable amount of protons from the
solvent ($>$ 10^4 fold molar excess with respect to the Pfr
peptide) leads to a very strong solvent line in the NMR
spectrum. This can principally be coped with by using a
computer with a sufficiently high dynamic range, but only
if this line is well separated from the resonance lines of
the solute. Since the chemical shift of water is strongly
increased in the presence of acid, the amount of acid
added to the system can be used to adjust the water line to
a position where it does not interfere with the solute
lines of interest.

With all the conditions adjusted accordingly, it was then
possible to obtain the NMR spectrum of a Pfr chromopeptide

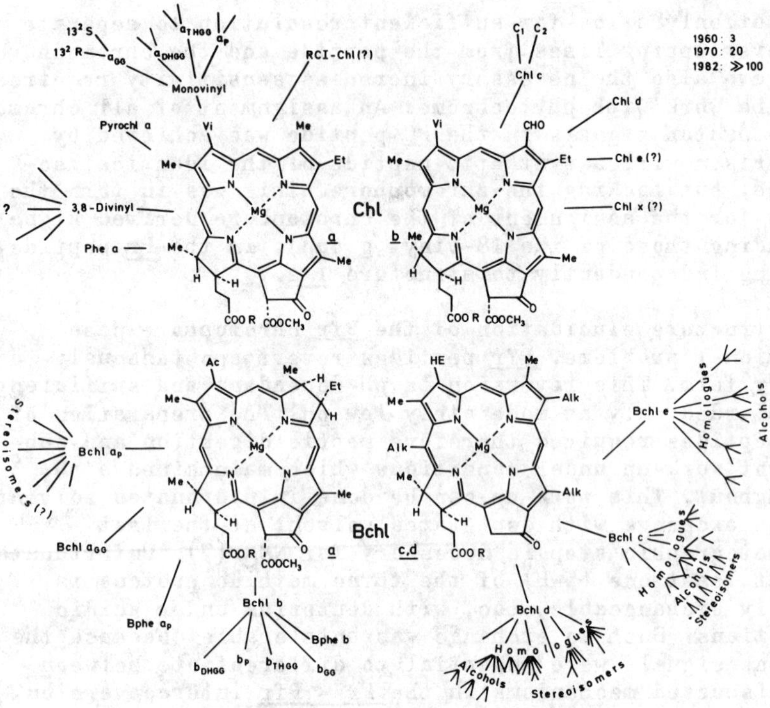

Fig.2: The "chlorophyll explosion". The time table
in the upper right gives the number of
reasonably well established structures.

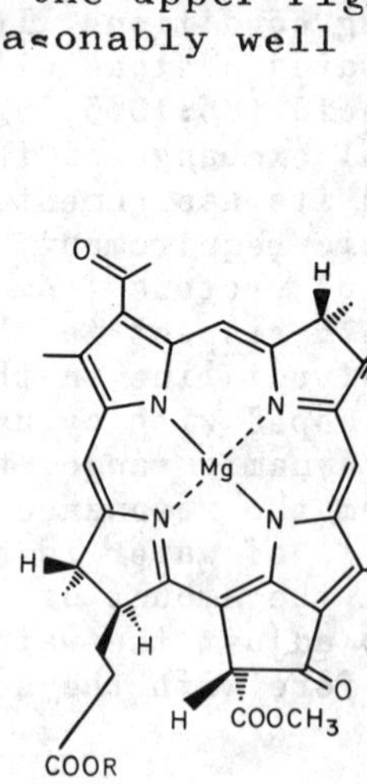

7: R = phytyl

at the first shot (7). Phytochrome accumulated over a
period of several months was phototransformed to Pfr, de-
graded and purified over night to the peptide by the bio-
chemist, taken over by the NMR crew and rushed in the early
morning hours to the 500 MHz NMR spectrometer 300 km away
in the laboratories of the BRUKER company. After measuring
first the spectrum of the Pfr peptide, the latter was
photochemically converted in the NMR tube to the Pr
peptide by irradiation with white light.

The analysis of the spectra clearly established the Pfr
chromophore as the 5Z, 10Z, 15E-isomer 1b of the all-Z
Pr chromophore 1a. This structure follows mainly from the
analysis of the methine proton signals. Studies with the
Z,E-isomers of PC and of the dihydrobilindiones 4a,b (8)
and 6a,b,c (9) have shown, that the chemical shifts of the
methine protons are somewhat solvent dependent. However,
the incremental shifts upon Z, E-isomerisation are solvent
independent, and characteristic for the site of isomerisation (table 1). The largest incremental shift is always observed at the site of isomerisation. In the Pfr
chromopeptide, the position of the methine signals of
the chromophore agree reasonably well with those of the
5Z, 10Z, 15E-isomers 2b and 4 b (fig. 1). There is a
second set of lines corresponding to the Pr chromophore,
which is always present and continously formed in Pfr
peptides and amounts to ≈ 50 % in the actual sample. This
second set of lines increases accordingly upon photoconversion at the expense of the first set,
thus proving directly the relationship of the two species.
The largest incremental shift is observed for the 15-H
signal, and the Pfr chromophore must then contain the
15E-isomer 1b rather than the 5E-isomer 1c (or the for
other reasons unlikely 10 E-isomer) (7).

B.b.: NEW CHLOROPHYLLS

The number of chlorophylls has been rather limited until
recently, when an "explosion" of new chlorophyllous pigments raised their number well beyond into account. It
becomes still larger if all suggested or suspected
structural variations are counted (fig. 2). A sound
structure analysis of even the minor pigments seems not
only urgent, but also possible with the recent advances
in spectroscopy. It should help to separate the chlorophylls proper from metabolites and artifacts, and to
provide a structural basis for their functional studies.

There are two general approaches for the structure
elucidation of chlorophylls. The first is the direct

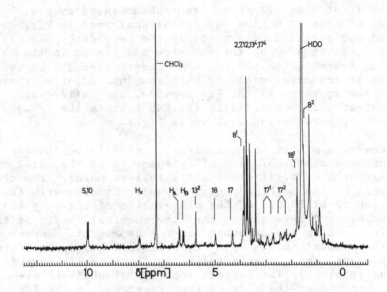

Fig.3: ^{1}H-NMR-spectrum (500 MHz) of methylpheophor-
bide-RC I (40 μg in CDCl$_3$). The spectrum is
rather similar to that of Mephe a (3) at the
same concentrations, but the 15-H is lacking
and the 13^2-H is shifted to higher field.
The spectrum is compatible with structure 11
for Mephe-RC I.

10   R = CH$_3$ 11

analysis of the isolated pigments. It can be limited by the
instability of the pigments, but is always necessary if the
molecule cannot be modified without extensive and
potentially unknown structural changes (see Bchl b as an
example (10). The second is the controled conversion to
derivatives like the methylpheophorbides which are more
stable and thus amenable to a broader variety of analytical
techniques, and which also provides already a means of
chemical structure relations (see e.g. the identification
of pyropheophetin as a degradation product (11) or of
"divynil" Chl a as a biosynthetic precursor (12). NMR
spectroscopy can usually be   performed with both the
genuine pigments and their derivatives. It gives a wealth
of information and is interpreted in a straightforward
fashion due to the inherent magnetic properties of the
tetrapyrrole macrocyle (13). The latter serves as a built-
in shift reagent which spreads the proton NMR spectrum over
a range from δ =-6 to 12 ppm. The only major source of
error is the pronounced solvent and concentration dependence, which is related to the facile aggregation of
these pigments. The second key method, mass spectroscopy,
has been widely used for the demetalated pigments (14)
but not been suitable for the chlorophylls proper, viz.
magnesium complexes of pheophorbides. The latter cannot
be analyzed by conventional mass spectroscopy with electron
impact ionization. Several different ionization techniques
have now been applied successfully to circumvent this
problem. BRERETON et al. (15) have used the "in beam"
ionization technique to determine the structure of the
allomerization product 7 of bacteriochlorophyll a (8).
KATZ and others (15,16) applied field and ^{252}Cf plasma
desorption as well as fast atom bombardment to a variety
of the "conventional" chlorophylls.

The latter mass spectroscopic technique has also been
applied to one of the new chlorophylls, e.g. Chl-RCI (17).
This pigment occurs only in trace amounts in organisms
capable of oxygenic photosynthesis, where it has been
quantitatively related to the contents of P700, the
reaction center of photosystem I(PS I) (17). The latter
is generally believed to contain the "normal" chlorophyll
a(9) in a special environment and/or aggregation state (see
section Cd). The suggestion that the red shift and other
changed properties of P700 may be due to a modified molecular structure (viz. Chl-RCI, is therefore of considerable interest.

Although the structure of Chl-RC I is at present not yet
fully elucidated, the known data are included here as an
example for the general approach to new structures in the

chlorophylls which are accessible only in small quantities.
DOERNEMANN and SENGER (17), who first described and isolated the compound from a Scenedesmus mutant enriched in
PS I, used the first aforementioned approach and studied
the pigment proper. They obtained a satisfactory 252Cf
plasma desorption mass spectrum, which gave a molecular
ion which is 35 mass units higher than that of Chl a
(9). They also obtained NMR evidence, that the C-20
methine proton is missing. With a sample provided to us
by the authors, we could show that the red-shift (as
compared to the respective derivatives of 9) is maintained when Chl-RC I is demetalated to its pheophytin and
transesterified to the methylpheophorbide, Mephe-RC I.
The uv-vis-nir absorption of the latter is furthermore
characteristic of a pigment similar to methypheophorbide
a (10), but carrying a substituent other than a hydrogen
atom at C-20. This interpretation was also supported by the
CD spectrum. It showed an increased ratio of the ellipti-
cities of the red and the near-uv ("SORET) band, which
is characteristic for steric hindrance due to substitution
at methine position(s). The convertibility of CHl-RC I
to its methylpheophorbide without an apparent loss of the
important structural variations was then used by us to
obtain larger ($\approx$ 100 μg) amounts of methylpheophorbide-
RC I from a different organism. The blue-green alga,
Spirulina geitleri is available commercially as a spray-
dried powder (Behr, Bonn, FRG). The material is extracted
exhaustively with methanol, and the crude extract containing mostly Chl a is immediately demetalated. The
resulting pheophytin mixture is chromatographed on silica
and transesterified to the methylpheophorbides. Further
repeated chromatography of the latter on silica, and
finally HPLC on a reverse phase adsorbent yielded the
pure Mephe-RC I. The 500 MHz proton NMR spectrum (fig. 3)
was again obtained at the BRUKER facilities. It fully
supported the substitution of C-20, and showed otherwise
all signals present in Mephe a (10). Variations of the
chemical shift can be accounted for by the low concentration of the former, with the possible exception of the
rather large shift of the 13^2-H signal. The "conventional"
electron impact mass spectrum gives a molecular ion which
is 50 mass units higher than that of 10 and shows an
unusally intense (M + 2) ion, which may relate to the
presence of Cl. Taken together with chemical and chromatographic evidence not discussed here, these results are
compatible (but not proof yet) of the 13 -oxy-20-chloro-
structure 11 for Mephe-RC I. Recent field desorption data
support this structure.

B.c.: BIOSYNTHESIS OF TETRAPYRROLS

Radioactive labeling is the most sensitive method to study
metabolic pathways, but it has several shortcomings which
can be overcome by heavy labeling with stable isotops. The
latter are detected by either mass or NMR spectroscopy. A
particular advantage is the easy distinction of double
labeling within a subset of molecules from a statistical
distribution at two or more sites. BROWN et al. have used
^{17}O labeling in conjunction with masspectroscopy to establish a twofold oxygenation with O_2 during bile pigment formation from heme in red algae. Spin-spin coupling in NMR gives
even a more detailed history on the biosynthetic pathway,as
demonstrated elegantly for the sequence of reactions involved
in the cyclisation of four porphobilingens to uroporphyrin
III (21a).
^{13}C labeling in conjunction with NMR has recently also been
applied to plants, where it provided conclusive evidence for
the predominance of the so-called C_5-pathway for chlorophyll
formation in maize (21 b,c). The key intermediate δ-amino-
levulinate (Ala) is formed in this pathway from C_5 compounds
like glutamate or ketoglutarate, rather than from succinate
and glycine. With the ^{13}C-NMR spectrum of chlorophyll a fully
assigned, the lable in any of the 35 C-atoms of the macrocycle   can be quantitated in a single experiment. The
technique requires, however, a substantial incorporation of
^{13}C above the 1.1% level of natural abundance. Since the
uptake of organic substrates by plants is low, and their
metabolism more complex than that of animals, it may be
difficult to attain such incorporation economically, ^{14}C
incorporation can then be used to test the feasibility of the
experiment. PORRA et al. incorporated after such tests 15oo
excised primary leaves of maize seedlings with selectively
labeled ^{13}C-glutamate. The contribution of the C_5-pathway to
the chlorophyll biosynthesis in maize is from these experiments $\geqslant 95\%$, and similar data are obtained for beans by
OHAMA et al. The only position labeled by 2-^{13}C glycine is
C-13^4, the methylester group at the isocyclic ring. Feeding
with 1-^{13}C glutamate yieled a label in all eight positions
originating from C-5 of Ala, and feeding 5-^{13}C-glutamate brings
the lablel to the two ester group C-13^4
Labeling in some other positions can furthermore be accounted
for quantitatively by a partial conversion of 5-^{13}C to
1-^{13}C-glutamate via the citric acid cycle.

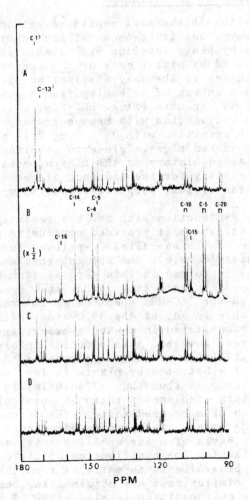

Fig. 3 A: Proton-coupled ^{13}C-NMR spectra from 90 ppm to
180 ppm of chlorophyll a produced by illumination of
excised etiolated maize leaves in the presence of 5-^{13}C
glutamate (A). 1-^{13}C glutamate (B) and 2-^{13}C glycine (C).
Spectrum D is the natural-abundance ^{13}C-NMR spectrum of
chlorophyll a. The vertical axes of spectra A. C and D
are expanded twofold relative to that of spectrum B. From
PORRA et al. (21b).

C: TETRAPYRROLES IN COMPLEX SYSTEMS

The functions of tetrapyrroles can only partly be unter-
stood on the basis of their molecular structure. The
properties of e.g. the biliproteins are quite different
(absorption, fluorescence) from what might be expected
from the structures of the chromophores. Also, a particular
pigment e.g. Chl a serves often rather different functions
(light harvesting, electron donor, electron acceptor) in
different environments. These changes are brought about
by chromophore-protein and chromophore-chromophore interactions. To unterstand the function of tetrapyrroles, their
structure has thus to be known in a more general sense
which encompasses these interactions. These investigations
are generally most effective by the combination of biochemical and spectroscopic techniques. The former allow a
study of the photosynthetic apparatus by topological studies and the isolation of functional subunits. Spectroscopic techniques serve as a control to ensure that these
subunits have retained their properties, and as a means for
their structure analysis in the aforementioned general
sense. Neither one of the two standard methods for the
analysis of molecular structures is currently applicable
to complex systems like tetrapyrrole-protein complexes,
whole organelles or organisms, or just larger aggregates.
Proton NMR can generally not cope with the line broadening
caused by the comparably slow tumbling of these systems,
and mass spectroscopy fails for the difficulties involved
in the ionisation process. Furthermore are both methods
principally difficult to apply to mixtures. The need for
both selectivity and sensitivity favors applications of
two- or more dimensional methods, of which only three
shall be discussed here. The first is fluorescence spectroscopy, which permits among others the analysis of
complex mixtures and of energy transfer. The second is
RAMAN resonance spectroscopy (RR), which adds a second
dimension to the conventional absorption spectroscopy,
and which is furthermore accessible to rather precise
computer modeling. The third of the selected methods is
ENDOR and related magnetic double resonance techniques,
which are selective for paramagnetic species, and also
well suited for theoretical analysis.

C.a: BIOSYNTHETIC STUDIES USING FLUORESCENCE SPECTROSCOPY

Fluorescence is, within the limits of photostability of
the investigated structures, a non-invasive technique and
thus applicable to complex systems. By recording the
excitation spectra for a number of excitation wavelengths
(or vice versa), it is possible to obtain a minimum set of

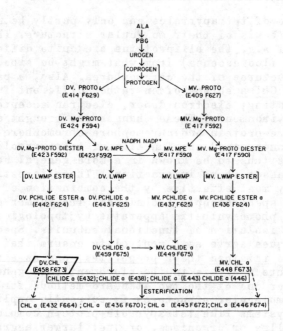

Fig.4: Branched pathway for the biosynthesis of
Chl a and homologues, as suggested by REBEIZ
et al. (ref.10). The classical pathway is the
second one from the right. The structure of
divinyl-Chlorophyll a (12,boxed) has been confirmed by NMR, MS and chemical correlation
(12,22).
From REBEIZ (ref.20).

12: R = phytyl

components contributing to the fluorescence. The excitation
spectra are furthermore similar to the absorption spectra
of the individual components, if no energy transfer among
the pigments (section C.b.) or other nonlinear effects take
place. The two major limitations are the necessity of
sufficient fluorescence, which has been improved e.g. by
photon counting, but also the difficulty to establish
a full structure from the absorption spectrum alone.

The metabolism of chlorophylls has long been known to
involve a number of cyclic tetrapyrroles which are fluore-
cent (19). REBEIZ and coworkers (20) have exploited the
fluorescence analysis of greening tissue to obtain more
details on the biosynthetic pathway. In addition to the
previously known chlorophylls and their metabolites they
were able to characterize both in vivo and in extracts a
variety of pigments with shifted absorption fluorescence
spectra. From a careful comparison of the excitation spectra
with the absorption spectra of known pigments, they suggested
in particular a branched pathway leading to chlorophylls
which carry not only the common vinyl group at C-3, but
have also still kept the second one at C-8 all the way from
protoporphyrin IX (fig. 4). The major argument for these
structures was, that their excitation spectra showed like
semisynthetic 3,8-divinyl-pigments, a considerably ($\approx$ 10nm)
red-shifted SORET-, but only minimally shifted (2-3 nm)
long wavelength band. This proposed structural principle
has now been proved for at least few of these pigments by
the more informative methods described in section B. BAZZAZ
et al. (21) have isolated 8-deethyl-8-vinyl-chlorophyll a
(12) from a mutant accumulating large amounts of this
pigment, and established its structure by NMR and mass
spectroscopy, and REBEIZ et al. (22) have furthermore performed chemical correlations with this and other related
pigments.

It is nonetheless desirable to increase the information
content of the in situ fluorescence analysis, because the
necessary reference material may not always be at hand. One
such possibility is its extension in the form of fluorescence detected magnetic resonance (FDMR). WOLF et al., and
others (23) have recently applied this technique to study
the pigments in photosynthetic bacteria. The biosynthetic
pathway of bacteriochlorophyll a has been known in many
details from studies with mutants (19). The conventional
fluorescence analysis of Rhodopseudomonas spheroides and
other photosynthetic bacteria made now most of these pigments visible in whole cells, and indicated that some
additional ones are present. In FDMR, the fluorescence at
a selected pair of excitation/emission wavelengths is

Chromophore Type	Absorption λ_{max}	Fluorescence λ_{max}
3	500	515
2 (s)	545	560
2 (f)	570	585
1 (s)	590	605
1 (f)	620	635
1	650	665
Chlorophyll a$_{II}$	686	
		Photooxidation

Fig.5: Models for the Rhodella violacea
Phycobilisome structure (a, from refs.24c)
and energy transfer scheme (b, from refs.3).
The outer "plates" (λ max=500-570 nm, light)
in (a) contain B-PE, the intermediate ones
(dark in (a), λ max=590-640 nm) contain C-PC,
and the inner core (gray in (a), λ max=650-
670 nm) contains APC.

monitored while a microwave field is swept, which gives the
zero-field splitting parameters of the pigments in their
triplet states. When applied to the bacteria, these parameters could be obtained for most of the pigments previously identified by fluorescence alone, and they agreed
well with the expected values. The full potential of this
technique is probably not yet exploited. First examples
have demonstrated its use where unusual zero-field
splittings can be expected. This does not only include
chlorophylls complexes with unusual metals and biosynthetic
precursors of the lesser investigated (bacterio)chloro-
phylls, but also perturbations due to the particular environment of a given pigment, viz. to the study of pigment-
protein and pigment-pigment interactions. (23 a, b d)

C.b.: ENERGY TRANSFER STUDIES USING FLUORESCENCE SPECTROS-COPY

Another important application of fluorescence spectroscopy
has been the investigation of energy transfer processes.
The majority of plant tetrapyrroles is involved in the
light collection process of photosynthesis. They are bound
to proteins and arranged in large light harvesting arrays.
Light energy captured by one of the chromophores is
efficiently transferred to neighboring ones and by
repetition eventually migrates to the reaction center(s),
in which the conversion to redox energy takes place. Several such antenna systems have been studied,but probably none
of them in such detail as the phycobilisomes. These are
electron microscopically visible aggregates consisting
mainly of biliproteins, which are attached to the outer
surface of the photosynthetic membranes of blue-green and
red algae. Their participation in the light harvesting process to photosystem II had been inferred from the classical
biochromatic experiments of EMERSON and coworkers, and
detailed studies became possible with the successful
isolation of entire phycobilisomes and their more or less
dissociated components (24). Phycobilisomes contain at
least two, generally three chromoproteins bearing bile
pigment chromophores. These are the green allophycocyanins
(APC), the blue phycocyanins (PC) and -optionally- the red
phycoerythrins (PE) (see fig. 5 for the spectral data).

All phycobiliproteins are highly fluorescent. Fluorescence
studies of intact phycobilisomes had shown, however, that
only the minor (generally 5 %) and red-most absorbing APC
is responsible for the fluorescence, irrespective of PE,
PC or APC being the absorber. The energy is thus transported with high efficiency from the higher energetic (PC,
PE) to the lower energetic component (APC). This transfer

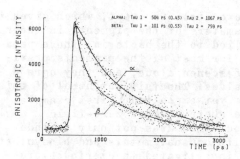

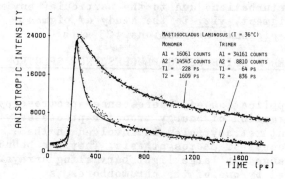

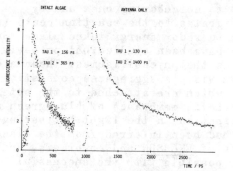

Fig.6: Fluorescence decay curves (points) and best fit
(Marquardt algorithm) for a biexponential
decay (solid lines). The maximum amplitudes
and the time constants are given in the insets.
(a) α-and β-subunits of PC from Mastigocladus
laminosus.(b) The same pigments in its tri-
meric state (α_3, β_3) and in its monomeric state
(α_1, β_1) induced by dissociation of the trimer
with KSCN. (d) Phycobilisomes from Mastigocladus laminosus (right) and whole cells
(left). Figs.(a) and (b) give only the anisotropic component of the fluorescence. Excitation at 580 nm, pulse intensity 10^{14} photons
.cm^{-2}, repetition rate 80 MHz.

is uncoupled after the dissociation of the phycobilisomes, a process which can be monitored by the gradual increase of the fluorescence of the fragments. The biochemical and fluorescence analysis of the fragments yielded a model in which the morphology of the pigments as well as their absorptions and emissions are optimized for an efficient energy transfer (fig. 5). Most if not all of the chromophores are weakly coupled and the energy is transfered by a Förster type process, which is also supported by circular dichroism of the isolated pigments (25).

Whereas this static analysis is well suited to study the energy transfer between chromophores with well separated absorptions, the individual transfer steps between neighboring and more alike chromophores required more sophisticated methods. Fluorescence depolarization revealed a further subdivision of the chromophores (26). Even in pigments like PC and many of the PE's bearing up to six chromophores of the same molecular structure, is the absorption (and fluorescence) of the individual chromophores fine-tuned for an efficient energy transfer. The three chromophores of C-PC absorb, for example, around 590, 610 and 620 nm, which allows a graded energy transfer even on this level. The static fluorescence experiments have been complemented by time-resolved studies. PORTER et al (27) monitored the small amount of leakage fluorescence emitted by PE, PC, and APC in intact algae. After excitation with green light absorbed mostly by PE, the PE fluorescence rises first, followed by the fluorescence from PC and finally from APC. Excitation with laser pulses of varying photon fluence has also been used to study the migration of excitation energy via singlet-singlet annihilation, which gives information on the number of coupled chromophores (28). It could e.g. be shown, that bacterial/antenna preparations with different detergents may have rather different numbers of coupled chromophores, in spite of their identical absorptions (28 b). These studies showed on the other hand, that rather low fluence rates are required to prevent interference from the latter process on the fluorescence decay rates obtained by picosecond time-resolved spectroscopy (see ref. 29 for a critical discussion).
Picosecond time-resolved spectroscopy has now been applied by several groups to investigate the energy transfer both in intact phycobilisomes and their componenents. (41,43). The is rather complex . Each one of the applied methods of measurement and data handling has certain advantages and shortcomings, which are complicated by preparative problems. A consistent and detailed model can thus not yet be given and will require considerably more data.

Table 2: Fluorescence decay (in psec) of phycocyanin
from Spirulina platensis

	T (°C)	Trimer		Monomer	
		τ_2 (%)	τ_1 (%)	τ_2 (%)	τ_1 (%)
Isotropic	18.6	1300(47)	320(53)	2200(51)	300(49)
	39.2	1500(49)	270(51)	2200(54)	290(46)
	51.8	1600(36)	260(64)	3200(51)	400(49)
Anisotropic	18.6	-	70(100)	1050(37)	150(63)
	39.2	-	70(100)	1000(38)	250(62)
	51.8	350(54)	70(46)	3500(47)	380(53)

The isotropic decay curves are calculated from $I_{||} + 2I_{\perp}$,
the anisotropic ones from $I_{||} - I_{\perp}$. In a different set of
experiments with C-PC from Mastigocladus laminosus, the
calculated isotropic decay curve has been shown to agree
satisfactorily with the experimental curve obtained with
the emission polarizer adjusted to 55°.

We have recently started a systematic investigation of
isolated phycobiliproteins and their aggregates by time-
resolved fluorescence depolarization. These studies use a
repetitive streak camera for the detection,which allows for
reduced photon fluence in the individual pulses. In a study
with PC from two different algae, it could be shown that
there is always a complex decay of fluorescence, which can
be fit satisfactorily by a biexponential (fig. 6). The
analysis of the components in the isotropic and anisotropic
fluorescence yielded different amplitues and rate constants,
but the slow component was always less pronounced and in the
aggregated pigments even absent in the polarized fluorescence
(table 2). Since rotational depolarization can be excluded
for proteins of this size on the picosecond time-scale, this
depolarization has been assigned to energy transfer among
the individual chromophores. It is accordingly more pronounced
in the aggregates and appears to be absent in the $\propto$-subunit
bearing only a single chromophore (fig. 6a). The two rather
long-lived components are here assigned to two species, the
shorter lived being probably an artifact from the denaturation
-renaturation necessary to separate the subunits. Additional
support for the interpretation of the short lived component
to reflect energy transfer comes from the increased difference
between isotropic and anisotropic fluorescence with increasing
aggregation (table 2). The decreased lifetimes in whole algae
(fig. 6c) are finally understandable on the basis of a
further transfer to the chlorophylls which acts as an efficient quencher for the biliprotein fluorescence. It should
be pointed out, however,that a quantitative analysis of this
process is at present rather difficult and depends not on
the signal-to-noise ratio of the data, but also critical on
the model. It was thus not possible to distinguish a biex-
ponentialfrom a higher exponential or a more complex decay
law (e.g. $e-\sqrt{t}$), which might be expected for a FOERSTER
type energy transfer.

High-resolution spectroscopy in the time domain can also
be supplemented by high-resolution spectroscopy in the frequency domain. Hole-burning experiments have been performed
with chlorophylls and biliproteins (fig. 7) to yield high-
resolution absorbtion spectra (42). The process involves
site-selective photochemistry at liquid helium temperatures,
most likely protein transfer reactions, which leave a "hole"
in the absorbtion band corresponding to the natural linewidth as determined by the kinetics of the process. In the
case of C-phycoerythrin, these holes are acompanied by
slightly wider satellite holes at the long-wavelenght side
of the primary hole. They may be related to energy transfer
between the different chromophores in this pigment (42a).

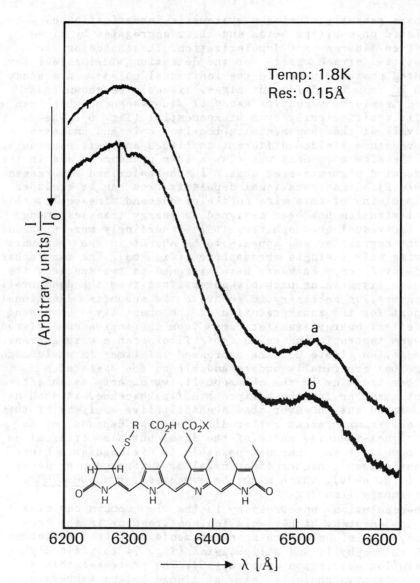

Fig.7: Hole burning experiment with a mixture of
        C-PC and APC from Spirulina platensis.
        Absorption spectrum at 1.7 K before (a) and
        after irradiation (b) with a narrow-band
        laser. The sharp hole develops at the laser
        frequency. It is accompanied by a broader
        phonon side-band.

C.c: NATIVE STATE OF CHROMOPHORES

The properties of tetrapyrole chromophores are usually
considerably changed when they are bound to the native
apoproteins. Spectroscopy has become in many ways a
mediator between the native structures and models mimicking certain aspects of this state . Chlorophyll aggega-
tion has played an important role to evaluate its donor-
acceptor properties for coordinative bonding (44). Although
the self-aggregation of chlorophylls appears currently less
important in vivo, the interactions with the protein are governed by essentially the same factors. It is also well possible, that the forme. have been an older evolutionary
feature which has later on been replaced by the latter, which
may be suggested from the antenna system of green bacteria.
        In the biliproteins, conformational changes have been
made most likely as an important factor determining the
properties of the native state. The uv-vis spectral differences between native and denatured biliproteins, e.g. PC or
Pr, can be accounted for by an extended and cyclic conformation,respectively, of the chromophore (3). This has been
supported both by the properties of synthetic bile pigments
with restricted conformational freedom, and by MO calculations.
The comparable large red-shift of native P_{fr} can, however,
not been rationalized on this basis alone. There are only two
experimentially studied processes leading to similarly large
red-shifts, e.g. deprotonation and cation radial formation.
Since the latter is unlikely from ESR data, the extended
anionic 15-E structure 1c has been suggested for native P_{fr}
(1,3).More recently,we have obtained evidence that the high
fluorescence of native biliproteins can be explained in
terms of a restricted conformational freedom. This interpretation comes from the high fluorescence yield of rigid
bilirubins, which we investigated in cooperation with
A.R. Holzwarth from Mülheim (Isr. J. Chem., in press).

1c

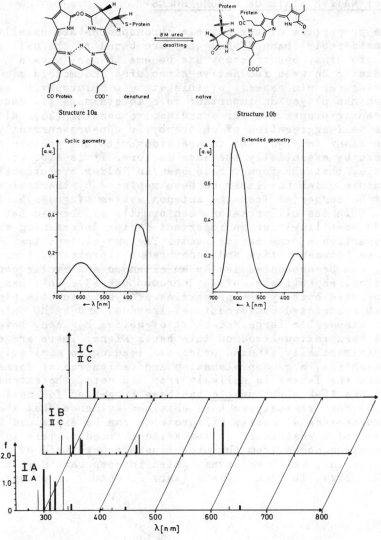

Fig.8: Suggested geometry for the <u>PC</u> chromophore in
its native (right) and denatured state, together with the corresponding experimental
absorption spectra (bottom of (a)) and the
calculated spectra (I,II A,C, respectively).
I and II denote the tautomer, carrying a
proton at N_{22} and N_{23}, respectively. B is
the calculated spectrum for a semi-open
geometry.

C.d: ENDOR SPECTROSCOPY OF THE PRIMARY DONOR IN PHOTO-
SYNTHESIS

Electron spin resonance (ESR) is a sensitive and highly
selective technique for the investigation of paramagnetic
species. It has also a comparably large information content
and its theory is well worked out to allow the investigation of structures without too stringent requirements
with regard to model compounds. The two major groups of
paramagnetic structures in biological systems are heavy
metal ions and organic radicals. Both participate –
among other functions – in biological oxidations and
reductions, and important information has been obtained
on these systems by the application of ESR. Major limitations are the overlap of signals from organic radicals,
which are usually clustered within a small spectral range
and the limited time resolution, which has only recently
been extended to the sub-nanosecond time scale.

One of the first studies of ESR in biology showed, that
an unstructured line can be observed in photosynthetic
tissue upon illumination. This signal (fig. 9a) has later
been assigned to the cation radical of the primary donor
of bacterial reaction centers, or of photosystemIof green
plants (29). The primary step during the photosynthetic
energy conversion is the electron transfer from an
electronically excited chlorophyll molecule to an acceptor
which is probably also a chlorophyll or pheophytin. This
leaves the primary donor as a cation radical, which is
re-reduced on the μsec or msec time scale. Since the latter
process is in most reaction centers rate limiting in strong
light, the primary donor can be fully converted to the
oxidized form during saturating irradiation. This results
in the reversible, light induced formation of a strong
ESR signal, and in the concomitant bleaching of absorption
bands associated with the primary donor chlorophyll.

The optical changes are usually monitored around 700, 870
and 960 nm in the reaction centers of PS I, Bchl a and
Bchl b containing bacteria, respectively. They are rather
small, however, in all but highly enriched reaction center
fractions, because they overlap with the absorption of
antenna pigments present in large excess. Since the latter
are ESR inactive, the ESR signal does not suffer from
this interference. After the kinetic correlation of the two
signals   and its assignment to a chlorophyll cation
radical, has ESR therefore been used extensively to monitor
and to obtain structural information on the primary donor.
The most data have currently been accumulated on bacterial
systems, because the chemical identity of its origin is

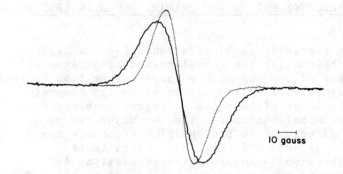

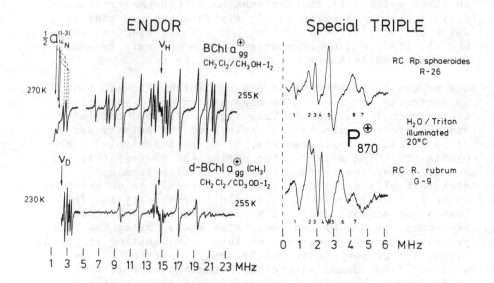

Fig.9: (a) ESR spectrum of the cation radical
Bchl a$^{+\bullet}$ in methanol /CH_2Cl_2 (heavy line)
and of P 870$^{+\bullet}$ in isolated reaction centers
from Rhodopseudomonas spheroides (thin line).
(b) ENDOR spectrum of fully protonated (upper
trace) and selectively deuterated Bchl a$^+$ in
the same solvent system. (c) Special TRIPLE
spectrum of P 870$^+$ in reaction centers from
Rhodopseudomonas spheroides R26 (upper trace)
and Rhodospirillum rubrum G9. In this technique, two radiofrequencies are applied
simultaneously which are symmetric to the free
proton frequency (=ν_H in b).

well established (Bchl a or Bchl b depending on the
species), and reaction centers can be isolated which are
essentially free from other components of the photosynthetic apparatus (30).

The ESR spectra of chemically oxidized bacteriochlorophyll a in solution and of the light induced in situ
signal of the primary donor in Bchl a containing bacteria
are very similar (fig.9a). Thenly exception is the line
width, which is reduced in the natural systems by a factor
of 1.4. NORRIS and coworkers were the first to relate this
narrowing to aggregation, and arrived at an aggregation
constant of two from a second moment analysis ("special
pair" model) (31). This interpretation was later on
supported by electron-nuclear double-resonance (ENDOR)
spectroscopy. This allowed the resolution of at least
five coupling constants in the inhomogeneously broadened
ESR line. All resolved ENDOR couplings were reduced in the
in situ signal by a factor of about two as compared to the
signal of free Bchl b, which strongly supported the above
model (32).

A drawback of earlier ENDOR studies had been their
restriction to low temperatures ($\leqslant$ 77K). The development
of ENDOR at ambient temperatures by MOEBIUS and coworkers
prompted us a few years ago, to study in cooperation with
them bacterial reaction centers in liquid solution (33).
This is not only an advantage with respect to the more
physiological temperatures, but gives also a considerably
increased resolution due to the averaging of dipolar
interactions (fig. 9   ). The ENDOR lines were assigned by
a combination of isotope labeling, chemical correlations
and spectroscopic techniques (fig. 9,b,c) and the results
are summarized in table 3. for reaction centers from two
photosynthetic bacteria, Rhodopseudomonas spheroides and
Rhodospirillum rubrum. There are some distinct and preparation independent differences between the species (fig.
9) but the general picture is identical: All coupling constants in the reaction centers are reduced in comparison
to the respective couplings from oxidized Bchl a in solution. The reduction factors are, however, considerably
different for different proton couplings, which indicates
a redistribution of the spin densities in the oxidized
primary donor. The average reduction factor is, however,
again close to two, thus indicating again a dimeric
structure for P870.[+] There is furthermore only one set of
lines for each coupling in the in situ species, e.g. the
individual lines are not split. Since such a doubling of
lines would be expected for an unsymmetric spin distribution over the two halves of the special pair, this argues

Table 3: Comparison of isotropic hfc's a (MHz) of
Bchl $\underline{a}^{+}$ (255 K) and P_{870}^{+} (20°C) cation radicals

position	Bchl $a_{gg,P}^{+}$ a	P_{870}^{+} a	R-26 RF	P_{870}^{+} a	G-9 RF
CH$_3$ (1a)	+ 4.85	+4.00	1.21	+3.40	1.43
CH$_3$ (5a)	+ 9.50	+5.60	1.70	+4.85	1.96
ß-H	+11.61	+3.30	3.52	+3.95	2.94
ß-H	+13.00	+4.45	2.92	+5.28	2.46
ß-H	+13.59	+8.60	1.58	+7.50	1.81
ß-H	+16.43	+9.50	1.73	+8.50	1.93

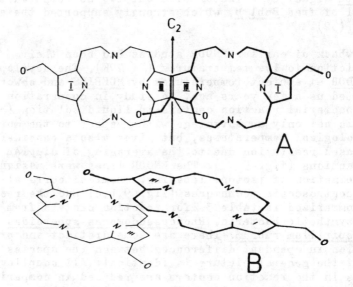

Fig. 10: Proposed geometry for P_{870}.t, the oxidized
primary donor of bacterial (Bchl a) photosynthesis. The plane-to-plane distance of the
molecules is 3.5-4A. The fit is obtained from
simulations of the ESR line shown in fig. 9
with the couplings from table 3.

for a dimer with C2 symmetry.

The experimental spin densities can now be used to simulate the ESR spectrum. A satisfactory fit is again only obtained under the assumption of a dimeric species for the oxidized primary donor. Based on these data has the Berlin group recently started molecular orbital calculations to assess the geometric relationship of the two molecules in the special pair. The two Bchl a molecules are treated as a super-molecule in these calculations. The preliminary results make the model shown in fig.10 most likely. The two molecules have a C2 axis, and they are essentially parallel ($\leqslant 5°$) to each other at a distance of 3.5 to 4 A. The two macrocyles are, however, shifted against each other and overlep only in the region of ring three. This model fits all the data, it accounts in particular for the symmetry of the entire system, and also for the observed spin density redistribution in the two halves.

Whereas it is rather nicely self-consistent, some constraints should be pointed out. It is firstly only applicable to Bchl a containing reaction centers from the Rhodospirillales, and the available ENDOR data indicate a rather different situation for Bchl b containing reaction centers and for the photosystem I and II of oxygenic photosynthesis. This is also evident from recent ESR experiments with organisms highly enriched in both ^{13}C and ^{2}H, which rather support a monomeric structure for the latter (35), and from the recent indication that a pigment structually different from Chl a is present in photosystem I (section B.b). There is finally a suggestion that severe distortions of the highest occupied molecular orbital (by mixing in of the energetically close lowest unoccupied MO), could principally explain the experimental data also with a monomeric species (36). Although none of our data give indications that this suggestion is correct, the model predicts some characteristic features (high spin density at the central nitrogen and methine positions) which are at least principally accessible to conclusive and direct measurements (e.g. ^{15}N-labeling). Such tests are currently under way.

C.e.: RAMAN RESONANCE STUDIES IN PHOTOSYNTHESIS

RAMAN resonance (RR) is probably one of the most general
techniques for the study of complex pigment systems, because it is very sensitive, widely applicable, quite
selective, provides an extremely high time resolution and
is accessible to a rather straightforward theoretical
analysis. The technique makes use of the strong enhancement
of the RAMAN spectrum of pigments, if the excitation occurs
in one of its absorption bands. This reduces the background
from ubiquitous components like proteins on water
sufficiently for a selective study. The major problem –
besides technology – is the interference of even low
levels of fluorescence. Unfortunately are most photosynthetic pigments highly fluorescent, but this problem
can be circumvented by excitation into higher excited
states, e.g. into the SORET band. Another problem is then,
however, the overlap of the absorption lines of many
pigments in this spectral range, which can partly be
circumvented by appropriate biochemical manipulations (isolation of complexes, redox changes) and by spectroscopic
means (variation of the excitation wavelength). Two
laboratories have concentrated in the past few years on RR
spectroscopy of photosynthetic pigments, in particular
the chlorophylls (37,38). COTTON et al. (38) have begun
with a systematic study of in vitro systems, e.g. free
chlorophylls in different states of aggregation and
oxidation, to obtain an interpretational basis for
photosynthetic complexes. LUTZ et al. (37) have started
with biological systems (whole cells, chlorophyll-protein
complexes), and based their interpretation strongly on
such complexes for which good structural data are available (e.g. the water-soluble antenna bacteriochlorophyll-
protein from Prostecochlorus aestuarii).

There are two spectral regions in the RR spectra of chlorophyll (-protein) which are well accessible to analysis.
One is the carbonyl region, which shows the C=O groups
which are conjugated to the macrocycle π-system. Their
binding state and conformation can be assesed from shifts
and intensity variations, respectively, in the carbonyl
stretching bands. The major conclusion from these data was,
that a differentiation is possible among most of the individual carbonyl groups   to be expected from the number
of chlorophylls present in a particular chromoprotein, and
that most if not all carbonyl groups are hydrogen bonded,
probably to the apoprotein. The second most informative
region is around 300 cm^{-1}, which has been assigned to the
central Mg-N vibrations. There is again one general result,
e.g. the central Mg is five coordinated in most of the pro-

tein bound pigments.

Studies with bacterial reaction centers showed changes in the carbonyl region upon oxidation of the primary donor, but few changes in other regions. LUTZ has used this as an argument against any close association among pigments, and in particular against the special pair model, because more pronounced changes are expected in aggregated systems (37a). The full meaning of this discrepan'cy is currently not yet clear. ENDOR and RR work at rather different time scales, which means that a hopping rate in the region of 10^{12} sec^{-1} would be seen as a delocalized electron by the former, but a localized one by the latter technique. BOWMAN and (39) NORRIS have recently narrowed this time gap by spin-echo experiments, but it is still in the range of two orders of magnitude.

It is also desirable, that the RR method becomes more selective for the different pigments present in reaction centers, since only one or two of the six tetrapyrroles are part of the primary donor. Two promising extensions of the conventional technique have recently been applied for the first time to chlorophyll proteins, which may be useful in this respect. One is the surface enhanced RR, the complexes are deposited on a silver electrode. COTTON et al. (38a) have shown with bacterial reaction centers, that this enhances strongly the sensitivity and quenches the interfering fluorescence, and that it is also possible to enhance selectively the RR signal from different pigments within the reaction center by variation of the potential applied to the electrode. Another technique which is at least principally related to RR is the coherent anti-STOKES RAMAN scattering. Its most important advantage over conventional RR is its greatly reduced sensitivity to fluorescence, because the vibrational lines are observed at the high-rather than the low-frequency side of the excitation line. HÖXTERMANN et al. (4o) have demonstrated, that this permits the measurement of chlorophylls in the red spectral region, which is much more sensitive to variations in the chlorophyll environment and thus allows more selective studies. Although the method seems to be somewhat less sensitive in the carbonyl region, it allowed like the RR technique a distinction of several bands assigned to the carbonyl groups of Chl a and Chl b molecules differing by their different environments. This site-specific distinction is supported by the analysis of the more intense spectral region between 1600 and 100 cm^{-1}. It is dominated by the double bond stretching vibrations of the the conjugated system, which also yield multiple lines in the chlorophyll proteins. The decreased interference of

fluorescence in <u>CARS</u> renders its application to
other pigments possible, e.g. the biliproteins.
There is only a limited amount of RR data on their
chromophores (40, and <u>CARS</u> may help to get more
details about their native chromophores.

Conclusions

It has been tried to sketch some recent examples of spectroscopic techniques to investigate plant tetrapyrroles,
either isolated or in their more or less intact natural
environment. It should be emphasized that all the selected
examples required the cooperation of spectroscopists and
biochemists although the latter have been neglected in this talk.
None can do without the other in such research, which poses
some problems in timing, a multiplied chance of breakdowns,
and last not least sample transfer. The examples show, that
these shortcomings are sometimes compensated by the results,
which have led to an improved picture of the structure and
function of tetrapyroles on the molecular level.

References

1. a) Rüdiger, W. 1980, Struct.Bond. 40, pp. 101,
   b) Rüdiger, W. and Scheer, H. 1983, Encyclopedia
   of Plant Physiology, 16, in press
2. Klein, G., Grombein S. and Rüdiger, W. 1977,
   Hoppe-Seyler's Z.Physiol.Chem. 358, pp. 1077-1079
3. Scheer, H. 1981, Angew.Chem. 93, pp. 230-250 and
   Angew.Chm.Int.Ed. 20, pp. 241-261
4. a) Gossauer, A., Hinze, R.P. and Kutschan, R. 1981
   Chem.Ber. 114, pp. 132-146, b) Placha-Puller, M.
   1979, Dissertation Techn.Universität Braunschweig
5. Rüdiger, W., Brandlmeier, T., Blos, I., Gossauer,
   A. and Weller, J.P. 1980, Z.Naturforsch. 35c,
   pp. 763-769
6. a) Lagarias, J.C. and Rapoport, H. 1980, J.Am.
   Chem.Soc. 102, pp. 4821-4828, b) Lagarias, J.C.,
   Glazer, A.N. and Rapoport, H. 1979, J.Am.Chem.Soc.
   101, pp. 5030
7. Rüdiger, W., Thümmler, F., Cmiel, E. and Schneider,
   S., Proc.Natl.Acad.Sci. USA in press
8. Kufer, W., Cmiel, E., Thümmler, F., Rüdiger, W.,
   Schneider, S. and Scheer, H. 1982, Photochem.
   Photobiol. 36, pp. 603-607
9. Falk, H., Kapl, G. and Müller, N., Monatsh.Chem.
   in press
10. a) Scheer, H., Svec, W.A., Cope, B.T., Studier,
    M.H., Scott, R.G. and Katz, J.J. 1974, J.Am.Chem.
    Soc. 96, pp. 3714, b) Steiner, R., Cmiel, E. and

Scheer, H., Z.Naturforsch. in press

11. Schoch, S., Scheer, H., Schiff, J.A., Rüdiger, W. and Siegelman, H.W. 1981, Z.Naturforsch. 36c, pp. 827-833

12. Bazzaz, M.B., Bradley, C.V. and Brereton, R.G. 1982, Tetrahedron Lett. 23, pp. 1211-1214

13. Scheer, H. and Katz, J.J. 1975 in K.M.Smith (ed.) "Porphyrins and Metalloporphyrins" 2nd ed. Elsevier, NewYork.

14. H. Budzikiewicz 1978 in D.Dolphin (ed.) "The Porphyrins",VIII, chapter 9, Academic Press, New-York;

15. Brereton, R.G., Rajanada, V., Blake, T.J., Sanders, J.K.M. and Williams, D.H. 1980, Tetrahedron Lett. 21, pp. 1671-74

16. a) Constantin, E., Nakatani, Y, Teller, G., Hueber, R. and Ourisson, G. 1981, Bull.Soc.Chim. France, pp. 303-305, b) Hunt, J.E., MacFarlane, R.D., Katz, J.J. and Dougherty, R.C. 1981, J.Am. Chem.Soc. 103, pp. 6775-6778

17. Dörnemann, D. and Senger, H. 1982, Photochem. Photobiol. 35, pp. 821-826

18. Wolf, H. and Scheer, H 1973, Ann.N.Y.Acad.Sci. 206, pp. 549

19. Jones, O.T.G. in Dolphin (ed.) 1978 "The Porphyrins" VI, chapter 3, Academic Press, NewYork

20. a) Rebeiz, C.A., Belanger, F.C., Freyssinet, G. and Saab, D.S. 1980, Biochim.Biophys.Acta 590, pp. 234-247, b) Rebeiz, C.A., Daniell, H., and Mattheis, J.R., Proceedings of the IVth symposium on biotechnology in energy production and conversation, in press

21. a) Battersby, A.R. and McDonald, E. 1979, Acc. Chem. Res. 12, pp. 14, b) Porra, R.J., Klein, O. and Wright, P.E., Eur.J.Biochem, 130,pp.509-516 c) Ohhama, T., Seto, H, Otake, N. and Miyachi, S. 1982, Biochem.Biophys.Res.Comm.105,pp.647-652

22. Belanger, F.C. and Rebeiz, C.A. 1982, J.Biol.Chem. 257 , pp. 1360-1371

23. a) Beck, J., v. Schütz, J.U. and Wolf, H.C. 1983 Z.Naturforsch. 38, pp. 220-229, b) Beck, J., v. Schütz, J.U. and Wolf, H.C., Chem.Phys.Lett. in press, c) Beck, J., v.Schütz, J.U. and Wolf, H.C., Chem.Phys.Lett., in press, d) Searle,G.F.W., Koehorst, R.B., Schaafsma, T.J., Moller, B.L., and v.Wettstein,D. 1981, Carlsberg Res 46 pp.183-194

24. a) Gantt, E. 1981, Ann.Rev.Plant Physiol. 32, pp. 327-347, b) Glazer, A.N. 1980 in D. Sigman, M.A.B. Brazier (eds.) "The Evolution of Protein Structure

and Function", Academic Press, NewYork, pp.221-
244, c) Mörschel, E., Koller, K. and Wehrmeyer,
W. 1980, Arch.Microbiol. 125, pp.43-51

25. Holzwarth, A.R., Wendler, J., and Wehrmeyer, W.
Biochim. Biophys.Acta in press

26. a) Grabowski, J. and Gantt, E. 1978, Photochem.
Photobiol. 28, pp. 39-45, b) Teale, F.W.J. and
Dale, R.E. 1970, Biochem.J. 116, pp. 161-169,
c) Zickendraht-Wendelstadt, B., Friedrich, J.
and Rüdiger, W. 1980, Photochem.Photobiol. 31,
pp. 367-376

27. a) Searle, G.F., Barber, J., Porter, G. and
Tredwell, C.J. 1978, Biochim.Biophys.Acta 501,
pp. 246, b) J. Breton and Geacintov, N.E. 1980,
Biochim.Biophys.Acta, 594, pp. 1-32

28. Doukas, A.G., Stefancic, V., Buchert, J. and
Alfano, R.R. 1981, Photochem. P. 34, pp. 505-510

29. Hoff, A.J. 1982, Biophys.Str. 8, pp. 107-150

30. Feher, G. and Okamura, M.Y. 1978 in R.K.Clayton
and W.R.Sistrom(eds.) "The Photosynthetic
Bacteria," Chapter 19, Plenum Press, NewYork

31. Norris, J.R., Uphaus, R.A., Crespi, H.L. and
Katz, J.J. 1971, Proc.Natl.Acad.Sci 68, pp.625

32. a) Wasielewski 1982 in F.K. Fong (ed.) "Light
Reaction Path of Photosynthesis, Chapter 7,
Springer-Verlag, Heidelberg, b) Katz, J.J.,
Shipman, L.L., Cotton, T.M. and Janson T.R. 1978,
in D. Dolphin (ed) "The Porphyrins", V Chapter 9,
Academic Press, NewYork

33. Möbius, K., Plato, M. and Lubitz, W. 1982,
Physics Report 87, pp. 171-208

34. a) Lendzian, F., Lubitz, W., Scheer, H.,
Bubenzer, C. and Möbius, K. 1981, J.Am.Chem.Soc.
103, pp. 4635-4637, b) Lubitz, W., Lendzian, F.,
Scheer, H., Gottstein, J., Plato, M. and Möbius,
K., Proc.Natl.Acad.Sci. in press

35. Wasielewski, M.R., Norris, J.R., Crespi, H.L. and
Harper 1981, J.Am.Chem.Soc.103, pp. 7664-7665

36. O'Malley, P.J. and Babcock, G.T., Proc.Natl.
Acad.Sci., submitted for publication

37. a) Lutz, M. Brown, J.S. and Rémy, R. 1979 in
"Chlorophyll Organisation and Energy Transfer
in Photosynthesis" (Ciba Foundation Symp.61 eds.
G. Wolstenholme, D.W.Fitzsimmons), Excerpta
Medica, Amsterdam, pp. 105-125, b) Lutz, M. 1981
in "Photosynthesis" III (ed. G.Akoyonoglu),
Balaban, Philadelphia, pp. 461-476

38. a) Cotton, T.M. and Vanduyne, R.P. 1981, J.Am.
Chem.Soc. 103, pp. 6020-6026, b) Cotton T.M. and
Vanduyne, R.P. 1982, Febs.Lett. 147, pp. 81-84

39. Bowman, M.K. and Norris, J.R. 1982, J.Am.Chem.
     Soc. 104, pp. 1512-1515
40. a) Höxtermann, E., Werncke, W., Stadnich, I.N.,
     Lau, A. and Hoffmann, P. 1982, Stud.Biophys.92,
     pp. 169-175, b) Höxtermann, E., Werncke, W.,
     Stadnich, I.N., Lau, A. and Hoffmann, P. 1982,
     Stud.Biophys. 92, pp. 159-168
41. Pellegring, F., Wong, D., Alfano, R.R., Zilinskas, B.
     PAB 34, 691-696 (1981).
42. Friedrich, J., Scheer, H., Zickendraht-Wendelstadt, B.,
     Haarer, D., J. Chem. Phys. 74, 2260-2266 (1981)
     Friedrich, J., Scheer, H., Zickendraht-Wendelstadt, B.,
     Haarer, D., J. Am. Chem. Soc. 103, 1030-1035 (1981)
     Rebane, K.K., Avarmaa R.A., Chem Phys 68(1-2): 191-2oo,
     (1982)
43. Pellegrino, F., Wong, D., Alfano R.R., Zilinskas, B.,
     PAB 34, 691-696 (1981).
     Kobayashi, T., Degenkolb,E. O., Behrson, R., Rentzepis,
     P. M., MacColl, R., Berns, D. S., Biochemistry 18,
     5073 (1979).
44. Katz, J.J., Norris, J.R., Shipmann, L.S., Thurnauer, M.C.,
     Wasielewski, M.R., Ann. Rev. Biophys. Bioeng. 7, 393-
     434,(1978).

PREPARATION OF ^{2}H AND ^{13}C LABELED RETINALS

Johan Lugtenburg

Department of Chemistry, State University Leiden,

The Netherlands

ABSTRACT. In the field of visual pigments and bacteriorhodopsin isotopic modification in the chromophores of these pigments is indispensable for the interpretation of spectroscopic data. These modified systems are accessible by regeneration of the apoprotein with the isotopically labeled retinal. Our strategy to prepare ^{2}H and ^{13}C labeled all-*trans* retinals and their 13-*cis*, 11-*cis* and 9-*cis* isomeric forms is discussed.

INTRODUCTION

Light plays a fundamental role in many biological processes. The absorption of a light quantum by the target molecules is the first act in these processes. Both in visual pigments and in bacteriorhodopsin (the light-driven proton pump in Halobacterium halobium) the chromophoric part is derived from retinal, Schiff's base linked to the peptide chain[1]. In visual pigments the chromophore has the 11-*cis* structure. Upon excitation with light, rhodopsin (the rod visual pigment) is converted into the labile primary intermediate bathorhodopsin. It has a half life of about 10^{-8} sec at physiological temperatures, it is converted into lumirhodopsin and via further intermediated into opsin and free all-*trans* retinal (in the case of vertebrate rhodopsin). Opsin combines with 11-*cis* retinal to regeneration rhodopsin.

Bacteriorhodopsin, the sole protein of the purple membrane of the halophilic bacterium Halobacterium halobium, occurs in a light- and a dark-adapted form. Bacteriorhodopsin functions as a light-driven proton pump, resulting in a vectorial transport of protons across the cell wall. The energy of this proton gradient is used by Halobacterium halobium to synthesize A.T.P. Light con-

C. Sandorfy and T. Theophanides (eds.), Spectroscopy of Biological Molecules, 447–455.

verts light-adapted BR into the energy rich photoproduct K. Via
thermal intermediates K is converted back into light-adapted BR.
During this cycle protons are transported from the inside of the
membrane to the outside. The energy for this proton pump action
is provided by the free energy of K. As a first approach for an
understanding of the molecular basis of the photochemistry of
visual pigments and bacteriorhodopsin it is necessary to elucidate the structure of the retinal chromophoric part in these pigments, their primary photoproducts and the further intermediates
in the photochemical reaction sequence. Resonance Raman spectroscopy has been used to obtain *in situ* vibrational information
about the chromophores in visual pigments and BR and the chromophore in the labile intermediates obtained by light excitation.[2]
Recently, FT IR difference spectroscopy has also been used to
acquire vibrational difference spectra between the intermediates
and the starting materials.[3]

The classical method of obtaining structural information is by
use of isotopic substitution. The vibrational spectra of isotopic
derivatives allow the assignment and interpretation of vibrational spectra whence the necessary structural information can be
derived. The only way to obtain visual pigments and BR with a
specific isotope label in their chromophores is total synthesis
of the thus labeled retinal and combining this with the apoprotein in question. The apoproteins of rhodopsin and bacteriorhodopsin are accessible. Reaction of opsin with labeled 11-*cis*
retinal gives rhodopsin with labeled chromophore, similarly
labeled bacteriorhodopsins are obtained by reacting BO with the
labeled all-*trans* retinal.

It is expected that the isotopic modification does not introduce ambiguities due to changes in electronic and steric interaction between chromophore and peptide part in any of these
systems. In fact, the native pigments contain each of the mono
labeled (^{2}H and ^{13}C) chromophores at the natural abundance level.
In the present paper the total synthesis of ^{2}H and ^{13}C labeled
retinals in all *trans*, 13-*cis*, 11-*cis* and 9-*cis* form is described.
These labeled retinals have been used by Prof. Richard Mathies
and coworkers at the University of California at Berkeley for the
assignment and interpretation of R.R. spectra of visual pigments
and bacteriorhodopsin, their photoproducts, and for the assignment and interpretation of the vibrational features of retinal
isomers.[4]

THE SYNTHESIS OF ^{2}H RETINALS

The structure of all-*trans* retinal ($C_{20}H_{28}O$) is depicted in
Figure 1. As an illustration of our strategy of synthesizing ^{2}H
labeled retinal, the synthesis of 10-, 11-mono and 10,11-dideute-
ro retinal will be discussed. The incorporation should be 95% or
better in order to ensure the required vibrational information.

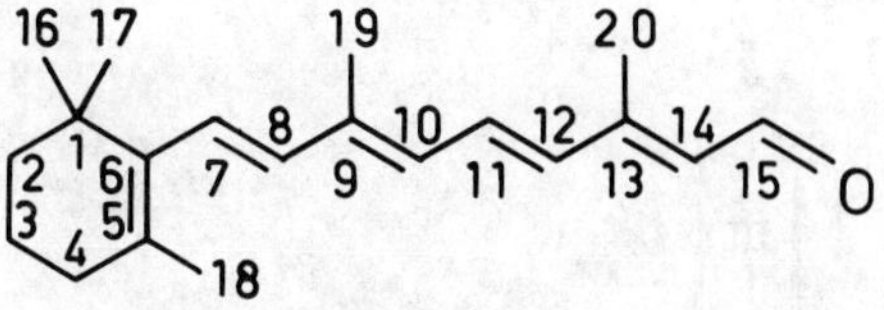

Figure 1. All-*trans* retinal.

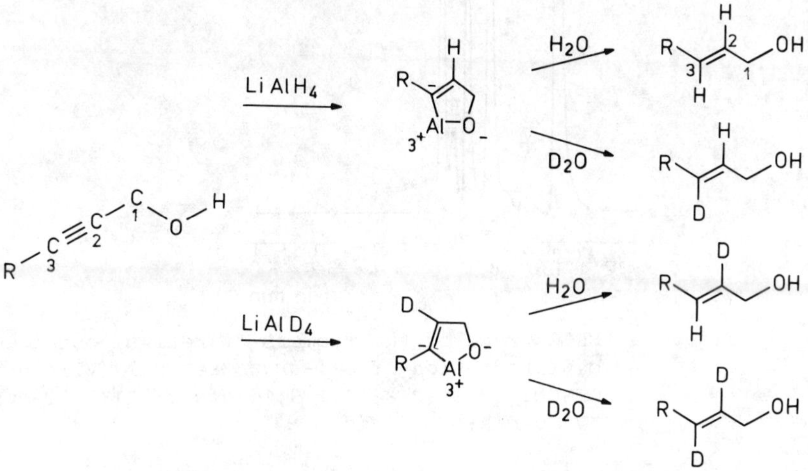

Figure 2. The conversion of a propynol into ^{2}H labeled propenols.

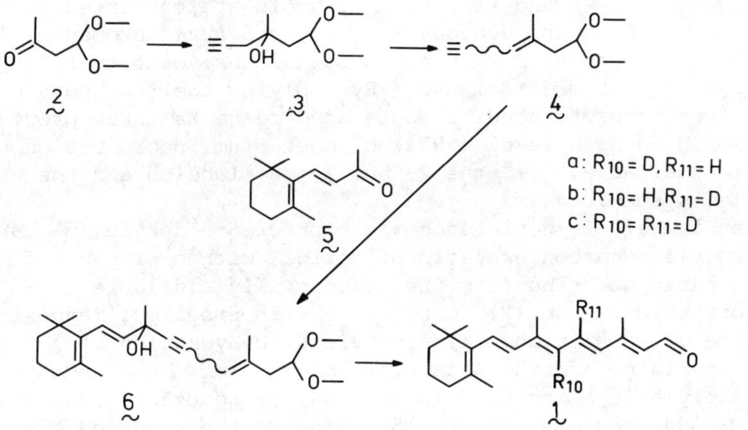

a: $R_{10} = D$, $R_{11} = H$
b: $R_{10} = H$, $R_{11} = D$
c: $R_{10} = R_{11} = D$

Figure 3. Reaction scheme for the preparation of retinal deuterated on positions 10 and 11.

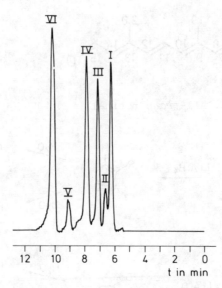

Figure 4. HPLC trace of the isomers of retinal obtained
by irradiation of all-*trans* retinal (VI) in
CH_3CN: I = 13-*cis*; II = 9,13-*dicis*; III = 11-*cis*;
IV = 9-*cis*; V = 7-*cis*.

With deuterated organic materials there is a possibility that at
later stages of the reaction the deuterium will partially or completely vanish. Therefore, we have chosen to introduce the ^2H
label as late as possible in the reaction sequence. Many inexpensive reagents are available for the introduction of the ^2H label.
We have used ^2H_2O and $LiAl^2H_4$. Propynols are converted by reaction with $LiAlH_4$ and aqueous work-up unto *trans* propenals (Figure
2). The hydrogen atom 2 in the propenal derives from $LiAlH_4$ and
the hydrogen atom 3 from H_2O.[5] By applying the four possible combinations of protonated or deuterated reagents three propenol
systems with high level (>99%) of deuterium incorporation have
become available, i.e. the 2- and 3-mono labeled and the 2,3-di-
labeled system.[6]

   In Figure 3 is depicted how a protected aldehyde (6) having
the complete carbon skeleton of retinal with a propynol function,
is prepared starting from the commercially available 1,1-di-
methoxy butan-3-one (2). Coupling 2 with propargyl magnesium
bromide gives 3 in quantitative yield. Dehydration of 3 with
$POCl_3$/pyridine affords 4 as a 1:1 mixture of the E and Z isomers.
The acetylenic anion of 4 is obtained by reaction with n-BuLi at
-60°C; addition of β ionone (5) leads to the required synthon 6.
Using the reactions with $LiAlH_4$ or H_2O and their deuterated forms
followed by treatment with 85% H_3PO_4 gives the labeled retinals.
The retinals were purified by Silicagel chromatography. This
tricky acid treatment affords high incorporation: 1a ≥ 96% ^2H,

1b ⩾ 95% ^{2}H and 1c ⩾ 97.5% ^{2}H, determined by mass spectrometry.
The various deuterated all-*trans* retinals were irradiated in
dilute acetonitrile solution. The HPLC trace of the mixture of
retinal isomers is shown in Figure 4. By preparative HPLC isolation the deuterated 13-*cis*, 11-*cis* and 9-*cis* forms together with
the all-*trans* form were obtained in 99% chemical purity in milli-
gram quantities.

THE SYNTHESIS OF ^{13}C LABELED RETINALS

For the synthesis of ^{13}C labeled retinals the exchange of
label is not expected to occur during the synthesis. However, the
use of labeled synthons more complicated than one or two carbon-
containing systems becomes prohibitively expensive. We have
therefore based our syntheses on ^{13}CH$_3$I, ^{13}CH$_3$CN, CH$_3$^{13}CN and
^{13}CH$_3$^{13}CN with 90% incorporation[7] (if necessary, these molecules
are also available in the 99% incorporation level). Other groups
have used different synthons.[8] The ^{13}CH$_3$I is used to label atoms
19 and 20 in retinal; the labeled acetonitriles are used for the
introduction of ^{13}C into various positions in the conjugated
chain (10, 11, 12, 13, 14 and 15).

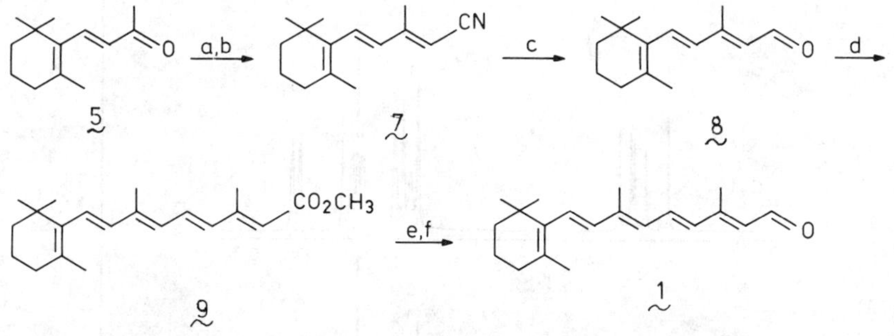

Figure 5. Preparation of retinal ^{13}C labeled on 10
and/or 11. a) CH$_3$CN, n-BuLi; b) cat. NBS;
c) (i-Bu)$_2$AlH; d) (EtO)$_2$P(=O)CH$_2$(CH$_3$)C=CH-
CO$_2$CH$_2$CH$_3$, NaH; e) LiAlH$_4$; f) MnO$_2$.

In Figure 5 is indicated how β ionone (5) is converted into retinal with ^{13}C label on positions 10 and 11. The anion of the
labeled acetonitrile is prepared by reaction with n-BuLi in THF
at -60°C; addition of β ionone (5) gives the aldol type addition
product. Dehydration with a catalytic amount of NBS leads to β
ionylidene acetonitrile (7).[9] Reaction with diisobutyl aluminium
hydride gives β ionylidene acetaldehyde (8).[10] The methylester of
retinoic acid (9) is obtained by the Wittig type reaction of 8

with diethyl 3-(methoxycarbonyl)-2-methyl-2-propenyl phosphonate[11]. A reduction oxidation sequence of the ester 9 yields
retinal (1)

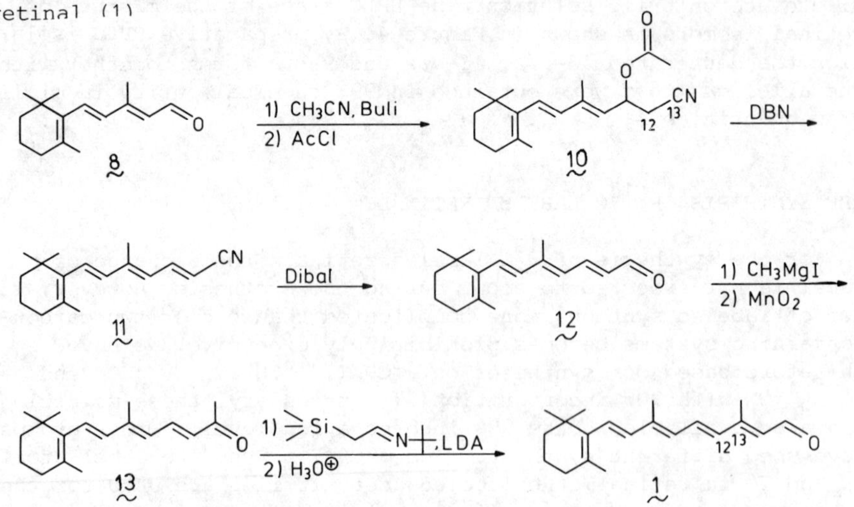

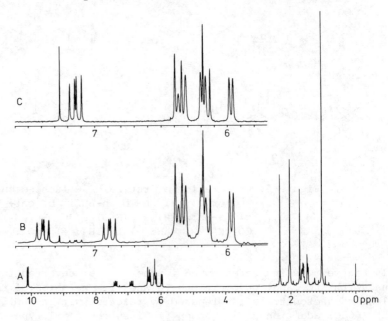

Figure 6. Preparation of retinal ^{13}C labeled on
positions 12 and/or 13 and/or 20.

Figure 7. A) 300 MHz ^{1}H NMR spectrum of 11-^{13}C retinal;
B) the vinylic region expanded;
C) ibid. of all-*trans* retinal.

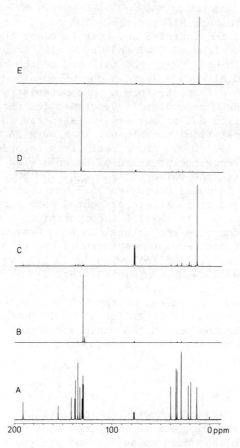

Figure 8. 75.5 MHz ^{13}C NMR spectra in CDCl$_3$ (TMS as reference) of:
A) all-*trans* retinal;
B) 10-^{13}C all-*trans* retinal;
C) 19-^{13}C all-*trans* retinal;
D) 11-^{13}C all-*trans* retinal;
E) 20-^{13}C all-*trans* retinal.

In Figure 6 the sequence for the preparation of all-*trans* retinal with ^{13}C label on positions 12, 13 and 20 is given. β ionylidene acetaldehyde (8) is reacted with the anion of aceto-nitrile and the resulting alcoholate is converted into the ace-tate 10. Base-catalysed acetic acid elimination gives the tetra-enenitrile 11. Diisobutylaluminiumhydride converts this into the corresponding tetraeneal (12). The Grignard reaction with (labeled) methyl magnesium iodide and subsequent oxidation gives the C$_{18}$ ketone (13). Peterson olefination with the anion of the tert. butylamine Schiff base of 2-trimethylsilyl acetaldehyde[12] and subsequent hydrolysis affords the ^{13}C labeled retinal.

From the schemes in Figures 5 and 6 it is clear that the judicious application of labeled CH_3I and CH_3CN will lead to any ^{13}C labeled retinal with the ^{13}C in the tail end of the retinal molecule. These labeled all-*trans* systems can be obtained in the 13-*cis*, 11-*cis* and 9-*cis* isomeric form by photochemistry and subsequent preparative HPLC procedure as described for the deuterated retinals. The position and amount of ^{13}C label can be checked with the various analytical techniques. In Figure 3A the 300 MHz ^{1}H NMR spectrum of 11-^{13}C all-*trans* retinal is recorded. The signals at δ 7.14 ppm show the presence of 8% unlabeled and 92% ^{13}C labeling at C 11 with the value for $J_{C\ 11-H\ 11} = 150.4$ Hz.

In Figure 8 the ^{1}H noise decoupled 75.5 MHz spectra of four ^{13}C labeled all-*trans* retinals are presented each showing distinctly the position of the 92% ^{13}C incorporation. Each of the labeled compounds contains the natural abundance amount of ^{13}C on all the other positions. From the spectra in Figure 8 at the natural abundance intensity the following $^{1}J\ 13_C-13_C$ values could be collected: $J_{9-10} = 70.4$ Hz, $J_{9-11} = 43.4$ Hz, $J_{10-11} = 58.7$ Hz, $J_{11-12} = 69.8$ Hz and $J_{13-20} = 40.4$ Hz. These $^{1}J\ 13_C-13_C$ values correlate well with the bond values of the C-C bond between the carbon atoms in question, about 70 Hz for a double bond, about 60 Hz for a single bond in the conjugated chain and about 42 Hz for a methyl carbon with the atom of the conjugated chain.

SUMMARY

Up to now 31 different isotopic labeled retinals have been made by our group and more are in the process of being prepared. These retinals have been used to assign and analyse resonance Raman spectra of pigments with a retinylidene chromophore and their photoproducts. It is to be expected that, analogously, they will give essential information in this field with FTIR spectroscopy and with magic angle spinning ^{13}C NMR spectroscopy. There are great expectations for the use of tailor-made isotopically modified systems prepared by organic synthesis in the field of spectroscopy of biological molecules.

ACKNOWLEDGEMENT

Thanks are due to the many persons without whose efforts this work would not have been realized, especially my coworkers Albert Broek, Hans Pardoen, Henk Nijenesch, Patrick Mulder, Marcel van der Brugge and Chris Winkel. I like to thank Prof. Rich Mathies and his coworkers for a fruitful and intense collaboration in the field of visual pigments and bacteriorhodopsins. Thanks are due to Sygna Amadio for her help in preparing this manuscript. This research was supported by the Netherlands Foundation for Chemical

Research (SON) with financial aid from the Netherlands Organisation for the Advancement of Pure Research (ZWO).

REFERENCES

1. M. Ottolenghi, Advances in Photochemistry *12*, 97 (1980);
   R.R. Birge, Annu. Rev. Biophys. Bioeng. *10*, 315 (1981);
   W. Stoeckenius, R.A. Bogomolni, Annu. Rev. Biochemistry *51*,
   587 (1982).
2. L. Packer, ed.,   Methods in Enzymology   , pp. 561-666 (1982).
3. K. Bagley, G. Dollinger, L. Eisenstein, A.K. Singh and L.
   Zimangi, Proc. Natl. Acad. USA *79*, 4972 (1982);
   K.J. Rothschild, H. Marrero, Proc. Natl. Acad. Sci. USA *79*,
   4045 (1982);
   F. Siebert, W. Mäntele, Eur. J. Bioch. *130*, 565 (1983);
   K.J. Rothschild, W.A. Cantore, H. Marrero, Science *219*, 1333
   (1983).
4. G. Eyring, B. Curry, R.A. Mathies, M.R. Fransen, H.A.M.
   Palings, J. Lugtenburg, Biochemistry *19*, 2410 (1980);
   G. Eyring, B. Curry, R.A. Mathies, A.D. Broek, J. Lugtenburg,
   J. Am. Chem. Soc. *102*, 5390 (1980);
   G. Eyring, B. Curry, A.D. Broek, J. Lugtenburg, R.A. Mathies,
   Biochemistry *21*, 384 (1982);
   B. Curry, A.D. Broek, J. Lugtenburg, R.A. Mathies, J. Am.
   Chem. Soc. *104*, 5274 (1982).
5. R.A. Raphael, Acetylenic Compounds in Organic Synthesis 29
   (1955);
   B. Grant, C. Djerassi, J. Org. Chem. *39*, 968 (1974).
6. A.D. Broek, J. Lugtenburg, Recl. Trav. Chim. Pays-Bas *99*, 363
   (1980);
   A.D. Broek, J. Lugtenburg, ibid. *101*, 102 (1982).
7. J.A. Pardoen, H.N. Neijenesch, P.P.J. Mulder, J. Lugtenburg,
   Recl. Trav. Chim. Pays-Bas *102*, 341 (1983).
8. G.D. Mateescu, W.G. Copan, D.D. Muccio, D.V. Waterhouse, E.W.
   Abrahamson, Proceedings of an International Symposium Kansas
   City, Elsevier 123 (1983);
   A. Yamaguchi, T. Unemoto, A. Ikegami, Photochem. Photobiol.
   33, 511 (1981);
   J. Shriver, G. Mateescu, R. Fager, D. Torchia, E.W. Abrahamson, Nature *270*, 271 (1977).
9. K. Eiter, E. Truscheit, U.S. Patent 2,974,155 C.A. 55 17541d.
10. A. Kini, H. Matsumoto, R.S.H. Liu, J. Am. Chem. Soc. *101*,
    5078 (1979).
11. H. Pommer, Angew. Chem. *72*, 811 (1960).
12. E.J. Corey, D. Enders, M.G. Bock, Tet. Lett. *7* (1976).

AN INTRODUCTION TO TWO-PHOTON SPECTROSCOPY

Robert R. Birge

Department of Chemistry
University of California
Riverside, CA 92521 U.S.A.

ABSTRACT: The basic theory and selected experimental methods of
two-photon spectroscopy are presented. Two-photon absorption
probabilities and the effect of photon polarization on the
absorptivities are analyzed using a single intermediate state
approximation. We demonstrate, however, that electronic transitions which involve large changes in dipole moment will have
enhanced two-photon absorptivities due to contributions from the
initial and final states. The two-photon excitation, thermal
lens and double resonance techniques are briefly described.

1.   INTRODUCTION

      Although two-photon spectroscopy has become a routine
experimental technique for studying atomic and small molecule
electronic levels, the increasing availability of reasonably
priced tunable pulsed lasers has encouraged a growing number of
researchers to adopt this technique for studying polyatomic
molecules (1-6). The purpose of this chapter is to briefly
introduce the basic theory and selected experimental techniques
associated with two-photon spectroscopy. An emphasis is placed
on describing experimental methods that are currently in use for
studying biological molecules. This discussion is neither complete nor sufficiently rigorous to provide a satisfactory
appreciation for all the advantages and experimental problems.
Nonetheless, it should prove useful to those readers unfamiliar
with this technique and will help direct the reader to the available literature on the subject. A number of recent reviews
should be consulted for a more detailed theoretical and experimental discussion of two-photon spectroscopy (1-9).

C. Sandorfy and T. Theophanides (eds.), Spectroscopy of Biological Molecules, 457–471.
© 1984 by D. Reidel Publishing Company.

2.   TWO-PHOTON SELECTION RULES

The electric dipole approximation provides a useful (though not rigorously accurate) approach to the analysis of one-photon and two-photon selection rules. One-photon transitions are allowed when the transition length has a component which transforms under x, y, or z symmetry operations. In contrast, two-photon transitions are allowed for transition-length products which have components that transform under x^2, y^2, z^2, xy, yz, or xz symmetry operations.

The selection rules for molecules with inversion symmetry are summarized in Fig. 1. The complementary nature of one-photon and two-photon spectroscopy is evident, but it should be emphasized that an affirmative selection rule does not guarantee an observable transition. A g←g two-photon transition can have negligible absorptivity due to orbital overlap or spin restrictions. In other words, one- and two-photon selection rules describe the possibilities, but not the probabilities of an absorption process. McClain has reviewed the two-photon selection rules for all 32 crystallographic point groups and the two linear molecule groups (10). The next section reviews the theoretical background necessary to understand the molecular-electronic parameters that determine the probability of two-photon absorption.

3.   THEORETICAL BACKGROUND

The allowedness of a one-photon transition is characterized by a dimensionless parameter called the oscillator strength f. This parameter is related to the transition length of the f←o electronic transition by the expression

$$f_{fo} = (8\pi^2 m_e/3h^2)\Delta E_{fo}|<f|\underset{\sim}{r}|o>|^2 = (1.0847\times10^{-5})\Delta\tilde{\nu}_{fo}|<f|\underset{\sim}{r}|o>|^2 \quad (1)$$

where m_e is the mass of an electron, h is Planck's constant, $\Delta\tilde{\nu}_{fo}$ is the transition energy in wavenumbers (cm^{-1}) and $|<f|\underset{\sim}{r}|o>|$ is the transition length in angstroms. The oscillator strength is related to the experimentally observed absorption spectrum by the expression

$$f_{fo} = \frac{10^3(\ln 10)m_e c^2}{N_A \pi e^2}\int\varepsilon_{\tilde{\nu}}^{fo}d\tilde{\nu} = (4.3190\times10^{-9})\int\varepsilon_{\tilde{\nu}}^{fo}d\tilde{\nu} \quad (2)$$

where c is the speed of light in a vacuum, N_A is Avogadro's number, e is the charge on an electron, and $\varepsilon_{\tilde{\nu}}^{fo}$ is the molar absorbtivity at wavenumber $\tilde{\nu}$ in liters per mole-centimeter. The one-photon cross section $\sigma_{\tilde{\nu}}^{fo}$ is related to the molar absorptivity

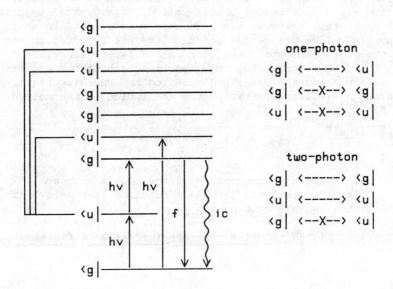

Figure 1. A two-photon transition from the ground state to the
lowest-lying ⟨g| state is shown at far left for a molecule with
inversion symmetry. This transition proceeds via preparation of
a virtual level which is a superposition of all the ⟨u| excited
states. A second photon arrives before the virtual level decays
($\sim 10^{-15}$ sec) to generate a final state of g symmetry. Accordingly,
an allowed two-photon transition can be viewed as a sequence of
two allowed one-photon transitions ⟨g|←⟨u|←⟨g| which leads to the
selection rules shown at right. The final excited state returns
to the ground state via fluorescence (f) and/or internal conversion (ic). If the fluorescence quantum yield is larger than 0.1,
the two-photon excitation method is usually the preferred technique. Molecules with low or negligible fluorescence quantum
yields are best studied using two-photon thermal lens, photo-
acoustic or double resonance techniques.

by the expression

$$\sigma_{\tilde{\nu}}^{fo} = 1000(\ln 10)\,\varepsilon_{\tilde{\nu}}^{fo}/N_A = (3.8235\times10^{-21})\,\varepsilon_{\tilde{\nu}}^{fo} \tag{3}$$

where $\sigma_{\tilde{\nu}}^{fo}$ is in units of square centimeters per molecule.

The one-photon oscillator strength has no direct analogy in
two-photon spectroscopy because the allowedness of a two-photon
process is dependent upon both the molecular properties and the

experimental configuration. This observation does not preclude
the definition of an appropriate parameter based on the integrated
two-photon cross section for a well-defined experimental condition.
No such parameter, however, has gained general acceptance and,
consequently, all experimental measurements are reported in terms
of the two-photon absorptivity (also called two-photon cross
section) $\delta_{\tilde{\nu}}^{fo}$. This parameter is a function both of the properties
of the molecule and the laser polarization and energies required
to generate a simultaneous two-photon absorption in the molecule
(9):

$$\delta_{\tilde{\nu}}^{fo} = \frac{(2\pi e)^4}{(ch)^2} \, \tilde{\nu}_\lambda \tilde{\nu}_\mu \, g(\tilde{\nu}_\lambda + \tilde{\nu}_\mu) \, |S_{fo}(\lambda,\mu)|^2 \tag{4}$$

where $\tilde{\nu}_\lambda$ and $\tilde{\nu}_\mu$ are the frequencies of the two laser beams,
$g(\tilde{\nu}_\lambda + \tilde{\nu}_\mu)$ is the normalized lineshape function (see below) and
$S_{fo}(\lambda,\mu)$ is the two-photon tensor:

$$S_{fo}(\lambda,\mu) = \sum_k \frac{(\lambda \cdot <k|\underline{r}|o>)(<f|\underline{r}|k> \cdot \lambda)}{\tilde{\nu}_k - \tilde{\nu}_\lambda + i\Gamma_k} + \frac{(\mu \cdot <k|\underline{r}|o>)(<f|\underline{r}|k> \cdot \mu)}{\tilde{\nu}_k - \tilde{\nu}_\mu + i\Gamma_k} \tag{5}$$

where λ and μ are the unit vectors defining the polarization of
the two photons and $\tilde{\nu}_k$ and Γ_k are the transition frequency and
the linewidth of the system origin of state k, respectively.
The summation is over all electronic states of the molecule
including the ground and final states (11,12).

Each molecule will have a unique lineshape (or bandshape)
function which will be very complicated in those instances where
vibronic structure can be resolved. It is important to note that
this function is solute, solvent, and temperature dependent and
therefore a maximum two-photon absorptivity measured under a
given set of experimental conditions will not, in general, be
meaningful for different solvent or temperature environments.
The normalized lineshape function takes on a particularly simple
form when the two-photon excitation profile for the final electronic state is Gaussian in shape (2,13):

$$g(\tilde{\nu}_\mu + \tilde{\nu}_\lambda) = g_{max} \exp\{[-4 \ln 2/(\Delta\tilde{\nu})^2](\tilde{\nu}_\mu + \tilde{\nu}_\lambda - \tilde{\nu}_f)^2\}, \tag{6}$$

$$g_{max}(sec) = (3.13363 \times 10^{-11})/\Delta\tilde{\nu}, \tag{7}$$

where $\Delta\tilde{\nu}$ is the full width at half maximum in wavenumbers. The
units of g(sec) derive from the arbitrary convention that this
function is normalized in frequency space so that

$$\int g(\nu)\,d\nu = 1. \tag{8}$$

The two-photon absorptivity $\delta_{\tilde{\nu}}$ has units of (centimeters)^4-second per molecule-photon. This factor has also been called the "two-photon cross section" because of its direct phenomenological relationship to the one-photon cross section $\sigma_{\tilde{\nu}}$ (Eq. 3).

A rough estimate of the two-photon absorptivity for a two-photon allowed state can be obtained by using a single intermediate-state approximation (2,13):

$$\delta_{max}^{f\leftarrow o} = \frac{8\pi^4 e^4}{15c^2 h^2} \tilde{\nu}_\lambda^2 g_{max} \frac{(a+b)[<k|\underset{\sim}{r}|o>\cdot<f|\underset{\sim}{r}|k>]^2}{(\tilde{\nu}_k - \tilde{\nu}_\lambda)^2} +$$

$$\frac{b|<k|\underset{\sim}{r}|o>|^2|<f|\underset{\sim}{r}|k>|^2}{(\tilde{\nu}_k - \tilde{\nu}_\lambda)^2} \qquad (9)$$

where $\tilde{\nu}_\lambda$ is the wavenumber of the laser excitation, a and b are photon propagation-polarization variables (see Section 4), and k represents a single (one-photon allowed) intermediate state at $\tilde{\nu}_k$. The above equation is a good approximation for nonpolar molecules with a low-lying, strongly (one-photon) allowed singlet $\pi\pi^*$ state, $<k|$, provided $<f|\underset{\sim}{r}|k>$ is also large. For example, the two-photon properties of the low-lying $^1A_g^{*-}$ state in linear polyenes are accurately described using the single intermediate state (SIS) approximation where $<k|$ is the strongly allowed $^1B_u^{*+}$ state (13,14). However, if the molecule is polar, the SIS approximation is often inadequate (see below). Equation 9 can be simplified further by assuming that the intermediate and final states (k and f) are close in energy [$\tilde{\nu}_f \sim \tilde{\nu}_k$ and therefore $\tilde{\nu}_\lambda = 1/2(\tilde{\nu}_k)$], the transition lengths $<k|\underset{\sim}{r}|o>$ and $<f|\underset{\sim}{r}|k>$ are similarly polarized, and the laser excitation is linearly polarized (a = b = 8):

$$\delta_{max}^{f\leftarrow o} \stackrel{\sim}{=} (1.682\text{x}10^{-3}) g_{max}|<k|\underset{\sim}{r}|o>|^2|<f|\underset{\sim}{r}|k>|^2. \qquad (10)$$

An allowed one-photon transition will have a transition length of $\sim 1\text{Å}$. A very broad absorption band will have a value of g_{max} on the order of 10^{-15} sec (see Eq. 7). Accordingly, a broad excitation band of a "typical" two-photon allowed state will exhibit a maximum two-photon absorptivity on the order of 10^{-50} cm^4 sec molecule^{-1} photon^{-1}. This absorptivity value is informally known as 1 Göppert-Mayer (GM) in honor of Maria Göppert-Mayer's pioneering theoretical treatment of the two-photon absorption phenomenon (15).

If the molecule undergoes a change in dipole moment upon excitation, then the initial and final states in the summation

in Eq. 5 will also contribute to the two-photon absorptivity
(11,12):

$$\delta^{f\leftarrow o}_{max} = \delta^{f\leftarrow o}_{max}(f\leftarrow k\leftarrow o) + \delta^{f\leftarrow o}_{max}(\Delta\mu) \tag{11}$$

where $\delta^{f\leftarrow o}_{max}(f\leftarrow k\leftarrow o)$ is the absorptivity due to the single intermed-
iate state, $\langle k|$, given by Eq. 9 and

$$\delta^{f\leftarrow o}_{max}(\Delta\mu) = \frac{8\pi^4 e^2}{30c^2 h^2}\, g_{max}\{(a+b)S_{ab} + b\,S_b\} \tag{12}$$

where

$$S_{ab} = (\underset{\sim}{\mu}_o \cdot \langle f|\underset{\sim}{r}|o\rangle)^2 + (\langle f|\underset{\sim}{r}|o\rangle \cdot \underset{\sim}{\mu}_f)^2$$
$$- 2(\underset{\sim}{\mu}_o \cdot \langle f|\underset{\sim}{r}|o\rangle)(\langle f|\underset{\sim}{r}|o\rangle \cdot \underset{\sim}{\mu}_f) \tag{13}$$

and

$$S_b = \underset{\sim}{\mu}_o^2|\langle f|\underset{\sim}{r}|o\rangle|^2 + \underset{\sim}{\mu}_f^2|\langle f|\underset{\sim}{r}|o\rangle|^2 - 2\underset{\sim}{\mu}_o\underset{\sim}{\mu}_f|\langle f|\underset{\sim}{r}|o\rangle|^2 \tag{14}$$

The terms in Eqs. 12-14 are defined as for Eq. 9 and $\underset{\sim}{\mu}_o$ and $\underset{\sim}{\mu}_f$
are the ground and excited state dipole moments. Note that S_{ab}
and S_b vanish if $\underset{\sim}{\mu}_o = \underset{\sim}{\mu}_f$. Accordingly, a change in dipole moment
is required. Furthermore, if the molecule to be studied is in a
solvent, then the reaction field of the solvent contributes to
the total dipole moment of the ground and excited states:

$$\underset{\sim}{\mu}_i^{tot} = \underset{\sim}{\mu}_i + \underset{\sim}{\alpha}_i \cdot \underset{\sim}{R}_{oo} \tag{15}$$

where $\underset{\sim}{\alpha}_i$ is the polarizability of the ith state and $\underset{\sim}{R}_{oo}$ is the
reaction field of the solvent in the ground state.

An examination of Eqs. 12-14 indicates that the one-photon
allowedness of the f←o transition ($\propto \langle f|\underset{\sim}{r}|o\rangle^2$) is an important
parameter in determining the two-photon absorptivity due to dipole
moment changes upon excitation. Accordingly, a one-photon allowed
state can have a large two-photon absorptivity. For example, if
$\langle f|\underset{\sim}{r}|o\rangle = 1\overset{\circ}{A}$, $g_{max} = 10^{-15}$ sec, $|\underset{\sim}{\mu}_o| = 5D$ and $|\underset{\sim}{\mu}_f| = 10D$, then Eq. 12
yields $\delta_{max}(\Delta\mu) \sim 0.5$ GM provided $\underset{\sim}{\mu}_o$, $\underset{\sim}{\mu}_f$ and $\langle f|\underset{\sim}{r}|o\rangle$ are identi-
cally polarized. This value is only half the value predicted for
the previous calculation of the absorptivity for a strongly two-
photon allowed state. Accordingly, the electronic states of a
highly polar molecule may be very difficult to analyze because
both the "allowed" and "forbidden" states will have similar two-
photon absorptivities. This observation will be important for
the interpretation of the two-photon spectra of the protonated
Schiff base retinyl polyenes (see following chapter).

4. POLARIZATION EFFECTS

One-photon and two-photon spectroscopy differ markedly with respect to polarization phenomena. The use of one-photon spectroscopy to assign the symmetry of an excited state requires the use of samples in which the solute molecule is oriented. Although this restriction can be partially circumvented by using photoselection techniques to determine the relative polarization of an absorbing versus an emitting state, the symmetry information is relative and not absolute. Two-photon spectroscopy can provide absolute symmetry assignments for many electronic states using randomly oriented samples (2,5,6,10). This unique characteristic of two-photon spectroscopy derives from the requirement that two photons must be absorbed simultaneously by the molecule. Hence, the relative polarization of the two photons will affect the two-photon absorption probability.

The ratio of the two-photon absorptivity for circularly polarized light versus linearly polarized light can be easily measured by using a Fresnel Rhomb or a quarter-wave plate to convert linearly polarized to circularly polarized light. The experimental accessibility combined with the diagnostic utility of this parameter has led to its definition as the "two-photon polarization ratio"

$$\Omega = \delta(\text{circ})/\delta(\text{lin}).$$ (16)

The single-intermediate-state approximation can be used to examine the range to be expected for this parameter (Eq. 9). The photon propagation-polarization variables, a and b in Eq. 9 are (13):

$$a = b = 8 \quad \text{(linearly polarized)},$$

and

$$a = -8, \; b = 12 \quad \text{(circularly polarized)}$$

where parallel propagation is assumed in both cases. Note that Eq. 9 assumes both photons have the same energy. If the transition length vectors $\langle k|\underset{\sim}{r}|o\rangle$ and $\langle f|\underset{\sim}{r}|k\rangle$ are identically polarized, then

$$\Omega = (a + 2b)_{\text{circ}}/(a' + 2b')_{\text{lin}} = \frac{16}{24} = 0.667.$$

If the transition length vectors $\langle k|\underset{\sim}{r}|o\rangle$ and $\langle f|\underset{\sim}{r}|k\rangle$ are orthogonal, then

$$\Omega = b_{\text{circ}}/b'_{\text{lin}} = \frac{12}{8} = 1.5.$$

The polarization ratios for molecules probed using a single laser source (parallel propagation, $E_\lambda = E_\mu$) therefore fall in the range $\frac{2}{3} \le \Omega \le 1.5$. A measured value of Ω can often help assign the origin of electronic states with the same absolute symmetry. For example, polyene-like $^1A*^-_g$ $_{\pi\pi}*$ states typically have polarization ratios in the range 0.6-0.8 whereas phenyl-like $^1A*^-_g$ $_{\pi\pi}*$ states typically have polarization ratios in the range 1.3-1.5 (14). Both of the above types are low-lying $^1A*^-_g$ singlet states in diphenylpolyenes and the observed two-photon polarization ratio has proved to be useful in assigning the origin of the observed resonances (2,14).

The above example serves to illustrate the diagnostic utility of the polarization ratio. This parameter, however, is only one of many experimentally measurable polarization variables. The review by McClain and Harris (6) and the symmetry tables of McClain (10) should be consulted for a more extensive treatment of the use of two-photon polarization measurements to assign excited-state electronic symmetries.

5.   THE TWO-PHOTON ABSORPTION PROCESS

The number of photons absorbed at time t during the time increment Δt via two-photon excitation, $^2N_{abs}^{\Delta t}(t)$, is given by:

$$^2N_{abs}^{\Delta t}(t) = N_{\Delta t}(t) A_2 \tag{17}$$

where $N_{\Delta t}(t)$ is the number of photons in the excitation beam that pass through an imaginary plane normal to the light vector and centered at the beam waist, and A_2 is the wavelength dependent two-photon absorption parameter. If a single mode laser is used for the excitation source, A_2 is given by (2):

$$A_2 = 1 - \exp[-P(t)S(^2E_{\ell m}^{TEM})\delta_\lambda C z_{02}^{eff}/(\pi w_{02}^2)] \tag{18}$$

$$\cong P(t)S(^2E_{\ell m}^{TEM})\delta_\lambda C z_{02}^{eff}/(\pi w_{02}^2) \quad \text{(small absorption)} \tag{19}$$

where $P(t)$ is the laser power (photons/sec) at time t, S is the dimensionless two-photon correlation parameter equal to 1 for laser excitation ($S_{coherent} = 1$, $S_{chaotic} = 2$), δ_λ is the wavelength dependent two-photon absorptivity (cm^4 sec molecule^{-1} photon^{-1}), C is the concentration of the two-photon absorber (molecule cm^{-3}), z_{02}^{eff} (cm) and w_{02}^2 (cm^2) are defined in Fig. 2 and $^2E_{\ell m}^{TEM}$ is the dimensionless mode efficiency parameter approximately given by (2):

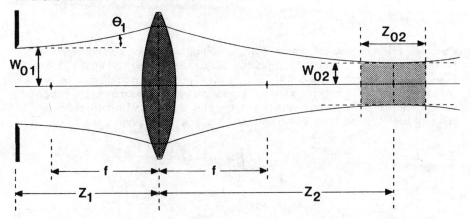

Figure 2. A single "thin" lens at Z_1 of focal length f is used
to focus a laser beam with an initial beam radius of W_{01} and
divergence half-angle of θ_1 to produce a focussed beam radius at
Z_2 of W_{02} with a confocal distance of Z_{02} ($= \pi W_{02}^2 n/\lambda$, where n =
refractive index and λ = laser wavelength). The effective
confocal distance, Z_{02}^{eff}, is equal to Z_{02} provided $Z_{02} \leq L$, where
L is the pathlength of the cell and the center of the cell is
placed at Z_2. If $Z_{02} > L$ then $Z_{02}^{eff} = L$.

$$^2E_{\ell m}^{TEM} \cong [(\ell + 1)(m +1)]^{-0.4} \tag{20}$$

where ℓ and m are the transverse excitation mode quantum numbers.
One of the major experimental complications of two-photon spec-
troscopy is the fact that $P(t)$, $^2E_{\ell m}^{TEM}$, Z_{02}^{eff} and W_{02} are all
potentially wavelength dependent. Furthermore, all of these
parameters are not trivial to experimentally measure.

    The efficiency of two-photon absorption is very sensitive to
mode structure. For example, $^2E_{\ell m}^{TEM}$ equals 1 for TEM_{00} but drops
to $\sim$0.6 for TEM_{11}. Complex modes often have efficiencies less
than 0.3 (see Table III of Ref. 2). Accordingly, it is important
to prevent higher order modes from lasing, and more importantly, to
maintain the same mode structure at all wavelengths to prevent
spurious signal changes. Spatial filtering can be used to select
TEM_{00} and we will assume TEM_{00} laser excitation for the remainder
of the discussion.

    If we assume a laser pulse with a Gaussian temporal profile,
the following relationships result:

$$P(t) = P_0 \exp[-4\ell n 2 \, t^2/\Gamma^2] \tag{21}$$

$$N = 1.06447 \, P_0 \, \Gamma \tag{22}$$

where P_0 is the laser power (photons/sec) at maximum intensity (time $t = 0$), Γ is the full-width at half-maximum (FWHM) of the laser pulse (sec) (i.e. Γ = the laser "pulse width") and N is the total number of photons in the laser pulse. Accordingly, the total number of photons absorbed by the solute via two-photon absorption during the laser pulse, $^2N_{abs}$, is given by:

$$^2N_{abs} = [\delta_\lambda C \, z_{02}^{eff} / (\pi w_{02}^2)] \int_{-\infty}^{\infty} N(t) P(t) \, dt \tag{23}$$

$$= 2^{-\frac{1}{2}} \, N \, P_0 \, \delta_\lambda \, C \, z_{02}^{eff} / (\pi w_{02}^2) \tag{24}$$

$$= 2^{-\frac{1}{2}} \, N \, P_0 \, \delta_\lambda \, C \, (n/\lambda) \qquad\qquad z_{02}^{eff} < L \tag{25}$$

where the last equation is only applicable if a single short focal length lens is used to focus the laser into the sample cell such that Z_{02} is less than the path length of the cell, L, and the confocal beam parameter can be described assuming Gaussian optics [$Z_{02} = n\pi w_{02}^2/\lambda$ (see Fig. 2)].

6.   THE TWO-PHOTON EXCITATION METHOD

     The two-photon excitation technique is the method of choice for the study of forbidden states in molecules that fluoresce efficiently. This technique has been reviewed recently by Birge (2) and its application to the study of the visual chromophores is discussed in Ref. 4 and in the next chapter. A schematic diagram of our two-photon excitation spectrometer is shown in Fig. 3a. [Note that the gas laser (GL) is not a part of the excitation apparatus, but is used for observing double resonance spectra (see Section 8).]

     A two-photon excitation spectrum is taken by monitoring the fluorescence (f in Fig. 1) induced by the absorption of two photons by the molecule. The number of photons emitted by the solute is likely to be 4-10 orders of magnitude less than the number of photons in the laser excitation pulse. A combination of interference and chemical filters (IF1, CF in Fig. 3) is often required to pass the fluorescence while simultaneously preventing scattered laser light from reaching the detector.

     The following equations give the two-photon absorptivity as a function of experimental observables assuming TEM$_{oo}$ laser

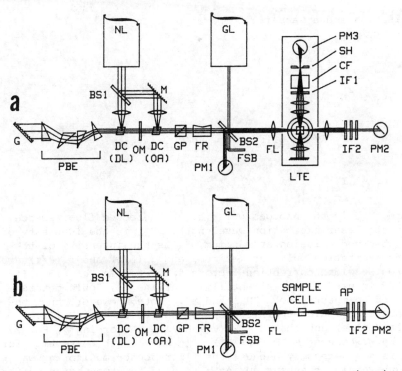

Figure 3. Schematic diagrams of typical two-photon excitation (a)
and two-photon thermal lens (b) spectrometers. The excitation
apparatus (a) can also be used to simultaneously observe the two-
photon double resonance spectrum for certain molecules (see text).
The following symbols describe the individual components of the
pulsed dye laser: G = grating, PBE = prism beam expander,
DC(DL) = dye cell (dye laser), OM = output mirror, DC(OA) = dye
cell (optical amplifier), BS1 = beam splitter, M = dielectric
mirror, NL = nitrogen (or YAG) laser. The following symbols
describe the remaining common components: GP = glan polarizer
to produce linearly polarized light, FR = Fresnel Rhomb which can
be rotated to generate linearly or circularly polarized light,
BS2 = quartz plate beam splitter which reflects ∿3% of pulsed
laser into PM1 = reference photomultiplier, FSB = field stop
baffle, GL = continuous wave gas laser (HeNe, HeCd or Argon ion),
FL = focussing lens, IF2 = interference filters which pass the
gas laser line, PM2 = gas laser monitoring photomultiplier. The
following components are unique to each configuration; a) excitation (top): LTE = light tight enclosure containing sample in
cryogenic dewar, IF1 = interference filter, CF = chemical filter
(IF1 and CF pass sample fluorescence and rigorously exclude laser
light), SH = shutter, PM3 = fluorescence monitoring photomultiplier; b) thermal lens (bottom): AP = small aperture which is
placed far down-field from sample cell.

excitation and a Gaussian temporal profile:

$$\delta_\lambda \propto {}^2N_{em}\lambda/(NP_0) \tag{26}$$

$$\delta_\lambda \propto {}^2N_{em}\lambda\Gamma/N^2 \tag{27}$$

$$\delta_\lambda \propto {}^2I_{em}\lambda/[N^2 \, g_2(\Gamma/\tau)] \tag{28}$$

$$\delta_\lambda \propto {}^2I_{em}\lambda/[I_0^2 \, f_2(\Gamma/\tau)] \tag{29}$$

where ${}^2N_{em}$ is the photons/pulse of the detected fluorescence, λ is the laser excitation wavelength, ${}^2I_{em}$ is the intensity of the observed emission at maximum, I_0 is the intensity of the laser excitation pulse at maximum, and $g_2(\Gamma/\tau)$ and $f_2(\Gamma/\tau)$ are the pulse width correction factors tabulated in Ref. 16. If Eq. 27 is used to experimentally determine δ_λ, Joule meters must be used to measure both ${}^2N_{em}$ and N^2. Note, however, that the output of a Joule meter is proportional to energy, E, (i.e. Joules/pulse) and that $N \propto E\lambda$ ($E = hc/\lambda$). Accordingly, Eq. 27 can be written $\delta_\lambda \propto {}^2E_{em}\Gamma/(\lambda E^2)$. (A wavelength correction for E_{em} is not necessary because the fluorescence monitoring wavelength remains constant during a scan of the two-photon excitation spectrum.) Equations 28 and 29 are the most experimentally useful because the peak fluorescence intensity (${}^2I_{em}$) is easily measured using boxcar averaging techniques whereas measurement of the total integrated photon emission ($\propto {}^2N_{em}$) is experimentally difficult. It should be noted that the pulse width correction factors, g_2 and f_2, are a function of the laser pulsewidth (Γ) and the molecular fluorescence lifetime (τ). If the pulsewidth is constant as a function of wavelength or $\Gamma/\tau > 3$ at all wavelengths of excitation, these correction factors can be ignored (16).

A plot of log(S) versus log(R) at a fixed wavelength, where S is the signal (${}^2N_{em}$ or ${}^2I_{em}$) and R is the reference (N or I_0), should yield a slope of 2 within experimental error.

7. THE THERMAL LENS TECHNIQUE

The thermal lens technique is a very sensitive method of observing two-photon spectra of molecules that do not fluoresce. The technique has recently been reviewed by Fang and Swofford (3). A schematic diagram of our two-photon thermal lens spectrometer is shown in Fig. 3b. Thermal lensing arises in an irradiated liquid sample when a fraction of the laser beam is absorbed and subsequently converted into thermal energy via internal conversion back to the ground state (ic in Fig. 1). The irradiated volume changes temperature radially from the beam axis, with the temperature increase being greatest on the axis. In response to the radial change in temperature there is a radial change in the refractive index in the irradiated volume. A transient "thermal lens" is formed which changes the intensity of the CW monitoring gas laser beam passing through a small aperture (AP, Fig. 3b) down-field, provided the center of the sample cell is displaced from the position of maximum focus of the CW gas laser beam.

The signal, $\underline{S}$, observed in pulsed laser thermal lensing is $\Delta I_{bc}/[I_{bc}(t = 0)]$, where $I_{bc}(t = 0)$ is the intensity of the monitoring laser at the beam center at maximum deflection (just after the dye laser pulse has passed through the cell) and $\Delta I_{bc} = I_{bc}(t = 0) - I_{bc}(t = \infty)$, where $I_{bc}(t = \infty)$ is the base line intensity observed after decay of the thermal lens. The two photon absorptivity is related to the observed thermal lens signal using the following formula:

$$\delta_\lambda \propto \lambda^2 S\Gamma/E^2 \tag{30}$$

The recent review by Fang and Swofford (3) should be consulted for a more detailed discussion of the two-photon thermal lens technique.

8. THE DOUBLE RESONANCE TECHNIQUE

One of the potentially most sensitive methods of observing two-photon spectra is the double resonance technique. Few molecular systems, however, are amenable to study via this technique, because the method of monitoring the absorption of two-photons requires that the molecule undergo a wavelength independent photochemical transformation following excitation to produce a stable or meta-stable product with a red-shifted absorption maximum. Furthermore, the product must be photochemically reconvertable to the initial starting material after each absorptivity measurement. For the purposes of discussion, we will examine the use of this technique to measure the two-photon spectrum of light adapted bacteriorhodopsin. A schematic diagram of our two-photon double resonance spectrometer is shown in Fig. 3a. Although the LTE assembly is used, the shutter (SH) to the photomultiplier (PM3) is closed (fluorescence is not monitored). The two-photon induced transition is measured by using the continuous wave gas laser (GL) to monitor the formation of the photochemical product. The two-photon spectrum of light adapted bacteriorhodopsin (bR_{568}) can be taken by using a helium-neon laser to monitor the formation of K_{610}. At 77K (liquid nitrogen), the photocycle stops at K_{610} and the monitoring laser will not only measure the formation of K_{610} but photochemically drive the $K_{610} \rightarrow bR_{568}$ back reaction (illumination at 6328Å produces a 77K photostationary state containing 98% bR).

The excellent sensitivity of this technique derives from the ability to "molecularly integrate" the signal. The CW monitoring gas laser can be blocked using a shutter and the dye laser fired for a specified number of shots ($\sim$100) to build up photoproduct. The dye laser is then blocked from entering the sample and the gas laser shutter opened. The "signal," S, is then measured as the difference between the initial value of the gas laser intensity subtracted from the intensity measured after a well defined time delay. The time delay is chosen to roughly coincide with $\sim$95% photoconversion back to the initial starting material. This entire sequence is best performed under computer control so that a number of such measurements can be averaged. A dual input multichannel scaler can be employed to average the signals. The two-photon absorptivity is proportional to $S\lambda/(nNP_0) = S\lambda^3\Gamma/(nE^2)$ where n is the number of dye laser excitation shots, λ is the wavelength of laser excitation, Γ is the pulsewidth at λ, n is the number of shots and E is the average laser excitation pulse energy measured using a Joule meter. The dye laser must have a shot to shot repeatability of better than 5% for reliable averaging. The above equation assumes a TEM$_{oo}$ laser excitation pulse with a Gaussian temporal profile (Eqs. 21-22) and "Gaussian" focussing (Fig. 2).

ACKNOWLEDGMENTS

This work was supported in part by grants from the National Institutes of Health (EY-02202), the National Science Foundation (CHE-7916336) and a Grant-in-Aid from the Standard Oil Company of Ohio (SOHIO). I am grateful to Professors David Kliger, George Leroi and Martin McClain and Drs. James Bennett, Howard Fang, Brian Pierce and Robert Swofford for interesting and helpful discussions.

REFERENCES

1. "Ultrasensitive Laser Spectroscopy," (1983), D. S. Kliger, ed., Academic Press, New York.
2. Birge, R.R. (1983), Chapter 2 of Ref. 1, pp. 109-174.
3. Fang, H.L. and Swofford, R.L. (1983), Chapter 3 of Ref. 1, pp. 221-232.
4. Birge, R.R. (1982), In "Methods of Enzymology," Vol. 88 (L. Packer, ed.), pp. 522-532, Academic Press, New York.
5. Friedrich, D.M., and McClain, W.M. (1980), Annu. Rev. Phys. Chem. 31, pp. 559-577.

6. McClain, W.M., and Harris, R.A. (1977), Excited States 3, pp. 2-56.
7. Yariv, A. (1975), "Quantum Electronics," 2nd Ed., Wiley, New York.
8. Gold, A. (1970), Proc. Scott. Yniv. Summer Sch. Phys. 10, pp. 397-420.
9. Peticolas, W.L. (1967), Annu. Rev. Phys. Chem. 18, pp. 233-260.
10. McClain, W.M. (1971), J. Chem. Phys. 55, pp. 2789-2796.
11. Mortensen, O.S., and Svendsen, E.N. (1981), J. Chem. Phys. 74, pp. 3185-3189.
12. Dick, B., and Hohlneicher, G. (1982), J. Chem. Phys. 76, pp. 5755-5760.
13. Birge, R.R., and Pierce, B.M. (1979), J. Chem. Phys. 70, pp. 165-178.
14. Bennett, J.A., and Birge, R.R. (1980), J. Chem. Phys. 73, pp. 4234-4246.
15. Göppert-Mayer, M. (1931), Ann. Phys. (Leipzig) [5] 9, pp. 273-285.
16. Pierce, B.M., and Birge, R.R. (1983), IEEE J. Quantum Electron QE-19, pp. 826-833.

TWO-PHOTON SPECTROSCOPY OF BIOLOGICAL MOLECULES

Robert R. Birge, Brian M. Pierce and Lionel P. Murray

Department of Chemistry
University of California
Riverside, CA 92521 U.S.A.

ABSTRACT: The use of two-photon excitation and thermal lens spectroscopy to study the "forbidden" excited states of selected biologically relevant chromophores is discussed. Spectroscopic studies of the following molecules and proteins are described: iso-tachysterol (an analog of Vitamin D), all-trans retinol (Vitamin A), all-trans retinal (Vitamin A aldehyde), the protonated Schiff base of all-trans retinal, and rhodopsin. All of these molecules contain a substituted polyene chromophore. We observe that a forbidden "$^1A_g^{*-}$" state is the lowest-lying $\pi\pi^*$ state in all of these chromophores except the protonated Schiff base of all-trans retinal and the chromophore in rhodopsin. The latter two chromophores have a lowest-lying, strongly allowed "$^1B_u^{*+}$" $\pi\pi^*$ state. The implications of this observation are briefly discussed.

1. INTRODUCTION

The application of two-photon spectroscopy to the study of the electronic properties of biological molecules is a relatively recent endeavor. The first two-photon spectrum of a biological chromophore was reported in 1978, and this study of Vitamin A was carried out using laser equipment which was unsophisticated by today's standards (1). Significant advances in the performance and reliability of tunable pulsed dye laser excitation sources and in the quality of commercially available electronic signal collection equipment have dramatically improved experimental capabilities (2). If the advances during the last five years continue at the present rate, two-photon spectroscopy may well become a routine spectroscopic technique in biophysics during the next decade. This development would be very important to a number

C. Sandorfy and T. Theophanides (eds.), Spectroscopy of Biological Molecules, 473–486.
© 1984 by D. Reidel Publishing Company.

of areas of study, because optically "forbidden" excited states
are believed to be of importance to the biological function of
many photon receptor chromophores in animals, plants and bacteria
(3-8). For example, Thrash et al. have proposed that a forbidden
lowest-lying $\pi\pi^*$ excited state in β-carotene acts as the energy
donor in Förster-type excitation transfer to antenna chlorophyll
molecules following β-carotene's light harvesting function in
photosynthesis (4). The barrierless excited state potential surfaces for the primary photoisomerizations in rhodopsin (6) and
bacteriorhodopsin (8) are believed to derive from the interaction
between an allowed and a forbidden excited state during double
bond isomerization (3,6,8).

     The purpose of the present chapter is to review the appli-
cation of two-photon spectroscopy to the study of the following
biologically relevant chromophores and proteins: iso-tachysterol
(an analog of Vitamin D), Vitamin A (all-trans retinol), Vitamin
A aldehyde (all-trans retinal), the protonated Schiff base of
all-trans retinal and rhodopsin. The previous chapter has
reviewed the theory and selected experimental techniques of two-
photon spectroscopy. All of the above chromophores are substituted polyenes, and the two-photon selection rules for polyenes
are briefly reviewed in the next section.

2. TWO-PHOTON SELECTION RULES IN POLYENES

     It is conceptually useful to analyze the two-photon absorp-
tion process as a combination of two, essentially simultaneous,
one-photon absorption events. The first photon is "absorbed" by
an extremely short-lived virtual state that has a lifetime of
approximately 10^{-16} sec. (The virtual state does not actually
absorb the photon. The process is a special case of Dirac's
dispersion theory and represents a type of scattering phenomenon.)
A second photon must "arrive" before the virtual state scatters
the first photon. If the sum of the photon energies corresponds
to an excited state vibronic level of the correct symmetry, both
photons are absorbed. The extremely short lifetime of the virtual
state requires the use of an intense radiation field to induce
experimentally a two-photon absorption. The virtual state also
plays an important role under lower light intensities, where its
interaction with the radiation field is responsible for Raman
scattering.

     A simplified analysis of the two-photon excitation process
in terms of two allowed one-photon absorption processes is useful
for assigning selection rules. One proceeds by recognizing that
the virtual state has definite symmetry properties and must be
"prepared" by an allowed one-photon transition from the ground
state. Linear polyenes are represented by the C_{2h} point group,

and the $\pi\pi^*$ states of these molecules can be classified under one
of four possible symmetries: $^1A_g^{*-}$, $^1A_g^{*+}$, $^1B_u^{*-}$, or $^1B_u^{*+}$. (The
superscript "+" and "-" labels derive from orbital pairing rela-
tionships as discussed in Refs. 6 and 7). The ground state has
$^1A_g^{*-}$ symmetry and couples via one-photon selection rules only
with $^1B_u^{*+}$ states. Accordingly, the virtual state must have $^1B_u^{*+}$
symmetry. The second photon must access the final state by a one-
photon allowed process, and therefore the final state must have
$^1A_g^{*-}$ symmetry. Accordingly, only $^1A_g^{*-}$ states are two-photon
allowed in linear polyenes. Retinyl polyenes belong to the C_1
point group. The fact that all electronic states have the same
symmetry (A) means that all the excited singlets are formally
allowed in both one-photon and two-photon spectroscopy. Nevertheless, the excited singlet states of the visual chromophores
maintain many of the characteristics of linear polyene excited
states and it is useful to describe these states by reference
to the C_{2h} point group (e.g., "$^1A_g^{*-}$", "$^1A_g^{*+}$", "$^1B_u^{*+}$", "$^1B_u^{*-}$").
The approximate symmetry classifications are given in quotation
marks and are derived by correlating the properties of a given
electronic state with those of the analogous state in a linear
polyene of C_{2h} symmetry. In other words, "$^1A_g^{*-}$" should be
interpreted as $^1A_g^{*-}$-like. The lowest-lying "$^1A_g^{*-}$" states of
both all-<u>trans</u>-retinol and all-<u>trans</u>-retinal are calculated to
be sufficiently more two-photon allowed than the nearby "$^1B_u^{*+}$"
states to make two-photon spectroscopy well suited for studying
these compounds (see below).

3. ISO-TACHYSTEROL

The excited state level ordering in polyenes has been the
subject of considerable experimental and theoretical study.
While it is now generally accepted that all long chain linear
polyenes have a lowest-lying forbidden $^1A_g^{*-}$ $\pi\pi^*$ state, the level
ordering in the shorter chain polyenes such as butadiene and
hexatriene remains a subject of debate. Molecular orbital
studies including extensive configuration interaction do not
agree on the level ordering in all-<u>trans</u> 1,3,5-hexatriene
(hereafter abbreviated all-<u>trans</u> hexatriene). Experimental
studies of all-<u>trans</u> hexatriene employing uv-visible absorption,
electron impact, multiphoton ionization and two-photon thermal
lensing techniques have failed to directly observe the valence
$2^1A_g^{*-}$ $\pi\pi^*$ state (see Ref. 9 and references therein).

We studied the two-photon excitation spectrum of the all-
<u>trans</u> hexatriene polyene system of iso-tachysterol to help
resolve this controversy. This compound (synthesized from
Vitamin D_3 (10)) was chosen because it fluoresces with reasonable
efficiency ($\phi_f \sim 0.01$) and its one-photon absorption spectrum in
EPA (77K) displays well-resolved vibronic structure (Fig. 1).

ISO-TACHYSTEROL

Furthermore, the one-photon fluorescence excitation spectrum
matches the one-photon absorption spectrum which indicates a
wavelength independence in the (one-photon) fluorescence quantum
yield. We assume this wavelength independence is maintained
under two-photon excitation. The fluorescence and two-photon
excitation spectra are also shown in Fig. 1. All spectra were
taken using EPA (ethyl ether:isopentane:ethanol; 5:5:2 v/v)
solvent at liquid nitrogen temperature (77K).

     Simulations of the one-photon absorption, the two-photon
excitation and the fluorescence spectral contours using semi-
classical Gaussian wave packet propagation theory lead to the
following conclusions (9): 1) the fluorescence spectrum and the
one-photon absorption spectrum cannot be fit to the same electronic transition, 2) the fluorescence spectrum and the two-
photon excitation spectrum can be fit to the same electronic
transition using a common system origin of 32,200+500 cm^{-1},
3) a better fit to the fluorescence spectrum is obtained by
assuming that a ^1b_u promoting mode of 300+200 cm^{-1} provides a
false vibronic origin. We assign the two-photon excitation and
the fluorescence spectra to the $2"^1A_g^{*-"}$ electronic state and the
one-photon absorption spectrum to the $1"^1B_u^{*+"}$ state. The above
simulations predict that the system origin of the $2"^1A_g^{*-"}$ state
at $\sim$32,200+500 cm^{-1} lies below the $1"^1B_u^{*+"}$ system origin at
$\sim$32,700+200 cm^{-1}. [Although the error bars provide for the
possibility that the $1"^1B_u^{*+"}$ state could lie below the $2"^1A_g^{*-"}$
state, the fact the fluorescence spectrum and the one-photon

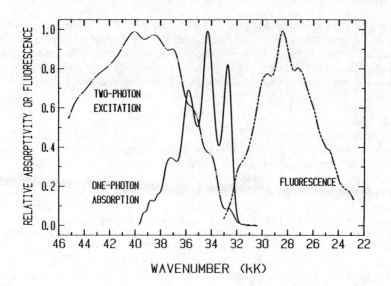

Figure 1. The one-photon absorption (——), two-photon excitation (-•-•) and uncorrected fluorescence (---) spectra of
iso-tachysterol in EPA (77K). The two-photon excitation spectrum
was taken using linearly polarized light and is plotted versus
the combined energy of the two photons (9,10).

absorption spectrum cannot be fit to the same electronic transition (see above) precludes this possibility.]

The large geometry change upon excitation into the $2"^1A_g^{*-}"$
state of iso-tachysterol [$\Delta r_{C=C}$ = 0.093+0.018Å, Δr_{C-C} = 0.050+
0.014Å] is associated with significant bond order reversal in
the polyene π-electron system upon excitation. This observation
is consistent with the importance of double excitations in the
configurational description of polyene $2^1A_g^{*-}$ and $2^1A_g^{*-}$-like ππ*
states (1,3,5-9).

We conclude that the system origin of the $2"^1A_g^{*-}"$ ππ* state
in iso-tachysterol lies slightly below (∿500 cm^{-1}) the $1"^1B_u^{*+}"$
system origin in EPA at 77K. Extrapolation to gas phase condi-

tions suggests that the system origin of the $2"^1A_g^{*-}"$ state will lie $\sim$3,500 cm^{-1} below that of the $1"^1B_u^{*+}"$ state assuming a $\sim$3,000 cm^{-1} dispersive red shift of the allowed state in EPA (77K). This observation implies that the $2^1A_g^{*-}$ state is also below the $1^1B_u^{*+}$ state in all-<u>trans</u> hexatriene (9).

The observation that the $"^1A_g^{*-}"$ and $"^1B_u^{*+}"$ states are very close in energy in iso-tachysterol and presumably all Vitamin D compounds in solution suggests that the wavelength dependence of the photochemistry of many Vitamin D compounds may be associated with this near degeneracy (see Ref. 11 for a recent review).

4.   THE VISUAL CHROMOPHORES

The level ordering of the low-lying excited singlet states of the visual chromophores has been the subject of extensive spectroscopic study and some controversy (3). The controversy is characterized by numerous conflicting assignments in the litera- ture and is associated with the inherent difficulty of spectro- scopically assigning the excited states of molecules, like the visual chromophores, which are subject to severe inhomogeneous broadening in their electronic spectra. The excited singlet state manifold of retinal provides a particularly challenging system to study because there are three low-lying states $["^1B_u^{*+}"$ $(\pi\pi*)$; $"^1A_g^{*-}"$ $(\pi\pi*)$, and $^1n\pi*]$ that are very close in energy (3). Furthermore, the relative level ordering appears to be solvent dependent (12).

The location of the low-lying "forbidden" $^1A_g^{*-}$-like $\pi\pi*$ state in the visual chromophores is believed to be of significant importance in defining the photochemical properties of these molecules (3,6,7,12). Recent calculations indicate that the interaction of this state with a lowest-lying $^1B_u^{*+}$ state may be directly responsible for producing a barrierless excited state potential energy surface for <u>cis-trans</u> isomerization of the chromophore in rhodopsin (6). The extremely rapid formation of bathorhodopsin may therefore be a consequence of the photochemical lability of the $"^1A_g^{*-}"$ state (6).

The pioneering spectroscopic investigations of Hudson, Kohler, Christensen, and co-workers indicate that the forbidden $^1A_g^{*-}$ state is the lowest-lying excited singlet in long-chain linear polyenes (13). It is not possible, however, to generalize this level ordering to the visual chromophores because this covalent state is predicted to be highly sensitive to conformation and polarity (3).

Two-photon spectroscopy has proved to be one of the most versatile methods of studying forbidden $^1A_g^{*-}$ and A_g^{*-}-like states

(1-3,5,7,9,10). Two-photon spectroscopy is particularly useful for studying the visual chromophores (1,3,7) because inhomogeneous broadening prevents the use of the high-resolution, low-temperature matrix techniques that have successfully been used to study the $^1A_g^{*-}$ state in the linear polyenes (13).

4.1. All-trans Retinol (Vitamin A)

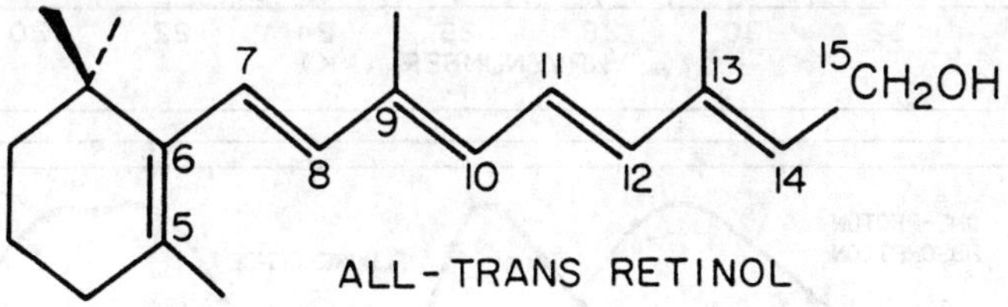

ALL-TRANS RETINOL

The first two-photon spectrum of a visual chromophore was reported in 1978 (1). This investigation of all-trans-retinol used the two-photon excitation technique to demonstrate that this chromophore has a lowest-lying, strongly two-photon allowed ($\delta = 2 \times 10^{-49}$ cm^4 sec molecule^{-1} photon^{-1}) "$^1A_g^{*-}$" excited state. The two-photon excitation maximum was observed $\sim$1600 cm^{-1} to the red of the one-photon ("$^1B_u^{*+}$") absorption maximum in EPA at 77K (Fig. 2). A comparison of maxima cannot be used as a reliable indicator of system origin energy separations because Franck-Condon factors associated with the "$^1A_g^{*-}$" state are expected to shift the energy vibronically of the inhomogeneously broadened two-photon excitation maximum to higher energy than that characteristic of the "$^1B_u^{*+}$" state contour. Accordingly, the system origin of the "$^1A_g^{*-}$" state in all-trans-retinol is probably >1600 cm^{-1} below the "$^1B_u^{*+}$" state system origin in EPA at 77K.

4.2. All-trans Retinal (Vitamin A Aldehyde)

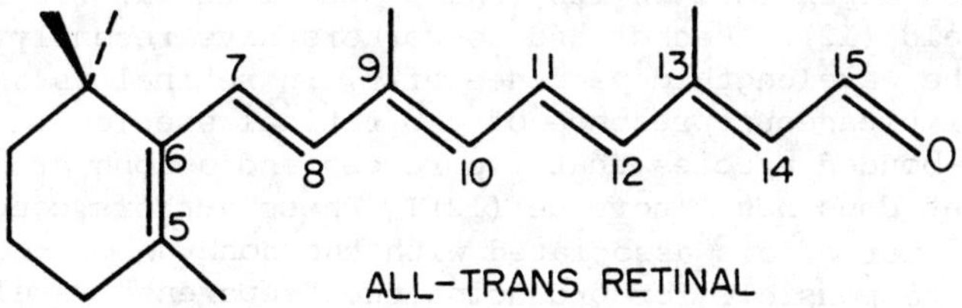

ALL-TRANS RETINAL

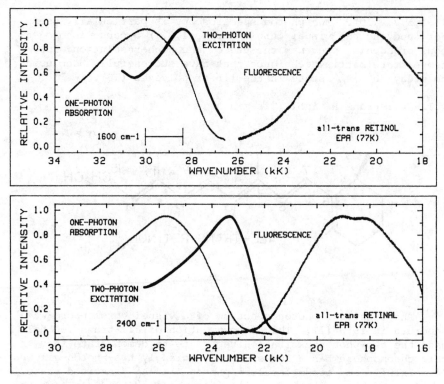

Figure 2. The one-photon absorption, two-photon excitation and uncorrected fluorescence spectra of all-trans retinol (top) and all-trans retinal (bottom) in EPA (77K). The increase in the two-photon excitation spectrum of all-trans retinol above ~30,300 cm^{-1} is primarily due to incomplete removal of scattered laser light by the monochromator (see Ref. 1). The two-photon excitation spectra were taken using linearly polarized light and are plotted versus the combined energy of the two photons (1,7).

The two-photon excitation spectrum of all-trans-retinal is shown in Fig. 2. The interpretation of the excitation spectrum is complicated by the observation that all-trans-retinal displays a wavelength dependence in its (one-photon induced) fluorescence quantum yield (12). Becker and co-workers have recently suggested that the wavelength dependence of ϕ_f in retinal is associated with the simultaneous presence of two retinal species in solution, a hydrogen-bonded species that fluoresces and a nonhydrogen-bonded species that does not fluoresce (12). These authors conclude that an inner-filter effect associated with the nonhydrogen-bonded species is responsible for producing the "apparent" wavelength

dependence (12). This inner-filter effect would not be observed
in two-photon spectroscopy because a negligible amount of the
laser irradiation is absorbed during a two-photon excitation
experiment (see discussion in Ref. 7). Accordingly, if the
inner-filter effect hypothesis is correct, no quantum yield
correction should be made to the observed two-photon excitation
spectrum. The spectra shown in Fig. 2 indicate that all-trans-
retinal has a "$^1A_g^{*-}$" $\pi\pi*$ excitation maximum roughly 2400 cm^{-1}
below the "$^1B_u^{*+}$" $\pi\pi*$ absorption maximum in EPA at 77°K. Molecular orbital calculations indicate that the two-photon absorptivity of the low-lying $n\pi*$ state is extremely small (7).
Accordingly, the $n\pi*$ state is definitely not associated with the
two-photon excitation spectrum shown in Fig. 2. The solvent
effect data reported by Becker and co-workers, however, suggest
that the $n\pi*$ state is nearly degenerate with the "$^1A_g^{*-}$" state and
that its level ordering relative to the "$^1A_g^{*-}$" state is solvent
dependent (12). The experimental results for all-trans-retinal
can be summarized as follows:

$$"^1B_u^{*+}" > "^1A_g^{*-}" > n\pi* \qquad \text{(nonpolar solvents)}$$

$$\left.\begin{array}{l} "^1B_u^{*+}" > n\pi* > "^1A_g^{*-}" \\[2mm] n\pi* > "^1B_u^{*+}" > "^1A_g^{*-}" \end{array}\right\} \text{(hydrogen bonding solvents)}$$

or

4.3. Protonated Schiff Base of all-trans Retinal

When 11-cis retinal is incorporated into opsin to form
rhodopsin (see Section 5), the chromophore is covalently bound
to the protein via a protonated Schiff base linkage with lysine
(3). The chromophore in light adapted bacteriorhodopsin is the
protonated Schiff base of all-trans retinal (3). Molecular
orbital calculations predict that the protonated Schiff bases of
retinal have an excited state level ordering which places the
"$^1A_g^{*-}$" state above the "$^1B_u^{*+}$" state (3,6,8). This level ordering
is believed to be responsible for producing a barrierless excited
state potential surface for double bond isomerization (3,6,8).
However, until the two-photon investigations reported below were
carried out, no direct experimental evidence was available to
support the theoretical predictions concerning level ordering.

A preliminary two-photon thermal lens spectrum of all-trans-
N-retinylidene-n-butylimine-HCl (ATRPSB) in carbon tetrachloride
at room temperature is shown in Fig. 3. Although the two-photon
thermal lens spectrum is very broad and an accurate assignment of
the two-photon absorptivity maximum is very difficult, it is
clear from a close examination of Fig. 3 that the two-photon
maximum is at a higher energy than the one-photon absorption
maximum. We assign the two-photon spectrum to contributions from
both the "$^1A_g^{*-}$" and "$^1B_u^{*+}$" states with the former dominating.

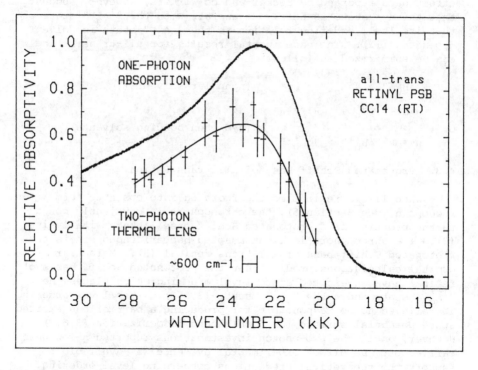

all-<u>trans</u>-N-Retinylidene-n-butylimine·HCl

Figure 3.   The one-photon absorption and two-photon thermal lens
spectra of all-<u>trans</u>-N-retinylidene-n-butylimine-HCℓ (ATRPSB) in
carbon tetrachloride at room temperature.   The two-photon thermal
lens spectrum was taken using linearly polarized light and is
plotted versus the combined energy of the two-photons (14).

As noted in Section 3 of the previous chapter, the $"^1B_u^{*+}"$ state may have a large two-photon absorptivity because of large dipole moment changes upon excitation (see Eqs. 12-15 of previous chapter). Furthermore, the symmetry labels $"^1A_g^{*-}"$ and $"^1B_u^{*+}"$ are highly qualitative due to significant state mixing due to the large polarity of the protonated Schiff base compound. Accordingly, the actual $"^1A_g^{*-}"$ state absorptivity maximum may be much higher in energy than the ∿600 cm^{-1} separation shown in Fig. 3 suggests. We conclude that the Franck-Condon maximum of the $"^1A_g^{*-}"$ state is at least 600 cm^{-1} above the $"^1B_u^{*+}"$ state maximum in ATRPSB in ambient temperature CCl_4 solution (14).

5. RHODOPSIN

The experimental assignment of the excited state level ordering of the chromophore in rhodopsin is very important for a number of reasons. As previously noted, a number of theoretical simulations of the primary event in rhodopsin have been carried out (3,6,8; for a recent review see 15). The reaction path for the photoisomerization in the primary event was generated using semiempirical molecular orbital theory. It is important to experimentally verify that the level ordering predicted by the theoretical procedures is, in fact, correct. Furthermore, the relative energies of the $"^1A_g^{*-}"$ and $"^1B_u^{*+}"$ states can be used to probe the nature of the protein active site. The stabilizations of these two states are affected in very different degrees by the counter-ion environment and the extent of protonation of the chromophore. In particular, protonation of the Schiff base linkage is predicted to preferentially stabilize the $"^1B_u^{*+}"$ state so that $"^1A_g^{*-}" > "^1B_u^{*+}"$ whereas hydrogen bonding of the Schiff base will produce $"^1B_u^{*+}" > "^1A_g^{*-}"$. The energetic separation of the $"^1A_g^{*-}"$ and $"^1B_u^{*+}"$ states of a protonated Schiff base is highly sensitive to the separation of the primary counterion from the $C_{15}=NHR$ group.

Our preliminary two-photon studies of rhodopsin have been carried out in collaboration with Dr. Valeria Balogh-Nair and Professor Koji Nakanishi at Columbia University (16). These researchers provided us with a rhodopsin analog containing a non-bleachable 11-cis-locked chromophore (Fig. 4). The synthesis of this chromophore and evidence that the chromophore occupies the same opsin binding site as with 11-cis-retinal is described in Ref. 17. The obvious advantage of using a non-bleachable 11-cis-locked rhodopsin for the present spectroscopic study is that the rhodopsin is not bleached during the spectroscopic study. This is particularly important for a two-photon investigation because of the long time periods required to obtain the spectrum. A molecule can be promoted into the excited state hundreds of times during an experiment and the use of samples that are photochem-

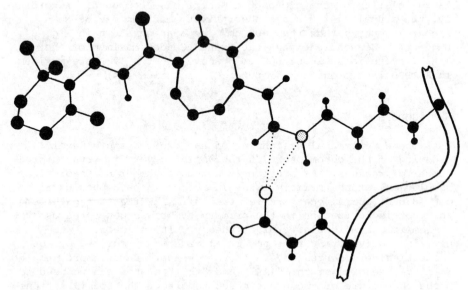

Figure 4. A possible geometry of the 11-<u>cis</u>-locked retinyl
chromophore in rhodopsin showing the lysine residue (the protonated lysine nitrogen atom is shaded) and the primary counterion
which is assumed to be a glutamic acid residue separated $\sim 3\overset{\circ}{A}$
(dotted lines) from the nitrogen and C_{15} carbon atoms. The
actual geometry of the rhodopsin binding site is not known, but
the above geometry conforms with the known spectroscopic observables (see text).

ically labile is rarely possible.

     The resulting (preliminary) two-photon thermal lens spectrum
is shown in Fig. 5. The two-photon maximum is ~ 2000 cm^{-1} higher
in energy than the one-photon absorption maximum. We conclude
for reasons identical to those discussed in Section 4.3 that the
"$^1A_g^{*-}$" state maximum is at least 2000 cm^{-1} above the "$^1B_u^{*+}$" state
maximum. Accordingly, the Schiff base nitrogen is protonated,
and the primary counterion in rhodopsin is separated further from
the C_{15}=NHR moiety in rhodopsin than for ATRPSB in solution (16).
The level ordering observed experimentally is in agreement with
the theoretical predictions (3,6).

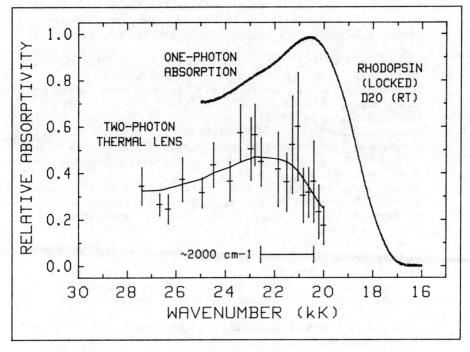

Figure 5. The one-photon absorption and two-photon thermal lens spectra of rhodopsin in D_2O with 2% digitonin and 67mM phosphate buffer (pH ∿7) containing the 11-cis-locked retinyl chromophore depicted in Fig. 4. The two-photon thermal lens spectrum was taken using linearly polarized light and is plotted versus the combined energy of the two photons (16).

ACKNOWLEDGMENTS

This work was supported in part by grants from the National Institutes of Health (EY-02202), the National Science Foundation (CME-7916336), The Committee on Research, University of California, Riverside, and a Grant-in-Aid from the Standard Oil Company of Ohio (SOHIO). RRB thanks Professors Barry Honig, David Kliger, George Leroi, Koji Nakanishi and Camille Sandorfy; Drs. James Bennett and Valeria Balogh-Nair for interesting and helpful discussions.

REFERENCES

1. Birge, R.R., Bennett, J.A., Pierce, B.M., and Thomas, T.M. (1978), J. Am. Chem. Soc. 100, pp. 1533-1539.

2. Birge, R.R. in "Ultrasensitive Laser Spectroscopy," (1983),
    D.S. Kliger, ed., Academic Press, New York, pp. 109-174.
3. Birge, R.R. (1981), Annu. Rev. Biophys. Bioeng. 10, pp. 315-
    354.
4. Thrash, R.J., Fang, H.L., and Leroi, G.E. (1979), Photochem.
    Photobiol. 29, pp. 1049-1050.
5. Friedrich, D.M., and McClain, W.M. (1980), Annu. Rev. Phys.
    Chem. 31, pp. 559-577.
6. Birge, R.R., and Hubbard, L.M. (1980), J. Am. Chem. Soc.
    102, pp. 2195-2205; (1981), Biophys. J. 34, pp. 517-534.
7. Birge, R.R., Bennett, J.A., Hubbard, L.M., Fang, H.L.,
    Pierce, B.M., Kliger, D.S., and Leroi, G.E. (1982), J. Am.
    Chem. Soc. 104, pp. 2519-2525.
8. Birge, R.R., and Pierce, B.M. in Proceedings of the Inter-
    national Conference on Photochemistry and Photobiology,
    Alexandria, Egypt, January 5-10, 1983 (in press) and Laser
    Chem. (in press).
9. Pierce, B.M., Bennett, J.A., and Birge, R.R. (1983),
    J. Chem. Phys. 77, pp. 6343-6344.
10. Pierce, B.M., Bennett, J.A., Gerdes, J.M., Lewicka-Piekut,
    S., Okamura, W.H., and Birge, R.R. (to be published).
11. Jacobs, H.J.C., and Havinga, E. (1979), Adv. Photochem. 11,
    pp. 305-373.
12. Takemura, T., Das, P.K., Hug, G., and Becker, R.S. (1978),
    J. Am. Chem. Soc. 100, pp. 2626-2630.
13. Hudson, B.S., Kohler, B.E., and Schulten, K. in "Excited
    States," (1982), E.C. Lim, ed., Vol. 6, Academic Press,
    New York.
14. Murray, L.P., Pierce, B.M., and Birge, R.R. (to be published).
15. Birge, R.R. in "Biological Events Probed by Ultrafast Laser
    Spectroscopy," (1982), R. Alfano, ed., Academic Press, New
    York, pp. 299-317.
16. Birge, R.R., Murray, L.P., Pierce, B.M., Findsen, L.,
    Balogh-Nair, V., and Nakanishi, K. (to be published).
17. Akita, H., Tanis, S.P., Adams, M., Balogh-Nair, V., and
    Nakanishi, K. (1980), J. Am. Chem. Soc. 102, pp. 6370-6372.

MOLECULAR STRUCTURE OF THE GRAMICIDIN TRANSMEMBRANE CHANNEL: UTILIZATION OF CARBON-13 NUCLEAR MAGNETIC RESONANCE, ULTRAVIOLET ABSORPTION, CIRCULAR DICHROISM AND INFRARED SPECTROSCOPIES

Dan W. Urry

Laboratory of Molecular Biophysics, University of
Alabama in Birmingham, School of Medicine, Birmingham,
Alabama 35294, USA

ABSTRACT

The phenomenology of Gramicidin A transmembrane channel transport is briefly reviewed. The problem of the conformational characterization of the channel state using absorption and circular dichroism spectroscopies is noted and the proposed structures for the channel are reviewed. The case is presented for the channel state of Gramicidin A being obtained on heat incubation of Gramicidin A with L-α-lysolecithin. Then ion induced carbonyl carbon chemical shifts are obtained using nuclear magnetic resonance and Gramicidin A syntheses in which one carbonyl carbon at a time was 90% carbon-13 enriched. The result is the definition of the channel structure to be the head-to-head dimerized, single-stranded, left-handed β^6-helix. It is further shown that this structure has an amide I frequency in the infrared spectrum at 1633 cm^{-1}, a frequency previously considered to be limited to multi-stranded pleated sheet structures. Importantly, the ion induced carbonyl carbon chemical shifts also locate two ion binding sites within the channel. Finally the dimerization process is considered in terms of the steric effects at the head-to-head junction and the destabilization of the six intermolecular hydrogen bonds. It is noted that the common anesthetic, halothane, also destabilizes the dimeric structure, as measured by its effect on decreasing channel mean lifetime.

I. INTRODUCTION

The selective transport of ions across a cell membrane is an element essential to cell viability; the capacity of cell

487

C. Sandorfy and T. Theophanides (eds.), Spectroscopy of Biological Molecules, 487–510.
© *1984 by D. Reidel Publishing Company.*

membranes to achieve rapid selective ion fluxes of the order of 10^7 ions/sec by each of a collection of transmembrane channel structures is fundamental to cellular excitability and motility, exemplified in higher organisms by nerve conduction and muscle contraction. The process, that spectroscopic methods are to be employed to describe, is the selective flow of ions across a planar lipid bilayer membrane resulting from the presence of Gramicidin A, $HCO-L-Val^1-Gly^2-L\cdot Ala^3-D\cdot Leu^4-L\cdot Ala^5-D\cdot Val^6-L\cdot Val^7-D\cdot Val^8-L\cdot Trp^9-D\cdot Leu^{10}-L\cdot Trp^{11}-D\cdot Leu^{12}-L\cdot Trp^{13}-D\cdot Leu^{14}-L\cdot Trp^{15}HN-CH_2CH_2OH$ (1). The three dimensional molecular structure is first required and then the detailed ionic mechanism is to be determined. The former is considered in the present manuscript (the first lecture) and the latter in the accompanying manuscript (the second lecture).

In applying spectroscopic methods to the characterization of the molecular structure of the Gramicidin transmembrane channel, there are two immediate problems. One is that the channel state of interest occurs within a lipid bilayer structure and the second is the extreme, the ultimate, sensitivity of the transport characterization where single channels may be observed and characterized. In common spectroscopic methods millimolar concentrations of molecular species are required i.e. greater than 10^{20} molecules rather than the one or few channels that are characterized by the electrical measurements on planar bilayer membranes.

II. PHENOMENOLOGY OF GRAMICIDIN CHANNEL TRANSPORT

The analytical system for monitoring conductance across lipid bilayer membranes, represented in Figure 1, is due to Mueller and colleagues (2). Two aqueous chambers are separated by a thin septum containing a small hole. The hole is coated or otherwise filled with lipid in a suitable solvent. The lipid droplet then thins to a planar lipid bilayer as evidenced by the appearance of a black area in the hole. This gave rise to the nomen of black lipid membrane studies. The black lipid membrane results from the interference of light reflecting from the two surfaces of the bilayer separated by a distance shorter than the wavelength of light. As shown by Hladky and Haydon (3), with sufficient sensitivity it is possible on addition of Gramicidin at very low concentrations to see conductance steps during the period of an applied potential across the membrane.

It was early suggested that the observed conductance was second order with respect to Gramicidin concentration (4,5). This was substantiated by the covalent dimer, malonyl-bis-desformyl Gramicidin A (6), by the kinetic studies of Bamberg and Läuger (7), by the noise analysis of Kolb et al. (8) and significantly by studies on the water soluble derivative O-pyromellityl Gramicidin

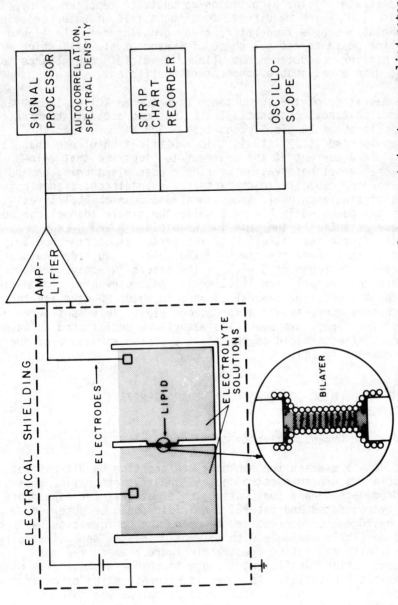

FIGURE 1: Planar lipid bilayer system for studying channel transport due to an applied potential or to a concentration gradient. Reproduced with permission from D. W. Urry, In: The Enzymes of Biological Membranes, Plenum Publishing Corporation, New York, (in press).

(9). Accordingly, the conducting channel is a dimer. That the conductance steps were due to a transmembrane spanning structure was demonstrated by the high proton conductance reported by Myers and Haydon (10) which required a continuous file of water crossing the membrane and by a temperature study crossing from the liquid crystalline phase to the gel phase of Krasne et al. (11) which blocked carrier conductance but allowed Gramicidin conductance to continue with a reasonable temperature coefficient.

The magnitude of the conductance step for 2M KCl at 100 mV and 28°C for diphytanoyl phosphatidyl choline membranes as shown in Figure 2, is about 28 picoSiemens (pS) which is more than 10^7 potassium ions/sec (12). It is this amplification of one channel resulting in a current of the order of 10^7 ions/sec that allows for the phenomenal observation of single channels turning on and off and for the capacity to characterize channel conductance both in terms of single channel conductances and mean channel lifetimes. As may be noted in Figure 2 while the single channel conductances are essentially the same for Gramicidin A and D·Leu^2 Gramicidin A, the mean lifetime of the latter is shorter due to the destabilization of the dimer by the bulk of Leu side chain replacing the hydrogen of Gly^2. In the case of N-acetyl desformyl Gramicidin the channel mean lifetime is reduced even more dramatically from the order of several seconds to about 50 msec and the single channel currents are also reduced (13). This subtle replacement of the formyl hydrogen by a methyl was anticipated to have significant effects based on the early proposed structure of the dimeric channel (6,14).

III. THE PROBLEM OF CONFORMATIONAL CHARACTERIZATION OF THE
     CHANNEL STATE

A. Studies in Organic Solvents

In order to examine the backbone conformation by ultraviolet and circular dichroism spectroscopies, the tryptophan side chains were hydrogenated and a qualitative conformational correspondence between hydrogenated and natural Gramicidin could be observed. For the hydrogenated derivative, however, the conformation was found to be highly variable with solvent. This is demonstrated by the ultraviolet absorption spectra of Figure 3 where the mean residue extinction coefficient is seen to change dramatically with the organic solvent (15). As shown in Figure 4 using circular dichroism, not only is the backbone conformation variable with similar solvents, it is very concentration dependent in each solvent system (16), but there is a systematic direction of change on increasing concentration. Interestingly, raising the temperature has the same qualitative shift as increasing concentration

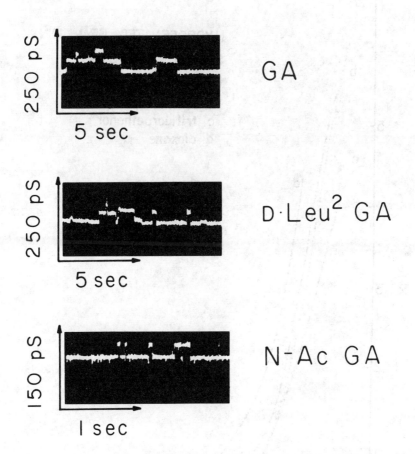

FIGURE 2: Current tracings at 100 mV applied potential showing
conductance steps in picoSiemens (pS) for channels due to A.
Gramicidin A (GA), B. D·Leu[2] Gramicidin A, and C. N-acetyl Gramicidin A. Note that the height of the steps in A and B are similar
but that the duration of the channels is less in B. In C where
both the scales are expaned, it is clear that replacement of the
formyl hydrogen by a methyl moiety, i.e. N-acetyl instead of N-
formyl, dramatically decreases the mean lifetime from several
seconds, to the order of 50-60 msec. Also the magnitude of the
current is reduced to about 60%. A height of 32 pS at 100 mV is
equivalent to 2×10^7 ions/sec passing through a single channel.
It should be again emphasized that these are single channel events
being observed.

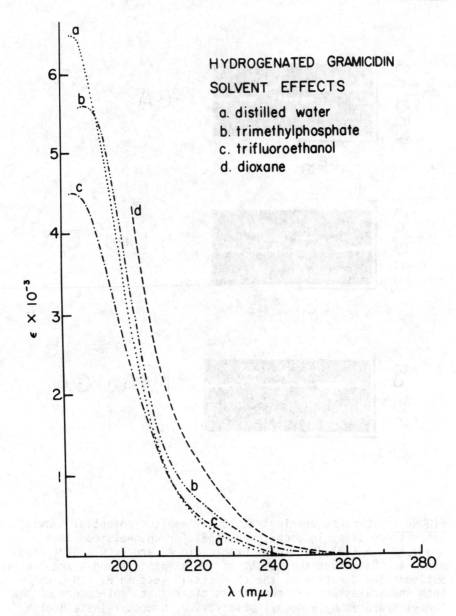

FIGURE 3: Absorption spectra of hydrogenated Gramicidin A. Hydrogenation was carried out to remove the side chain absorbances and leave only the peptide chromophores, the absorbance of which reflect the backbone conformation of the polypeptide. The different absorption intensities near 190 nm reflect different polypeptide conformations in the different solvents. Reproduced with permission from Reference 15.

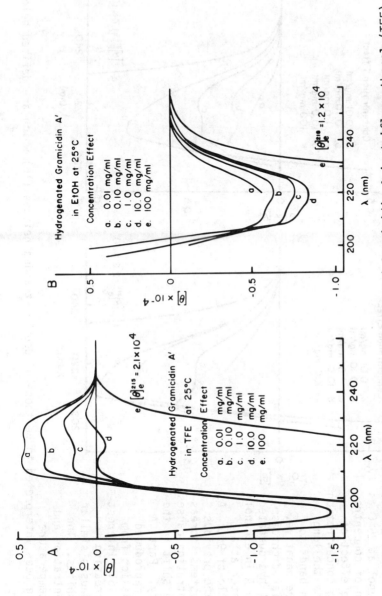

FIGURE 4: A. Circular dichroism spectra of hydrogenated Gramicidin A in trifluroethanol (TFE) at 25°C as a function of concentration. As the concentration increases, the long wavelength bands shift from positive to negative. B. Circular dichroism spectra of hydrogenated Gramicidin A in ethanol (EtOH) at 25°C as a function of concentration. Again as the concentration is raised the ellipticity of the long wavelength bands becomes more negative.

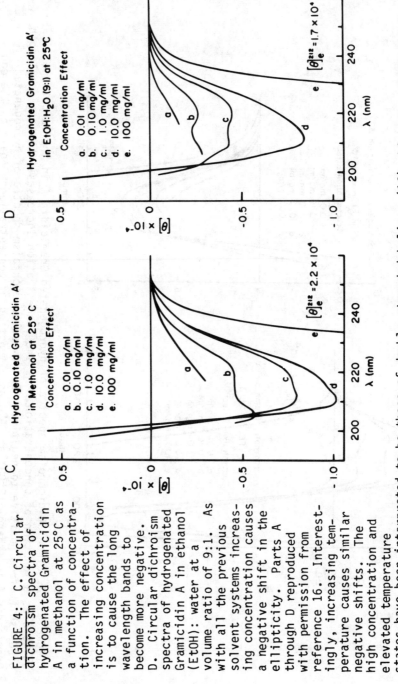

FIGURE 4:   C. Circular dichroism spectra of hydrogenated Gramicidin A in methanol at 25°C as a function of concentration. The effect of increasing concentration is to cause the long wavelength bands to become more negative. D. Circular dichroism spectra of hydrogenated Gramicidin A in ethanol (EtOH): water at a volume ratio of 9:1. As with all the previous solvent systems increasing concentration causes a negative shift in the ellipticity. Parts A through D reproduced with permission from reference 16. Interestingly, increasing temperature causes similar negative shifts. The high concentration and elevated temperature states have been interpreted to be those of double stranded helices (16) which are defined below.

(16). When using the organic solvent system which was the better candidate for the channel conformation, however, the ion binding properties did not correlate with the channel transport properties; Ca^{2+} was found to bind strongly to the Gramicidin though it is not transported by the channel and its effects on transport of monovalent ions is a secondary effect. On the other hand Na^+ and K^+ which are well transported by the channel show little or no interaction (17). Accordingly studies on Gramicidin in organic solvents are of limited relevance to the transport process.

B. Incorporation of the Channel State into Phospholipid Structures

In a configuration suitable for spanning a lipid bilayer, the structure requires hydrophobic sides and hydrophilic ends and interior. Characterization of a derived channel state would be facilitated in a lipid incorporated state when there is an equivalence of both entrances to the channel and when the resulting particle size is small. The latter property would be to avoid the particulate problems in optical spectroscopy and to give rise to high mobilities for improving resolution in nuclear magnetic resonance studies of resonances within the channel containing particle. A micelle forming lipid would seem to provide for such possibilities.

On addition to aqueuous solutions of L-α-lysolecithin micelles, Gramicidin is found to be immediately taken up by the micelles with limited sonication and heating, but the product does not significantly perturb the lipid core nor does it give rise to significant sodium ion interaction as monitored by sodium-23 nuclear magnetic resonance (NMR) resonance chemical shifts (ν), by longitudinal relaxation times (T_1) nor by line widths (ν_{1_2}). On heating, however, the lipid core mobility can be seen to decrease; the sodium-23 T_1 is found to decrease; an increasingly upfield sodium-23 chemical shift is found and the CD spectrum is markedly changed. With extended heating, these changes result in a state where further heating has no effect. The state is stable for very long periods of time (several weeks) and there is no evidence of significant L-α-lysolecithin breakdown as measured by carbon-13 NMR and thin layer chromatography. For details see references 18 and 19. The sodium ion interaction is competitively reduced by Ag^+ and by Tl^+ which competitively blocks transport through the channel (20,21). Using temperature dependence of T_1 and of ν_{1_2}, an activation energy for exchange with the sodium ion binding site is obtained ($E_a \approx 6.8 - 7.4$ kcal/mole) which is essentially the same ($E_a \approx 7.3$ kcal/mole) as for Na^+ transport through the channel (22). On the basis of these correlations, it is argued that the channel state is obtained. Further correlations in this and the accompanying paper demonstrate conclusively that the channel state is indeed obtained.

Examination by electron microscopy using the negative staining technique of the product of heating L-α-lysolecithin with Gramicidin shows that the latter induces formation of planar bilayers (23,24) as shown in Figure 5. While many of the structures appear vesicle-like, many appear to be folded bilayer sheets such that it would appear that the problem of sealed vesicles developing an electrochemical potential during an ion titration is not expected. As considered above, this is not the ideal phospholipid incorporated state for spectroscopic characterization. It will be seen below, however, that it is quite adequate for determining conformation by means of ion induced carbonyl carbon chemical shifts, and as will be shown in the accompanying paper, this channel state serves well for the determination of ion binding and rate constants by sodium-23 NMR. It is particularly well-suited for determining intrachannel ion translocation rates by means of dielectric relaxation studies as in these studies, it is desirable that the whole structure have as slow reorientation correlation time.

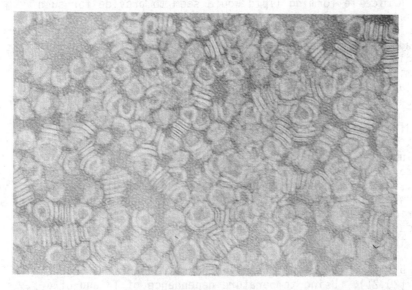

FIGURE 5: Electron micrograph of negatively stained phospholipid structures formed on heating Gramicidin A with L-α-lysolecithin. Note that many of the vesicle-like structures are seen on edge as having a single bilayer thickness and give the appearance of folded bilayer sheets. Reproduced with permission from Reference 24.

IV. PROPOSED STRUCTURES FOR THE CHANNEL

This laboratory (6,14) and independently Ramachandran and Chandresekharan (25,26) proposed new conformations for polypeptides with alternating L-D sequences. The initial terminologies for the new conformations were $\pi_{L,D}$-helices and L-D helices, respectively. The name was subsequently changed to single stranded β-helices to reflect the hydrogen bonding patterns between turns of the helix which was the same as for the β-pleated sheet conformations (27). In the one set of studies (6,14), it was proposed that the channel state of the Gramicidins was specifically a state of two helical molecules hydrogen bonded head-to-head (amino or formyl end to formyl end) by means of six intermolecular hydrogen bonds (6) and that the preferred helix sense was left-handed (6,14). The general set of single-stranded β-helices are represented in Figure 6A. Considering different end to end hydrogen bonded associations and both left- and right-handed helix senses gives six possible structures. Possible helix pitches with different numbers of residues per turn, e.g. 4.4, 6.3, 8.2 and 10 expands the possible number of structures to twenty four. Structure I with about 6.3 residues per turn was early proposed (1971) as the dominant con-formational state of the channel (6). This is shown in terms of CPK and wire models in Figure 7.

In 1974, Veatch et al. (28) proposed the double stranded struc-tures schematically represented in Figure 6B. These may be either antiparallel or parallel aligned chains and they may have either left or right handed helical sense to give the four schematic representations. In addition, there may be 5.6, 7.2, 9.0 or 10.8 residues per turn (29). This gives some sixteen possible double stranded structures. Thus in all there are between 30 and 40 dif-ferent reasonable conformational states possible for the channel. The antiparallel double stranded β-helix with 7.2 residues per turn is given in space filling and wire models in Figure 8.

In addition, very recently a new antiparallel β-structure has been suggested for the channel state of Gramicidin (30,31) as being consistent with the data of Weinstein et al. (32). All of these conformations can be placed with respect to the planar bilayer and the relative positions of their carbonyl carbons can be plotted with respect to a transmembrane axis. In the following section a means is utilized to determine which structure is the state of Gramicidin A in L-α-lysolecithin packaged phospholipid bilayers.

V. CARBON-13 NUCLEAR MAGNETIC RESONANCE OF ION INDUCED CARBONYL CARBON CHEMICAL SHIFTS: DETERMINATION OF CHANNEL STRUCTURE AND LOCATION OF ION BINDING SITES

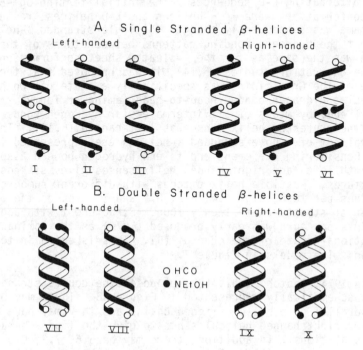

A. Single Stranded β-helices

B. Double Stranded β-helices

○ HCO
● NEtOH

FIGURE 6: A. Schematic representation of single stranded β-helical dimers with different end to end alignments and with different helix senses. Also, these structures can occur with different numbers of residues per turn. Structure I with 6.3 residues per turn was early proposed to be the channel structure (6). It is shown in molecular detail in Figure 7. B. Schematic representation of double stranded β-helices with different chain directions and helix senses. These may also occur with different numbers of residues per turn. A detailed structure of X with 7.2 residues per turn is shown in Figure 8. It is an antiparallel, right-handed double stranded β-helix. Reproduced with permission from Reference 39.

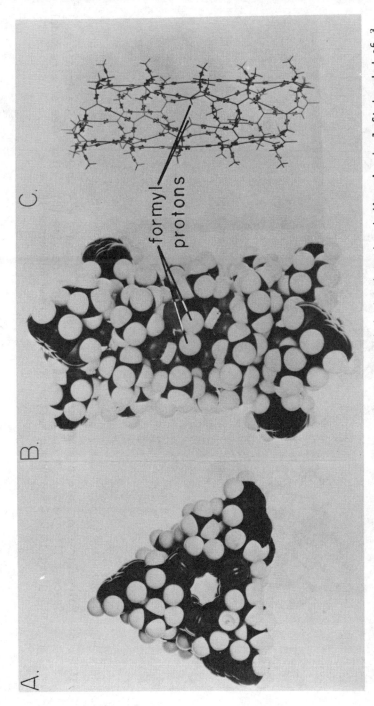

FIGURE 7: Molecular structure of the single-stranded, head-to-head dimerized, left-handed-$\beta^{6.3}$-helical structure. This is the Gramicidin A transmembrane channel. A. Channel view, B. Side view, C. side view of wire model. Note the juxtaposition of formyl protons at the center of the structure in B and C. In the malonyl-bis-didesformyl Gramicidin A the formyl protons are replaced by a single CH$_2$ moiety. In the N-acetyl desformyl GA derivative, each formyl proton is replaced by a CH$_3$ moiety. Reproduced with permission from Reference 16.

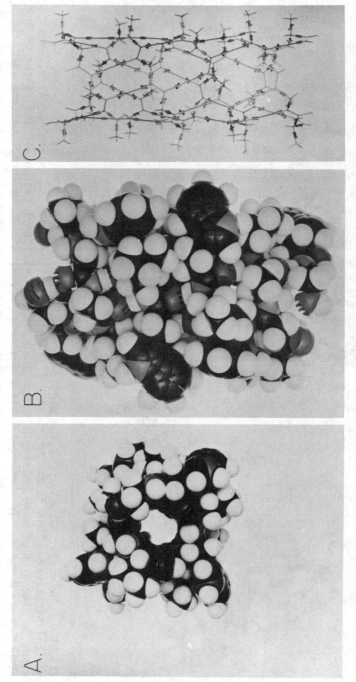

FIGURE 8: Molecular structure of the antiparallel, double stranded $\beta^{7.2}$-helix of Gramicidin A. Space filling model in channel view (A) and in side view (B). Side view of wire model is given in C. Reproduced with permission from Reference 16.

A. Ion Induced Carbonyl Carbon Chemical Shifts Using a Cyclic
    Conformational Correlate of Gramicidin

Enniatin B, which is cyclo (N-Methyl-L·Valyl-D·Hydroxyiso-
valeryl)$_3$ (33), may on the basis of its primary structure be considered a cyclic conformational correlate of the linear Gramicidins. This cyclohexapeptide with alternating L and D residue
configurations functions as a carrier of monovalent cations across
lipid bilayers. The structure of the monovalent cation complex
has been determined both in the crystal (34) and in solution (35).
As such it is an especially relevant model for obtaining an estimate of the direction and magnitude of carbonyl carbon chemical
shifts resulting from complexation with alkali metal cations. In
Figure 9 is the plot of the carbonyl carbon chemical shifts
resulting from a KCl titration of enniatin B in per-
deuteromethanol. The magnitude of the monovalent cation induced
chemical shifts on complex formation is in the range of 1.5 to 2.0
ppm and the direction is downfield (36). This provides a useful
calibration of the sign and magnitudes of chemical shifts that can
be expected to result on complexation of certain monovalent
cations by Gramicidin A.

B. Ion Induced Carbonyl Carbon Chemical Shifts Exhibited by the
    Carbonyl Carbon Resonances of Gramicidin A in the Channel
    State

One of the effects of the relatively large lipid bilayer
sheets formed on heat incubation of Gramicidin A with L-α-lysolecithin is the broadening of the carbon resonances of the channel. This may be overcome by obtaining spectra at elevated
temperatures of 90% enrich carbonyl carbons. The different
single-site enriched Gramicidin A (GA) syntheses utilized were the
follow-ing: ^{13}C-formyl-GA; (1-^{13}C)L·Val^{1}GA; (1-^{13}C)L·Ala^{3}GA;
(1-^{13}C)L· Ala^{5}GA; (1-^{13}C)L·Val^{7}GA; (1-^{13}C)D·Val^{8}GA; (1-^{13}C)L·Trp^9
GA; (1-^{13}C) L·Trp^{11}GA; (1-^{13}C)L·Trp^{13}GA; (1-^{13}C)D·Leu^{14}GA; and
(1-^{13}C)L·Trp^{15}GA. Each of these syntheses were individually
incorporated into phospholipid bilayers, and the carbon-13 NMR
spectra were obtained. As the lipid carbonyl carbon resonance
chemical shift was unchanged by the ion additions, this resonance
can be used as a convenient reference. The thallium ion induced
carbonyl carbon chemical shifts for residues 1, 3, 5, 7, 9, 11, 13
and 15 are given in Figure 10A through H (37,38). These chemical
shifts may be plotted as a function of the different classes of
proposed structures for the channel. This is done in Figure 11.

In Figure 11A is a representative structure for the antiparallel double stranded β-helices. Here it is seen that carbonyl
carbons of residues 11 and 13 show large chemical shifts whereas

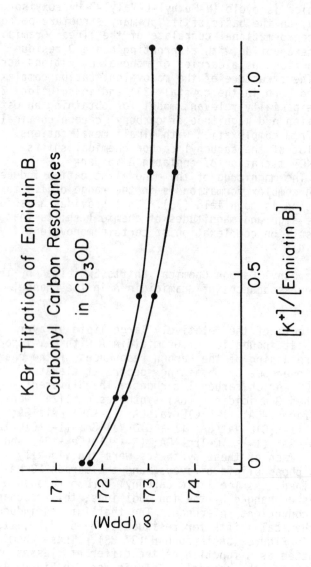

FIGURE 9:   A plot of the carbon-13 nuclear magnetic resonance carbonyl carbon chemical shifts of Enniatin B as a function of KBr concentration in CD_3OD. In this cyclic conformational correlate of Gramicidin A (see text), the carbonyl carbon resonances shift downfield by up to 2.0 ppm on direct coordination of the monovalent cation by the carbonyl moieties. This provides an effective calibration of the sign and magnitude of chemical shift that can be expected on coordination of monovalent cations by carbonyls of Gramicidin A. Reproduced with permission from Reference 36.

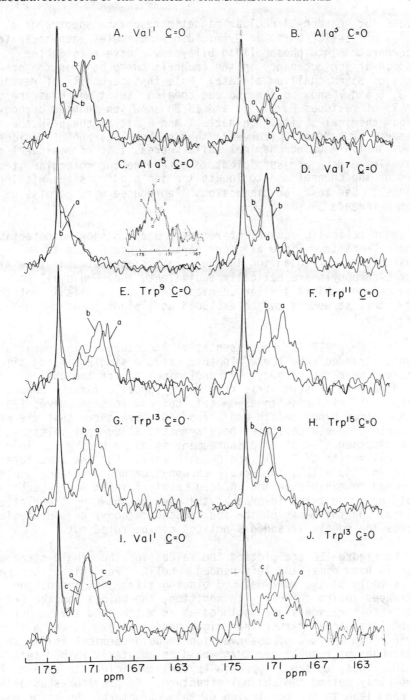

Figure 10 legend on next page.

FIGURE 10: Carbon-13 nuclear magnetic resonance spectra of a series of (1-^{13}C) enriched Gramicidin A molecules each separately incorporated into phospholipid bilayers. Curve a is in the absence of ion occupancy in the channel; curve b is in the presence of 83 mM thallium acetate. Note that carbonyls of residues 1, 3, 5 and 7 show no ion induced chemical shift where as the carbonyls of residues 9, 11, 13 and 15 do show ion induced carbonyl carbon chemical shifts. In parts I and J, it is shown in the presence of 1 M BaCl$_2$ (curve c) that carbonyl carbon of residue 13 shows an ion induced chemical shift whereas that of residue 1 does not. These data are sufficient to determine the molecular structure of the channel and to locate the ion binding sites within the channel. See text for discussion. Reproduced with permission from Reference 38.

those of helically equivalent residues 3 and 5 show no detectable chemical shifts. Since all four carbonyls would have to be equally proximal to an ion in the putative channel and since all four L-carbonyls are helically equivalent, they would all be expected to exhibit similar chemical shifts. As this is not the case, this structure can be excluded as a significant channel structure.

In Figure 11B is a representative structure of the parallel double stranded β-helices. In this case, a single binding site is observed near the carboxyl (ethanolamine) end of the chains. This would require an asymmetric free energy profile for the channel which would give rise to asymmetric current-voltage curves (39). On the other hand, Busath and Szabo (40) have shown that the most probable conductance states have symmetrical current voltage curves. Another element of disagreement is given by the ion induced chemical shifts of D·Val^8 and D·Leu^{14} (41). For the structure shown in Figure 11B, the D·Val^8 carbonyl carbon resonance would be expected to exhibit an ion induced chemical shift and D·Leu^{14} would not. Just the inverse is the case as shown by the plotted x values. On the basis of the ion induced carbonyl carbon chemical shifts the double stranded β-helices can be ruled out.

In Figure 11C are plotted the values for the single-stranded, head to head dimerized left-handed β-helix. For this structure, there would be two well-defined binding sites just inside the entrances to the channel. In addition, the points for the D·Val^8 and D·Leu^{14} residues are included as xs which show the structure to be left-handed rather than right handed (41). It can also be argued from the set of observed ion induced chemical shifts that the number of repeats per turn would be about six rather than 4, 8 or 10. Thus, this one approach appears sufficient to essentially completely define the channel structure. As previous studies had already lead to the conclusion of this structure, the major new

A. Antiparallel, double stranded β-helix

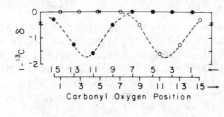

B. Parallel, double stranded β-helix

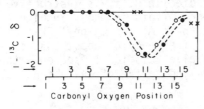

C. Single stranded, head-to-head β-helix

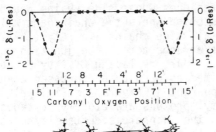

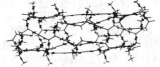

FIGURE 11: Ion induced carbonyl carbon chemical shifts of Figure 10 plotted with respect to carbonyl locations of several, proposed channel structures. A. Antiparallel, double stranded β-helix. The contradiction that the carbonyls of residues 11 and 13 show large ion induced chemical shifts whereas those of helically equivalent residues 3 and 5 shown no chemical shifts although they would have to be equally proximal to the ion argues that this is not the channel structure (see text for discussion). B. Parallel double stranded β-helix. An asymmetric channel would be required with a resulting non-symmetric current voltage curve as this is not the case, this structure is eliminated (see text for discussion). The ion induced chemical shifts for D-residues 8 and 14 also indicate a contradiction with this structure. C. Single-stranded, head-to-head dimerized, left-handed β-helix. The ion induced chemical shifts indicate that this is the correct structure. See text for discussion. Reproduced with permission from Reference 38.

point that is drawn from the study of the ion induced carbonyl carbon chemical shifts is the locations of the ion binding sites.

It should also be mentioned that the considered antiparallel structure or cross β-structure (30,31) can be shown to be inconsistent with the Ba^{2+} induced carbonyl chemical shifts as this ion would exhibit ion induced chemical shifts all the way across the membrane but it is not transported by the channel. With this considered structure, there are also problems of data arguing for a dimeric channel and of steric crowding of side chains.

VI. INFRARED SPECTRUM OF THE SINGLE-STRANDED, HEAD-TO-HEAD DIMERIZED, LEFT-HANDED β^6-HELIX: THE GRAMICIDIN A TRANSMEMBRANE CHANNEL

Since it has just been demonstrated that the single stranded, head-to-head dimerized, left-handed β^6-helix is the channel state obtained on heat incorporation of Gramicidin A into phospholipid bilayers comprised of L-α-lysolecithin, it is instructive to determine the amide I frequency. This is shown in Figure 12 where it is seen that Gramicidin A (part A) and malonyl Gramicidin A (part B) both give an amide I frequency of about 1633 cm^{-1}. Thus the singlestranded β-helices give an amide I frequency that had previously been considered as indicative of the absence of such a conformation (30,32,43,44).

VII. THE HEAD-TO-HEAD JUNCTION AND THE "ON-OFF" MECHANISM: ELEMENTS OF THE DIMERIZATION PROCESS

Another element of molecular structure that may be briefly mentioned is that of the head-to-head junction. It is noteworthy that the amino blocking group is a formyl moiety rather than the more common acetyl or larger blocking moieties and that the amino acid side chains tend to be smaller at the amino end of the molecule. Of the first five residues whose side chains would interact across the head to head junction where the chains are aligned antiparallel three have small side chains i.e. Gly^2, L·Ala^3, L·Ala^5. These side chains arrange across the junction so as to limit side chain crowding. Thus, replacing Gly^2 by D·Leu^2 led to a destablization of the dimer as seen by a decreased channel mean lifetime (12). This is apparent in Figure 2. Similarly as seen in Figure 7B, the two formyl protons are closely juxtaposed. Again the dimeric state of the channel is destabilized as shown in Figure 2C (13). Thus even these minor structural modifications provide detailed verification of the structure. It is of further interest to note, when the N·Ac desformyl GA is studied in the presence and absence of halothane, that the dimer is further destabilized (45). As the dimer results from the formation of six

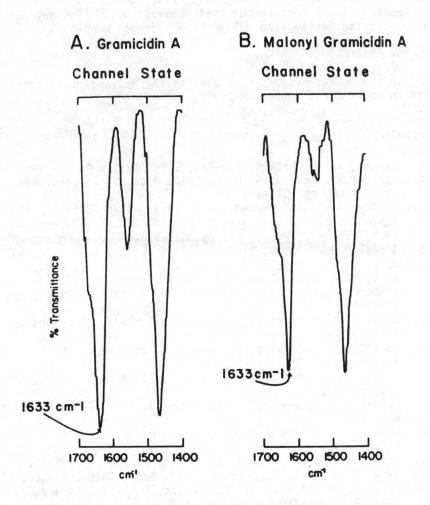

FIGURE 12:   Infrared spectrum in the 1400 to 1700 cm^{-1} range for
Gramicidin A (A) and malonyl Gramicidin A (B) heat
incorporated with lysolecithin to form channels in
lipid bilayers. In Figures 10 and 11, it was shown
that the structure is that of Figures 11C and Figure
7. The structure gives an amide I frequency of 1633
cm^{-1} for the single-stranded, head-to-head dimerized,
left-handed β^6-helix. This value was previously considered to be indicative of double-stranded structures
and β-pleated sheets. Reproduced with permission from
Reference 42.

intermolecular hydrogen bonds, this raises the point stressed by
C. Sandorfy (this conference) that general anesthetics may act in
a capacity to destabilize polypeptide hydrogen bonding.

ACKNOWLEDGEMENT

This work was supported in part by the National Institutes of
Health under Grant Number GM-26898.

REFERENCES

1. Sarges, R. and Witkop, B. 1965, Biochemistry 4, pp. 2491-2494.
2. Mueller, P. and Rudin, D. O. 1967, Biochem. Biophys. Res.
    Commun. 26, pp. 398-404.
3. Hladky, S. B. and Haydon, D. A. 1970, Nature 225, pp. 451-453.
4. Tosteson, D. C., Andreoli, T. E., Tieffenbery, M. and Cook, P.
    1968, J. Gen. Physiol. 51, pp. 373-384.
5. Goodall, M. C. 1970, Biochim. Biophys. Acta 219, pp.471-478.
6. Urry, D. W., Goodall, M. C., Glickson, J. D. and Mayers, D. F.
    1971, Proc. Natl. Acad. Sci. USA 68, pp. 1907-1911.
7. Bamberg, E. and Läuger, P. 1973, J. Membrane Biol. 11, pp.
    177-194.
8. Kolb, H.-A, Bamberg, E. and Läuger, P. 1975, J. Membrane
    Biol. 20, pp. 133-154.
9. Apell, H.-J., Bamberg, E., Alpes, H. and Läuger, P. 1977, J.
    Membrane Biol. 31, pp. 171-188.
10. Myers, V. B. and Haydon, D. A. 1972, Biochim. Biophys. Acta
    274, pp. 313-322.
11. Krasne, S., Eisenman, G. and Szabo, G. 1971, Science 174, pp.
    412-415.
12. Bradley, R. J., Prasad, K. U. and Urry, D. W. 1981, Biochim.
    Biophys. Acta 649, pp. 281-285.
13. Szabo, G. and Urry, D. W. 1979, Science 203, pp. 55-57.
14. Urry, D. W. 1971, Proc. Natl. Acad. Sci. USA 68, pp. 672-676.
15. Urry, D. W, Glickson, J. D., Mayers, D. F. and Haider, J.
    1972, Biochemistry 11, pp. 487-493.
16. Urry, D. W., Long, M. M., Jacobs, M. and Harris, R. D. 1975,
    Ann. NY Acad. Sci. 264, pp. 203-220.
17. Urry, D. W. 1978, Ann. NY Acad. Sci. 307, pp. 3-27.
18. Urry, D. W., Spisni, A. and Khaled, M. A. 1979, Biochem.
    Biophys. Res. Commun. 88(3), pp. 940-949
19. Urry, D. W., Spisni, A., Khaled, M. A., Long, M. M. and
    Masotti, L. 1979, Int. J. Quantum Chem.:Quantum Biology Symp.
    No. 6, pp. 289-303.
20. McBride, D. and Szabo, G. 1978, Biophys. J. 21, pp. A25.
21. Neher, E. 1975, Biochim. Biophys. Acta 401, pp. 540-544.
22. Bamberg, E. and Läuger, P. 1974, Biochim. Biophys. Acta 367,
    pp. 127-133.

23. Pasquali-Ronchetti, I., Spisni, A., Casali, E., Masotti, L. and Urry, D. W. 1983, Bioscience Reports 3, pp. 127-133.
24. Spisni, A., Pasquali-Ronchetti, I., Casali, E., Lindner, L., Cavatorta, P., Masotti, L. and Urry, D. W., Biochim. Biophys. Acta, (in press).
25. Ramachandran, G. N. and Chandrasekharan, R. 1972, In: Progress in Peptide Research Vol. II, Lande, S., Ed., Gordon and Breach, Science Publishers, Inc. New York, pp. 195-215.
26. Ramachandran, G. N. and Chandrasekaran, R. 1972, Indian J. Biochem. Biophys. 9, pp. 1-11.
27. Urry, D. W. 1973, In: Conformation of Biological Molecules and Polymers - The Jerusalem Symposia on Quantum Chemistry and Biochemistry, V, Jerusalem, Bergmann, E. D. and Pullman, Eds, Israel Academy of Sciences, pp. 723-736.
28. Veatch, W. R., Fossel, E. T. and Blout, E. R. 1974, Biochemistry 13, pp. 5249-5256.
29. Lotz, B., Colonna-Cesari, F., Heitz, F. and Spach, G. 1976, J. Mol. Biol. 106, pp. 915-942.
30. Sychev, S. V., Ivanov, V. T. 1982, In: Membranes and Transport Volume 2, Martonosi, A. N., Plenum Press, New York, pp. 301-307.
31. Ovchinnikov, Y. A., Ivanov, V. T. 1982, In: Conformation in Biology, Srinivasan, R. and Sarma, R. H., eds., Adenine Press, Guilderland, New York, pp. 155-174.
32. Weinstein, S., Wallace, B. A., Blout, E. R., Morrow, J. S. and Veatch, W. 1979, Proc. Natl. Acad. Sci. USA 76(9), pp. 4230-4234.
33. Plattner, Pl. A., Vogler, K., Studer, R. O., Quitt, Pl. and Keller Schierlein, W. 1963, Helv. Chim. Acta 46, 927- .
34. Dobler, M., Dunitz, J. D. and Krajewski, J. 1969, J. Mol. Biol. 42, pp. 603-606.
35. Shemyakin, M. M., Ovchinnikov, Y. A., Ivanov, V. T., Antonov, V. K., Vinogradova, E. I., Shkrob, A. M., Malenkov, G. G., Eustratov, E. V., Laine, I. A., Melnik, E. I. and Ryabova, I. D. 1969, J. Membr. Biol. 1, pp. 402-430.
36. Urry, D. W. 1976, In: Enzymes of Biological Membranes, Vol. 1, Martonosi, A., Ed., Plenum Publishing Corp., New York, New York, pp. 31-69.
37. Urry, D. W., Prasad, K. U. and Trapane, T. L. 1982, Proc. Natl. Acad. Sci. USA 79, pp. 390-394.
38. Urry, D. W., Trapane, T. L. and Prasad, K. U., Science (in press).
39. Urry, D. W., In: The Enzymes of Biological Membranes, Martonosi, A. N., Ed., Plenum Publishing Corporation, New York, New York (in press).
40. Busath, D. and Szabo, G. 1981, Nature 294, pp. 371-373.
41. Urry, D. W., Walker, J. T. and Trapane, T. L. 1982, J. Membr. Biol. 69, pp. 225-231.

42. Urry, D. W., Shaw, R. G., Trapane, T. L. and Prasad, K. U.,
    Biochem. Biophys. Res. Commun., (in press).
43. Spach, G., Trudelle, Y. and Heitz, F. 1983, Biopolymers 22,
    pp. 403-407.
44. Sychev, S. V., Nevskaya, N. A., Jordanov, S., Shepel, E. N.,
    Miroshnikov, A. I. and Ivanov, V. T. 1980, Bioinorganic Chem.
    9, 121-151.
45. Urry, D. W., Venkatachalam, C. M., Prasad, K. U., Bradley, R.
    J., Parenti-Castelli, G. and Lenaz, G. 1981, Int. J. Quantum
    Chem.:Quantum Biology Symp. No. 8, pp. 385-399.

IONIC MECHANISMS AND SELECTIVITY OF THE GRAMICIDIN TRANSMEMBRANE
CHANNEL: CATION NUCLEAR MAGNETIC RESONANCE, DIELECTRIC RELAXATION,
CARBON-13 NUCLEAR MAGNETIC RESONANCE, AND RATE THEORY CALCULATION
OF SINGLE CHANNEL CURRENTS

Dan W. Urry

Laboratory of Molecular Biophysics
University of Alabama in Birmingham
School of Medicine
Birmingham, Alabama 35294, USA

ABSTRACT

Cation nuclear magnetic resonance of lithium-7, sodium-23,
potassium-39, cesium-133 and thallium-205 is utilized to determine binding constants and rate constants for ion interactions
with the channel state. A dielectric relaxation study is used to
provide the rate of ion jump within the channel from one binding
site to the other. It is demonstrated that five experimentally
derived rate constants obtained in the sodium-23 NMR and in
dielectric relaxation studies can be used to calculate the single
channel currents of the electrical studies on planar bilayers.
The approach is to use Eyring rate theory to introduce voltage
dependence into the five rate constants, expanding the five to the
ten rate constants required for a two-fold symmetric, two site
channel in the presence of an electric field. Having arrived at a
clear description of ionic mechanism the physicochemical basis for
ion selectivity is discussed. The molecular structure of the
channel itself is the basis for anion vs cation selectivity.
Carbon-13 NMR is utilized to demonstrate divalent cation binding
in the channel and rapid exchange with the binding site. This
leads to the conclusion that divalent cation impermeability is due
to the lipid proximity. Finally selectivity among monovalent
cations is discussed in terms of the energetics and dynamics of
peptide librational type motion. It is noted that non-linear
Arrhenius type plots can be expected in which a degree of non-
linearity would be proportional to the conformational energy
change required to go from the preferred conformation in the
absence of ion to the librated state in the presence of ion.

C. Sandorfy and T. Theophanides (eds.), Spectroscopy of Biological Molecules, 511–538.
© 1984 by D. Reidel Publishing Company.

I. INTRODUCTION

In the previous lecture (1), it was demonstrated that Gramicidin A (HCO-L·Val^1-Gly^2-L·Ala^3-D·Leu^4-L·Ala^5-D·Val^6-L·Val^7-D·Val^8-L·Trp^9-D·Leu^{10}-L·Trp^{11}-D·Leu^{12}-L·Trp^{13}-D·Leu^{14}-L·Trp^{15}HNCH$_2$CH$_2$OH) could be incorporated into phospholipid bilayers on heat incubation with L-α-lysolecithin. During the time course of heat incubation, the relative mobility of lipid CH$_2$ moieties with respect to choline CH$_3$ moieties decreases; the circular dichroism progressively changes to a terminal unique pattern, and there is a progressive increase in sodium ion interaction as measured by decreases in the longitudinal relaxation time (T$_1$) and increases in the sodium-23 resonance line width ($\nu_{1/2}$). Once the final state is reached, temperature dependence of T$_1$ and $\nu_{1/2}$ gave activation energies for exchange with the binding site which were the same as obtained for transport through the channel as determined by planar bilayer studies of temperature dependence of single channel currents. In addition, silver ion and thallium ion, which competitively block sodium or potassium ion transport through the channel, compete with sodium ion binding to the lysolecithin packaged Gramicidin A as measured by increases in T$_1$ and decreases in $\nu_{1/2}$. Furthermore, Ca^{2+} and Ba^{2+} interactions (see below) also quantitatively parallel effects found for these divalent cations on the magnitude of monovalent cation single channel currents. Accordingly a strong case is developed that the channel state is obtained on heat incubation of Gramicidin A with lysolecithin. Then carbon-13 NMR ion induced carbonyl carbon chemical shifts were shown only to be consistent with the single-stranded, head-to-head dimerized, left-handed β$^{6·3}$ helix. Thus the structure of the lysolecithin packaged GA channel state was determined and importantly two binding sites separated by about 22 Å were observed to occur just inside each entrance to the channel.

Using an entirely independent line of reasoning, one of employing Gramicidin A derivatives and determining their effects on channel formation and properties, the same channel structure is deduced. For example, the malonyl dimer (malonyl-bis-dideformyl Gramicidin A) forms channels at very low concentrations, of very long lifetimes and with apparent first order concentration dependence (2,3). This is a specific test for head-to-head dimerization. The O-pyromellityl Gramicidin A, which places the triply negatively charged group at the ethanolamine oxygen, at high ionic strength produces channel conductances that are almost indistinguishable from regular Gramicidin A when placed on both sides of the planar bilayer and results in no channels when placed only on one side (4). The N-pyromellityl desformyl Gramicidin A forms no measureable channels whether placed on one or both sides or when placed on one side with O-pyromellityl GA on the other side. In general, any modification at the formyl (head) end markedly alters properties in a manner explicable by the head-to-head dimerized, single-stranded β-helix whereas modifications at the ethanolamine end of the molecule has lesser effects again reasonably explained by the head-to-head dimerized, single stranded β-helix (6-9). Thus two entirely separate lines of argument, those directly using the planar bilayer observations of channel transport and those using spectroscopic methods on suspensions of the channel state in phospholipid bilayers, arrive at the same channel structure.

In the present lecture, cation nuclear magnetic resonance is utilized to determine binding and where possible, rate constants for cation interaction with the channel state obtained by heat incubation of Gramicidin A with L-α-lysolecithin. Data will be considered for lithium-7, sodium-23, potassium-39, cesium-133 and thallium-205. Dielectric relaxation studies will be used to obtain estimates of rates of intrachannel ion movements. Using the cation NMR and dielectric relaxation derived rate constants derived in the absence of an applied potential, Eyring rate theory will be used to introduce voltage dependence. Then single channel currents will be calculated and compared with experimental values. A useful result is the plot of the free energy profile for passage of sodium ion through the Gramicidin A channels at a single temperature. Finally the physicochemical basis for ion selectivity will be discussed in terms of anion versus cation, of monovalent versus divalent cation (using carbon-13 NMR) and of the set of alkali metal ions.

II. CATION NUCLEAR MAGNETIC RESONANCE AND THE DETERMINATION OF BINDING AND RATE CONSTANTS

When certain nuclei are placed in a magnetic field, they exhibit magnetic moments which tend to align and precess about the field direction. The small increase in numbers of nuclei aligned parallel, as opposed to antiparallel to the field, results in a macroscopic magnetic moment parallel to the magnetic field that can be pulsed with a radio frequency field and caused to orient perpendicular to or even antiparallel to the magnetic field. The relaxation rate of the macroscopic magnetic moment so driven to a non-equilibrium state can be determined as the macroscopic moment returns to the equilbrium distribution. The relaxation rate of nuclei with spins greater than $1/2$, i.e. quadrupolar nuclei, depends dominantly on the immediate electric field gradient in which the nucleus is found. The alkali metal ions, lithium-7 (3/2), sodium-23 (3/2), potassium-39 (3/2) and cesium-133 (7/2), exhibit relatively slow relaxation rates in aqueous solutions due to the symmetric field of the solvation sphere. The slow relaxation that does occur is considered to result from the transient exchange of a water molecule in the first hydration shell (10). When bound to

a site in a peptide, the relaxation can be very fast. This large
difference in relaxation rates between solution and a peptide
binding site can, in favorable circumstances, be used to determine
the binding constant for the site and the off-rate constant from
the site. This effect will be shown below to be very effective
for studies of cation interactions with the Gramicidin A
transmembrane channel.

A. Sodium-23 Nuclear Magnetic

The sodium-23 nucleus is a particularly favorable nucleus with
which to study ion interactions with the Gramicidin A channel
state. This is because it exhibits large chemical shifts on
binding, large changes in longitudinal relaxation rates, bi-
exponential transverse relaxation rates as well as more general
line broadening effects due to exchange with the channel binding
site. In the folllowing review, the malonyl-bis-didesformyl
Gramicidin A is utilized to ensure that the dimeric state is
achieved essentially quantitatively.

1. Longitudinal Relaxation Studies for Determination of
Binding Constants. When a cation is exchanging rapidly between
solution and a binding site, the experimental relaxation rate, R_1,
may be written,

$$R_1 = 1/T_1 = P_f R_{1f} + P_b R_{1b} \tag{1}$$

where T_1 is the experimental longitudinal relaxation time; P_f and
P_b are the mole fractions of free and bound ions, respectively;
and R_{1f} and R_{1b} are the relaxation rates when the ion is free in
solution and when it is at the binding site, respectively. The
excess longitudinal relaxation rate due to binding is simply

$$(R_1 - R_{1f}) = \Delta R_1 = P_b(R_{1b}-R_{1f}) \tag{2}$$

For the situation where the total ion concentration is greater
than the site concentration, James and Noggle (11) have shown that
the reciprocal of the excess longitudinal relaxation rate, ΔR^{-1},
is approximated as,

$$\Delta R_1^{-1} = (R_1-R_{1f})^{-1} = \frac{([Na]_T + K_b^{-1})}{(R_{1b}-R_{1f})[Ch]_T} \tag{3}$$

where $[Na]_T$ and $[Ch]_T$ are the total sodium ion and total site
(channel) concentrations, respectively and K_b^{-1} is the reciprocal
of the binding constant. Since the product $(R_{1b}-R_{1f})[Ch]_T$ is a
constant for a given study, an extrapolation to the negative x axis

intercept gives the reciprocal of the binding constant. The data for the sodium ion titration of malonyl GA channels plotted in this way is given in Figure 1A (12) where at least two binding processes are seen. The apparent binding constants, $K_{(app)}$, are indicated on the negative $[Na]_T$ axis for a tight and a weak binding site. Analysis by Eqn. 3, however, assumes for conditions of multiple binding that R_{1b} is the same for each binding site. This is not necessarily the case even for a two-sited, two-fold symmetric channel because the presence of the first ion can be expected to change, reasonably to increase, the electric field gradient experienced by each of the two ions in the channel. The relaxation rates for one and for two ions in the channel should each be considered explicitly, i.e.,

$$R_1 = P_f R_{1f} + P_t R_{1t} + P_w R_{1w} \qquad (4)$$

On using the relationship that the sum of the mole fractions is one, Eqn. 4 may also be placed in the James and Noggle form

$$\Delta R_1 = (R_1 - R_{1f}) = P_t(R_{1t} - R_{1f}) + P_w(R_{1w} - R_{1f}) \qquad (5)$$

It should be noted that a term $P_\ell R_{1\ell}$ for interaction with lysolecithin molecules is negligible and therefore has not been included in the sodium-23 binding study. The complete analysis of the data in Figure 1A (12,13) gives rise to the following values for the binding constants and relaxation rates, $K^t \simeq 70M^{-1}$; $R_{1t} \simeq$ 1.9/msec; $K^w_b \simeq 1.3$ M^{-1} and $R_{1w} \simeq 6.2$/msec. As anticipated, the relaxation rate for two ions in the channel, R_{1w}, is faster than for one ion in the channel, R_{1t}. This is part of the reason why the apparent binding constant in Figure 1A for the weak site is a larger value. The rest of the effect is due to the proper separating of tight and weak binding site effects rather than the simple extrapolation to the intercept.

It may also be noted in Figure 1B that a plot of excess line width as a function of ion concentration gives similar apparent binding constants as in Figure 1A. This approach could also be used to obtain estimates of binding constants. In this case, the correction for the differing character of the process for one and for two sites in the channel would involve differences in the exchange terms.

2. Transverse Relaxation Studies and the Determination of Rate Constants. For a spin 3/2 nucleus the decay of the transverse magnetization, M_t, is bi-exponential (14-16),

$$M_t = M_0[0.6 \exp(-t/T_2') + 0.4 \exp(-t/T_2'')] \qquad (6)$$

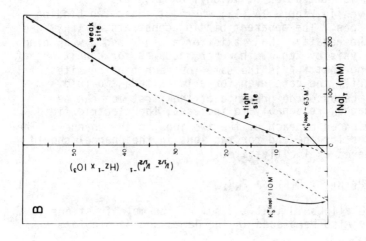

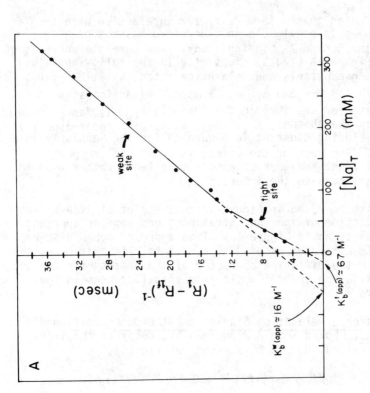

FIGURE 1: A. Excess longitudinal relaxation rate plot for sodium-23 interaction with malonyl Gramicidin A (3 mM) in phospholipid bilayers. Extrapolations to negative x-axis intercepts give apparent binding constants. Two sites are apparent. B. Excess line width plot using sodium-23 NMR to follow sodium interaction with malonyl Gramicidin A channel (3 mM) in phospholipid bilayers. Again two binding sites are apparent. See text for discussion. Reproduced with permission from Reference 12.

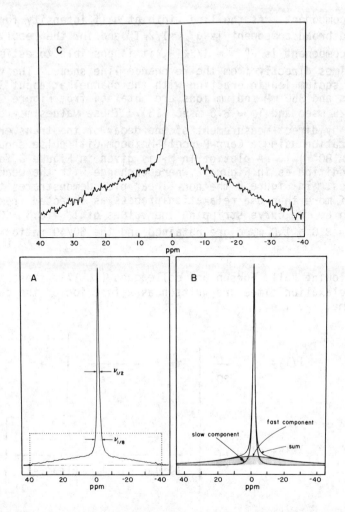

FIGURE 2: Sodium-23 NMR resonance line showing both broad and nar
row components. From the line widths at half intensity for the
resolved compnents (part B) the transverse relaxation times T_2' and
T_2'' are obtained for the fast and slow processes, respectively.
Reproduced with permission from Reference 13.

where T_2' and T_2'' are the relaxation times for the fast and slow
components of the transverse relaxation. Under favorable conditions the two components can readily be seen in the resonance
signal of the quadrupolar nucleus. At high sodium ion concentration in the presence of channels, the sodium-23 resonance
line, shown in Figure 2, is seen to be composed of a broad and a

narrow component. As the line width at half intensity for the
resolved broad component is $\nu_{\frac{1}{2}}' = 1/2\pi T_2'$ and for the resolved
narrow component is $\nu_{\frac{1}{2}}'' = 1/2\pi T_2''$, it is possible to estimate
these times directly from the resonance line shape. The values
for the sodium ion interaction with the channel at about 3 mM
channels and 350 mM sodium ions are obtained from Figure 2 to be
$T_2' \simeq 0.25$ msec and $T_2'' \simeq 8.6$ msec (13). These values may be
checked by direct measurement of the decay of the transverse
magnetization using a Carr-Purcell-Meiboom-Gill pulse sequence,
$(90°-(\tau-180°\tau)_n)$. A plot of ln M_t is given in Figure 3 for the
same condition as in Figure 2 where exchange with the weak site
dominates. In Figure 3 the non-linear plot demonstrates the presence of more than one relaxation process as expected from
Equation 6. By curve stripping the values of $T_2' = 0.4 \pm 0.15$ msec
and $T_2'' = 8.0 \pm 1.0$ msec are obtained and the 60/40 ratio is also
found.

Following Bull, Forsén and colleagues (16-17), these transverse relaxation times are written as a function of the correlation time, τ_C, i.e.,

$$1/T_2' = 1/T_{2f} + P_W \frac{\chi^2}{20} \left[\tau_C + \frac{\tau_C}{(1+\omega^2\tau_C^2)} \right] \qquad (7)$$

$$1/T'' = 1/T_{2f} + P_W \frac{\chi^2}{20} \left[\frac{\tau_C}{(1+4\omega^2\tau_C^2)} + \frac{\tau_C}{(1+\omega^2\tau_C^2)} \right] \qquad (8)$$

where χ is the quadrupolar coupling constant for the weak site and
is equal to e^2qQ/h with eQ being the electric quadrupole moment of
the sodium-23 nucleus and eq being the mean electric field gradient
experienced by the ion when there is double occupancy in the channel. By transposing $1/T_{2f}$ and dividing Eqn. 7 by Eqn. 8, the
ratio, which is the experimentally derived quantity, is a function
only of τ_C as $\omega = 2\pi\nu$ where ν is the observation frequency.

$$\frac{1/T_2' - 1/T_{2f}}{1/T_2'' - 1/T_{2f}} = \frac{1 + 1/(1+\omega^2\tau_C^2)}{1/(1+4\omega^2\tau_C^2) + 1/(1+\omega^2\tau_C^2)} \qquad (9)$$

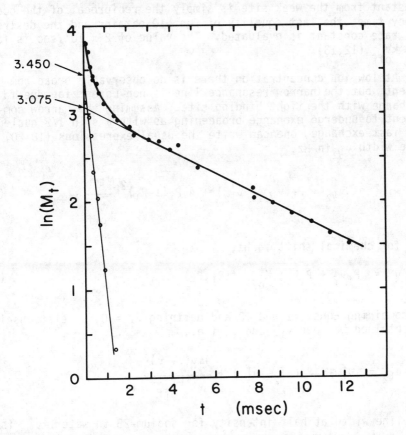

FIGURE 3: Relaxation of the macroscopic transverse magnetization obtained using the Carr-Purcell-Meiboom-Gill pulse sequence. Two relaxation processes are apparent and can both be resolved to obtain T_2' and T_2'' of Eqn. 6. These values compare favorably with those of Figure 2 and can be used to obtain the off rate constant for the weak site (see text for discussion). Reproduced with permission from Reference 13.

Of the two possible solutions for τ_c only one value makes physical sense and the inverse of this is the sum of two quantities i.e.

$$\frac{1}{\tau_c} = \frac{1}{\tau_r} + \frac{1}{\tau_b} \simeq k_{off}^w \qquad (10)$$

The quantity τ_r is the reorientation correlation time for the channel bilayer system which is very long compared to τ_b, the occupancy time for the ion in the channel. Since the off rate constant from the weak site is simply the reciprocal of the occupancy time, the last equality of Eqn. 10 obtains and the desired off rate constant is evaluated. The value of 2×10^7/sec is found for k^w_{off} (12,13).

At low ion concentration there is no observable broad component, but the narrow resonance line is non-Lorentzian due to exchange with the tight binding site. Assuming the narrow component to undergo exchange broadening as with a spin 1/2 nucleus, for fast exchange, one can write the usual expressions (18-20) for line width $\nu_{1/2}$ in Hz,

$$\nu_{1/2} = \nu^f_{1/2} + P_t(\nu^t_{1/2} - \nu^f_{1/2}) + 4\pi P_t(1-P_t)^2 \frac{\nu_t^2}{k^t_{off}} \qquad (11)$$

and for chemical shift in Hz,

$$\nu = P_f\nu_f + P_t\nu_t; \quad K^t_b = f(P_f,P_t) \qquad (12)$$

By combining Eqns. 11 and 12 and defining $\nu_f = 0$, a relationship is obtained between $\nu_{1/2}$ and ν, i.e.,

$$\nu_{1/2} = \nu^f_{1/2} + \frac{\nu}{\nu_t}(\nu^t_{1/2} - \nu^f_{1/2}) + \frac{4\pi\nu(\nu_t-\nu)^2}{\nu_t k^t_{off}} \qquad (13)$$

The line width at half intensity for sodium-23 in water, $\nu^f_{1/2}$ is, of course, known. The chemical shift at the tight binding site, ν_t, can be independently obtained from a plot of ν versus [Na] in which K^t_b is also found to be consistent with that obtained from the longitudinal relaxation studies (see Eqn. 3 and 5 and Figure 1A). The value of K^t_b so obtained assuming one tight site per empty channel gives a value of 146 M^{-1} with $\nu_t = 1850$ Hz, but since there are two tight sites in an empty channel this quantity should be halved, i.e. about 70 M^{-1}. Now in a plot of $\nu_{1/2}$ versus ν the unknowns are $\nu^t_{1/2}$ and k^t_{off} and it is k^t_{off} that gives rise to non-linearity. As shown in Figure 4, reasonable fit is obtained for $k^t_{off} = 3 \times 10^5$/sec. A similar value is obtained when intermediate exchange (18-20) is employed (13). With $K_b = k_{on}/k_{off}$, four rate constants have been obtained; $k^t_{on} \simeq 2.1 \times 10^7$/Msec; $k^t_{off} \simeq 3 \times 10^5$/sec; $k^w_{on} \simeq 2.6 \times 10^7$/Msec and $k^w_{off} \simeq 2 \times 10^7$/sec. For a two site, two-fold symmetric channel in the absence of an applied field, there are five rate constants. The one remaining rate constant is for the rate over the central barrier, k_{cb}, i.e., the jump between sites in a singly occupied channel. This value may be estimated in favorable cases from dielectric relaxation studies as outlined below in Part III.

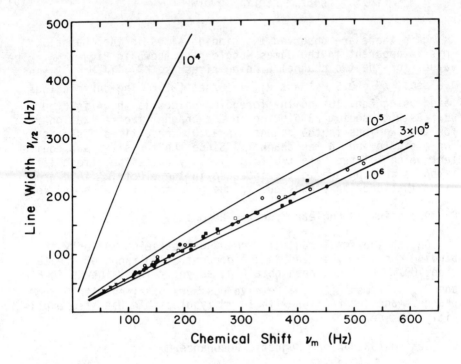

FIGURE 4: Plot of half line width versus chemical shift using Eqn. 13. The best fit gave k_{off}^t of 3 x 10^5/sec. Reproduced with permission from Reference 12.

B. Lithium-7 Nuclear Magnetic Resonance

The use of lithium-7 NMR to determine binding constants for the malGA channel in phospholipid bilayers formed on interaction with malGA is complicated by two additional considerations (21). One is that T_{1f} in D_2O is not independent of the concentration of LiCl. The dependence can be expressed by the relationship

$$T_{1f} = 39.4 \text{ sec} - 4.2 \text{ [LiCl] sec} \tag{14}$$

Second the interaction of lithium ion with L-α-lysolecithin cannot be neglected. The binding constant, K_b^{ℓ}, and the relaxation rate at

the binding site, $R_{1\ell}$, were determined for a 100 mM concentration
of L-α-lysolecithin to be 0.2 M^{-1} and 2/sec, respectively. This
allows evaluation of the term, $P_\ell R_{1\ell}$, in the required expression,

$$R_1 = P_f R_{1f} + P_\ell R_{1\ell} + P_t R_{1t} + P_w R_{1w} \tag{15}$$

as again there are observed two binding sites in the channel.
This is apparent in the James-Noggle plot shown in Figure 5A. The
values for the two channel binding sites are $K^t_b = 14$ M^{-1}, $R_{1t} =$
4.8/sec., $K^w_b = 0.5$ M^{-1} and $R_{1w} = 16$/sec (21). The curve calculated using Eqn. 15 and the foregoing values is shown in Figure
5B. As with sodium-23 binding in the channel the relaxation rate
for the two ions in the channel is about three times faster than
for a single ion in the channel. Since the two sites are equivalent being related by a two-fold symmetry axis, the effect is
taken as arising due to the increase in the electric field gradient that one ion senses due to the presence of the second ion.

C. Potassium-39 Nuclear Magnetic Resonance

Due to the small magnetic moment of potassium-39, only the
binding constant observed in the concentration range of greater
than 100 mM [K^+] has been obtained, as yet. The value K^w_b is 5 M^{-1}
and $R_{1w} = 37$/msec (22). These values were obtained with 0.3 mM
channels and interaction with 10 mM lysolecithin was also explicitly considered.

D. Cesium-133 Nuclear Magnetic Resonance

Another alkali metal ion of interest for the Gramicidin A
transmembrane channel is the cesium ion. As this is a spin 7/2
nucleus, both the longitudinal and transverse relaxation processes
are multi-exponential. This is true also for spin 3/2 nuclei, e.g.
sodium-23. In that case the bi-exponential transverse relaxation
was used to estimate the off-rate constant from the weak site but
in the longitudinal relaxation studies, the coefficient of the
first term is 0.2 whereas that for the slower term is 0.8 such
that it was only necessary to consider the slower term. For a
spin 7/2 nucleus the transverse relaxation process is too complex
to allow evaluation of τ_c from an expression of the form of Eqn.
9. As with sodium-23, however, the slow term dominates the
measurable longitudinal relaxation time (23) and a James-Noggle
type of analysis still obtains. The result is $K^t_b = 55M^{-1}$, $R_{1t} =$
21/msec, $K^w_b = 4M^{-1}$ and $R_{1w} = 86$/msec (24). It is interesting
to note here that with the large, polarizable cesium ion $R_{1w} > R_{1t}$
but it is not much larger than obtained for Li^+ and Na^+ binding.

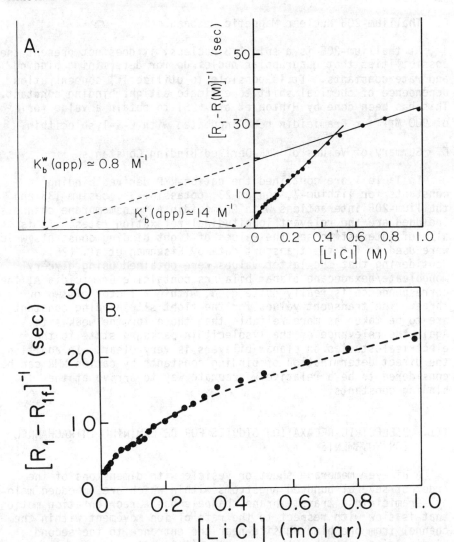

FIGURE 5: Excess longitudinal relaxation rate plot of lithium-7 interactions with the malonyl Gramicidin (malGA) channel in phospholipid bilayers. The excess rate is plotted with respect to the relaxation rate in the presence of 100 mM L-α-lysolecithin at the appropriate ion concentration. Two binding processes are seen and the apparent binding constants are obtained from the negative x-axis intercepts. B. Excess longitudinal relaxation rate plot using an equation of the form of Eqn. 15 and plotting the excess with respect to the relaxation rate in D_2O, R_{1f}. The binding constants and relaxation rates are given in the text and in Table I. Reproduced with permission from Reference 21.

E. Thallium-205 Nuclear Magnetic Resonance

As thallium-205 is a spin 1/2 nucleus, it does not present the possibilities that quadrupolar nuclei do for determining binding and rate constants. It is possible to utilize Tl^+ concentration dependence of chemical shift to estimate a tight binding constant. This has been done by Hinton et al. (25) to obtain a value for K_b^t of 900 M^{-1} for Gramicidin heat incubated with L-α-lysolecithin.

F. Summary of Values of NMR-Derived Binding Constants

In Table I are contained the cation-NMR derived binding constants for lithium-7, sodium-23, potassium-39, cesium-133 and thallium-205 interactions with Gramicidin A transmembrane channels. Included are the relevant longitudinal relaxation rates. It is also of interest to see the values of tight binding constants which were used to fit the transport data by Eisenman et al. (26). Considering that the latter values were obtained using glyceryl monooleate/hexadecane planar bilayers containing Gramicidin A, the correspondence is really quite good, within a factor of two or three. The transport values for the tight site binding constant are to be taken as more reliable than those for the weak site. Again the relevance of the lysolecithin-packaged state to the electrical studies on planar bilayers is very clear. In addition, the direct determination of binding constants by cation NMR can be considered to be a relatively accurate way to arrive at the binding constants.

III. DIELECTRIC RELAXATION STUDIES FOR DETERMINING INTRACHANNEL ION MOVEMENTS

A bilayer membrane sheet or vesicle with dimensions of the order of several hundred Angstroms within which are embedded malonyl Gramicidin A transmembrane channels has a reorientation motion that is slow with respect to the rate of ion movement within the channel from one binding site near the entrance to the second binding site near the other entrance (1,27,28). Such an ion movement is equivalent to a large dipole moment change as depicted in Figure 6. The magnitude of the dipole moment change is of the order of 100 Debye. Accordingly dielectric relaxation studies provide an opportunity, under favorable circumstances, for determining the rate constant for passage of an ion over the central barrier. A required condition is that the channel be singly occupied at relatively low ion concentration. This condition is met by thallium ion because of its high binding constant with K_b^t being of the order of $10^3 M^{-1}$.

TABLE I: NMR-DERIVED PARAMETERS FOR ION BINDING TO
LYSOLECITHIN-PACKAGED GRAMICIDIN CHANNELS

Parameter	7Li[a]	23Na[b]	39K c	133Cs c	205Tl d
K_b^t	14/M (9/M)e	63/M (20/M)e		55/M (100/M)e	900/M (950/M)e
K_b^w	0.5/M (1.2/M)e	1.4/M (2.0/M)e	8.0/M^{-1} (1.5/M)e	4.0/M (0.8/M)e	
R_{1f} (D$_2$O)	0.025/sec	0.017/msec	0.017/msec	0.086/msec	
$R_{1\ell}$	0.064/sec	0.020/msec	0.019/msec	0.12/msec	
R_{1t}	4.8/sec	1.9/msec		21.0/msec	
R_{1w}	16.0/sec	6.20/msec	37.0/msec	86.0/msec	

a. Reference 21
b. References 12 and 13
c. Reference 24
d. Reference 25
e. The binding constants in parentheses were obtained by Eisenman et al. (26), utilizing single channel conductance data from electrical measurements on glyceryl monooleate/hexadecane planar bilayers containing Gramicidin A.

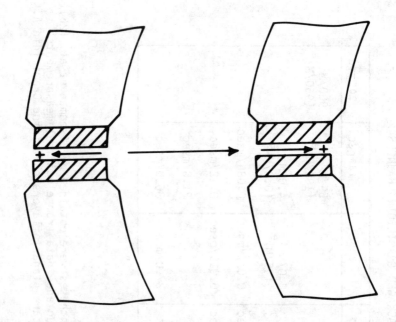

FIGURE 6: Schematic representation of the dipole moment change
when an ion jumps from one binding site to the other. This dipole
moment change is detectable in the real part of the dielectric
permittivity as shown in Figure 7 and provides the rate constant
for the central barrier, k_{cb}.

The real part of the dielectric permittivity is given in
Figure 7 over the frequency range of 0.005 to 900 MHz at 25°C.
The conditions were approximately equimolar channels and salts of
near 10 mM. The curve for thallium ion (the solid dots) show a
relaxation centered between 1 and 10 MHz with a relaxation time of
120 nsec. This translates to a rate constant for the central
barrier, k_{cb}, of 4×10^6/sec (29). In addition, the temperature
dependence of the relaxation indicates an activation energy of
less than 6.7 kcal/mole which is reasonable given that exchange
with the channel binding site gives an activation energy of just
over 7 kcal/mole and that transport through the channel is found
to be about 7.3 kcal/mole (30). The implication is that the rate
limiting barrier is at the month of the channel and that the
central barrier is less. Because of lower binding constants,
there is not sufficient occupancy in the channel for Na^+ and Li^+
to exhibit a relaxation. In what follows it is assumed, however,
that binding is a normalization process, since the backbone structure is uniform across the channel, and that the ions would all
exhibit a similar free energy height of central barrier relative
to the free energy at the binding site. In addition for the
sodium calculation of single channel currents, the fitted rate

constant for the central barrier of mal GA using the other four
rate constants derived from NMR was found to be 3.2×10^6/sec
(12). This gives an experimental basis for the assumption that
the thallium ion k_{cb} could be used for the Na^+ transport and that
the required five rate constants have been estimated directly and
independent of the electrical measurement of single channel
currents.

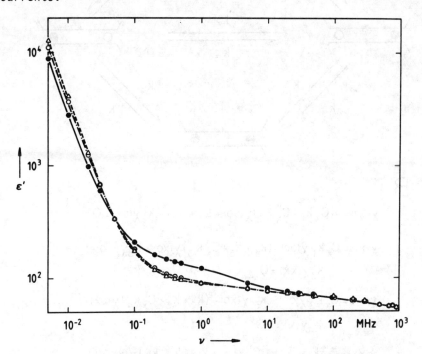

FIGURE 7: A log-log plot of the real part of the dielectric permittivity as a funciton of frequency for malonyl Gramicidin A
channels (10 mM) in lysolecithin (300 mM) phospholipid bilayers
with different salts. Solid dots, thallium acetate (10 mM); open
dots, lithium chloride (10 mM); triangles, sodium chloride (10 mM).
At these concentrations of salts only thallium ion occupancy in
the channel is significant. The relaxation observed for Tl^+ has a
relaxation time of 120 nsec giving a rate constant for the central
barrier, k_{cb}, of 4×10^6/sec. Reproduced with permission from
Reference 29.

IV. CALCULATION OF SINGLE CHANNEL CURRENTS USING RATE CONSTANTS
    DETERMINED IN THE ABSENCE OF AN APPLIED ELECTRICAL FIELD

    The steady state scheme for a two site channel is given in
Figure 8 (31,32). There are ten required rate constants but due

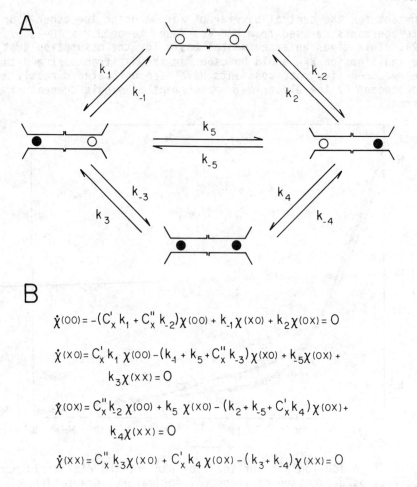

A.

B.

$$\dot{\chi}(oo) = -(C_x' k_1 + C_x'' k_{-2})\chi(oo) + k_{-1}\chi(xo) + k_2\chi(ox) = 0$$

$$\dot{\chi}(xo) = C_x' k_1 \chi(oo) - (k_{-1} + k_5 + C_x'' k_{-3})\chi(xo) + k_{-5}\chi(ox) + k_3\chi(xx) = 0$$

$$\dot{\chi}(ox) = C_x'' k_{-2} \chi(oo) + k_5 \chi(xo) - (k_2 + k_{-5} + C_x' k_4)\chi(ox) + k_{-4}\chi(xx) = 0$$

$$\dot{\chi}(xx) = C_x'' k_{-3}\chi(xo) + C_x' k_4 \chi(ox) - (k_3 + k_{-4})\chi(xx) = 0$$

FIGURE 8: A. Scheme of rate processes for a single-filing, two-site channel. Ten rate constants are defined. The plus subscript keeps track of net ion movements to the right and the negative subscript indicates net ion movement to the left. These ten rate constants may be defined in terms of five experimentally derived rate constants expanded to ten on inclusion of the effect of a transmembrane potential as shown in Eqn. 19. B. Steady state equations for a single-filing, two-site channel. These equations are used to solve for the probability χ, of each of the four occupancy states; $\chi(oo)$, $\chi(xo)$, $\chi(ox)$ and $\chi(xx)$. C_x' and C_x'' are the ion activities on the left and right sides, respectively. Reproduced with permission from Reference 31.

to the two-fold symmetry of the Gramicidin A channel these ten
rate constants are the five experimentally derived rate constants
(see above) modified by the effect of the applied potential. A
given experimentally derived rate constant, for example, becomes
two rate constants, one k^w_{off} modified for the ion moving up the
potential gradient and one for movements down the potential gradient. Using Eyring rate theory, k^t_{on}, for example, is written

$$k^t_{on} = \frac{kT}{h} e^{-\Delta G^t_{on}/RT} \tag{16}$$

As early shown by Eyring and colleagues (33,34), k_1 of Figure 8
would be written

$$k_1 = \frac{kT}{h} e^{-\Delta G^{\ddagger}_1/RT} \tag{17}$$

where

$$\Delta G^{\ddagger}_1 = \Delta G^t_{on} + (d-b_1)\ zFE/2dRT \tag{18}$$

and z is the charge on the ion, F is the Faraday constant, E is
the applied potential and d, b_1 and a_1 are defined in Figure 9.
Thus, the rate constants can be written

$$k_1 = k^t_{on}\ X^{\ell_1} \quad ; \quad k_{-1} = k^t_{off}\ X^{-\ell_2}$$

$$k_2 = k^t_{off}\ X^{\ell_2} \quad ; \quad k_{-2} = k^t_{on}\ X^{-\ell_1}$$

$$k_3 = k^w_{off}\ X^{\ell_2} \quad ; \quad k_{-3} = k^w_{on}\ X^{-\ell_1} \tag{19}$$

$$k_4 = k^w_{on}\ X^{\ell_1} \quad ; \quad k_{-4} = k^w_{off}\ X^{-\ell_2}$$

$$k_5 = k_{cb}\ X^{a_1} \quad ; \quad k_{-5} = k_{cb}\ X^{-a_1}$$

where $\ell_1 = (d-b_1)$, $\ell_2 = (b_1-a_1)$ and $X \equiv \exp(zFE/2dRT)$. The set of
four steady state equations in Figure 8 constitute four simulta-
neous equations with four unknowns, namely $\chi(oo)$, $\chi(xo)$, $\chi(ox)$ and
$\chi(xx)$ which are the probabilities for the four possible occupancy
states. Thus each occupancy state probability can be expressed in

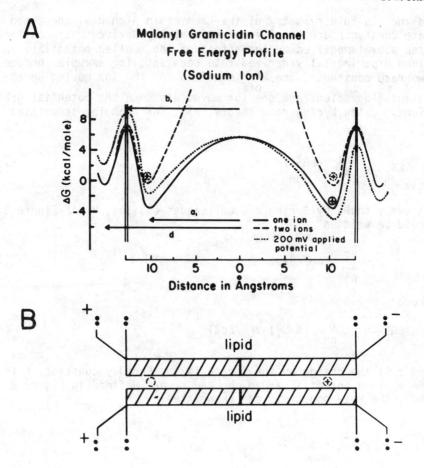

FIGURE 9: A. Free energy profile for sodium ion passage through the malonyl Gramicidin A transmembrane channel. The free energy of the binding sites were determined from the expression ΔG = $-RT\ln K$. The positions of the binding sites were determined using ion induced carbonyl carbon chemical shifts (1). The free energies of the barriers were determined using Eqn. 16. The lengths a_1, b_1 and d are defined in the figure. B. Schematic representation of the channel. An applied potential shows the left side to be positive and the right side to be negative. The effect of a 200 mV applied potential is shown in part A. Reproduced with permission from D. W. Urry, In: Methods in Enzymology, (L. W. Cunningham and D. W. Frederiksen, Eds.) Academic Press, Inc., New York, 82, 673-716, 1982.

terms of the ten rate constants. If we know the shape of the
potential gradient, which is taken to be linear to a good approximation, then with the five experimental rate constants everything
is known for calculation of the single channel currents.
Recognizing that at steady state the net current over each barrier
encountered on passage through the channel is the single channel
current, the simplest expression for the single channel current,
i_x, is the net current over the central barrier, i.e.,

$$i_x = ze[k_5\chi(xo) - k_{-5}\chi(ox)] \tag{20}$$

Using this equation, it should be realized that $\chi(xo)$ and $\chi(ox)$
are each a function of the ten rate constants of Eqn. 6 and i_x of
Eqn. 20 is not just a function of the rate constants for the
central barrier.

Using the five experimentally determined rate constants, the
single channel current at 100 mV can be calculated to a particularly satisfying extent (12,31) as shown in Figure 10A. If
the rate constants are allowed to vary to obtain a "best fit", the
values do not change by more than a factor of two or three and the
single channel currents can be very well fitted with a single set
of five rate constants over several decades of ion activity and
for 50, 100 and 150 and 200 mV applied potential as shown in
Figure 10B. Thus, the ionic mechanism has reached a well-
described stage for a single temperature. The corresponding
free energy profile is given in Figure 9.

VI. PHYSICOCHEMICAL BASIS FOR ION TRANSPORT SPECIFICITY (31)

The Gramicidin A transmembrane channel is essentially imper-
meable to anions and to multivalent cations and it has a reasonable capacity to discriminate among monovalent cations. Now that
the molecular structure is well-established and the ionic mechanisms are at an advanced stage of description, it becomes possible
to discuss the physicochemical basis for ion transport specificity with some degree of confidence.

A. Anion vs Cation Selectivity

A stereo view of the channel structure in the calculated
lowest energy in vacuo conformation is given in Figure 11(35).
The L-residue carbonyl oxygens are seen to be slightly librated
inward such that these C-O axes make an angle of about 10° with
the channel axis. The carbonyl oxygen carries about a 0.4 negative charge and librates easily into the channel, whereas the NH
hydrogens carry less than half the positive charge, i.e. about
+0.17, and move into the channel at a substantially greater cost in

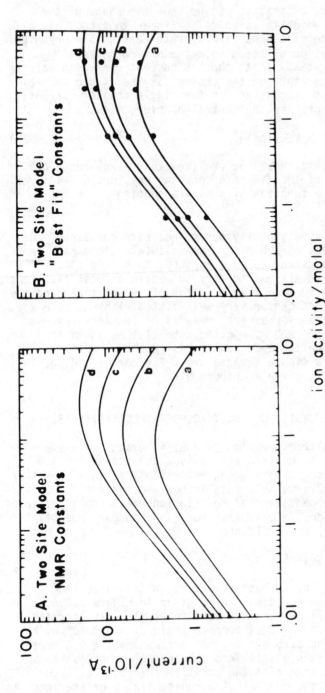

FIGURE 10: A. Calculation of single channel currents for a (50 mV), b (100 mV), c(150 mV) and d (200 mV) applied potentials using four ^{23}Na-NMR derived rate constants and with k_{cb} = 3.2 x 10^6/sec which is essentially the same as k_{cb} of 4 x 10^6/sec obtained for thallium ion from the dielectric relaxation studies. B. Comparison of experimental single channel current values, dots, with calculated single channel currents starting with the experimentally derived rate constants and allowing these to vary to obtain, with a single set of five rate constants, a best fit to the data for the four applied potentials and over the indicated ion activity range. In obtaining this best fit the rate constants did not vary by more than a factor of two or three from the experimentally derived values. Reproduced with permission from Reference 12.

A

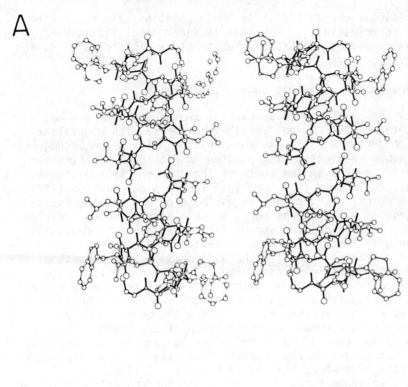

B

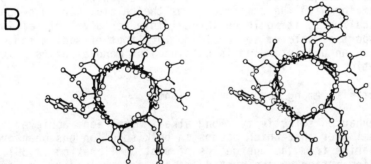

FIGURE 11:   Stereoview, set for cross-eye viewing, of the lowest energy, in vacuo conformation of the Gramicidin A transmembrane channel.   A. side view, B. channel view of one-half of channel. Note that the L-residue carbonyls are librated slightly into the channel.   It is the carbonyls of L-residues 9, 11, 13 and 15 that coordinate the cations in the binding site (1).   The charge distributions of the peptide moiety and the ease of the L-residue carbonyl oxygens to librate into the channel are responsible for the cation selectivity of the channel.   Reproduced with permission from Reference 35.

conformational energy (36). As these moieties have to provide
lateral coordination for any ions in the channel, it is clear that
the physicochemical nature of the structure dictates high cation
selectivity.

B. Monovalent vs Divalent Cation Selectivity

Divalent cations such as calcium and barium ions are not
transported by the channel yet they interfere with monovalent
cation conductance (37). As shown by the calcium ion induced carbonyl carbon chemical shifts plotted in Figure 12, it is clear
that Ca^{2+} can bind in the mouth of the channel, though the site is
pressed 1 to 2Å away from the lipid (38). A calcium ion titration
of the chemical shift allows an estimate of the binding constant
to be about 1 M^{-1}. Furthermore, since a single carbonyl carbon
resonance is obtained throughout the titration, this means that
exchange with the binding site is faster than 10^6/sec. The divalent ion exchanges fast with the binding site in the mouth of the
channel but it does not pass through the channel. This requires
that the central barrier be rate limiting. A reasonable explanation is the repulsive image force of the lipid. By the Born
solvation energy expression, there is a charge squared, a z^2,
dependence to the repulsive image force of the lipid. Thus the
portion of the barrier in Figure 9 that is due to the proximity of
the lipid surrounding the channel would have to be multiplied by z^2
or by 4 for a divalent ion and by 9 for a trivalent ion. Taking
the positive image force contribution to the central barrier for
the sodium ion with the same radius as the calcium ion to be only
about 3 kcal/mole, it would be 12 kcal/mole for divalent calcium
and transport of calcium ion would be 10^{-6} that of sodium ion.
Accordingly an understanding is obtained for monovalent vs multivalent cation selectivity.

C. Selectivity Among Monovalent Cations

The enhanced selectivity among alkali metal ions achieved by
the channel over that which occurs in bulk solution has been proposed to arise from the energetics of peptide libration (2,39).
This rocking motion of the peptide moiety is required to bring the
carbonyl into the channel to achieve direct coordination of the
laterally bare cation. A general expression for the single channel current may be written (31)

$$i_x = ze \sum_r x_r \sum_s \overline{\lambda}_{rs} \, q_{rs} \qquad\qquad (21)$$

where x_r is the probability of a given occupancy state, $\overline{\lambda}_{rs}$ is the
distance the ion moves as a result of the rate process q_{rs}. The
line is above the distance moved to indicate that this is a signed

quantity, positive when moving with the potential gradient and negative when moving against the potential gradient. The rate process, q_{rs}, may be either the rate constant or a concentration times the on rate constant. If a ratio of the probability of all other states of the channel to the librated state required for the local coordination of the ion for an individual rate process is defined as K°_{rs}, then the expression is modified to give

$$i_x = ze \sum_r xr \sum_s \overline{\lambda}_{rs} \, q_{rs} \, (1 + K^{\circ}_{rs})^{-1} \tag{22}$$

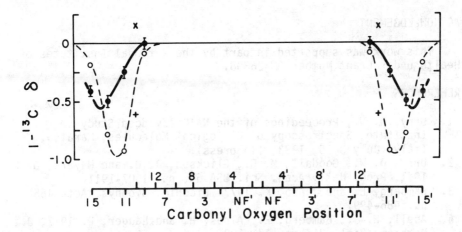

FIGURE 12: Comparison of the on induced (carbon-13) carbonyl carbon chemical shifts due to Na$^+$(o) and Ca^{2+}(•) which have nearly identical ionic radii. This demonstrates that calcium ion binds in the channel. It can also be argued that Ca^{2+} exchanges rapidly with the channel binding site. Therefore, the lack of calcium ion transport by the channel must be due to the height of the central barrier, the so-called repulsive image force due to the lipid with a charge squared dependence. See text for discussion. Reproduced with permission from Reference 38.

In a plot of ln i_x vs reciprocal temperature in °K the term -ln $(1+K^{\circ}_{rs})$ introduces a non-linearity in the Arrhenius type plot.

This has been observed for K$^+$ transport (40). The degree of non-linearity is expected to increase as the librational energy required to achieve coordination increases. The size of the channel in the lowest energy configuration is closely that required for coordination of Rb$^+$. Some slight outward libration might be required for Cs$^+$ and inward librations are increasingly required as the ionic radius decreases though this is not necessarily a monotonic increase in energy with decreasing radius (36). It can be anticipated that a complete and most accurate description will require consideration of a dynamic channel not just in the sense of distributions between states but rather in terms of the rates of peptide librations being induced by the ion. Work is in progress to achieve such descriptions. It does seem evident, however, that a qualitative picture of selectivity among monovalent cations involves a dynamic peptide libration mechanism. It is an accurate description of this process that is currently the most challenging.

ACKNOWLEDGEMENT

     This work was supported in part by the National Institutes of Health under Grant Number GM-26898.

REFERENCES

1. Urry, D. W., Proceedings of the NATO Advanced Study
    Institute, Spectroscopy of Biological Molecules, Maratea,
    Italy, July 3-15, 1983, (in press).
2. Urry, D. W., Goodall, M. C., Glickson, J. D. and Mayers, D. F.
    1971, Proc. Natl. Acad. Sci. USA 68, pp. 1907-1911.
3. Bamberg, E. and Janko, K. 1977, Biochim. Biophys. Acta 465,
    pp. 486-499.
4. Apell, H.-J., Bamberg, E., Alpes, H. and Läuger, P. 1977, J.
    Membrane Biol. 31, pp. 171-188.
5. Bamberg, E., Apell, H.-J. and Alpes, H. 1977, Proc. Natl.
    Acad. Sci. USA 74, pp. 2402-2406.
6. Szabo, G. and Urry, D. W. 1979, Science 55, pp. 55-57.
7. Bradley, R. J., Prasad, K. U. and Urry, D. W. 1981, Biochim.
    Biophys Acta 649, pp. 281-285.
8. Bradley, R. J., Urry, D. W., Okamoto, K. and Rapaka, R. S.
    1978, Science 200, pp. 435-437.
9. Bamberg, E., Apell, H.-J., Bradley, R., Härter, B., Quelle,
    M.-J. and Urry, D. W. 1979, J. Membr. Biol. 50, pp. 257-270.
10. Abragam, A. 1961, In: The Principles of Magnetism, Chap. 8,
    Oxford Univ. Press, London.
11. James, T. L. and Noggle, J. H. 1969, Proc. Nat. Acad. Sci.
    USA 62, pp. 644-649.
12. Urry, D. W., Venkatachalam, C. M., Spisni, A., Bradley, R. J.,
    Trapane, T. L. and Prasad, K. U. 1980, J. Membr. Biol. 55,
    29-51.
13. Venkatachalam, C. M. and Urry, D. W. 1980, J. Magn. Resonance
    41, pp. 313-335.
14. Hubbard, P. S. 1970, J. Chem. Phys. 53, pp. 985-997.
15. Bull, T. E. 1972, J. Mag. Res. 8, 344-353.

16. Bull, T. E., Andrasko, J., Chiancone, E. and Forsén, A. 1973, J. Mol. Biol. 73, pp. 251-259.

17. Norne, J. E., Gustavsson, H., Forsén, S., Chianocone, E., Kuiper, H. A. and Antonini, E. 1979, Eur. J. Biochem. 98, pp. 591-595.

18. Pople, J. A., Schneider, W. G. and Bernstein, H. J. 1959, In: High-Resolution Nuclear Magnetic Resonance, McGraw-Hill, New York, pp. 218-230.

19. Binsch, G. 1968, In: Topics in Stereochemistry, E. L. Eliel and N. L. Allinger, Eds., Wiley, New York, pp. 97-192.

20. Feeney, J., Batchelor, J. B., Albrand, J. P. and Roberts, G.C.K. 1979, J. Magn. Reson. 33, pp. 519-529.

21. Urry, D. W., Trapane, T. L., Venkatachalam, C. M. and Prasad, K. U., J. Chem. Phys. (in press).

22. Trapane, T. L. and Urry, D. W. (unpublished).

23. Bull, T. E., Forsén, S. and Turner, D. L. 1979, J. Chem. Phys. 70, pp. 3106-3111.

24. Urry, D. W., Venkatachalam, C. M. and Trapane, T. L. (unpublished data).

25. Hinton, J. F., Young, G. and Millett, F. S. 1982, Biochemistry 21, pp. 651-654.

26. Eisenman, G., Sandblom, J. and Neher, E. 1978, Biophys. J. 22, pp. 307-340.

27. Pasquali-Ronchetti, I., Spisni, A., Casali, E., Masotti, L. and Urry, D. W. 1983, Bioscience Reports 3, pp. 127-133.

28. Urry, D. W., In: The Enzymes of Biological Membranes, Martonosi, A. N., Ed., Plenum Publishing Corp., New York (in press).

29. Henze, R., Neher, E., Tapane, T. L. and Urry, D. W. 1982, J. Membr. Biol. 64, 233-239.

30. Bamberg, E. and Läuger, P. 1974, Biochim. Biophys. Acta 367, pp. 127-133.

31. Urry, D. W., In: Topics in Current Chemistry, F. L. Boschke, ed., Springer-Verlag, Heidelberg, Germany, (in press).

32. Urban, B. W., Hladky, S. B., Haydon, D. A. 1978, Fed. Proc. 37, 2628-2632.

33. Zwolinski, Bi. I., Eyring, H. and Polissar, M. I. 1949, J. Phys. Chem. 53, pp. 1426-1453.

34. Parlin, B. and Eyring, H. 1954, In Ion Transport Across Membranes, H. T. Clarke, Ed., Academic Press, New York, pp. 103-118.

35. Venkatachalam, C. M. and Urry, D. W., J. Comput. Ch., (in press).

36. Venkatachalam, C. M. and Urry, D. W., J. Comput. Ch. (in press).

37. Bamberg, E. and Läuger, P. 1977, J. Membr. Biol. 35, pp. 351-375.

38. Urry, D. W., Trapane, T. L., Walker, J. T. and Prasad, K. U. 1982, J. Biol. Chem. 257, pp. 6659-6661.

39. Urry, D. W. 1973, In: Conformation of Biological Molecules and Polymers - The Jerusalem Symposia on Quantum Chemistry and Biochemistry, V, Jerusalem, Israel Academy of Sciences, E. D. Bergmann and B. Pullman, Eds., V, Jerusalem, Israel Academy of Sciences, pp. 723-736.
40. Urry, D. W., Romanowski, S. and Bradley, R. J., unpublished.
41. Läuger, P. (private communication).

IN SITU MONITORING OF MEMBRANE-BOUND REACTIONS BY KINETIC LIGHT-SCATTERING

Andreas Schleicher and Klaus Peter Hofmann

Institut für Biophysik und Strahlenbiologie
Albertstr.23, 7800 Freiburg/Brsg. FRG

Introduction

In vertebrate rod cells absorption of a light quantum by rhodopsin is followed by intramolecular reactions of rhodopsin (photoproducts batho to meta II, ps to ms (1,2)). After milliseconds, photoexcited rhodopsin is able to trigger membrane bound enzymatic reactions. The first step thereby is stoichiometric association of the G-unit (G-protein, "transducin") to the activated rhodopsin molecule (3).

The work of our laboratory is dedicated to these coupling processes between rhodopsin photochemistry and enzymatic activation. By means of absorption flash photometry and kinetic light-scattering these first enzymatic reactions can be monitored in real time. This abstract shortly introduces to the problems. The main goal is to describe the methods, emphasizing the kinetic light scattering technique.

The rhodopsin G-protein complex

In rod outer segment suspensions binding of the peripheral G-protein to the intrinsic membrane protein rhodopsin was observed as a light scattering change termed 'binding signal' (3). This light scattering transient was later found to be basically identical to the so-called "signal P" (4) described by Hofmann et al. (5). It is a positive scattering change with complex ms kinetics and a saturation at a rhodopsin

C. Sandorfy and T. Theophanides (eds.), Spectroscopy of Biological Molecules, 539–545.

turnover of about 10%. As the content of G-protein
in rod outer segments is only about 10% of the one of
rhodopsin the P-signal saturates after all of the
available G-protein is bound. Kinetic investigations
have shown that binding occurs after the reaction of
metarhodopsin I (M I) to metarhodopsin II (M II).
The following reaction scheme was deduced (4):

$$Rh \rightsquigarrow \ldots \ldots \longrightarrow M\ I \rightleftharpoons M\ II \rightleftharpoons M\ II\ G$$

The especially high affinity to G of the M II-state
becomes evident by the fact that the M I-M II equilibrium is shifted - at low rhodopsin bleaching - by
the protein binding reaction towards higher M II concentrations (4,6,7). After further bleaching (⩾ 10%)
excess G-protein is no longer present and the rhodopsin
from further flashes runs into the normal M I-M II
equilibrium. The equilibrium shift can therefore be
used as another indicator of binding. However, com-
parative kinetic and stoichiometric analysis has shown
(4) that the equilibrium shift monitors directly the
binding of G, whereas the scattering transients represent structural changes which are subsequent to binding
and not the binding reaction itself. It is our hope
that a physical investigation of the scattering changes
will allow to describe the underlying processes and the
enzymatic mechanism.

Principle of the method

     Kinetic light scattering and absorption photo-
metry are analogous to each other (8). In both methods
the sample suspension is irradiated by a beam of monochromatic light. In contrast to absorption photometry,
in kinetic light scattering not the light passing
through the sample but the scattered light intensity
is measured. The scattering angle is a measuring parameter whereas the wavelength is chosen outside the
absorption bands of the sample (fig. 1).
In analogy to the absorption spectrum which gives information about the photochemical state of the sample,
the scattering curve is characterized by structural
properties of the scattering particles in the sample
suspension. The scattering changes can be described in
form of difference scattering curves $\Delta I\ (\theta)/I(\theta)$ in
analogy to difference spectra. Although a general interpretation of the scattering curve is often not
possible for complex biological structures, the additional kinetic information contained in the triggered

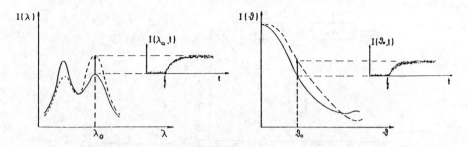

Fig.1 Analogy of kinetic absorption and
light-scattering photometry.
Left: absorption spectrum; right: scattering
curve. In both methods, transitions are measured
in form of intensity changes.

scattering changes considerably reduces the interpretation problems.

Current apparative and theoretical approach

        Hofmann and Emeis (8) developed a measuring de-
vice which allows simultaneous time resolved measurements of flash induced absorption and scattering
changes on the same sample. This device is shown
schematically in fig.2a. It consists of a fast (20μs)
two wavelength double beam absorption photometer with
semiconductor detectors and a scattering photometer
using a near infrared measuring beam. The nearly parallel measuring light comes from a NIR emitting diode.
The scattered light is collected by two Fresnel lenses
and focused on a semiconductor photodiode. Using
different diaphragms between both Fresnel lenses the
scattering angle is selected. The scattering photometer has an amplitude resolution of $\Delta I\,(\theta)/I(\theta) =
2\times 10^{-5}$ (single flash).
        In view of the labile samples scattering signals
at different scattering angles can be compared much
more reliably when measured simultaneously on the
same sample. Therefore a second light scattering
photometer was developed (fig.2b) which monitors with
several detector arrays the scattered intensity at
different angles at the same time (9).
A parallel near infrared beam is radiated on the entrance of a half spherical cuvette. The scattered light
is focused by the optical properties of the cuvette on

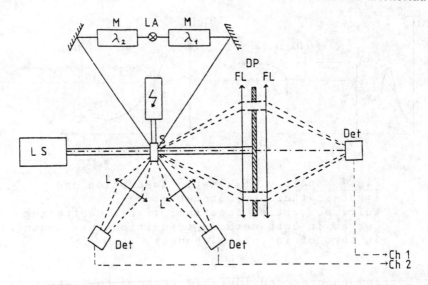

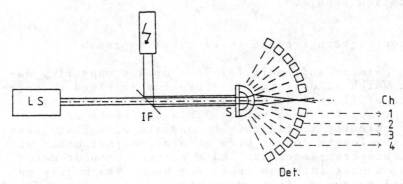

Fig.2 Simplified schematic drawing of the optics for
time-resolved measurement of:
2a) Simultaneous absorption and scattering changes
    (after (8))
2b) Simultaneous scattering changes at several angles
    (after (9))
Symbols are: LA: lamp;   M: monochromator; S:
sample; Det: detector; ⚡ : flash lamp with collimator and blocking filters; LS: light source, light
emitting diode (λ=880nm) with collimator; FL: Fresnel
lenses; DP: diaphragm; IF: interference filter
used as spectrally dependent mirror.

photodiode arrays arranged centrosymmetrically around
the sample. The different circular detector arrays
corresponding to certain scattering angles produce
parallel photocurrents which are amplified and recorded. It is possible to measure at 4 scattering
angles with an amplitude resolution of $\Delta I(\theta)/I(\theta) =$
5×10^{-4} and a time resolution of 35 µs.

An interpretation of the difference scattering
curves was attempted on the basis of the classical
Raleigh-Gans approximation (10). This theory neglects
the phase shift of waves passing through the particle
due to the higher refractive index of the particle
compared to the surrounding medium. This approximation
is valid for particles with small refractive index
and sizes ($< \lambda/2$).
For investigations on isolated disc vesicles this
approximation can be applied in the whole angular
range. For rod outer segments which are much longer
than the wavelength but have a rather small refractive
index this approximation is still valid for small
scattering angles. More reliable results are obtained
on magnetically oriented rod outer segments with incident light normal to the axis (10). For such oriented
circular cylinders exact calculations based on the
Maxwell equations can be made (11) and compared with
the Raleigh-Gans approximation (12). These calculations
show that the scattering of particles with size and
refractive index of rod outer segments is correctly
described in the angular range $\theta \leqq 40°$.

For unpolarized incident light, the Raleigh-Gans
approximation writes the scattering intensity as a
product of two terms (11)

$$I = (1+\cos^2\theta)\ I_\alpha\ P(\theta)$$

The first term I_α depends on the polarisability of the
particle. $(1 + \cos^2\theta)I_\alpha$ is the sum of the scattering
of the Raleigh dipoles contained in the particle
where I_α is independent of the scattering angle. The
term $P^\alpha(\theta)$, particle scattering function, describes
the intraparticle interference of the scattered light
and contains geometry and orientation of the scattering
mass within the particle. Because no destructive
interference of the scattered light occurs in the
forward direction, $P(\theta=0)$ is equal to unity independent of the particle shape.

According to these terms one can distinguish
between two principally different light scattering
transformations. One of them is due to changes of the
polarisability caused by changes of the individual

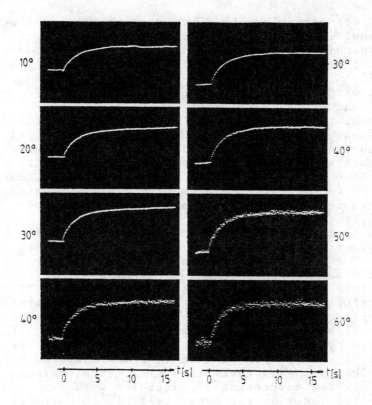

Fig.3 Two simultaneous four-angle records
of the signal P in isolated disc membranes.
The two measurements at 30° and 40° fit to
each other without normalization.

Raleigh dipoles. The other one belonging to the
second term includes particle swelling shrinkage
reorientation or any other shift of the scattering mass
or parts thereof.
These two scattering contributions are rather typical
in the angular dependence and may be distinguished
from each other.
Convenient feature is the behaviour at small angles
where for a shape change the scattering difference
approximates to zero whereas it remains constant for
a change in polarisability.
Quantitative approximations start from numerical calculations of the difference scattering curves. Analytical solutions for the particle scattering functions
$P(\theta)$ of several particle shapes are found in literature.

The physical interpretation of the structural processes involved in the P-signal in rod outer segments are still under investigation in different laboratories. There is now general agreement that this signal is due to complex gross-structural shifts of the scattering mass in axial and radial direction with different kinetics. A change of particle polarisability is not seen in the P-signal but in a slower light scattering transient, the so-called Ps-signal (10).

In washed disc membranes which no longer form membrane stacks, after recombination with the peripheral proteins a P-termed P_D-is observed, which is much slower than the triggering binding reaction (4). Two records of this signal at four angles each (using the device fig. 2b) is shown on fig.3. The same relative intensity change is observed at all angles indicating a homogeneous polarisability change without essential gross-structural contributions.

References

1. Wald, G. 1968, Nature 219, pp.800-807.
2. Abrahamson, E.B. 1983, this volume, p. 385.
3. Kühn, H., Bennett, N., Michel-Villaz, M. and Chabre, M. 1981, Proc. Natl. Acad. Sci. USA 78, pp.6873-6877.
4. Emeis, D., Kühn, H., Reichert, J. and Hofmann, K.P. 1982, FEBS Lett.143, pp.29-34.
5. Hofmann, K.P., Uhl, R., Hoffmann, W. and Kreutz, W. 1976, Biophys. Struct. Mech.2, pp.61-77.
6. Emeis, D. and Hofmann, K.P. 1981 FEBS Lett.136, pp.201-207.
7. Bennett, N., Michel-Villaz, M. and Kühn, H. 1982, Eur. J. Biochem.127, pp.97-103.
8. Hofmann, K.P. and Emeis, D. 1981, Biophys. Struct. Mech.8, pp.23-34.
9. Schleicher, A. and Hofmann, K.P. 1983, J. Biochem. Biophys. Meth. in press.
10. Hofmann, K.P., Schleicher, A., Emeis, D., and Reichert, J. 1981, Biophys. Struct. Mech.8, pp.67-93.
11. Hulst, H.C. van de 1957, Light scattering by small particles. J. Wiley and Sons, New York; Chapman and Hall, London.
12. Kerker, M. 1969, The scattering of light and other electromagnetic radiation. Academic Press, New York, London.

FOURIER TRANSFORM INFRARED SPECTROSCOPY AS A PROBE OF BIOMEMBRANE STRUCTURE

Henry H. Mantsch

National Research Council Canada, Division of
Chemistry, 100 Sussex Drive, Ottawa, K1A OR6, Canada.

SYNOPSIS

Fourier transform infrared spectroscopy is becoming an
increasingly useful technique for studying the structural
organization and dynamics of cell membranes. An important
feature of this technique is that it can be applied to
biological membrane preparations in a straightforward manner
which does not involve introduction of exogenous probes or
potentially harmful preparative procedures.

INTRODUCTION

Membranes are the most common cellular structures in both
plants and animals. They are now recognized as being involved
in almost all aspects of cellular activity ranging from motility
or food entrapment in simple unicellular organisms, to such
complex functions as energy transduction, immunorecognition, or
biosynthesis in higher organisms. This functional diversity is
based on a corresponding structural diversity, which in turn is
reflected in the wide variety of lipids and proteins that
compose different membranes. Thus, an understanding of the
physical principles that govern the molecular organization of
membranes is essential for an understanding of their
physiological roles since structure and function are strongly
interdependent in membranes. The present commonly accepted view
of biological membranes is embodied in the fluid mosaic model of
Singer and Nicolson which envisages membranes as composed of

547

C. Sandorfy and T. Theophanides (eds.), Spectroscopy of Biological Molecules, 547–561.

proteins incorporated either wholly or partly in a fluid-like
sea of bilayer lipids; the latter provide the structural
framework of the membrane and allow the proteins to move about
more or less freely in the lipid matrix. An excellent
introduction to the subject of cell membranes can be found in
the book edited by Weissmann and Claiborne[1].

THE LIPID MATRIX OF MEMBRANES

Physical studies of biomembranes have concentrated mainly
on the organization and dynamics of the lipid matrix. Perhaps
the best studied property of membranes and lipids is their
thermotropic mesomorphism. This change of state induced by
temperature is reflected in an order-disorder phase transition
of the lipid matrix commonly referred to as the gel to liquid-
crystalline phase transition. This transition has been widely
studied with a variety of physical techniques. Spectroscopic
methods are a good choice for such studies because information
can be obtained about molecular motion and even about small
changes occurring in different molecular moieties; thus
molecular spectroscopy provides a submolecular look at the
membrane phase behavior.

Infrared spectroscopy has been a late addition to the field
of membrane spectroscopic research. The reason behind this late
arrival of infrared as a tool for the membrane biophysicist has
been the presence of water. Biological membranes, like most
biological structures, require an aqueous environment whereby
water is not only the solvent of choice, but often part of the
molecular structure itself. While water does not impair
spectroscopic measurements using the NMR, ESR, UV or Raman
techniques, both water and heavy water are strong infrared
absorbers, a fact which has precluded or severely limited the
application of conventional infrared spectroscopy to the study
of aqueous systems. However, the advent of interferometric
Fourier transform infrared spectroscopy, along with the recent
availability of sophisticated data handling techniques, has
opened up new avenues for the study of aqueous biological
systems, such as natural membranes.

The field of infrared instrumentation has changed
dramatically in the last few years. Fourier-transform infrared
(FT-IR) spectrometers are now widely available including
sophisticated low-cost models which are quite suitable for
studies of lipids and aqueous membrane preparations. The
advantages of FT instruments over conventional dispersive
instruments are well documented and shall not be discussed here;
the interested reader is referred to a number of publications
for an up to date discussion on this subject.[2-4]

Experiments directed towards studying the thermotropic phase behavior of membrane lipids involve collecting the infrared spectrum of the same system at various temperatures below, at and above the various phase transitions and monitoring changes in band parameters as a function of temperature. Variable temperature experiments are dependent on auxiliary apparatus which in general consist of a variable temperature circulating bath and thermostrated cell mount.[5] FT-IR spectra can be easily recorded at a large number of different temperatures, the range and temperature increment being selected according to the particular sample under study. In our experimental setup this process is completely under the control of the spectrometer computer, which records a spectrum, increments the temperature, waits for temperature equilibration, then records another spectrum.[6] The temperature is monitored by a copper-constantan thermocouple located against the edge of the cell windows. The thermocouple output is routed to a printer via a Newport digital pyrometer and the printer is triggered by the spectrometer at the end of each spectral collection, thus providing a complete temperature record.

The FT-IR spectrometer generates an immense amount of data in a very short time which requires adequate data reduction procedures. Due to recently improved data processing techniques changes in infrared frequencies and bandwidths can now be measured to within 0.1 cm^{-1} even when the original spectra are recorded only at 2 cm^{-1} resolution.[7] Furthermore, it takes only a few minutes to obtain a plot of frequency or bandwidth versus temperature for spectra recorded at, say 50 different temperatures. Among other data processing techniques, the band-narrowing procedure (deconvolution) has been particularly useful.[8-10] The procedure removes a known lineshape from the spectra and the resulting deconvoluted spectrum will have narrower bands but the correct integrated intensities and frequencies are retained.

When studying the thermotropic mesomorphism of membrane lipids the sample preparation becomes rather important to ensure reproducible infrared spectra. Artificial membranes are usually prepared using synthetic lipids, or lipids extracted from natural sources. Lipid dispersions for infrared measurements are generally prepared by mixing the desired amounts of solid lipid and solvent (H_2O, D_2O or buffer solution). When using D_2O as solvent a closed vessel should be used in order to minimize exchange with atmospheric H_2O. In order to ensure proper hydration the lipid:water mixture should be heated for a few minutes to some 10°C above the temperature of the gel to liquid crystalline phase transition of the corresponding lipid, and then allowed to cool before commencing the measurements. The necessity of well hydrated samples is paramount when studying

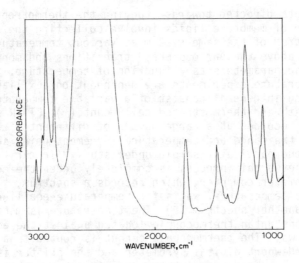

Figure 1. FT-IR spectrum of an aqueous dispersion of human
           erythrocyte phosphatidylethanolamine in excess water
           (D_2O).

phase transitions in lipid assemblies as several phenomena are
dependent on the amount of water associated with the lipid. A
recently discussed case is that of saturated
phosphatidylethanolamines.[11] In the case of saturated
phosphatidylcholines, the so called "subtransition" is known to
involve two lipid phases with different degrees of
hydration,[12,13] which again emphasizes the need for a well
proven hydration procedure for sample preparation.

     The study of natural membranes involves re-hydration only
if the membranes are isolated from the other cell components and
lyophilized for storage[14,15]. The experiments on bacterial
membranes currently under way in our laboratory utilize live
bacteria or isolated membranes which have not been lyophilized,
therefore there is no hydration procedure involved and one less
source of experimental error is introduced.[16]

     For the study of lipid-protein interactions it is necessary
to reconstitute the lipid and the protein under investigation.
Several studies describe methods for achieving proper
reconstitution. In general, care must be taken to ensure that
detergents or other chemical agents used in the purification of
lipids and proteins are completely removed prior to the
formation of the lipid: protein complexes.[17,18] Needless to
say, analytical methods are needed to prove that protein and
lipid are indeed associated. Of these, column chromatography
and electron microscopy are particularly well suited.

The infrared spectra of lipids have been studied in detail and most bands have been assigned.([19-21]) With regard to studies of the polymorphism in membrane lipids, the infrared spectrum can be separated to great advantage into spectral regions which originate from molecular vibrations of different molecular moieties. In this manner we can refer to "head-group" and/or "hydrocarbon tail" spectral regions. For the purposes of illustrating this, Figure 1 shows a typical infrared spectrum of an aqueous membrane preparation, namely that of human erythrocyte phosphatidylethanolamines in the presence of excess water, in this case D_2O.([22]) The vibrations of the acyl chains are readily assigned by comparison with infrared spectra of esters of fatty acids, and other methylene-chain compounds. Carbon-hydrogen stretching vibrations give rise to bands in the 2800-3100 cm^{-1} spectral region, and are in general the strongest bands in the spectra of lipids. The CH_2 antisymmetric and symmetric stretching modes are generally found around 2920 and 2850 cm^{-1} respectively, while vibrational modes due to the terminal CH_3 groups are found around 2960 cm^{-1} (asymmetric) and 2870 cm^{-1} (symmetric). A band at 3010 cm^{-1} arises from the olefinic =C-H stretching vibration, due to acyl chains which contain carbon-carbon double bonds.

The 1350-1500 cm^{-1} spectral region contains the bending vibrations of the methylene and methyl groups. A particularly useful band in this region is the CH_2 scissoring mode around 1470 cm^{-1}, which has been used extensively for the characterization of the acyl chain packing in lipids.([23-26]) The infrared spectra of long methylene chains also show a series of bands in the region 1200-1400 cm^{-1}, whose number and frequency are characteristic of the chain length. The observation of this CH_2 wagging band progression is made difficult because it overlaps the strong band due to the antisymmetric stretching of the PO_4 group.

The CH_2 rocking band progression appears in the 700-1200 cm^{-1} region. The rocking bands are extremely weak and their observation and study is difficult due to overlap with many other vibrational modes. However, the head band of this progression at 720 cm^{-1}, which is the most intense band in this region, can also be used to characterize the nature of the packing of long methylene chains.([27,28])

Figure 2 illustrates the temperature-induced changes in the infrared spectra of the phospholipid 1,2-dipalmitoyl phosphatidylcholine (DPPC). The solid traces show the CH_2 scissoring band (Figure 2A) and the CH_2 rocking band (Figure 2B) at temperatures below the so called "pretransition" at T_p, while the broken traces show the same spectral region at

temperatures above T_p. The splitting of the two bands observed
in the solid traces results from crystal field interaction which
occurs only when the acyl chains are packed in an orthorhombic
subcell with two molecules as shown in Figure 3A. The broken
traces with a single band correspond to a hexagonal crystal
packing with only one molecule per unit cell as shown in Figure
3B. As the splitting of these two vibrational modes arises from
intermolecular vibrational coupling due to the crystal field,
[29] it is temperature dependent; therefore, the frequency
separation of the two band components decreases with increasing
temperature.[12,25,30]

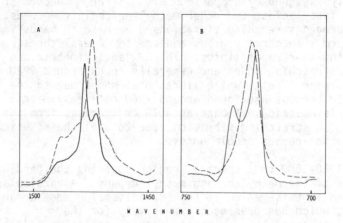

Figure 2. FT-IR absorbance spectra of fully hydrated DPPC
multibilayers at 5°C (solid traces) and at 30°C
(broken traces) for the CH_2 scissoring mode (part A)
and for the CH_2 rocking mode (part B).

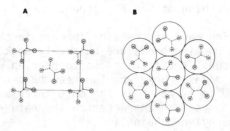

Figure 3. Acyl chain packing patterns showing the orthorhombic
(part A) and hexagonal (part B) crystal-packing. The
long axes of the chains are projecting from the page.
For the hexagonal packing the torsion about the long
axes is such that the chain orientations relative to
each other at a given moment are random.

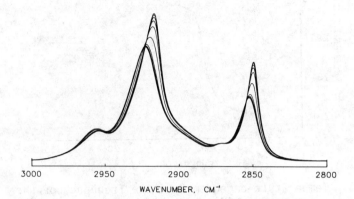

WAVENUMBER, CM⁻¹

Figure 4.   Temperature-induced changes in the region of the C-H
            stretching bands of palmitoyl lysolecithin. There
            are ten individual infrared spectra recorded between
            -4 and 14°C in 2°C intervals; the peak height
            decreases with increasing temperature.

The structural changes occurring in aqueous lipid
dispersions can be studied by monitoring various infrared
absorption bands as a function of temperature. Among the
vibrational modes of major diagnostic value are the methylene
stretching vibrations of the acyl chains. Figure 4 displays a
series of ten infrared spectra of a lysolecithin in the C-H
stretching region at temperatures between -4 and 14°C.[31]
Lysolecithins are natural intermediaries in lipid metabolism and
exist in membranes during phospholipid turnover. The strong
bands at 2920 and 2850 cm⁻¹ which are common to compounds with
long methylene chains are the antisymmetric and symmetric CH_2
stretching modes, respectively; the weaker bands at 2955 and
2872 cm⁻¹ are the asymmetric and symmetric CH_3 stretching modes
of the terminal methyl group. All of these bands exhibit
temperature-dependent variations in frequency and width which
can be related to structural changes at the molecular level.
Illustrated in Figure 5 is a detailed temperature profile
obtained for the frequency of the 2920 cm⁻¹ band in palmitoyl
lysolecithin. The frequency of this vibrational mode changes
slightly with temperature below 0°C and again above 8°C, but
shows a steep increase between 0 and 8°C. The frequency values
at temperatures below 0°C (~2917 cm⁻¹) are characteristic of
conformationally highly ordered acyl chains as found in solid
hydrocarbons, whereas the values at temperatures above 8°C
(~2923 cm⁻¹) are characteristic of conformationally disordered
acyl chains with a high content of gauche conformers such as

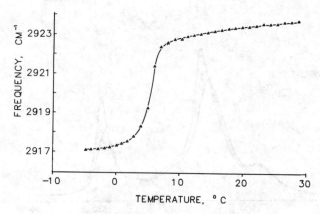

Figure 5.   Temperature-dependence of the frequency of the
             antisymmetric CH_2 stretching band of palmitoyl
             lysolecithin. The frequencies were obtained by
             computing the center of gravity of the 3 top most
             points of the band (8).

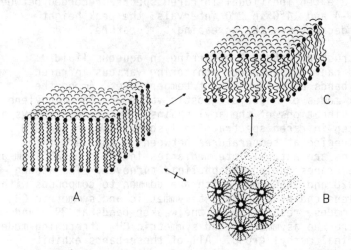

Figure 6.   Pictorial description of the lamellar bilayer
             structures as in the solid-like gel phase (part A)
             and in the liquid-like liquid-crystalline phase (part
             C). T_m indicates the acyl chain melting phase
             transition between structures A and C. Part B shows
             the non-bilayer structure referred to as the inverted
             hexagonal phase consisting of tubular inverted
             micelles. T_h indicates the bilayer to non-bilayer
             phase transition between structures B and C. Both
             phase transitions are reversible, however, structures
             A and B can only interconvert via structure C.

those found in liquid hydrocarbons. The midpoint of the steep
sigmoidal frequency curve can be taken as the midpoint of the
corresponding phase transition. There is ample evidence in the
literature that the frequencies of the CH_2 stretching bands of
acyl chains depend on the degree of conformational disorder and
hence can and have been used to monitor the average trans/gauche
ratio in such systems([12],[25],[30-33]); the higher this frequency
the higher the conformational disorder.

Certain lipids obtained from the egg yolk or from the human
erythrocytes show not only the characteristic gel to liquid
crystalline phase transition, but also undergo a more dramatic
structural re-arrangement to a non-bilayer assembly, as
indicated in Figure 6C. The structural changes occurring at
this thermally-induced phase transition have also been
characterized by FT-IR spectroscopy.([22],[32])

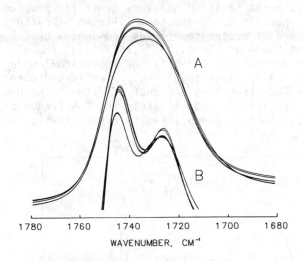

WAVENUMBER, CM⁻¹

Figure 7.   FT-IR spectra of 1,2-dipalmitoyl
            phosphatidylsulfocholine in the C=O stretching region
            at temperatures above and below Tm before (part A)
            and after (part B) Fourier self deconvolution with a
            Lorentzian line of 20 cm⁻¹ half width which results
            in a reduction of the bandwidth by a factor of 2.5.
            The midpoint of the C=O stretching band contour in
            part A is at a higher frequency in the gel phase
            compared to that in the liquid crystalline phase. In
            fact, as shown by the spectra in part B there are two
            component bands which do not change in frequency,
            however, the intensity of the high frequency band at
            1743 cm⁻¹ decreases above Tm, while that of the low
            frequency band at 1726 cm⁻¹ increases slightly.

Vibrational modes of the moieties found in the head-group of phospholipids and other membrane lipids give rise to strong infrared bands such as the ester carbonyl vibrations. The C=O stretching modes between 1700-1745 cm^{-1} are by far the most intense of the head-group bands. This region consists of at least two bands, in principle originating from the two C=O groups in the molecule([34]) however, in general, in hydrated samples only a broad band contour is found.([24],[32]) When resolution enhancement techniques are applied, which reduce the intrinsic width of individual bands, the two component bands become clearly visible as shown in Figure 7 for the case of phosphatidylsulfocholine bilayers.([30])

NATURAL BIOMEMBRANES

The complexity of natural membranes and studies involving addition of substances to lipid bilayers generally precludes the observation of specific bands due to well-known vibrations. In such cases the use of deuterated lipids provides a useful means of circumventing these problems.([14],[15],[18],[35-37]) The effect of deuteration, in general, is to shift the vibrational mode to a lower frequency. Vibrations involving primarily hydrogen motions such as C-H stretching are most affected. The C-D stretching frequencies are approximately the C-H frequencies divided by $(2)^{\frac{1}{2}}$. This is exemplified in Figure 8 which shows

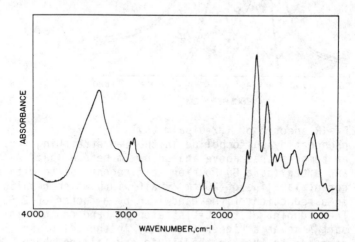

Figure 8.   FT-IR spectrum of the bacterial plasma membrane of
            Acholeplasma laidlawii B grown at 30°C in the
            presence of perdeuteromyristic acid.

the infrared spectrum of whole membranes of an organism which was grown on a diet containing deuterium labelled fatty acids. The spectrum contains a variety of bands from lipids and proteins as well. However the temperature-dependent C-D stretching bands of the acyl chains occur near 2200 cm^{-1}, a region free from interfering bands. Since the deuterated fatty acid is biosynthetically incorporated as a lipid acyl chain, these bands provide a specific probe of the membrane. Similar to the CH$_2$ stretching modes, the frequencies of the CD$_2$ stretching bands of acyl chains depend on the degree of conformational disorder and hence can be used to monitor the average trans-gauche ratio in the lipid ensemble. A frequency of 2090 cm^{-1} for the symmetric CD$_2$ stretching band is characteristic of a highly ordered gel phase with a low gauche population. Higher frequencies, the upper limit being about 2096 cm^{-1}, are indicative of an increased population of gauche rotamers, that is, a transition to the liquid crystalline phase([16],[38]). However, the relation between frequency and conformational disorder is not linear. This necessitates the use of a simple two-component overlapping band model to obtain the proportions of gel and liquid crystal phases.([18])

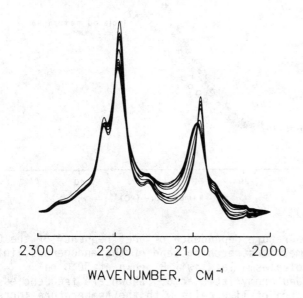

WAVENUMBER, CM^{-1}

Figure 9.   Temperature-dependence of the carbon-deuterium stretching bands of the infrared spectra of deuterium enriched plasma membranes of Acholeplasma laidlawii B.

Figure 9 shows the region of the C-D stretching bands as a function of temperature. The bands around 2194 and 2090 cm^{-1} are due to antisymmetric and symmetric CD_2 stretching vibrations respectively; these modes are the strongest features in this region and have been used to study the thermotropic phase behavior of intact biomembranes[14,15] as well as the temperature-response of deuterium-labelled plasma membranes in live bacteria.[16]

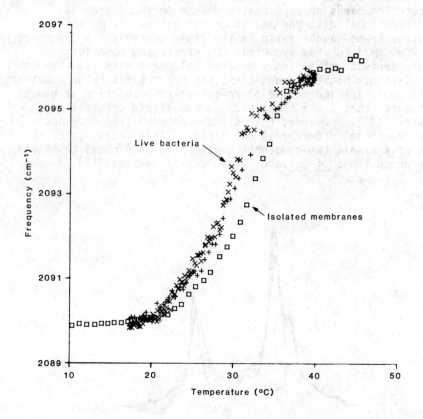

Figure 10. Temperature dependence of the frequency of the CD_2 symmetric stretching band of the endogenous lipids of Acholeplasma laidlawii B grown at 30°C on perdeuteromyristic acid. Shown are frequencies from spectra of live cells with the temperature increasing from 20 to 39°C (+) or decreasing from 39 to 16°C (x) and frequencies from spectra of isolated membranes with the temperature increasing from 5 to 45°C ( ).

Figure 10 shows temperature plotted against frequency of the CD_2 symmetric stretching band in the spectra of the live cells and the isolated membranes. Below 20°C, membranes of live cells are in the gel phase; between 20 and 34°C they undergo a transition to the liquid crystalline phase. On cooling the system exhibits slight hysteresis but reverts to the gel phase at low temperature. At the growth temperature (30°C) the frequency is 2093.3 cm^{-1}. Use of the two-component model indicates that at least 50 percent of the lipids are in the liquid crystalline phase at this temperature.

The isolated membranes also undergo a gel to liquid crystal phase transition. However, while the widths of the transitions are the same for isolated membranes and live bacteria, the transition of the former occurs at a temperature about 4°C higher. At the growth temperature the frequency of the CD_2 stretching band in the spectrum of the isolated membranes is 2092 cm^{-1}. That is, the liquid crystalline phase content is only about 20 percent, as compared to the 50 percent content of the live cell membranes.

These data indicate that, at the growth temperature, the isolated bacterial plasma membranes of Acholeplasma laidlawii B are highly ordered, and that only a small proportion of their lipids is in the liquid crystalline phase. However, in the membranes of live Acholeplasma laidlawii B a much higher proportion of the lipids is in the liquid crystalline phase at the growth temperature. Hence, data on conformational order obtained from isolated membranes are not always directly applicable to the live microorganism.

REFERENCES

1. G. Weissmann and R. Claiborne (Editors) "Cell Membranes", HP
    Publishing Co., Inc. New York, N.Y. 1975 (283 pp).
2. A.G. Marshall (Editor) "Fourier, Hadamard and Hilbert
    Transforms", Plenum Press, N.Y. 1982 (562 pp).
3. P.R. Griffiths (Editor) "Transform Techniques in Chemistry",
    Plenum Press, N.Y. 1978 (400 pp).
4. P.R. Griffiths, "Chemical Infrared Fourier Transform
    Spectroscopy", Wiley and Sons, N.Y. 1975 (340 pp).
5. D.G. Cameron and R.N. Jones (1981) Applied Spectrosc., 35,
    pp. 448.
6. D.G. Cameron and G.M. Charette (1981) Applied Spectrosc.,
    35, pp. 224-225.
7. D.G. Cameron, J.K. Kauppinen, D.J. Moffatt and H.H. Mantsch
    (1982) Applied Spectrosc., 36, pp. 245-250.
8. J.K. Kauppinen, D.J. Moffatt, H.H. Mantsch and D.G. Cameron
    (1981) Applied Spectrosc., 35, pp. 271-276.
9. J.K. Kauppinen, D.J. Moffatt, D.G. Cameron and H.H. Mantsch
    (1981) Applied Optics, 20, pp. 1866-1879.
10. J.K. Kauppinen, D.J. Moffatt, H.H. Mantsch and D.G. Cameron
    (1981) Analytical Chemistry, 53, pp. 1454-1457.
11. H.H. Mantsch, S.C. Hsi, K.W. Butler and D.G. Cameron (1983)
    Biochim. Biophys. Acta, 728, pp. 325-330.
12. D.G. Cameron and H.H. Mantsch (1982) Biophysical J., 38, pp.
    175-184.
13. J.M. Seddon, K. Harlos and D. Marsh (1983) J. Biol. Chem.,
    258, pp. 3850-3854.
14A H.L. Casal, I.C.P. Smith, D.G. Cameron and H.H. Mantsch
    (1979) Biochim. Biophys. Acta, 550, pp. 145-149.
14B H.L. Casal, D.G. Cameron, I.C.P. Smith and H.H. Mantsch
    (1980) Biochemistry, 19, pp. 444-451.
15. H.L. Casal, D.G. Cameron, H.C. Jarrell, I.C.P. Smith and
    H.H. Mantsch (1982) Chem. Phys. Lipids, 30, pp. 17-26.
16. D.G. Cameron, A. Martin and H.H. Mantsch (1983) Science,
    219, pp. 180-182.
17. R. Mendelsohn, R. Dluhy, J. Taraschi, D.G. Cameron and H.H.
    Mantsch (1981) Biochemistry, 20, pp. 6699-6706.
18. R.A. Dluhy, R. Mendelsohn, H.L. Casal and H.H. Mantsch
    (1983) Biochemistry, 22, pp. 1170-1177.
19. H. Akutsu, Y. Kyogoku, H. Nakahara and K. Fukuda (1975),
    Chem. Phys. Lipids, 15, pp. 222-242.
20. J.E. Fookson and D.H. Wallach (1978) Arch. Biochem.
    Biophys., 189, pp. 195-204.
21. E. Grell (Editor) "Membrane Spectroscopy" Springer Verlag,
    Berlin, 1981 (498 pp.).
22. H.H. Mantsch, A. Martin and D.G. Cameron (1981) Can. J.
    Spectroscopy, 26, pp. 79-84.

23. D.G. Cameron, H.L. Casal, E.F. Gudgin and H.H. Mantsch (1980) Biochim. Biophys. Acta, 596, pp. 463-467.
24. J. Umemura, D.G. Cameron and H.H. Mantsch (1980) Biochim. Biophys. Acta, 602, pp. 33-44.
25. D.G. Cameron, H.L. Casal and H.H. Mantsch (1980) Biochemistry, 19, pp. 3665-3672.
26. D.G. Cameron, E.F. Gudgin and H.H. Mantsch (1981) Biochemistry, 20, pp. 4496-4500.
27. H.H. Mantsch, D.G. Cameron, J. Umemura and H.L. Casal (1980) J. Mol. Structure, 60, pp. 263-268.
28. H.L. Casal, H.H. Mantsch, D.G. Cameron and R.G. Snyder (1982) J. Chem. Phys., 77, pp. 2825-2830.
29. R.G. Snyder (1979) J. Chem. Phys., 71, pp. 3225-3229.
30. H.H. Mantsch, D.G. Cameron, P.A. Tremblay and M. Kates (1982) Biochim. Biophys. Acta, 689, pp. 63-72.
31. H.H. Mantsch, A. Garg and D.G. Cameron (1983) Spectroscopy Int. J., 2, pp. 88-96.
32. H.H. Mantsch, A. Martin and D.G. Cameron (1981) Biochemistry, 20, pp. 3138-3145.
33. C. Huang, J.R. Lapides and I.W. Levin (1982) J. Am. Chem. Soc., 104, pp. 5926-5930.
34. I.W. Levin, E. Mushayakarara and R. Bittman (1982) J. Raman Spectrosc., 13, pp. 231-234.
35. I.C.P. Smith and H.H. Mantsch (1979) Trends Biochem. Sciences, 4, pp. 152-154.
36. S. Sunder, D.G. Cameron, H.L. Casal, Y. Boulanger and H.H. Mantsch (1981) Chem. Phys. Lipids, 28, pp. 137-148.
37. D.G. Cameron, H.L. Casal, H.H. Mantsch, Y. Boulanger and I.C.P. Smith (1981) Biophysical J., 35, pp. 1-16.
38. S. Sunder, D.G. Cameron, H.H. Mantsch and H.J. Bernstein (1978) Can. J. Chem., 56, pp. 2121-2126.

INVESTIGATIONS ON BIOLOGICALLY IMPORTANT HYDROGEN BONDS

C. Sandorfy, R. Buchet and P. Mercier

Département de chimie, Université de Montréal
C.P. 6210, Succ. A, Montréal, Québec CANADA H3C 3V1

Abstract

Anesthetic action appears to be a convenient point of entry into the study of biologically important hydrogen bonds. The possible role that hydrogen bonds play in the mechanisms of anesthesia is briefly outlined. Hydrogen bonding by cholesterol is examined. Hydrogen bond equilibria in nucleotide base pairs and the way in which they can be perturbed are important for studies on cell division and cancer. Spectroscopic investigations relating to these problems are reviewed.

Introduction

Hydrogen bonds are just as widespread and, possibly, just as important in nature as covalent chemical bonds. This is perhaps even more true for the living matter. It would be hard to find a biological system in which H-bonds are not present. This review cannot pretend to covering them all. Rather, we will concentrate on a few observations and phenomena in which the formation and breaking of H-bonds are likely to constitute a step in a series of events leading to a given biological happening and in whose study our laboratory has been involved. They will be treated in three brief sections: A) H-bonding and anesthesia; B) H-bonding by cholesterol; C) H-bonding in nucleotide base pairs.

A) Hydrogen Bonding and Anesthesia

General (inhalation) anesthetics have widely different chemical structures: Xenon, paraffins, ether, chloroform, nitrous oxide, alcohols, etc. all have anesthetic potency. (For

563

reviews see (1-5)) It is remarkable, however, that the most potent ones are halocarbons invariably containing a so-called acidic hydrogen. Examples are chloroform, halothane ($CF_3 CHClBr$), methoxyflurane ($CH_3 OCF_2 CHCl_2$). Furthermore, it is known that anesthetic action does not involve chemical reactions, only subchemical ones, that is changes in molecular associations. This immediately leads to the following assumption: a) since almost every molecule can have at least a weak anesthetic potency and since what every molecule can do is to interact with van der Waals forces, weak anesthetic action should be made possible by van der Waals forces; b) for strong anesthetic action polar interactions are also needed.

Several years ago, an observation was made in our laboratory which indicated that the polar interactions that are involved are likely to be the "breaking" and formation of H-bonds. (6-11) Let us consider a solution of a system containing an O-H...:O-, or N-H...:N, or N-H...:C=O etc. type H-bond. At a given concentration and temperature the infrared spectrum will exhibit a free and an association O-H or N-H stretching band with an intensity ratio corresponding to the given conditions. Normally, if concentration is increased, or if temperature is lowered the intensity of the free band decreases and that of the association band increases. If, however, an anesthetic like chloroform, is added to the solution, the exact opposite occurs: the association band decreases in intensity and the free band increases. This shows that in presence of the anesthetic a part of the H-bonds were dissociated. The generality of this phenomenon has been demonstrated on a number of different H-bonds with a number of different anesthetics by measurements of infrared spectra in both the fundamental and the overtone regions. (6-11) A semi-quantitative relationship was found between anesthetic potency and H-bond breaking ability of anesthetics. (11)

This, if applicable to in vivo conditions, would contradict the current unitary-hydrophobic theory of anesthesia. (1-5) According to the latter all anesthetics exert their action in hydrophobic parts of the lipid membrane and, qualitatively, in the same manner. This idea is based on the Meyer-Overton rule, a good smooth relationship between anesthetic potency and lipid solubility. In reality it only shows, however, that dissolution in lipids is needed to bring the anesthetic on the site of action; the site may be in the ion channel which is in the protein, or the lipid-protein interface in the membrane. We have to remember in this respect that the functioning of the nervous system depends on the permeability of the membrane of the neuron, to ions mainly Na^+ and K^+. The permeability, in turn, depends on the dimensions and other properties of the ion channels which are formed in proteins crossing the lipid membrane. Thus our suggestion amounts to postulating that strong anesthetics perturb the ion channel by direct polar interactions. Recently Urry and his Coworkers (12-14) constructed ion channels of known structure in gramicidine. These can be considered as dimers of peptides where the two monomers are bound together by H-bonds and the gate to the channel is also controlled by H-bonds. Urry has been able to show that if a strong anesthetic is brought to the gates it slows down the rate of their opening and closing by a factor of about a hundred! This is a striking demonstration of the role of H-bonding in the mechanisms of anesthesia.

If potent anesthetics are able to "break" H-bonds this means that another molecular association must be formed to replace it. Supposing that the H-bond was of the O-H...:O-type (like in a water dimer) or of the N-H...:O=C type (like in a formamide dimer), we have the following equilibria:

$$H_2 O...H_2 O+A \overset{K_T}{\rightleftharpoons} H_2 O...A+H_2 O$$

or

$$H_2 NCHO...HNHCHO+A \overset{K_T}{\rightleftharpoons} H_2 NCHO...A+H_2 NCHO$$

The equilibrium constant K_T is linked to the Gibbs free energy by:

$$\Delta G_T^o = -RT \ln K_T$$

where A means anesthetic, chloroform in our case. In two theoretical papers Hobza, Mulder, and Sandorfy (15,16) computed ΔG_T for both equilibria by ab initio quantum chemical and statistical thermodynamic methods. The total energies for all the systems on the two sides of the equilibria were calculated using the 4-31G basis set. The van der Waals contribution was calculated by a multiple expansion method and was added to the SCF energies. The vibrational frequencies needed to compute the partition fucntions were taken from experimental spectra or from a Wilson FG analysis. Details and justification for the various approximations that have been sued were given. (15,16) The results clearly indicate that the equilibrium is in favor of the water-chloroform or formamide-chloroform complex, that is under the effect of chloroform large fractions of the water-water or formamide-formamide H-bonds are dissociated. In other words the competitive interaction that makes a part of the original H-bonds break is the formation of another H-bond, of the C-H...O type in this case.

The case of the weak anesthetics is different. In most cases these cannot be expected to open H-bonds. (But see (15)) They can, however, perturb the ion channel by van der Waals interactions directly or indirectly, through the lipid. For chlorinated and brominated anesthetics, infrared spectra indicate that the interacting part of these is the halogene. The extent of the interaction can be monitored through the C-Cl stretching frequency. (11)

The whole range of intermolecular interactions has to be

invoked in order to understand anesthesia. H-bonds are needed
for the case of strong anesthetics.

B) Hydrogen Bonding Involving Cholesterol

Cholesterol is a major constituent of cell membranes and
plays a decisive role in determining their structure and
fluidity. The main constituent of membranes are phospholipids;
they form bilayers so that their polar heads are in contact with
an aqueous phase and their hydrophobic parts point towards the
interior of the cell. In-between there is a so-called hydrogen
belt which in most cases contains an ester (or amide) carbonyl
group and a number of water molecules. (17)

How is cholesterol bound in the lipids? In general, it is
believed that it is held by non-polar interactions between the
hydrocarbon chains. Chemical intuition, however, makes one
wonder for what purpose does cholesterol contain a hydroxyl
group? Unless severe steric hindrance prevents it from doing so,
the OH group must be H-bonded to either a carbonyl or to water
molecules. (18)

In order to explore these conditions the infrared spectra
of cholesterol and a number of saturated and unsaturated esters
of different chain lengths have been recorded. Cholesterol has
a strong tendency to self-association, just as any other
alcohol. (19) At a concentration of 0.2M in carbontetrachloride
solution we see a free band at 3622 cm^{-1} and a broad band
centered at 3340 cm^{-1}, typical of H-bonded multimers. a
shoulder at about 3470 cm^{-1} belongs to the H-bonded
dimer.Despite this strong tendency to self-association by
O-H...O-H...O-H... type H-bond formation, in a solution of 0.2M
cholesterol and 0.2M ethyl caprylate (or any other ester) a well
pronounced shoulder appears at 3530 cm^{-1} and the intensity of
the multimer band of cholesterol decreases considerably. At
higher ester concentrations the shoulder at 3530 cm^{-1} becomes a
band. This band can only belong to O-H...O=C H-bonds.
Despite their weakness, these H-bonds have considerable
influence on the association equilibrium involving cholesterol.
A carbonyl band at 1715 cm^{-1} corresponds to this species, while
the unassociated ester carbonyl gives a band at 1735 cm^{-1}.
These alkyl ester-cholesterol mixtures can be considered as
simplified models for the study of phopholipid-cholesterol
H-bonding.

Amide-cholesterol mixtures, on the other hand, can be
considered as models for sphingolipid-cholesterol H-bonding.
Nerve cells contain a large amount of sphingolipids. The
changes observed in the infrared spectra are very similar to
those encountered with esters showing that H-bonds are formed
between the OH group of cholesterol and the amide carbonyl group
of sphingolipids.

When an anesthetic, like chloroform or halothane is added
to the solutions containing ester-cholesterol mixtures both the
3340 cm^{-1} and 3530 cm^{-1} bands decrease in intensity and the free

band increases. This shows that halocarbon anesthetics can open both the OH...OH...OH... and O-H...O=C   H-bonds. The same is observed for amide-cholesterol mixtures. So it is seen that halocarbon anesthetics (containing the acidic hydrogen) can interfere with the associations of cholesterol with the constituents of the lipid membrane.

The simple models which have been sued are remote from actual in vivo conditions. The H-bonds which are involved are, however, the same type and of very similar strength. Now, it has been reported in several instances that anesthetics increase the fluidity of the cell membrane. Whether this is connected with the mechanism (20-22) of anesthesia or is just a phenomenon accompanying it, is not clear. It seems to be likely, however, that fluidization is due, at least in part, to the interference by anesthetics with the molecular associations which determine the position and distribution of cholesterol in the membrane. The H-bonds that are affected might be either cholesterol-carbonyl or cholesterol-water or both. (For a different view on this problem see (23) and (24).)

C) Hydrogen Bonding in the Nucleotide Base Pairs

Cell division, either normal or abnormal (cancer) is bound to begin with the "opening" of H-bonds linking the nucleotides together. During our work on anesthetics we encountered several molecules that are capable of "opening" H-bonds. It was tempting to inquire if they have an effect on nucleotide base pairs.

As a first model Buchet (25) studied the 1-cyclohexyluracil/9-ethyladenine (CU/EA) base pair by means of both IR and proton NMR spectroscopy. This pair has advantages from the point of view of solubility in organic solvents. Although this pair does not occur in nature, it can be considered as a reasonable model for the adenine-thymine pair. In the IR spectrum of the CU/EA pair the bands at 3537 and 3416 cm^{-1} are the free asymmetrical and symmetrical NH_2 stretching vibrations of EA. The corresponding association (H-bonded)bands are at 3485 and 3312 cm^{-1} respectively. However, a part of the 3416 cm^{-1} band is due to the stretching vibration of the free secondary NH group of CU. It is readily seen in the IR spectrum that the free bands increase in intensity while the association bands decrease when the proportion of chloroform increases in the solvent, a mixture of CCl_4 and $CDCl_3$. This clearly illustrates the fact that an important fraction of the H-bonds in the CU/EA pair has been dissociated by chloroform.

The next model, closer to biological reality, was the 8-bromo2',3',5'-tri-O-acetyl adenosine/1-methylthyine pair, a Watson-Crick type pair. (26) The H-bond breakers were barbiturates: phenobarbital, secobarbital, allobarbital, pentobarbital, and barbital. The result was that a large fraction of the adenine-thymine H-bonds were dissociated, the

barbiturate replacing thymine in the base pair. Infrared,
near-infrared (νNH overtones) and proton NMR spectra were used
to demonstrate the effect of the barbiturates on the H-bonds and
the equilibrium constants were determined in many cases. (26)

While the adenine-thymine base pair appears to be very
vulnerable to attack by H-bond breakers of both the halocarbon
and the barbiturate type, under the same conditions the
guanine-cytosine pair was not affected. This leads to the
proposal that if such H-bond breakers are able to penetrate to
molecular distance from the base pairs, "opening" of some of the
A-T H-bonds will occur. This then might cause irregular cell
division and cancer.

Under in vivo conditions the helix is surrounded by a
water layer and by a helix of metal ions. (27,28) Now, positive
ions, like Na^+ion, are expected to be more powerful H-bond
breakers than halocarbons and barbiturates. In their quantum
chemical study Hobza and Sandorfy (29) investigated the
equilibrium related to the opening of an adenine thymine H-bond
by a Na^+ ions. The computed equilibrium constants and Gibbs
free-energies actually show that Na^+ ions can easily distroy a
H-bond in an A-T pair but that even one water molecu'e between
the ion and the H-bond provides complete protection to the
latter. If, however, a carcinogen like, for example, an
aromatic hydrocarbon perturbs the water structure around the DNA
helix this can make it possible for Na^+ to move sufficiently
close to the H-bonds so that dissociation can occur.

The fact that halocarbons, barbiturates and positive ions
can dissociate H-bonds in DNA is to be taken into consideration
in studies on cell division, cancer, mutations and seems to us
of general importance.

Concluding remarks.

Hydrogen bonds are of general occurrence in the living
matter; their importance cannot be exaggerated. The H-bond
systems treated in this short review represent only a fraction
of them. H-bonds are important in pigments. For example, the
chromophore of vision, the Schiff-base of 11-cis-retinal is
usually assumed to be protonated. (Cf.(30,31)) This is needed,
in order to explain that while, in vitro, the chromophore
absorbs at 370nm, in the actual (human or bovine) pigment, the
absorption maximum is at about 500nm. The proton is also needed
in order to interpret proton transfer which has observed to be a
part of the photochemical event of vision. (32)(33) Now, if a
C=N group is protonated, the proton remains H-bonded to the
"original" proton donor in all stable systems. Thus instead of
stating that the Schiff base is protonated or is not, we have to
inquire about the shape of the potential that governs the
motions of the proton in the $C=N...H^+...X$ bridge. X, the proton
donor is very probably aspartic or glutamic acid. (34) Since
the pK_a value of these acids is about 4 we are in the

situation of the "hesitating proton": the Schiff base may or
may not be protonated and the potential in the bridge might well
be double well. The possible consequences of this on the
observed Raman spectra have been recently explored by Leclercq
et al. (35) This "H-bond of vision", being formed between
charged species must be a strong H-bond. It can presumed to
have appreciable stabilizing effect on rhodopsin. While the
weaker H-bonds are of more frequent occurence in living
organisms, such strong interionic H-bonds (50-100 kJM^{-1}) must be
formed between water and phosphate or ammonium or carboxylate
ions and be of considerable importance for the stability of
lipids and, possibly, proteins.

Proton relay networks received a great deal of attention
in recent years. They would require a separate review. (36,37)

It should be emphasized that many important events in the
mechanisms of life occur through changes in the patterns of
molecular associations. These can be as important as changes
requiring chemical reactions. H-bonds occupy an especially
important place among the molecular associations that are
involved with the mechanisms of life.

References

1.   K.W. Miller and E.B. Smith in "A Guide to Molecular
       Pharmacology-Toxicology", Part 2, R.M. Featherstone, Ed.,
       Marcel Dekker, New York, 1973, pp. 427-475.
2.   M.J. Halsey in "Anesthetic Uptake and Action", E.I. Eger,
       Ed., Williams and Wilkins, Baltimore, 1974, pp 45-76.
3.   J.C. Miller and K.W. Miller in "Physiological and
       Pharmacological Biochemistry", H.K.F. Blaschko, Ed.,
       Butterworths, London, 1975, pp 33-75.
4.   R.D. Kaufman, Anesthesiology, 46, 49 (1977).
5.   D.D. Denson, Chemtech., 8, 446 (1977).
6.   T. Di Paolo and C. Sandorfy, J. Med. Chem., 17, 809
       (1974).
7.   T. Di Paolo and C. Sandorfy, Can. J. Chem., 52, 3612
       (1974).
8.   R. Massuda and C. Sandorfy, Can. J. Chem., 55, 3211
       (1977).
9.   C. Sandorfy, Anesthesiology, 48, 357 (1978).
10.   G. Trudeau, K.C. Cole, R. Massuda, and C. Sandorfy, Can.
       J. Chem., 56, 1681 (1978).
11.   G. Trudeau, P. Dupuis, C. Sandorfy, J.M. Dumas, and M.
       Guérin, Top. Curr. Chem., 93 (1980).
12   D.W. Urry, K.U. Prasad, and T.L. Trapane. Proc. Nat.
       Acad. Sci. U.S.A. 79, 390 (1982).
12.   D.W. Urry, Proc. Nat. Acad. Sci. U.S.A., 68, 672 (1971).
13.   D.W. Urry, K.V. Prasad, and T.L. Trapane, Proc. Nat.
       Acad. Sci. U.S.A. 79, 390 (1982).
14.   D.W. Urry, A. Spisni, and M.A. Khaled, Biochem. Biophys.
       Res. Commun., 88, 940 (1979).

15. P. Hobza, F. Mulder, and C. Sandorfy, J. Am. Chem. Soc.,
     103, 1360 (1981).
16. P. Hobza, F. Mulder and C. Sandorfy, J. Am. Chem. Soc.,
     104, 925 (1982).
17. H. Brockerhoff, Lipids, 9, 645 (1974).
18. R.A. Demel and B. de Kruiff, Biochim. Biophys. Acta, 457,
     109 (1976).
19. F.S. Parker and K.R. Bhaskar, Biochemistry, 7, 1286
     (1968).
20. D.D. Shieh, I. Ueda, and H. Eyring in "Molecular
     Mechanisms of Anesthesia", Vol.1, B.R. Fink, ed. Raven
     Press, New York, 1975, pp. 307-12.
21. J.R. Trudell, W.L. Hubbell, and E.N. Cohen, Biochim.
     Biophys. Acta, 291, 321 (1973).
22. C.J. Mastrangelo, J.R. Trudell, N.H. Edmunds, and E.N.
     Cohen, Mol. Pharmacol., 14, 463 (1978).
23. S.F. Bush, H. Levin, and I.W. Levin, Chem. Phys. Lipids,
     27, 101 (1980).
24. S.F. Bush, R.G. Adams, and I.W. Lewin, Biochemistry, 19,
     4429 (1980).
25. R. Buchet and C. Sandorfy, J. Phys. Chem., 87, 275 (1983).
26. R. Buchet and C. Sandorfy, to be published.
27. E. Clementi and G. Corongiu in Biomolecular
     Stereodynamics. Vol.1. R.H. Sarma, ed. Adenine Press, New
     York (1981). pp. 209-259.
28. E. Clementi and G. Corongiu, Biopolymers 21, 763 (1982).
29. P. Hobza and C. Sandorfy, Proc. Nat. Acad. Sci. U.S.A.,
     80, 2859 (1983).
30. B. Honig, Ann. Rev. Phys. Chem., 29, 31 (1978).
31. R.R. Birge, Ann. Rev. Biophys. Bioeng., 10, 315 (1981).
32. V. Sundstrom, P.M. Rentzepis, K. Peters, and M.L.
     Applebury, Nature 267, 645 (1977).
33. M.L. Applebury, K. Peters, and P.M. Rentzepis, Biophys.
     J., 23, 375 (1978).
34. F. Siebert, in "Spectroscopy of Biological Molecules". C.
     Sandorfy   and T. Theophanides, eds. Reidel, Dordrecht,
     1984, p. 347.
35. J.M. Leclercq and C. Sandorfy, Croatica Chim. Acta, 55,
     105 (1982).
36. C.I. Bränden, H. Jörnvall, H. Eklund, and B. Furugren, in
     "The Enzymes". Vol.3, 3rd edition. P.D. Boyer, ed.
     Academic Press, New York, 1975. pp. 323-373.
37. O. Tapia and G. Johannin, J. Chem. Phys. 75, 3624 (1981).

A Review of the Fourier Transform Near Infrared Spectrometer for

the Determination of Non-Metals in Organic Compounds by Atomic Emission

from an Atmospheric Pressure Argon Inductively Coupled Plasma

A.J.J. Schleisman, John A. Graham, R.C. Fry, and W.G. Fateley

Department of Chemistry

Kansas State University

Manhattan, KS 66506, USA

I. ABSTRACT

The application of Fourier transform spectroscopy to argon inductively coupled plasma (ICP) atomic emissions of nonmetals, e.g. carbon, hydrogen, nitrogen, oxygen, fluorine, chlorine, bromine, and sulfur, is reported in the spectral region from 15,800 to 10,000 cm^{-1} (633 - 1000 nm). This type of Fourier transform near infrared (FT-NIR) investigation is potentially a very useful technique for the simultaneous monitoring of ICP excited atomic carbon [C(I)] emissions around 10,995 cm^{-1} and 10,632 cm^{-1} (909.5 nm), atomic hydrogen [H(I)] emission at 15,239 cm^{-1} (656.2 nm), and atomic sulfur (S(I)) emissions around 10,855 cm^{-1} (921.2 nm) deriving from organic molecules atomized in the hot ICP. Within the spectral region and reaction conditions used here, the ratio of atomic carbon to atomic hydrogen emission intensity (and got that matter other nonmetallic elements) is measured simultaneously from an interferometric scan. This gives potentially a very important technique for the determination of the chemical formula stoichiometry yet independent of the total number of hydrocarbon molecules originally injected. The technique appears to be capable of stoichiometry evaluating the empirical chemical formula of unknown organic compounds atomized in the ICP.

C. Sandorfy and T. Theophanides (eds.), Spectroscopy of Biological Molecules, 571–586.

II.   INTRODUCTION

Our previous work has reported the use of dispersive grating instruments for quantitative submicrogram to nanogram detection of the elements, N, O, F, Cl, Br, S, C, and H (1-8).   Element selective GC-ICP determination of N and O in organic compounds and mixtures has also been reported (4,5).   The present review involves the coupling of Fourier transform near infrared interferometry (FT-NIR) to the argon ICP as an alternative yet simpler means of simultaneously determining a variety of nonmetals of interest in synthetic chemistry (12).

FT-NIR allows a simultaneous multielemental measurement utilizing atomic emission analysis which Horlick, et al. have reported for monitoring flame and plasma excited ultraviolet-visible resonance line emissions of metals (9).   This review reports Fourier transform (FT) studies related to the NIR region of the argon ICP emission spectrum and to FT-NIR:ICP emission determinations of nonmetals pertinent to empirical and molecular formula elucidation in the field of synthetic chemistry.

The present paper reports the practicallity of using FT-NIR spectrometry for the simultaneous monitoring of C, H, O, N, F, Cl, Br and S and other nonmetals atomic emissions excited in the argon ICP.   The near infrared spectral region studied here is from 15,800-10,000 cm^{-1} (633-1000 nm).

II.   EXPERIMENTAL

Apparatus

The commercially available ICP and Fourier transform spectrometer along with experimental operating conditions are described in Table 1.   The light source housing of the IBM-98 instrument was modified so a periscope and a window could be attached (see figure 1).   An intermediate image (I_1) of the plasma was produced at unity magnification on the window in the top plane of the source housing.   The periscope included a 7.7 x 12 cm plane mirror (M_1) located +29 cm from the plasma.   This mirror directed the plasma emissions downward.   The periscope also included a single bi-convex glass lens of 18 cm focal length (f) and of 7.6 cm diameter located directly below M_1 at a total optical distance of +2f (36 cm)

Table I

Experimental System and Operating Conditions

A. Light Source	Plasma Therm ICP 2500 (Kresson, N. J.) with APCS-3 auto power control AMNPS-1 auto matching network
ICP, radio frequency (r.f.) power	Incident: 2.0 kW Reflected: <20 W
Quartz ICP torch	Plasma Therm
Pressure	Atmospheric
Argon flow rates (L/min)	Plasma: 20 Auxiliary: 2.0 Sample: See text
Plasma region viewed	A 10 mm vertical zone centered 5 mm above the load coils except where noted in text
Primary external optics	An f/2.4 mirror and glass lens periscope used to project an intermediate plasma image (I_1) of unity magnification at the top plane of the source housing (see text and figure 1).

B. Fourier Transform Spectrometer	IBM-IR98 interferometer and data system (Danbury, CT)
Entrance aperture	10 mm diameter, circular
Beamsplitter	(SiO_2) Quartz
Mirror velocity	0.118 cm/sec (velocity = 4)
Scan time	approx. 1.0 sec (velocity = 4 and 2 cm^{-1} resolution)
Resolution used	either 2.0 or 0.25 cm^{-1} (see text)
Minimum resolution of the IBM-IR98	0.03 cm^{-1}
Number of scans	1 or 2, unless otherwise noted
Reference laser	He-Ne (15,800 cm^{-1})
Optical reference freq.	15,800 cm^{-1}
Electronic reference freq.	31,600 cm^{-1}
Detector	Silicon photodiode, uncooled
Mode	Single beam (emission)

Table 1 (cont.)

Electronic filters	High pass filter approx. 8800 cm^{-1} Low pass filter approx. 35,000 cm^{-1}
Optical filters	Hoya 25A red filter placed at the image I_1
Effective spectral region	15,800-9,000 cm^{-1}
Spectral region plotted	15,800-10,000 cm^{-1}, unless otherwise noted in text or figure
Apodization	Triangular
Zero fill	none (1)

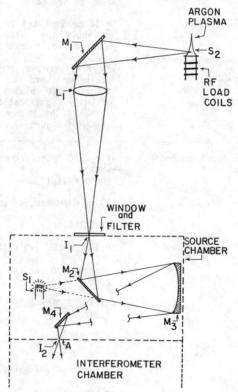

Figure 1. External Optics. Includes IBM-IR98 light source housing with mirros (M_3 and M_4). Also includes periscope (M_1 + L_1 + window and filter + M_2 (operationally removed and installed)) to view the argon plasma emission source (S_2). S_1 is the normal IBM-IR98 Nernst glower source for mid-IR molecular vibrational absorption spectrometry.

from the plasma and -2f from the top plane of the source housing. An additional

5.0 x 3.0 cm front surface plane mirror (M_2) was placed directly under I_1 to

reflect the diverging rays from intermediate image (I_1) toward the normal

focusing mirror (M_3) of the IBM-IR98 source housing. The mirror, M_2, is easily

removed or installed on an operational, reproducible basis by using a mount custom

fitted to the optical platform of the source housing. Figure 1 shows that when

M_2 is in place, the intermediate image (I_1) of the plasma (S_2) replaced the

normal IBM-98 Nernst glower source (S_1) and becomes the object of M_3 which

refocuses a second image (I_2) of the plasma on the normal entrance aperture mask

(A) to the interferometer chamber. M_3, M_4, S_1, and S_2 are now a permanent

part of the system. M_2 is operationally installed or removed to select either

souce (S_2 or S_1) as the need arises.

The limits of the spectral region studied here were determined by the spectral

response limit of the diode detector (about 9000 cm^{-1}) and the frequency of the

He-Ne reference laser (15,800 cm^{-1}). A Hoya 25A red camera filter prevented

radiation approximately 17,000 cm^{-1} from entering the interferometer and an

electronic filter in the data system was set to cut off any frequencies lower than

8,000 cm^{-1}.

Because the spectral range reported here lies between the optical frequency of

the reference laser (15,800 cm^{-1}) and one half that frequency (7,900 cm^{-1}), it

was necessary to take steps to avoid aliasing errors due to a 2-fold undersampling

of the interferometric waveform (11). To reduce the folding problem, the normal

rate of data collection was doubled in order to increase the effective reference

frequency to an equivalent 31,600 cm^{-1}.

Procedure

With the periscope and mirror M_2 installed, the ICP discharge was initiated.

Gaseous samples, e.g. CH_4, SF_6, and H_2, etc., were introduced continuously

into the ICP. The resulting C, H, O, N, Cl, Br, F and S atomic emissions were

simultaneously monitored with the interferometer and data system. This was done at

several different sample introduction rates, mirror velocities, scan distances and

corresponding resolving powers.   Generally one or two interferograms were

collected.   A background emission spectrum of the argon plasma was also recorded
for reference.

III. RESULTS AND DISCUSSION

Full Spectra Simultaneous Monitoring

Figure 2 shows the argon ICP background emission spectrum from 15,800-10,000
cm^1 (633-1000 nm).   This spectrum was obtained in 1 second with the spectral
resolution selected to be 2 cm^{-1}.   Comparison of this spectrum with the argon
emissions listed in reference 10 will indicate that there are about 15 more argon
lines here than one would expect.   These additional lines appear to be a minor
system artifact yet unidentified.   The instrumental origin of the folded lines may
be easily identified as known argon lines if the spectrum is assumed to be folded

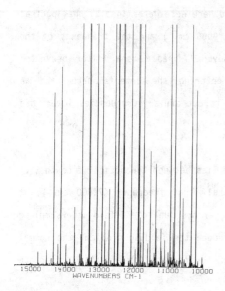 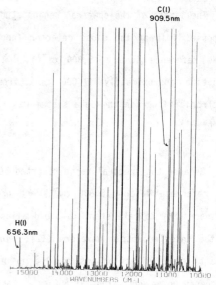

Figure 2.   Argon ICP Background
Emissions from 15,800-10,000 cm^{-1}.
Resolution = 2 cm^{-1}; scan and data
acquisition time = 1 second.

Figure 3.   Atomic Carbon and Hydrogen
Emissions in the ICP. Sample = CH_4;
resolution = 2 cm^{-1}; scan and data
acquisition time = 1 second.

about the following frequencies: 9875, 11,850, and 13,825 cm^{-1}.   The anomalously
folded argon lines observed here are a minor artifact we have encountered and these
lines have not affected our results of and the quantitative interpretation of the

data. That is to say the "extra" argon lines do not overlap any of the nonmetal analytical lines of interest and no spectral distortion has been observed. The extra argon lines therefore appear to be of little or no consequence for this application.

Figure 3 shows the ICP excited spectrum while introducing a sample of CH_4. The main difference from the earlier background spectrum (fig. 2) is that the emission of atomic hydrogen [H(I)] is now clearly visible at 15,239 cm^{-1} (656.2 nm) along with the atomic carbon [C(I)] emissions at 10,995 cm^{-1} (909.5 nm) and 10,632 cm^{-1} (940.5 nm). These elements were simultaneously monitored with a data acquisition time of 2 second and a resolution of 2 cm^{-1}.

Resolution

A. Sulfur Emissions

Figure 4 shows the S(I) emission lines around the 10,855 cm^{-1} (921.2 nm) region of the spectrum collected by the FT-NIR plasma emission system while a sample of SF_6 is continuously introducing during the scan. This spectrum was taken at medium interferometric resolution (0.25 cm^{-1}) with an acquisition time of 10 seconds. The same three atomic sulfur lines are seen in figure 4 as that which we reported earlier with a long focal length (3.4 m) dispersive grating spectrometer (8). The 0.25 cm^{-1} interferometer resolution employed here is similar to that of the 3.4 meter grating instrument used earlier. Under similar experimental condition, the grating instrument was approaching its slit diffraction limit on resolution; whereas, the interferometer used here could yield another order of magnitude improvement in resolution (theoretically to 0.03 cm^{-1}). This additional interferometer resolution was, however, not used because i) the S(I) spectral siimplicity and linewidth do not warrant it and ii) the scan time and data acquisition times would be excessive. In practice the resolutions of 2 to 4 cm^{-1} appear to suffice and will allow much smaller data acquisition times.

B. Carbon Emissions

Figure 3 was a low resolution scan (2 cm^{-1}) in which both H(I) and C(I) emissions is observed. Figure 5 is an expanded scale showing the C(I) emissions in the vicinity of 10,995 cm^{-1} (909.5 nm). This spectrum was taken at medium resolution of 0.25 cm^{-1} and required a longer acquisition time of 10 sec. A

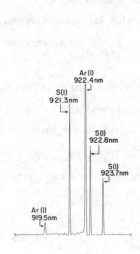

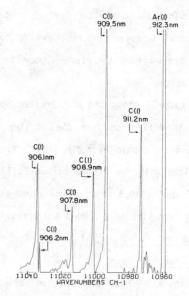

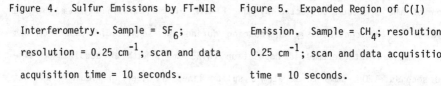

Figure 4. Sulfur Emissions by FT-NIR
Interferometry. Sample = SF$_6$;
resolution = 0.25 cm^{-1}; scan and data
acquisition time = 10 seconds.

Figure 5. Expanded Region of C(I)
Emission. Sample = CH$_4$; resolution =
0.25 cm^{-1}; scan and data acquisition
time = 10 seconds.

number of additional carbon lines are observed with this resolution which give
greater spectral detail. The resolution and spectral carbon features observed here
are similar to those previous observed by dispersive grating spectrometer (8).
From these FT-NIR atomic emission data it is not necessary to use the extreme
resolution capabilities (0.03 cm^{-1}) of the interferometer because i) no further
C(I) spectral features of interest need to be resolved and ii) the acquisition time
would increase unreasonably. In fact, plots of the C(I) emission (see figure 6)
demonstrated that the low resolution case (2 cm^{-1}) is adequate to isolate the
pincipal C(I) emission at 10,995 cm^{-1}) (909.5 nm). Of course, the benefit of
using low resolution is to shortened data the acquisition time for the experiment
(e.g. 1 sec. for one interferogram). This will be important in future testing with
the type of transient sample introduction encountered in chromatographic analysis
when empirical formula data is required. In a combined interferometric-
chromatographic application, the maximum allowed interferometer sampling time would
be about 1 sec. to maintain reasonable chromatographic resolution.

C. Quantitative Analysis

Figure 6 shows that the C(I) emission intensity is related to the amount of CH_4 sample introduced.   Figure 7 shows that H(I) emission inensity is also related to the amount of CH_4 introduced.   These relationships form the preliminary basis for quantitative carbon and hydrogen determination by FT-NIR atomic emission.

1. Formula Stoichiometry

It would be desirable if these C(I) and H(I) emission lines could reflect not only the percent C and H content of the sample, but also serve to directly evaluate the C/H chemical formula stoichiometry ratio of the compound originally introduced. To pursue this important goal, the C(I) and H(I) emissions were measured simultaneously from several samples containing varied amounts of CH_4 atomized in the plasma.

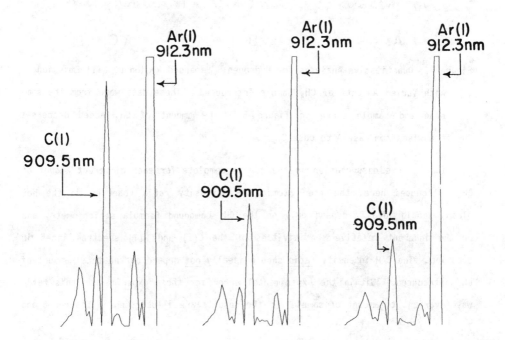

Figure 6. Quantitative Analysis for Carbon: Expanded Region of C(I) Emission with Varied Amounts of CH_4 Sample Introduced.   Resolution = 2 cm^{-1}; scan and data acquisition times = 1 second.   The amount of CH_4 added decreased stepwise from case A to case C.

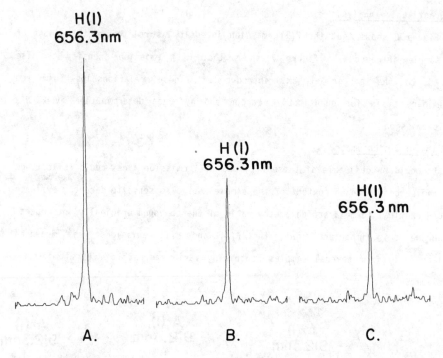

Figure 7. Quantitative Analysis for Hydrogen: Expanded Region of H(I) Emission with Varied Amounts of CH$_4$ Sample Introduced. These data were from the same scan and sample shown in figure 5. The amount of CH$_4$ added decreases stepwise from case A to case C.

If CH$_4$ bond dissociation in the ICP is complete for each different amount of CH$_4$ introduced here, the free atomic C/H intensity ratio observed in the hot plasma should ideally depend only on the CH$_4$ compound formula stoichiometry and on the inherent relative sensitivities of the C(I) and H(I) spectral lines in question. The C/H intensity ratio should ideally not depend on the total amount of CH$_4$ introduced. Within the experimental error for these experiments, this ratio was invariant to amount of sample and the results are illustrated in figures 6 and 7.

Once calibration and normalization for inherent line sensitivities has been performed, relative C/H intensity ratios measured by FT-NIR atomic emission in the ICP therefore appear to be very important in evaluation of the compound formula stoichiometry. The authors deem it fortuitous that this intensity ratio does not depend on the reproducibility, accuracy, or amount of sample introduction. In fact, several different injected amounts of the same compound appear to give an invariant C/H ratio. It therefore appears possible to accurately determine C/H ratios even for uncalibrated amounts of injected sample. This is a result of the simultaneous monitoring of C(I) and H(I) emissions inherent in a Fourier transform interferometric approach.

2. Spectral Emissions of Nonmetals and Argon

Table II lists the spectral emissions (633-1250 nm) detected by the photodiode of this FT-NIR:ICP system for Ar and the seven nonmetals investigated. Relative

Table II

Argon(I) emission lines observed by FTI

Wavelength (nm)	Wavenumber (cm^{-1})
641.631	15585.3
667.728	14976.2
675.283	14808.5
687.129	14553.3
693.767	14414.1
696.543	14356.6
703.025	14224.2
706.873	14146.8
706.723	14149.8
714.704	13991.8
720.698	13875.4
727.293	13749.6
735.332	13599.3
737.212	13564.6
738.398	13542.8
	13526.4
	13491.0
750.387	13326.5
751.465	13307.3
763.511	13097.4
772.376	12947.1
772.421	12946.3
794.818	12581.5
800.616	12490.4
801.479	12476.9
810.369	12340.1
811.531	12322.4
826.452	12099.9
840.821	11893.1
842.465	11869.9
852.144	11735.1
866.974	11534.4
912.297	10961.3
919.464	10875.9
922.450	10840.7
929.158	10762.4
935.422	10690.4
965.778	10354.3

Table II (continued)

Argon(I) emission lines observed by FTI

Wavelength (nm)	Wavenumber (cm^{-1})
978.450	10220.2
1047.005	9551.1
1047.810	9543.7
1050.647	9517.9
1067.355	9369.0
1068.178	9361.7
1088.096	9190.4
1095.074	
1107.887	9026.2
1139.3703	8776.8
1144.1832	8739.9
1146.7545	8720.3
1148.81076	8704.7
1158.0445	8635.2
1166.8709	8569.9
1171.9487	8532.8
1173.3235	8522.8

spectral intensities, uncorrected for instrumental response, were determined with the most intense spectral emission of each element normalized to 100. The spectral range allowed by the system resulted in detection of spectral emissions further into the near infrared (>1,000 nm) than had been possible with our previous photomultiplier-dispersive studies. The larger spectral converage did not reveal any ion lines of the nonmetals or of argon. Nonetheless the greater spectral range did allow the detection of nonmetal emissions previously unreported from the Ar-ICP. Figures 8 and 9 show the energy diagrams. The wavelengths (in Å) denote the first member of a transition series. A detailed listing of all spectral lines

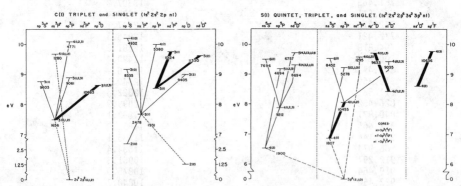

Figure 8. Triplet and Singlet energy   Figure 9. Quintet, Triplet, and Singlet
    diagrams for carbon (I).   energy diagram for Sulfur (I).

observed in this study is given in tables II and III. The dotted lines in figures 8 and 9 indicate some of the more prominent, ICP excited, vacuum UV lines of these elements while the thin solid lines are spectral emissions reported in our previous dispersive studies. The thick solid lines denote spectral emissions unreported

Table III

S(I) emission lines observed by the ICP-FTI

Wavelength (nm)	Wavenumber (cm^{-1})
921.291	10854.3
922.811	10836.5
923.749	10825.5
938.294	10657.6
941.346	10623.1
942.193	10613.5
943.711	10596.5
943.760	10595.9
945.543	10575.9

Cl(I) emission lines observed by the ICP-FTI

Wavelength (nm)	Wavenumber (cm^{-1})
837.597	11938.9
894.801	11175.7
912.112 obscured by an Ar(I) line	10963.6
919.167	10879.4
928.882	10765.6
945.206	10579.7
959.220	10425.1
970.235	10306.8

Br(I) emission lines observed by the ICP-FTI

Wavelength (nm)	Wavenumber (cm^{-1})
827.244	12088.3
844.655	11839.2
847.745	11796.0
863.866	11575.9
869.853	11496.2
881.996	11337.9
882.522	11331.2
889.702	11239.0
896.400	11155.7
916.606	10909.8
917.363	10900.8
917.816	10895.4
926.542	10972.8
932.086	10728.6
989.640	10104.7

Table III (continued)

Hydrogen (I) emission lines observed by ICP-FTI

Wavelength (nm)	Wavenumber (cm^{-1})
656.285	15237.3
656.273	15237.6

Carbon (I) emission lines observed by ICP-FTI

Wavelength (nm)	Wavenumber (cm^{-1})
906.147	11035.7
906.247	11034.5
907.828	11015.3
908.851	11002.9
909.483	10995.3
911.180	10974.8
940.573	10631.8
960.303	10413.4
962.079	10394.2
1012.387	9877.65
1068.308	9360.60
1068.534	9358.62
1069.125	9353.44
1070.733	9339.40
1072.953	9320.07
1133.028	8825.91

N(I) emission lines observed by the ICP-FTI

Wavelength (nm)	Wavenumber (cm^{-1})
746.8309	13389.9
821.6317	12170.9
868.027	11520.4
867.843	11522.8
868.340	11516.2
938.6805	10654.1
939.2789	10646.5

O(I) emission lines observed by the ICP-FTI

Wavelength (nm)	Wavenumber (cm^{-1})
844.625	11839.6
844.636	11839.4
844.676	11838.9
926.267	10796.0
926.277	10795.9
926.594	10792.2
926.601	10792.1

from an ICP. In addition there are several spectral lines of sulfur originating in the ICP at S(I), 941.3, 942.1, 943.7, and 945.5 nm that are unclassified and are reported here for the first time.

3. Summary

Figure 10 and 11 illustrates the quantitative capabilities of FT-NIR:ICP system. Units proportional to moles/sec of carbon or hydrogen are plotted along

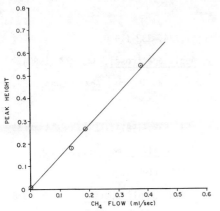

Figure 10. A plot of carbon atom emission at 909.4nm peak height verse CH_4 flow.

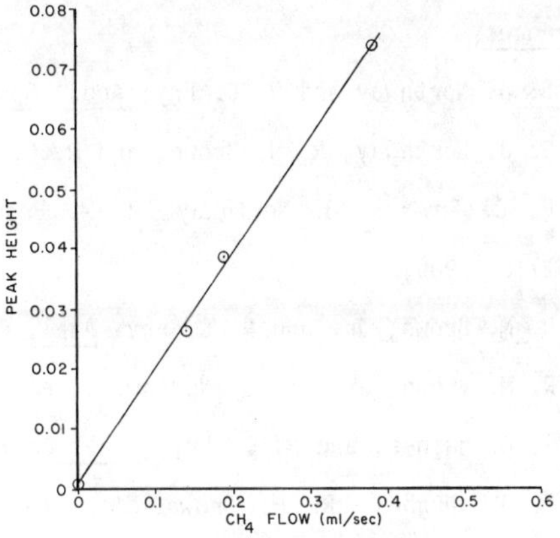

Figure 11. A plot of carbon atom emission at 656.2nm peak height verse CH_4 flow.

the abscissa and the peak height (units proportional to intensity) is potted along the ordinate. An approximate detection limit for carbon originating from methane is 5 micrograms of carbon per second. While an approximate detection limit for hydrogen is 0.07 micrograms of hydrogen per second. The detection limit for hydrogen has been divided by 4 to reflect the stoichiometric ratio of C to H in methand. These detection limits are preliminary and do not represent the minimum attainable values. On first inspection these detection limits seem to be an order of magnitude less sensitive than our disspersive studies; however, the dispersive detection limits were based on 8 microliter gas sampling loops that were purged with argon at a flow rate of 100 ml/min. This suggests that the dispersive detection limits were based on a significantly shorter time basis than the FT-NIR:ICP detection limits. The detection limits for the dispersive studies for H and C were detained by using two different photomultiplier tubes, while a single photodiode (unit gain) simultaneously the emissions of C and H.

Acknowledgment

One of the authors (WGF) wishes to acknowledge the National Science Foundation for partial support of this work through grant number CHE-8109570.

References

1. S. J. Northway and R. C. Fry, Appl. Spectrosc., 34, 332 (1980).

2. S. J. Northway, R. M. Brown, and R. C. Fry, Appl. Spectrosc., 34, 338 (1980).

3. R. C. Fry, S. J. Northway, R. M. Brown, and S. K. Hughes, Anal. Chem., 52, 1716 (1980).

4. R. M. Brown, Jr. and R. C. Fry, Anal. Chem., 53, 532 (1981).

5. R. M. Brown, Jr., S. J. Northway, and R. C. Fry, Anal. Chem., 53, 532 (1981).

6. S. K. Hughes, and R. C. Fry, Anal. Chem., 53, 1111 (1981).

7. S. K. Hughes, R. M. Brown, Jr., and R. C. Fry, Appl. Spectrosc., 35, 396 (1981).

8. Steven K. Hughes and Robert C. Fry, Appl. Spectrosc., 35, 493 (1981).

9. G. Horlick and M. K. Yuen, Anal. Chem., 47, 775A (1975).

10. M. L. Parsons, A. Forster, and D. Anderson, "An Atlas of Spectral Interferences in ICP Spectroscopy," (Plenum, New York, 1980).

11. Robert J. Bell, "Introductory Fourier Transform Spectroscopy," (Academic Press, New York, 1972).

12. A.J.J. Scheisman, W. G. Fateley, and R. C. Fry, J. Phys. Chem, accepted for 1984.

RAMAN MICROPROBE AND MICROSCOPE. TIME RESOLVED RAMAN TECHNIQUES

M. DELHAYE

L.A.S.I.R., C.N.R.S., Université de Lille, C.5
59655 Villeneuve d'Ascq (France) or B.P. 28
94320 Thiais (France).

ABSTRACT

The impact of recent technological improvements on the progress
of Raman spectroscopic techniques is reviewed and illustrated by
examples of applications. After a brief description of the conventional Raman apparatus, the emphasis is put on two recently
opened field of applications :
- Non destructive analysis of minute quantities of various samples by means of the laser-Raman optical microprobes ;
- Time resolved studies of spontaneous Raman spectra in the millisecond to picosecond time domain.

Raman scattering provides direct information about vibrational
spectra of molecules or crystals. This optical process meets the
desirable criteria to accomplish by a non destructive method, the
analysis of a variety of gaseous liquid or solid samples. Raman
spectra offer both qualitative and quantitative accurate data
which complement other physical methods to determine the structure
and conformation of molecules.

1. CONVENTIONAL LASER-RAMAN SPECTROMETERS

A block-diagram of the experimental setup which is used to record
the Raman spectra excited by a laser light source is shown in
Figure 1.

This arrangement corresponds to the most widely used type of
Raman spectrometers, both home-made or commercially available.

C. Sandorfy and T. Theophanides (eds.), Spectroscopy of Biological Molecules, 587–607.
© *1984 by D. Reidel Publishing Company.*

A single photoelectric detector is used to record spectra, by
sequential measurement of the spectral elements, by means of a
grating monochromator, whose mechanical scanning devise is syn-
chronized with a strip chart recorder or a digital data treatment
and storage.

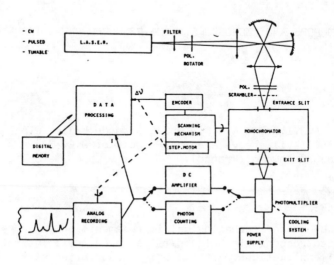

Figure 1. Block diagram of a Raman Spectrometer with single
          channel detection.

By considering the calculation of the light flux which is collected from the scattering sample and transmitted to the photodetector, it is possible to derive expressions of the "figure of merit"
of a Raman spectrometer (Figure 2).

In the grating monochromator, the rays which pass through the
optical system are limited by the area s of the rectangular entrance slit, and by the aperture stop S which limits the size of
the beam falling onto the grating. This aperture stop corresponds
to the projection of the grating G on the collimator mirror M,
placed at the focal distance F_c from the slit. The slit has to be
properly illuminated by the light coming from the sample, so as
to cover entirely the solid angle $\Omega = S/F_c^2$ in order to insure
full spectral resolution by the whole grating area.

This system of aperture stops is coupled by an optical system,
often called "transfer optics" to the scattering sample volume
irradiated by the laser beam. Evidently, the useful rays collected
by the transfer optics are limited by the images, real or virtual,
of the actual stops in the monochromator. There is a simple relationship between the sizes s' and S and separations L of aperture

stops images in various parts of the optical system :

$$\frac{s'S'}{L^2} = \frac{sS}{F_c^2} = U$$

In this expression, areas s, s', S, S' are normal to the optical
axis and L is the distance along the axis. The "lightgrasp" is
entirely defined by U, also called the "Etendue" of the optical
system.

The transfer optics introduces a magnification factor Γ between
the sample and the image projected on the slit. In considering
the sample volume illuminated by the laser beam, the question
arises as to the optimum magnification factor from Γ ≠ 1 corresponding to large sample cells (10 or 20 mm wide) with a small collection angle convenient for accurate polarization measurements,
to Γ > 10 for microsamples, which require a wide angle of collection, with aberration corrected transfer optics at low F number
(F/1 or better).

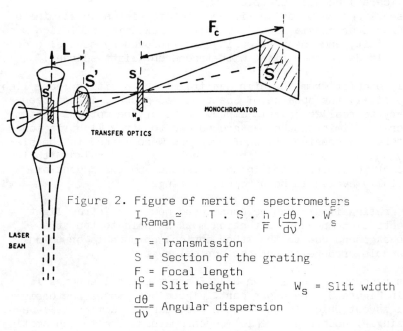

Figure 2. Figure of merit of spectrometers

$$I_{Raman} \simeq T \cdot S \cdot \frac{h}{F} \left(\frac{d\theta}{d\nu}\right) \cdot W_s$$

T = Transmission
S = Section of the grating
F_c = Focal length
h = Slit height W_s = Slit width
$\frac{d\theta}{d\nu}$ = Angular dispersion

Commercially available camera lenses are routinely used as collection and transfer optics in the visible spectrum. They offer excellent correction of aberrations and give sharp enlarged images,
with the low F number which is required to satisfy the above conditions when a high magnification factor is needed.

Ultraviolet excitation of resonant species requires special optical components both for laser focusing and scattered light collection. Mirror arrangements, of the Cassegrain type, or off axis
elliptical mirrors, or UV corrected lenses are required.

2. STRAY LIGHT IN MODERN MONOCHROMATORS

Obviously the basic properties and performances of single and
double monochromators have been studied a long time ago. Calculations and comparison of various Raman systems were reported ten
years before the development of the laser (1) and the first generation of commercial equipment (Cary 81) was perfectly adapted to
the excitation of the Raman scattering by mercury lamps.

The advent of the laser sources in the 60's has modified the situation in two respects :
- A coherent laser beam can be focused to the diffraction limit,
  resulting in extremely high irradiance on a very small volume
  of sample. As a consequence, any attempt to illuminate a large
  height of the entrance slit of the monochromator (i.e. by "image slicers") became obsolete.
- A laser light beam can be filtered out to virtually eliminate
  background or plasma lines, so that the exciting radiation mono-
  chromaticity and spectral purity are improved by several orders
  of magnitude in comparison with incoherent light sources.

To fully benefit from the high quality of the exciting laser light,
and especially the absence of spectral background, it has been
necessary to realize drastic improvements of the optical system
in order to eliminate the background due to stray light in monochromators. Successive steps of progress were marked in the six-
ties by the development of "ghost-free" gratings, ruled with interferometric control, and in the seventies by the advent of
commercially available holographic gratings.

If the grating itself can be considered now as a "stray light
free" optical component, attention must be paid to the stray
light background due to the other optical components or to an
unfavourable arrangement of the optical system.

To this respect, a comparison of different optical schemes is
given in Figure 3. The upper part, Figure 3 A, shows the basic
optical layout which is the most widely used in commercially available single, double or triple monochromators. The monochromatic
images of the entrance slit which are dispersed by the grating,
give rise to the spectrum, focussed in the plane of the exit slit
which isolates a narrow spectral element. Unfortunately, a large
part of the spectrum reaches the walls of the cabinet, and is
scattered,or falls onto the grating again and is re-diffracted.

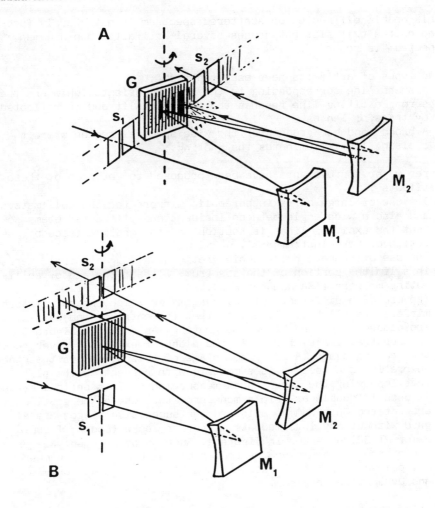

Figure 3. Stray light in monochromator and spectrograph

A. Double diffraction in conventional monochromators

B. The "tilted mirror" arrangement which avoids double
   diffraction

S_1 : entrance slit

S_2 : exit slit

G : plane grating

M_1, M_2 : spherical mirrors

This double diffracted or scattered spectrum is not sharply focused on the exit slit but causes several artifacts when a Raman spectrum is recorded.

Two types of artifacts have been reported :
- A sharp line corresponding to the grating orientation where the strong exciting line reaches the entrance slit and is reflected by transfer lenses.
- A broad band background for the wide spectral region where the exciting line illuminates the grating.

Three major improvements have been proposed to get reed of these artifacts :
- The use of large off-axis parabolic mirrors for the collimators and also a wide angle between incident and diffracted beams so that the exciting lines is kept out of the grating (second generation of Cary instruments).
- The use of concave holographic gratings, giving spectral images in a limited portion of the spectrum, without the aid of collimating mirrors (ISA-Ramanor HG2).
- The use of a different optical systems in which the collimators mirrors are slightly tilted, so that the entrance and exit slit are placed above and below the grating. As shown in Figure 3 B, the spectral images formed in the plane of exit slit never reach the grating again. Another advantage of this system is that the aberration due to coma is symmetric with respect to the slit, thus improving the line shapes measurements. This optical system is used in double or triple monochromators where the gratings are rotated on a common axis and the successive stages are aligned without any intermediate mirror or lense (Coderg T 800, Dilor RT 30) as shown in Figure 4. Such an arrangement has proven extremely stable and reproductible with a minimum number of optical surfaces which results in a high throughput for visible and UV spectral regions.

Testing the stray light figure of Raman spectrometers :

Modern monochromators exhibit very good performance, even for difficult measurements such as very low wavenumbers in the spectra of powered solids. Commercially available double or triple grating spectrometers are proposed with stray light rejection of 10^{-9} to 10^{-12} at $\Delta\nu = 20$ cm^{-1}, for a given slit width.

Attention should be called to the very difficult experimental procedure which is needed to achieve artifact-free stray light measurements at this level : (A factor of 10^{-12} is equivalent to the ratio of the peak height of a typical Raman line as recorded on a strip chart paper, to the radius of the orbit of our planet !) Practical testing requires a comparison of the background level, recorded at a given wavenumber by sending the laser beam focused

on the slit with a wide solid angle to fully illuminate the grating, to the intensity of the laser line itself as measured under
the same conditions, with a series of calibrated neutral filter
to attenuate the laser power. Light scattering by atmospheric air
along the light path in a monochromator is of the same order of
magnitude as the stray light due to optical components.

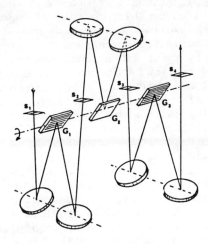

Figure 4. Low stray light triple monochromator with
tilted mirrors (Dilor RT 30)

3. RAMAN MICROSCOPE AND MICROPROBE

In principle the conditions of observation of Raman spectroscopy
are ideally suitable for extending its applications to the domain
of optical microscopy :
- Both excitation and observation are made in the visible or near
  UV range, where high quality optical components, light sources
  and photodetectors are available.
- Diffraction limited focussing and imaging can be performed by
  using the existing microscope objectives, which exhibit resolving power and aberration corrections close to the perfection.
- It is obvious that fluorescence has been widely used as an analytical tool in microscopy. In comparison, Raman spectra should
  provide a much more detailed information on the structure and
  conformation of molecular species observed in microscopic samples.

However, observing the Raman effect seems much more difficult for
evident reasons :
1) Normal Raman scattering is a very weak phenomenon compared to

fluorescence of the molecules used as "fluorescent markers" in
biology.
2) A better spectral resolution is required, both for monochromatic excitation and for scattered light analysis.
3) The irradiation of a small volume of sample by the laser beam
focused to the diffraction limit may cause irreversible damaging
by photodecomposition or thermal degradation.

In practice, the applications of micro-Raman to biological samples
are essentially limited by the threshold of degradation. In the
case of non resonant excitation where the amount of absorption of
laser light can be neglected, picograms of matter can be unambiguously identified by their vibrational spectra by using laser
power as low as 0.1 to 1 mW. This can be used for observing small
inclusions or local concentration of transparent compounds in
tissues, histological preparations, or even, under favourable
conditions, in living cells where the aqueous environment offers
the advantage of efficient heat transfer and cooling properties.

By resonant excitation, the detection limit could be improved by
several orders of magnitude. However, attention must be paid to
the drastic limitation of the maximum allowed power and/or irradiance which is a consequence of the absorption of laser light
by the sample. It is apparent that a prediction of the optimum
choice of laser power and spot diameter to be used in resonant
micro-Raman would require accurate data on local absorption coefficients, concentrations and thermal properties of the sample.
Such data are often not available for biological samples. It is
desirable to use some empirical approach and to verify in preliminary experiments that the level of laser irradiation is kept
low enough to avoid artifacts or local degradation.

Visible and UV excitation

Resonance enhancement of the molecular spectra associated with
chromophores offers a very promising approach to the problem of
localization and identification of biological markers in living
cells. The most interesting feature of resonant Raman microspec-
trometry is its specificity, which enables to observe in situ the
"labelled" compound, even when it is surrounded by a complex environment of non resonant molecules.

The extension of resonant excitation to the near ultra violet
spectrum is feasible by means of minor modifications of the microprobe system. These modifications involve the use of commercially
available microscope objective, specially corrected for the near
UV range, or Cassegrain type mirror objectives, or ellipsoïdal
mirrors. Also the coupling optics and all the lenses, mirrors and
gratings of the monochromator and spectrograph have to be optimized
for UV light. Photodetectors, single or multichannel, can be deli-

vered with silica or sapphire entrance window.

However, it seems that the instrument can be constructed with
high efficiency only for a limited spectral range, so that it
requires at least two sets of coated optical elements and blazed
gratings to cover the spectrum from 250 nm to the visible.

It is also apparent that, when excitation is made in the short
wavelength UV range, where most organic and biological molecules
strongly absorb, the enhancement of Raman intensity by resonance
or preresonance is less selective, and the limitations due to
laser damaging are more stringent.

The first generation of Raman microprobe:

More than ten years ago, the first Raman microprobes have been
independantly studied, at NBS in USA, and at CNRS in France, and
have been used to observe a variety of samples in geochemistry,
semiconductor and industrial analysis, as well as academic
research.
The principle of operation is given on fig. 5, which shows an
inexpensive setup which can be realized by coupling a commercially available microscope to an existing spectrometer.
A gaussian monomode (TEM$_{oo}$) Laser beam is focussed to the
diffraction limited spot by a standard microscope objective.
To improve the spurious light rejection, a spatial filter, made
of a pinhole placed at the common focus of two convergent lenses,
is often used in the incident beam. The same optical system,
beeing an afocal beam expander, matches the diameter of the
beam to the pupil of the microscope objective, by a proper choice
of the ratio of focal length.
For 90° scattering a variable angle observation, a second microscope objective can be used for light collection, but such an
arrangement is reserved for special experiments (i.e. oriented
crystals or thin films). Extremely stable holders and micuroposi-
tioners are needed. The magnification power and numerical aperture are limited to x 50 and 0,65 respectively, with long working
distance objectives.

Most commonly a single objective arrangement has prevailed as
the preferred technique for Raman spectroscopy at nearly 180°
scattering. This setup that we had proposed in our early experiments, circumvents the problem of difficult adjustment.   The
same objective serves to focus the Laser beam, partially reflected by a beam spletter, and to collect the scattered light
directed to the spectrometer. A beam splitter with 10 to 20%
reflexion, 90 to 80% transmission minimizes losses in scattered
radiation energy. The available Laser power has to be strongly
attenuated to prevent sample burning (typically 1 to 20 milli-
atts).

Microscope objectives corrected to give the primary image at infinite distance are preferred, so as to match the quasi-parallel
beam from the Laser filter unit. The scattered radiation is simply focused to the center of the entrance slit of the monochro-
mater. Different "coupling optics" have been described (5,7).
In the most simplified form, the "coupling optics" may be a
single lens, whose focal length is calculated so as to illuminate the whole width of the grating. With existing monochromators
apertures (typically F/10 to F/5) the geometrical "Etendue" U
is much larger than the one of microscope objectives (pupil diameter from 3 to 6 mm, normalized tube length 160 mm). Consequently, the focal length of the coupling lens is shorter than 160
mm, and the magnification factor of the microscope is reduced.
To improve the field coverage, for observing objects out of
optical axis, a "field lens" can be added in front of the slit.

Such an inexpensive experimental setup which can be realized in
any laboratory, exhibits performance levels equivalent to those
of commercially available instruments of the first generation,
using optical microscope attachments coupled to scannings monochromators with photomultiplier detection.

After some initial laboratory study of this kind of instrumentation in 1973, a joint development work made in collaboration
between CNRS and French industry resulted in a first generation
of commercial instruments, constructed under ANVAR license
by ISA-Jobin Yvon from 1976 to 1980, and called "MOLE".
This kind of microprobe has been widely used for various applications, in which the physical limitations of the technique
have been clearly established:
Available microscope objectives produce a 1 μm spot, with a depth
of focus of 3 to 10 μm. In the small irradiated volume the high
irradiance (typically 10^5 to 10^6 W cm^{-2}) may cause degradation
of samples during a long exposure time, when the Raman spectra
are recorded by means of a monochromator with a single photodetector.
To preserve the non destructiveness of Raman microanalysis,
it has been necessary to redesign entirely the microprobe, in
order to improve the sentivity, so as to enable recording the
spectra with a better signal-to-noise ratio and decreasing
the Laser power at the sample.

The second generation: Multichannel Microprobe.

A second generation of Raman microprobe has been studied and
developped since 1980 in close collaboration between CNRS and
french industry (DILOR). The new instrument now commercially
available under the name MICRODIL, takes fully benefit from the
most recent developments of multichannel photodetection systems.

The detectivity of this new generation is increased by a factor
of 10 to 100 compared to conventional systems. The gain can be
used both to observe very weak signals by long time integration
and data treatment by microcomputers, or to avoid sample damaging
owing to the short exposure to low power Laser beam.(Fig. 6).

The study and realization of this new instrument have been based
on the following principles:
- The available image intensifiers and photodiode arrays offer
the best compromise for sensitivity, wide dynamic range, geometrical stability, signal handling and treatment.
- The whole optical system has to be designed to fit the multichannel photodetector.
- Attention must be paid to optimize the coupling optics, from
the sample to the detector, so as to take fully benefit from the
"multichannel" advantage, both for spatial resolution and spectral requirements.
- All mechanical motions, spectral parameters and signal treatments, must be computerized or computer controlled.

The intensified detector is made by coupling by optic fibers an
image intensifier (LEP proximity focused microchannel plate
intensifier-)to a Reticon 512 elements photodiode array . The
spectral range is limited to 400-850 nm but the overall gain
enables single photon detection. Optional intensifiers are available for UV detection. The whole detector head filled with
argon and sealed is cooled to -28°C by Peltier elements. A part
of the electronic circuit, comprising the FET low noise pream-
plifiers and control is included in the head. To avoid extra
noise this unit is powered by batteries. A separate console
includes: Gauss filter, sampling, multiplexing A/D converter
and control logic. The signal output is compatible with a fast
computer, or with currently available microcomputer systems
(i.e. Apple II).

The time of integration varies from 20 ms to 990 ms by ms steps
or from 1 s to 99 s.

A new kind of transfer optics has been placed between the microscope and the spectrometer, which enables an optimized coupling.
Instead of moving the sample under the microscope objective the
exploration of the sample is performed by moving with a controlled scanner a couple of intermediate lenses, the one deflecting
the laser spot over the whole field of the microscope, the other
compensating for deviation of the image of the probed area to
bring it to the entrance slit of the spectrometer. A fixed sample can thus be analysed with high accuracy over a wide field,
and this optical arrangement has proven most convenient for a
variety of cumbersome samples (i. e. cryostat, variable tempera-

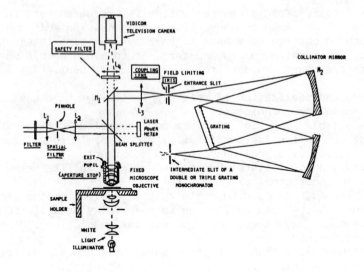

Figure 5. The "poor man's" Raman Microprobe

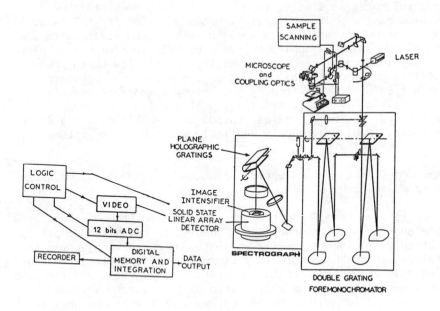

Figure 6.   The last version of multichannel Raman
             Microprobe : MICRODIL

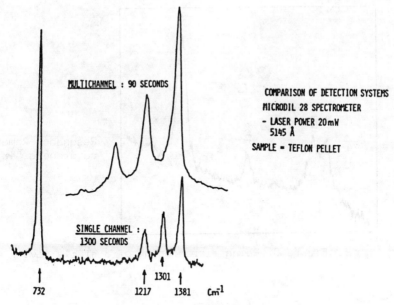

Figure 7. Comparison of Raman spectra recorded by
photomultiplier and multichannel detection systems.

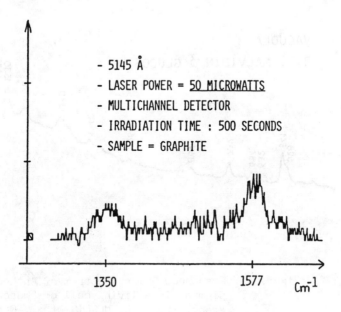

Figure 8. A good test of sensitivity for multichannel
detection : The Raman spectrum of graphite with
50 microwatts of Laser power.

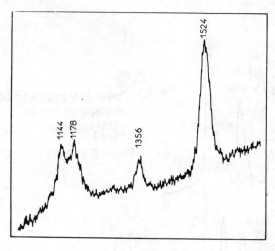

Resonance Raman of
carotenoïd "in vivo"

Figure 9. Raman microprobing of a single living cell :
Pyrocystis Lunula

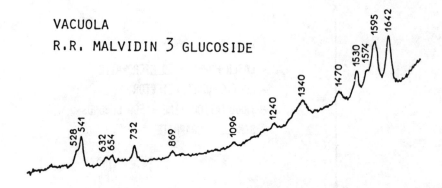

Figure 10. Resonance Raman spectrum of Anthocyanin
           pigment in a living cell of "Muscat noir"
           grape (from J.C. MERLIN et al. Ref (8)).

ture or pressure cells, pieces of art or industrial materials).

The spectrometer itself comprises a double holographic grating
foremonochromator, followed by a grating spectrograph. By using
a turret of gratings, combined with internal commutations which
realize either substractive or additive dispersions in the foremonochromator, a large choice of spectral resolution and spectral
band is offered.

As shown in fig.7 the multichannel system clearly exhibits a
better signal-to-noise ratio with a much shorter recording time.
A good test of the significant improvement in detectivity is
given in fig. 8, which shows an acceptable record fo very weak
Raman signals from a particle of graphite excited by only 50µw
of Laser power. Most chemical compounds can be identified from
subnanogram samples. In many cases, very good vibrational spectra
are obtained from particles of a few cubic micron, typically
10^{-11} grams of matter or less. In the case of Resonance Raman
excitation, the detection limit reaches 10^{-15} grams. (Fig.9, 10).

In addition to the Raman spectra, the same instrument can give
the distribution of the intensity of a characteristic Raman line
in an heterogeneous sample. By isolating this characteristic
Raman radiation, a selective "map" or "image" of the sample can
be rebuilt, which gives a precise distribution of a given polyatomic compound. By using powerful microscope objectives, with
a resolving power of the order of 10^{-3} millimetre, detailed
micrographic "Raman Images" can be obtained from a variety of
samples.

4. TIME RESOLVED RAMAN SPECTROSCOPY

Several time resolved experiments and applications are described
in the other chapters of the present book. We would like to
emphasize here some general features of Raman techniques, which
may complement the other spectroscopic methods to observe structural and kinetic evolution in time resolved studies. We shall
however restrict ourselves to some technical aspects which are of
interest for the possible applications to biological problems.
The most general method is called "Pump then Probe":
- The sample is first submitted to a rapid perturbation, (Rapid
mixing, flash photolysis, pulse radiolysis, jump of temperature,
electric field, current, pressure or strength). In many experiments, this perturbation can be produced by a Laser pulse (Pump).
- After a given delay, the evolution of the sample is observed
by means of a second Laser (Probe)whose wavelength is choosen so
as to excite the Raman scattering of transients or excited molecules, and to minimize the perturbing effects.
In principle, time resolved Raman spectroscopy is only limited
by Heisenberg's uncertainty principle, which means that the product of temporal resolution by spectral resolution is constant.
This would theoretically permit to observe Raman spectra with
a sufficient resolution for analytical and structural analysis
(1 to 10 cm^{-1}) in extremely short time, of the order of picoseconds to a few tens of picoseconds.
The photodetection of a spectral element requires to collect a
sufficient number of scattered photons to insure a good S/N ratio.
Owing to the low cross section of the Raman effect this condition
is only full filled by illuminating the sample by a very large
number of exciting photons (10^{14} to 10^{20} in typical experiments).
A question often raised is how the S/N ratio compares with a
pulsed Laser versus a CW Laser. To a first approximation , in
shot noise limited operation, an integrating photodetector should
exhibit the same S/N when the total number of photons is the
same during the time of measurement.
But if we consider the ability of gating the photodetector in
synchronism with light pulses, we found it advantageous.
When the gate is open during the light pulse, the photodetector
integrates the signal and the noise, but when it is closed, all
the noise originating from darkcurrent or from spurious light
emission of the sample is eliminated. The same advantage may be
taken to separate Raman scattering, which is pratically "instantaneous" from the long living fluorescence or luminescence.
A second question is how the sample degradation is modified by
pulsed operation. Again, to a first approximation, integrating
the same number of photons should result in the same thermal or
photochemical effects of the Laser beam which is used to probe
the sample by Raman spectrometry. But this approximation is
not valid in short pulse regime (ns to ps)when the life time

of the excited states are of the same order as the pulse width,
which results in important modifications of the populations
on electronic levels.

The experimentalist must also pay attention to the fact that the
same total energy E_t of exciting Laser light as needed in CW
experiments has to be delivered in a short pulse, with a very
high peak power.

In practice, the limitation of single pulse Raman experiments
is fixed by the threshold of non linear processes:

- Stimulated Raman, which produces a strong emission of light
and renders the photodetection of spontaneous Raman more difficult.

- Dielectric breakdown, which destroys the sample when the irradiance reaches 10^8 to 10^{10} W cm^{-2}.

To prevent these phenomena, the Laser beam can be defocused
to irradiate a larger area of the sample, at the expense of poo-
rer spatial resolution.

Another means to avoid destructive breakdown is to divide the
exciting pulse in a series of successive pulses with lower peak
power, and to perform pulse accumulation in the detection system.
The time-resolution may be preserved when the pumping and probing
processes are reproducibly repeated in the sample, in synchronism
with a gated photodetector. The cumulative photochemical effects
of the light pulses on the sample often require to continuously
renew the irradiated volume by a rapid mechanical motion of the
sample(rotating cell, flowing cell or flowing jet) or by a fast
deflection of the Laser beam. Attention must be paid to the
rate of relaxation phenomena. The thermal relaxation processes
of biological specimens, which are immersed in aqueous media,
are favourable. For instance, a micrometer sized particle in
water exhibits a thermal relaxation time of the order of 10^{-7}
second, which means that it is efficiently cooled in short pulse
regime even with megahertz repetition rate.

Commercially available Laser sources offer a large choice of
fixed or tunable wavelength and pulse duration from milliseconds
to picoseconds.

To establish a comparison of experimental conditions, we shall
consider in table 1 some typical values of the parameters, for
a given Raman spectrum, corresponding to a possible observation
of a biological sample with a modern highly sensitive multichannel spectrometer:

A = CW Laser:
total time of measurement T = 100 seconds
Laser power P_0= 1 milliwatt
                           E_t= 100 millijoules
B = Picosecond Mode locked Dye Laser, pumped by Ar$^+$ pulsed Laser
C = Microsecond Dye Laser, pumped by flash lamps.
D = Nanosecond Dye Laser, pumped by Q switched YAG or by excimer
Laser.
E = Picosecond Mode locked Dye Laser, pumped by YAG.

TABLE 1 : THE ORDER OF MAGNITUDE OF THE PARAMETERS OF INTEREST IN PULSED RAMAN EXPERIMENTS

		A	B	C	D	E
PULSE WIDTH (SECOND)		100	10^{-11}	10^{-6}	10^{-8}	10^{-11}
REPETITION RATE (PULSE PER SECOND)		CW	10^6	10^{-1}	10	10
NUMBER OF PULSES ACCUMULATED IN 100 s		1	10^8	10	10^3	10^3
ENERGY PER PULSE (JOULE)		10^{-1}	10^{-9}	10^{-2}	10^{-4}	10^{-4}
PEAK POWER (WATT)		10^{-3}	10^2	10^4	10^4	10^7
SAMPLE IRRADIANCE ($W CM^{-2}$)	DEFOCUSED LASER BEAM SPOT AREA = 10^{-2} CM^2	10^{-1}	10^4	10^6	10^6	10^9
	FOCUSED LASER BEAM SPOT AREA = 10^{-4} CM^2	10	10^6	10^8	10^8	10^{11}
	MICROSCOPE SPOT AREA = 10^{-8} CM^2	10^5	10^{10}	10^{12}	10^{12}	10^{15}

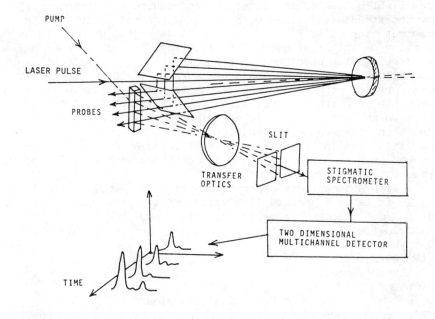

FIG.11 MULTIPASS OPTICAL DELAY LINE

Taking into account the available repetition rates of these pulsed lasers, we assume that the energy per pulse has been attenuated so as to normalize the total energy E_t = 0,1 J corresponding
to an average power of 1 milliwatt.
The limitation due to the effect of peak irradiance is clearly
apparent in Table 1, which shows that microprobing is not feasible with pulsed lasers, unless the peak power is attenuated at
the expense of accumulation of a large number of pulses.

Optical delay lines :

An elegan way to perform time measurements in the nanosecond to
picosecond range is offered by optical delay lines. We would like
to emphasis the properties of multiple path optical delay lines
which provide also a means to generate from a single laser pulse
a train of pulses with accurate time separation and calibrated
decreasing intensities. Such a pulse train is useful, either for
time resolved Raman interrogation of an evolutive process, or
to divide the energy of a laser pulse so as to avoid destructive
breakdown.
Fig. 11 shows an example of multipass delay line in which an exponentially decreasing series of pulses is generated by a simple
optical device. The edge of a 90° roof mirror system is placed
near the center of curvature of a spherical mirror, whose optical
axis is slightly tilted. One of the mirrors of the roof system is
partially transparent, in order to extract a known part of the
energy from successive roundtrips of the light pulse. The sample
to be observed by Raman spectroscopy receives the successive pulses of the train at different places, whose images are projected
on the slit of the multichannel spectrograph and are analysed as
separate spectra by means of a two dimensional photodetector
(Intensified vidicon or CCD matrix). Remembering the velocity of
light C = 3.10^8m s^{-1} it is obvious to adjust the length of optical path from i.e 3 cm (100 ps) to 30 m (100 ns). A set of parallel semi transparent mirrors (corresponding to the well known
Fabry Perot interferometer) can also be employed to generate an
exponential train of pulses in picosecond domain. We have recently proposed to make use of such a multiwave interferometer as a
temporal filter, to discriminate picosecond excited Raman spectra
from long living fluorescence emissions of the impurities often
present in the samples (16).

Multichannel spectrometers :

Although a high degree of sophistication has been attained with
photomultiplier detection, conventional spectrometers suffer in
time resolved experiments from a major drawback, inherent to the
scanning operation which is necessary to measure sequentially the
intensities of spectral elements with a single photodetector.
This result in a long time of exposure to the laser source, with
subsequent risk of decomposition of the sample.

The convenience and high detectivity offered by the last generation of multichannel spectrometer make it the experimentalist's
obvious choice for pulsed Laser time resolved studies. There
is in pratice no limitation of the sensitivity of multichannel
detectors in short pulse regime, from milliseconds to picoseconds,
and it is quite easy to integrate a number of successive pulses,
if necessary to obtain the desired S/N. However,many spectroscopists, experts in conventional Raman instrumentation, have been
disappointed by the poor results obtained in trying to adapt a
commercially available multichannel detector to an existing spectrometer.
The principles of multichannel Raman spectrometers were pioneered
by M. Bridoux and coworkers twenty years ago [10]. It has been
demonstrated that multichannel operation offered a substantial
amount of advantages, especially to detect simultaneously a large
number of spectral elements when the Raman spectrum is excited
by a pulsed Laser. It was also clear that single photon detection
was only possible under two conditions:
- The gain of the intensifier has to be high enough to give rise
to a measurable signal from a single electron emitted by the pho-
tocathode.
- The noise arising from the readout process of the secondary
photodetector (Photographic emulsion, Vidicon, diode array or CCD)
has to be reduced to a low level, compatible with the shot noise
regime of the primary photodetector.
It is obvious that these conditions were not fullfilled in commercially available multichannel analysers. Attention must also be
paid to the light collection efficiency and optimized coupling
of the optical part of the spectrometer to the photodetector.
As mentioned above, a special effort has been made in close
collaboration between CNRS and french industry to develop a
multichannel instrument which takes fully benefit of the poten-
tialities of this technique. The most important improvement,
introduced three years ago, consisted in a complete redesign
of the electronic circuitry associated with diode arrays, so that
the level of spurious noise related to the switching elements
became compatible with photon-counting regime (12-14). As a result,
a last generation of Raman spectrometer, based on the optical and
electronic schemes above described for the MICRODIL microprobe,
has been developped by DILOR under the name "OMARS". (16)

Conclusion

A good review of the time resolved Raman techniques and applications can be found in recent papers published by R.E. Hester (18)
and G.H. Atkinson (11). The first Conference on Time resolved
vibrational spectroscopy, organized in August 1982 at Lake Placid
(USA) by G.H. Atkinson, gave a brilliant demonstration of the po-
tentialities of Raman probing. One can predict a further development of the applications in various domains, including biological processes.

REFERENCES

(1) R.F. STAMM, C.F. SALZMANN, J. Opt. Soc. Am. 43 - 126 (1953)
                                             43 - 708 (1953)

(2) T. HIRSCHFELD, J. Opt. Soc. Am. 63 - 476 (1973)

(3) G.J. ROSASCO, E.S. ETZ, W.A. CASSATT - Appl. Sp. 29-396 (1975)

(4) M. DELHAYE, P. DHAMELINCOURT - J. Raman Sp. 3 - 33 (1975)

(5) P. DHAMELINCOURT, Doctoral Thesis - Lille (1979)

(6) Microbeam Analysis, K.F.J. Heinrich Edit. 1982 San Fransisco
    Press

(7) J. BARBILLAT, Doctoral Thesis - Lille (1983)

(8) J.C. MERLIN et al., C.R. Acad. Sc. Paris, 296 - 1397 (1983)

(9) A. DUPAIX et al. Biol. Cell.43. 157 (1982).

(10) M. BRIDOUX, M. DELHAYE Advances in Infrared and Raman Spec-
     troscopy Vol.2 (1976).
     (R.J.H CLARK and R.E HESTER, eds) Heyden. London.

(11) G.H. ATKINSON Advances in Infrared and Raman Spectroscopy
     vol. 10 (1982).

(12) H. SURBECK, W. HUG, M. GREMAUD, M. BRIDOUX, A. DEFFONTAINE,
     E. DA SILVA Optics communications 38,57 (1981).

(13) J.C MERLIN, J.L. LORRIAUX, R.E. HESTER.
     Jnal of Raman sp. 11-384 (1981).

(14) M. BRIDOUX, A. DEFFONTAINE, E. DA SILVA, M. DELHAYE, B. ROSE.
     Jnal of Raman Spectroscopy 11-515 (1981).

(15) A. DEFFONTAINE, B. ROSE, B. ROUSSEL, M. BRIDOUX, M. DELHAYE.
     C.R. Acad. Sc. Paris 292, 567 (1981).

(16) M. DELHAYE, M. BRIDOUX, E. DA SILVA. Spectra 2000- 77-10
                                              (1982)

(17) M. DELHAYE, A. DEFFONTAINE, A. CHAPPUT, M. BRIDOUX to be
     published in Jnal of Raman sp.

(18) R.E. HESTER The Spex Speaker. Aug. 1982.

FAR-INFRARED SPECTROSCOPY OF BIOMOLECULES

L. Genzel, L. Santo, S.C. Shen

Max-Planck-Institut für Festkörperforschung
Heisenbergstraße 1, D-7000 Stuttgart 80
Federal Republik of Germany

ABSTRACT

Far-infrared spectra in the 20 - 500 cm^{-1} region are presented for solid layers of L-alanine and several proteins. A special study is devoted to the spectroscopy of films of amino-acids, homopolypeptides and proteins which were cast from solutions of trifluoro-acetic acid for investigating the chemical binding of this strong denaturing agent to the biomolecules.

C. Sandorfy and T. Theophanides (eds.), Spectroscopy of Biological Molecules, 609–619.
© 1984 by D. Reidel Publishing Company.

INTRODUCTION

   A great number of infrared and Raman spectroscopic investigations and also normal mode calculations have been reported concerning amino acids and their polypeptides. These have included studies
of helical- or sheet-conformations and of the eigenfrequencies, ei-
genvectors and forcefields characteristic for the polypeptide
chains. Most of the work has been confined to the wavenumber region
above about $300 cm^{-1}$ because such spectra are easier to interpret and
also because of increasing experimental difficulties towards lower
frequencies. In general, the low-frequency infrared region contains less spectral features than the middle infrared but in special cases one might get there information which cannot be reached
otherwise as directly. An important example of this kind are the
vibrational modes which involve the hydrogen bonds as restoring
force constants. This example is demonstrated on the far-infrared
spectra of crystalline L-alanine. Another example is given by the
investigation of the influence of the strong denaturant trifluoro-
acetic acid (TFA).

EXPERIMENTAL

   Spectroscopy in the far-infrared $(10 - 400 cm^{-1})$ is nowadays
not more such a difficult experimental problem as it was still
15 to 20 years ago. This is due to the availability of several commercial Fourier-transform spectrometers for this spectral region
and rather sensitive low-temperature detectors. Other problems,
however, remain principally among which the high absorption of liquid water is particularly severe for studying biomolecules in
their natural environment. The absorption coefficient of water at
$190 cm^{-1}$ and 300 K reaches a value of about $1200 cm^{-1}$ /1/ while
strong vibrational bands of biomolecules have maximal absorption
coefficients of less than $400 cm^{-1}$. It is thus difficult, though
not completely impossible, to get reliable results from spectra
of biomolecules in aqueous solutions.

   The experimental results described below were, therefore,
made on solid films or layers of the biomolecules with thicknesses
in the 40 - 100 μm range. These films were cast from water or TFA
solutions onto a silicon substrate which is almost free of absorption in the far-infrared. For some polypeptides, free films
could be obtained by detaching the films from the highly polished
silicon plate. All the films were measured directly after preparation by transferring them as fast as possible into the spectrometer (IFS 114c Fourier-transform Spectrometer from Bruker-Physik,
Karlsruhe-Forchheim, Germany) which was either evacuated or flushed
with dried air.

Instead of films, cast from solutions, it is also possible to use the materials in the form of small particles (size << wavelength) which are embedded in transparent matrices like polyethylene with a low filling factor (smaller than 5%). This sample preparation might have some advantage for crystalline materials because cast films could have the tendency to be amorphous (see, however, the very sharp absorption bands in Fig.1). The technique of embedding small crystallites or particles in a matrix, on the other hand, may give wrong eigenfrequencies of strong vibrational bands. This is due to the depolarizing electric field of the surface polarisation (surface phonon mode) which shifts the band-frequencies slightly upwards /2/.

RESULTS AND DISCUSSION

Amino-Acids

We have first investigated various amino-acids, namely glycine, L-alanine, valine, leucine and tyrosine. Shown here are only the spectra of L-alanine (Fig.1) for various temperatures while the corresponding transmission spectra of the other amoniacids are published elsewhere /3/. As seen in Fig.1 for L-alanine and as evident from our full normal coordinate analysis /4/, one can distinguish the intramolecular vibrational modes (> 200 cm^{-1}) from the intermolecular lattice modes (< 200 cm^{-1}). It is astonishing that the line widths of the letter ones are almost generally smaller than of the former ones, and this even at low temperatures. A remarkable exception is the doublet band near 142cm^{-1}, which is seen at room temperature only as a shoulder but which sharpens up dramatically on going to lower temperatures. The eigenvector of this doublet is almost completely due to stretch vibrations of the three hydrogen bridges NH...OC which are placed on the NH_3^+-group of the zwitterion $NH_3^+ CH_3 CHCOO^-$ of the L-alanine molecule. Four of these zwitterions form the unit cell of the crystal /5/. If bands show such a strong anharmonicity as the 142cm^{-1} doublet then one should expect to see the overtone which here may be embedded in the strong doublet occuring around 284cm^{-1} which is assigned /4/ to modes with CH_3 torsions, $CC_\alpha N$ deformation and NC_α stretch. In contrast to the 142cm^{-1} doublet, one observes for the low-frequency doublet around 80cm^{-1} almost no anharmonicity. These modes contain hydrogen-bridge bending /4/ and may thus nearly be decoupled from acoustical phonon-modes of the lattice in which they could otherwise decay by a multi-phonon process.

The strong intramolecular bands at 326cm^{-1} which are assigned to skeletal deformation modes show striking temperature-dependent features. At 300 K they are seen as a single band which, however, splits with decreasing temperature into two components. The first component stays at 326cm^{-1} while the second component shifts up to

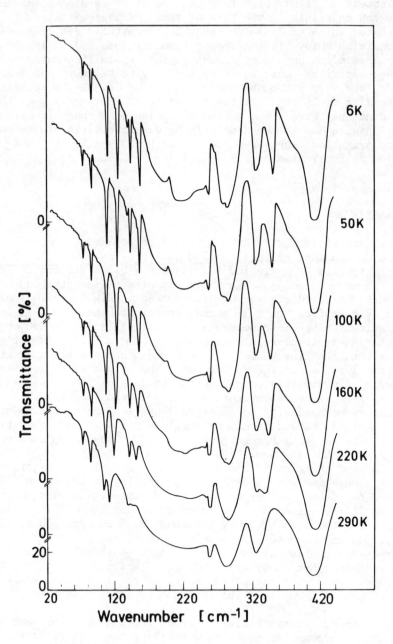

Fig.1: Transmission spectra for six temperatures of a layer of L-alanine, cast from water solution. Layer thickness was about 80μm. (Taken from /4/).

351cm^{-1} at 6 K. The very strong band at 410cm^{-1}, assigned also to
a skeletal deformation, does not change its eigenfrequency with
temperature and has an anomalously large width even at 6 K. At
present, one is not really able to understand such an anharmonic
behaviour which is a complicated consequence of the electronic
structure determining the potential.

Polypeptide And Proteins

A large number of spectroscopic investigations in the infrared
and with Raman scattering have been made especially on homo-polypeptides like poly-glycine and poly-L-alanine above 400 cm^{-1}/see
e.g.6-19/ which show that characteristic vibrational modes (amid
bands) are common to all polypeptides and which partly can be used
to distinguish between the possible conformations (helical, β-
sheeted). A selection of these characteristic modes with their
wavenumber regions is compiled in Table 1.

Table 1 : Characteristic Frequencies of Polypeptides.

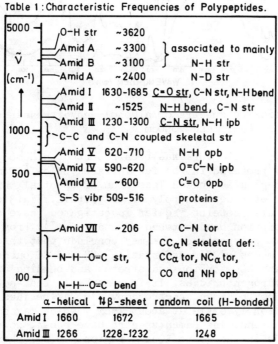

	O-H str	~3620	
	Amid A	~3300	} associated to mainly
	Amid B	~3100	N-H str
	Amid A	~2400	N-D str
	Amid I	1630-1685	C=O str, C-N str, N-H bend
	Amid II	~1525	N-H bend, C-N str
	Amid III	1230-1300	C-N str, N-H ipb
	C-C and C-N coupled skeletal str		
	Amid V	620-710	N-H opb
	Amid IV	590-620	O=C'-N ipb
	Amid VI	~600	C'=O opb
	S-S vibr	509-516	proteins
	Amid VII	~206	C-N tor
	-N-H⋯O≈C str,		CCαN skeletal def: CCα tor, NCα tor, CO and NH opb
	N-H⋯O≈C bend		

	α-helical	↑↓β-sheet	random coil (H-bonded)
Amid I	1660	1672	1665
Amid III	1266	1228-1232	1248

str = stretch, tor = torsion, ipb = in plane bend, opb = out of plane bend,
def = deformation.

It has been studied how these modes buildt up when going from the
pure amino-acid over the dimer, trimer etc. to the polymer /18/
showing that the knowledge of the spectra of the pure amino-acids
is not at all sufficient for an understanding of the vibrational
modes of the corresponding polymers. The characteristic modes of
the polypeptides (Table 1) are also very useful for the spectroscopic study of proteins as the example for the alcohol-dehydrogenase in Fig. 2 shows.

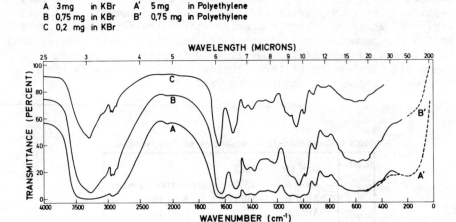

Alcohol-Dehydrogenase from Yeast

A 3 mg   in KBr   A' 5 mg   in Polyethylene
B 0,75 mg in KBr   B' 0,75 mg in Polyethylene
C 0,2 mg in KBr

Fig. 2: Transmission spectra of alcohol-dehydrogenase at 300 K
between $50 cm^{-1}$ and $4000 cm^{-1}$.

Comparably little work has been reported on homo-polypeptides
and especially proteins in the far-infrared below $400 cm^{-1}$/6,15,16,
20-23/. This region is indeed less rich of spectral information
characteristic of a certain polymer or protein. Fig.3 shows for
instance that all proteins, so far investigated, have a very similar broad absorption band between 50 and $250 cm^{-1}$. From comparable
studies of homo-polypeptides one may conclude that this band contains several modes - $CC_\alpha N$ deformation, NH and CO out-of-plane
bending, CC_α and NC_α torsion and stretch and bending vibrations of
the NH...OC hydrogen bridges. It is remarkable that nothing is seen
from these modes in Raman spectroscopy /19,24/ besides a weak broad
band occuring for lysozyme at about $75 cm^{-1}$ which may contain some
skeletal vibrations. The Raman spectra have shown, however, very
low lying bands in the 15 to $30 cm^{-1}$ region /19,24/. For lysozyme,
such a band has been observed at $25 cm^{-1}$ /24/ and was recently also
found by neutron time-of-flight spectroscopy /25/.

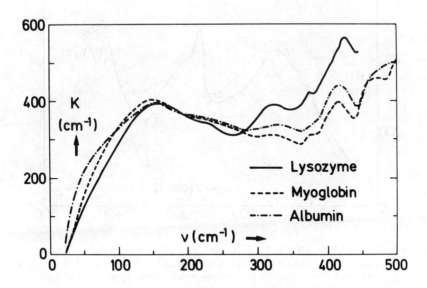

Fig. 3: Absorption spectra of films of lysozyme (full line), myo-
        globin (dashed line) and Albumin (dashed-dotted line);
        room temperature. The films were cast from water solutions.
        (Taken from /26/).

Spectra of Denatured Biomolecules

     The term denaturation is defined more so with respect to
biological activity of a protein than with respect to a definite
chemical change of the molecule. Thermal denaturation is clearly
different in this sense than denaturation by a chemical.

     We have performed now a far-infrared spectroscopic study of
the influence of the strong denaturant trifluoro-acetic acid (TFA)
on various biomolecules /26/. We came about this problem when we
produced films of several proteins which were cast from TFA-solu-
tions instead of films cast from water solutions. The resulting
spectra are shown in Fig.4 which should be compared with the spectra of Fig.3. The dramatic and characteristic change is the occurence of a strong band between 260 and 300cm^{-1}. In order to understand this band we first measured the far-infrared spectrum of pure,
liquid TFA (Fig.5) which yields indeed a strong structurized band

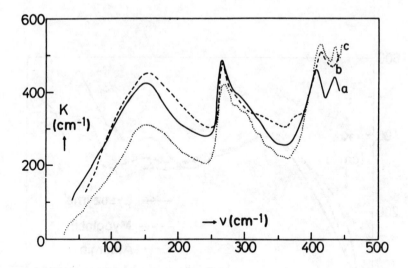

Fig. 4: Absorption spectra of films of lysozyme (full line),
myoglobin (dashed line) and albumin (dotted line); room
temperature. The films were cast from TFA solutions.
(Taken from /26/).

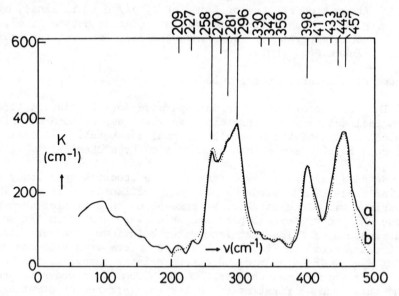

Fig. 5: Curve a (full line): Absorption spectrum of a liquid film
of pure TFA. Curve b (dotted line): oscillator fit of curve a with various oscillators defined on the top line (see
Fig.6); all damping constants are $21cm^{-1}$.(Taken from /26/).

between 258 and 300cm^{-1}. Interestingly, this band including its
structure is seen in the spectrum of a layer of glycine, cast from
TFA solution (Fig.6), but the whole band is shifted upwards by
9cm . Since TFA is a very strong acid we can assume that the

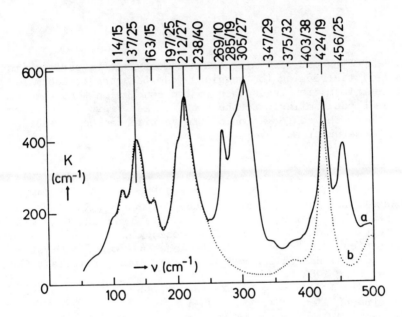

Fig.6: Curve a (full line): absorption spectrum of a layer of
glycine, cast from TFA solution; room temperature. Curve
b (dotted line): calculated absorption spectrum after subtracting several bands attributed to TFA. Numbers on top
line: frequencies and damping constants for an oscillator
fit of curve a. The vertical lines represent the oscillator strengths of the fitted oscillators.
(Taken from /26/).

complex of TFA and glycine forms a salt by means of protonation,
$F_3CCO_2^- \, ^+H_3NCH\,COOH$, which means that the characteristic TFA band
is a vibration (probably a rocking mode) of the CF_3-group which
remains almost free after bonding. We have also observed an additional band around 268cm^{-1} for films of polyglycine and poly-
L-alanine, cast from TFA solution. In these cases, protonation
cannot occur. But by a careful inspection of the spectra above
500cm^{-1}/26/ of polyglycine, cast from TFA solution and cast from
water solution and of the spectrum of pure TFA, we can conclude
that the TFA molecule binds to the backbone HN...OC bridges of the
polymer by means of hydrogen bonds in the following ways:

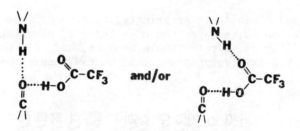

Thus, the CF_3-group of TFA is again almost free to perform its characteristic rocking (or wagging) mode occuring at 268cm^{-1}. All this information leads then to the conclusion that the TFA molecule binds similarly to the protein (see Fig.4) which then will deforme the conformation of the peptide backbone leading thus to denaturation.

REFERENCES

1. Afsar, M.N., Hasted, J.B., Infrared Physics 18, 835 (1978)
2. Genzel, L., Martin, T.P., Surface Science 34, 33 (1973)
3. Genzel, L., Kremer, F., Poglitsch, A., Bechtold, G., "Milli-meterwave and Far-infrared Spectroscopy on Biologial Macro-molecules", in "Coherent Excitations in Biological Systems", p. 58, Eds. H.Fröhlich and F. Kremer, Springer Verlag, Berlin, Heidelberg 1983.
4. Bandekar, J., Genzel, L., Kremer, F.,Santo, L., Spectrochim.Acta 39A, 357 (1983)
5. Lehmann, M.S., Koetzle, T.F., Hamilton, W.C. J.Am.Chem.Soc. 94, 2657 (1972)
6. Itoh, K., Nakahara, T., Shimanouchi, T., Biopolymers 6, 1759 (1968)
7. Gupta, V.D., Trevino, S., Boutin, H., J.Chem.Phys. 18, 3008 (1968)
8. Small, E.W., Fanconi, B., Peticolas, W.L., J.Chem.Phys. 52, 4369 (1970)
9. Lord, R.C., Yu, N.T., J.Mol.Biol. 50, 508 (1970)
10. Itoh, K., Shimanouchi, T., Biopolymers 9, 383 (1970)
11. Koenig, J.L., Sutton, P.L., Biopolymers 10, 89 (1971)
12. Fanconi, B., Small, E.W., Peticolas, W.L., Biopolymers 10, 1277 (1971)
13. Fanconi, B., Peticolas, W.L., Biopolymers 10, 2223 (1971)
14. Abe, Y., Krimm, S., Biopolymers 11, 1817 (1972)
15. Fanconi, B., Biopolymers 12, 2759 (1973)
16. Shotts, W.J., Sievers, A.J., Biopolymers 13, 2593 (1974)
17. Moore, W.H., Krimm, S., Biopolymers 15, 2439 (1976)
18. Fasman, G.S., Itoh, K., Liu, C.S., Lord, R.C., Biopolymers 17, 125 (1978)

19. Peticolas, W.L., Methods in Enzymology 61, 425 (1979)
20. Chirgadze, Y.N., Ovsepyan, A.M., Biopolymers 12, 637 (1973)
21. Buontempo, U., Careri, G., Ferraro, A.,
    Biopolymers 10, 2373 (1971)
22. Ataka, M., Tanaka, S., Biopolymers 18, 507 (1979)
23. Hasted, J.B., Husain, S.K., Ko, A.Y., Rosen, D., Nicol, E.,
    Birch, J.R., in "Coherent Excitations in Biological Systems",
    p.71, Eds. H. Fröhlich and F. Kremer, Springer Verlag,
    Berlin, Heidelberg, 1983.
24. Genzel, L., Keilmann, F., Martin, T.P., Winterling, G.,
    Yacoby, Y., Fröhlich, H., Makinen, W.M.,
    Biopolymers 15, 219 (1976)
25. Bartunik, H.D., Jollś, Berthou, J., Dianoux, A.J.,
    Biopolymers 21, 43 (1982)
26. Shen, S.C., Santo, L., Genzel, L.,
    Canad.J.Spectrosc. 26(13), 126 (1981)

PICOSECOND RELAXATIONS IN HAEMOGLOBIN, LYSOZYME, AND POLY-L-ALANINE OBSERVED BY MM-WAVE SPECTROSCOPY

L. Genzel, F. Kremer, A. Poglitsch, G. Bechtold

Max-Planck-Institut für Festkörperforschung,
Heisenbergstraße 1, D-7000 Stuttgart 80,
Federal Republic of Germany

ABSTRACT

Broadband measurements of the millimeter absorption of lyophilized haemoglobin and lysozyme are reported. Additionally, the absorption of poly-L-alanine and crystalline L-alanine at 70 GHz was measured for comparison. All measurements were extended over the temperature range from liquid helium to room temperature. For the millimeter range this was attained by using the novel oversized cavity technique. It was found that the millimeter wave absorption of the materials increased nearly exponentially with temperature and increased as $\nu^{1.2} - \nu^2$ with frequency. The frequency- and temperature dependence of the millimeter wave absorption is quantitatively described as due to three distinct relaxation processes on a picosecond timescale occuring in asymmetric double-well potentials. These processes are most probably assigned to the NH...OC hydrogen bonds of the peptide backbone. Hydrated lysozyme yields additionally above 150 K a nearly frequency independent contribution to the absorption indicating a slower relaxation process of the bound water molecules.

C. Sandorfy and T. Theophanides (eds.), Spectroscopy of Biological Molecules, 621–635.
© *1984 by D. Reidel Publishing Company.*

INTRODUCTION

Knowledge concerning the dielectric properties of biological macromolecules at millimeter (mm) wavelenghts is sparse with almost nothing known about the frequency- and temperature-dependence of the absorption coefficient in this regime. This lack of knowledge is caused by two reasons, namely firstly the large absorption of liquid water preventing meaningful studies of aqueous solutions and secondly experimental difficulties for solid state spectroscopy using the conventional single-mode waveguide and resonator technique over a broad range of frequencies and temperatures. A first step to overcome these difficulties was possible due to the rather novel oversized cavity technique /1-5/. We have conducted therefore, a broadband mm-wave study of haemoglobin (HE), lysozyme (LY) and poly-L-alanine $(Ala)_n$ over temperatures ranging from 4 K to 300 K.

EXPERIMENTAL

Various experimental reasons are responsible for the almost complete lack of spectroscopic information about the dynamical properties of biomolecules in the mm-region. There is firstly the low absorption of biomolecules in contrast to the high absorption of liquid water which prevents at present the possiblity of obtaining useful results from aqueous solutions. Secondly there is the difficulty of handling solid samples in the small-sized single-mode waveguide- and resonator-systems which are conventionally used in the mm-region. There is thirdly the demand of carrying out broadband measurements over a wide temperature range, at least as long as the general spectroscopic knowledge is sparse. This demand causes again severe technical problems if the singlemode waveguide systems are used. Finally, quasi-optical systems using mirrors, lenses and horn antennae suffer from frequency dependent diffraction effects and disturbing standing waves when the coherent microwave sources are used.

One rather novel technique, however, allows many of the afore-mentioned experimental problems to be overcome; it is the use of an oversized high-Q cavity with mode stirring /1-5/. The mm-radiation is coupled into the cavity from the source by the end of a waveguide and a small fraction of the cavity-field is coupled out to the detector. The large size of the cavity ($\sim$ 100 wavelengths) and the mode stirrer cause in time average an isotropic and homogeneous field. A helium cryostat, made by fused silica, can be placed inside the cavity and contains the absorbing sample which reduces the Q-value compared to the empty cryostat. This technique measures only the absorption (averaged over all angles of incidence) of the sample, but is essentially not disturbed by scattered and diffracted radiation which remains in the cavity.

The technique, however, results in the loss of the phase information. In order to get the absorption coefficient of the sample one has to assume a value of n, the real part of the refractive index. The weak mm-wave absorption of the biomolecules, however, has the consequence of a small index dispersion allowing thus a constant n to be used. For all samples under investigation, we used a value of n = 1.6 as it can be shown that a variation of n within limits of 0.2 will change the calculated value of the absorption coefficient α by less than 2% only.

The study presented here /6,7/ deals with the dielectric properties of anhydrous haemoglobin (HB), poly-L-alanine ($(Ala)_n$), polycrystalline α-L-alanine (Ala) and anhydrous as well as hydrous lysozyme (LY). The experiments with the oversized resonator require large amounts of sample material. The powdered chemicals were pressed into disc-shaped pellets of 50 mm in diameter and about 12 mm thickness which had then almost the bulk density of the solid. To prevent water adsorption in the anhydrous materials, they were dried over P_2O_5 for one week before the preparation of the pressed samples, and during the measurements they were kept under either dry nitrogen or helium. Anhydrous LY was additionally dried by keeping it for three days under 90^0C, a procedure which did not harm its enzymatic activity.

RESULTS

The respective absorption coefficients α as function of the temperature T of HB, $(Ala)_n$, and LY are shown in Figs.1-3. They all exhibit a similar temperature dependence, namely a nearly exponential increase of α with T from about 50 to 300 K. Below 50 K, α drops at first rapidly with decreasing T and then levels off at about 10 K. Furthermore, the absorption coefficient increases monotonic with frequency as ν^2 at low T and as $\nu^{1.2}$ at room temperature. We did not observe any sharp features in the mm-spectra which could be discussed in terms of vibrational modes. In addition to the results shown here we found quite similar $\alpha(T,\nu)$ curves for anhydrous keratin and for polyamid (nylon). A remarkable change of the temperature dependence of α was found with hydration (18% H_2O) as seen in Fig.4 for LY (compare with Fig.3). Above about 150 K an additional absorption process, also strongly temperature dependent, adds to the absorption already present in the dry material. This additional absorption is approximately proportional to the water content and is almost frequency independent.

In order to obtain a more complete survey for an understanding of the observed mm-wave absorption we performed a preliminary study of an amino-acid, α-L-alanine, which is a crystalline material with a water-free crystal structure. The absorption of Ala at 70 GHz is shown in Fig.5. It is found that α again increa-

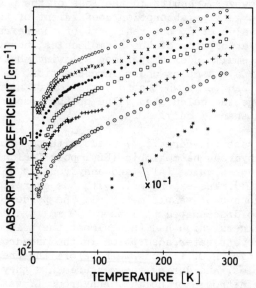

Fig.1: Absorption coefficient of dried haemoglobin from 4 K to
       300 K. Untuned cavity technique: O 148 GHz, x 111 GHz,
       • 100 GHZ, ⊡ 85 GHz, + 70 GHz, ⊙ 50 GHZ. Cavity perturba-
       tion technique: ⁎ 10 GHz (multiply by 10^{-1} to get actual
       value). (Taken from /6/).

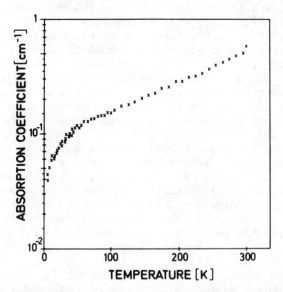

Fig.2: Absorption coefficient of dried poly-L-alanine from 4 K to
       300 K at 70 GHz using untuned cavity technique.
       (Taken from /6/).

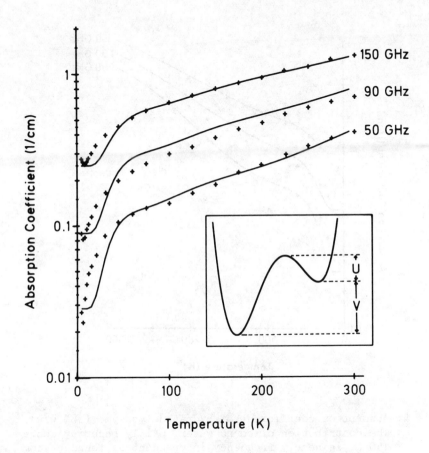

Fig.3: Absorption coefficient of dried lysozyme (water content
       < 0.5%). The solid line is a fit with Eq.(7). Inset:
       Asymmetric double-well potential used as relaxation model.
       (Taken from /19/).

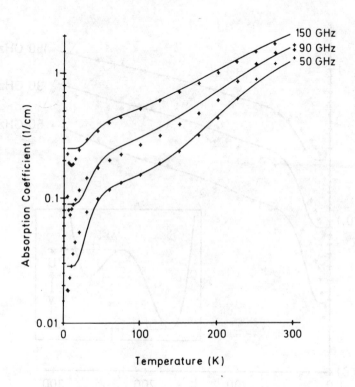

Fig.4: Absorption coefficient of hydrated lysozyme (18% w/w).
       The contribution of bound water, mainly occuring above
       150 K, is nearly frequency independent as found by com-
       parison with the dried sample. (Taken from /19/).

ses rapidly with temperature but with much lower values than the
biopolymers. This measurement appears to be rather interesting
because of the structure occuring in the T-dependence of α.

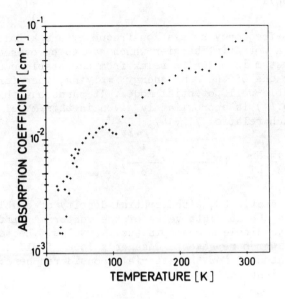

Fig.5: Absorption coefficient of poly-crystalline L-alanine from
       4 K to 300 K at 70 GHz. (Taken from /6/).

DISCUSSION

     The most significant result of this study resides in the
T- and ν-dependence of the mm-wave absorption coefficient α of
the various biomolecules and their respective similarity. We considered several known mechanisms which result in an increase of
α with temperature, such as two-level resonant absorption as observed at very low T in amorphous materials /8/, multi-phonon
difference processes /9-11/ and dielectric relaxation effects
/12-14/. Then it turned out that only the latter processes are
able to explain the observed ν- as well as T-dependence of α over
the whole range provided one uses an asymmetric double-well potential within which the relaxation occurs.

     To explain this in more detail we consider a Debye-type di-
electrical function

$$\varepsilon(\nu,T) = \varepsilon_\infty + \frac{\varepsilon_0 - \varepsilon_\infty}{1-i\nu\nu_\tau^{-1}} \qquad (1)$$

where ε_∞ is the high-frequency dielectrical constant (DC) and ε_0
the DC below the frequency regime of the considered relaxation
frequency ν_τ which is usually assumed to have the following temperature dependence for solids /12-14/:

$$\nu_\tau = \nu_\infty e^{-U/kT} \tag{2}$$

ν_∞ is the high-frequency relaxation frequency and k Boltzmann's
constant. U is a potential barrier which has to be overcome ther-
mally by the system in order to relax from one stable configuration to another one of the same energy implying, therefore, a
symmetrical double-well potential model. It turns out that
$(\varepsilon_o - \varepsilon_\infty)$ of Eq.(1) is approximately given in this model by the
Kirkwood-Fröhlich relation /12-14/:

$$\varepsilon_o - \varepsilon_\infty = \frac{4\pi}{3} \frac{N\mu^2}{kT} \; [\frac{3\,\varepsilon_s}{\varepsilon_\infty + 2\varepsilon_s}(\frac{\varepsilon_\infty + 2}{3})^2] \equiv \frac{C}{T} \tag{3}$$

Here, ε_s is the static DC, N the spatial density of double-well
systems and μ is the absolute value of the vectorial difference
of the respective dipole moments in the two wells. The term in
square brackets which represents Onsager's local field correction
has a value of about 2 for the materials considered here. The absorption coefficient follows from Eq.(1):

$$\alpha = \frac{2\pi\nu\varepsilon_2}{nc} = \frac{2\pi}{nc} \frac{(\varepsilon_o - \varepsilon_\infty)\nu^2\nu_\tau}{\nu^2 + \nu_\tau^2} \tag{4}$$

with c being the velocity of light and ε_2 the imaginary part of
ε, Eq.(1). Eq.(4) with Eqs.(2) and (3) will be discussed in two
limiting cases: a) $\nu^2 \ll \nu_\tau^2$, in this case we find α to be proportional ν^2/ν_τ T yielding thus a strong decrease of α with T
and an increase with ν; b) $\nu^2 \gg \nu_\tau^2$, here α is proportional to
ν_τ /T which means an increase of α up to a maximum at T = U/k,
but now the frequency dependence is omitted. Evidently, neither
of these cases is in accordance with the experimental results.

Staying with relaxation processes, one is left with the use
of an asymmetric double-well system as the next most simple model
which is, anyway, more probable (Fig.6). The treatment of this
dielectric model yields instead of Eq. (3)/6/

$$\varepsilon_o - \varepsilon_\infty = \frac{C}{T} \frac{e^{-V/kT}}{(1 + e^{-V/kT})^2} \tag{5}$$

and instead of Eq.(2)

$$\nu_\tau = \frac{\nu_\infty}{2} e^{-U/kT}(1 + e^{-V/kT}). \tag{6}$$

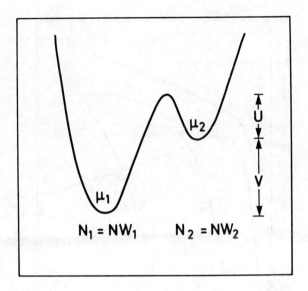

Fig.6: Asymmetric double-well potential used in the theoretical
model for the dielectric function. N = total number of
systems involved, W_i = occupation probability, μ_i = dipole
moment. $W_1 = \{1+ \exp(-V/kT)\}^{-1}$, $W_2 = 1 - W_1$

If $V > U$ one can obtain, even for $\nu^2 \lesssim \nu_\tau^2$, the desired increase
of α with T up to a certain T_{max} and from thereon a decrease with
T. This made it necessary to fit the data with a set of relaxation
processes, the minimal number of which turned out to be three –
a number which is also suggested from the result on polycrystalline Ala (Fig.5).

The observed levelling off of $\alpha(T)$ below 10 K together with
the observed ν^2-dependence is explained by a temperature independent wing of a strong and broad vibrational band found for all
proteins as well as for $(Ala)_n$ in the far-infrared around 150 cm^{-1}.
With this, our dielectric function for fitting the data is written
as

$$\varepsilon(\nu,T) = \varepsilon_\infty + \frac{S_o \nu_o^2}{\nu_o^2 - \nu^2 - i\nu\gamma_o} + \sum_{j=1}^{3} \frac{C_j}{T} \frac{1}{1-i\nu\nu\tau_j^{-1}} \frac{e^{-V_j/kT}}{(1+e^{-V_j/kT})^2} \tag{7}$$

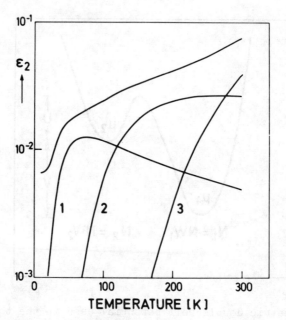

Fig.7: The temperature dependence of the imaginary part ε_2 of the
dielectric function for haemoglobin according to Eq.(7).
Curves 1-3 correspond to the relaxation processes 1-3,
respectively. The total dielectric function is composed
of the three relaxation processes plus the vibrational
term which causes the levelling off at low temperatures.
(Taken from /6/).

Relaxation	ν_∞ (GHZ)	U (kcal/mol)	V (kcal/mol)	$N\mu^2$ (cgs)
1	370 ± 20	0.04 ± 0.01	0.24 ± 0.05	2.9×10^{-16} $\pm\ 2 \times 10^{-17}$
2	350 ± 20	0.20 ± 0.05	0.91 ± 0.15	2.2×10^{-15} $\pm\ 2 \times 10^{-16}$
3	300 ± 20	0.40 ± 0.1	3.2 ± 0.5	8.3×10^{-14} $\pm\ 1 \times 10^{-14}$

Oscillator: $\varepsilon_\infty = 1.78$, $\nu_o = 4400$ GHz
$S_o = 0.385$, $\gamma_o = 3880$ GHz

Table 1: List of parameters for fitting the data of haemoglobin
with Eqs.(6), (7), and (3).

As an example of the final fit we show in Fig.7 for HB the three relaxation processes and their sum for the resultant $\varepsilon_2(T)$ curve. Table 1 compiles the various fitting parameters for HB, which are quite similar to those found for LY and $(Ala)_n$. Fig.8 demonstrates how the experimental $\alpha(T,\nu)$ of HB can be described by the model over the whole range of variables.

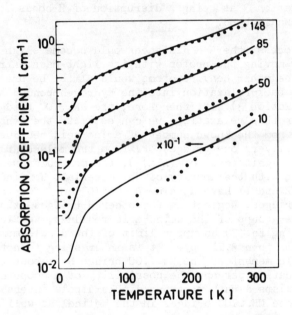

Fig.8: Comparison of the experimental values (dots) of the absorption coefficient for haemoglobin with the fit using Eq.(7)(full line). Parameter of the curves is the frequency in GHz. (Taken from /6/).

It should be emphasized, however, that we cannot exclude completely some small influence of multi-phonon difference processes /9-11/. Two- and three-phonon difference processes could yield a rather similar $\alpha(T)$ function above 100 K as found experimentally. These processes, though, would be frequency dependent at least as ν^2, which is in contradiction to our findings a higher temperatures. In addition, the difference phonon processes are expected to cause absorption coefficients being about one order of magnitude lower in our frequency range than those found for the biopolymers.

The question arises now about the microscopic nature of the assumed relaxation processes. Looking firstly at the values of the high-temperature relaxation frequencies ν_∞ (Table 1) one

finds them all of the same order of magnitude resulting in relaxation times of only several picoseconds. This is indicative of
the dynamics of small molecular submits with low mass. Secondly,
the depth H = U+V of well 1 for the relaxation 3 is found to be
3.6 kcal/mol ($\sim$ 1800 K) which is not much smaller than typical
binding energies of hydrogen bonds. In fact, hydrogen bonds of the
type NH...OC are the only molecular submits common to all of the
materials (including nylon) investigated here. We consider, therefore, the relaxation 3 as being a disruption of H-bonds to form
a weak Van-der-Waals bond.

In this process, the N-H distance would become slightly
smaller by only moving the proton with its light mass. A rearrangement of all the other heavier atoms would almost be prevented
due to the fast back-relaxation into the hydrogen-bonded situation
(well 1 of relaxation 3). If, therefore, the NH...OC bridges are
the location for this relaxation, we can estimate the density N
of the bridges from the known number of amino-acid residues per
protein (each delivers one bridge) or, from the molecular weights
of the repetitive unit (case of $(Ala)_n$), the molecular weight of
the bridge along with the density of the material. We find that
proteins like HE and LY have a value $N \simeq 6x10^{21} cm^{-3}$ for the peptide backbone H-bridges. Neglecting the other H-bridges existing
between the side-groups of the amino-acid residues, we find with
this N-value using Eq.(3) an upper limit of the dipole moment
change μ which is then 3.7 Debye. It is interesting to note that
the static dipole moment of the NH...OC bridge has about the same
value /14/ indicating perhaps the possibility of a cooperative
effect: There exists a rather strong dipole-dipole interaction between the backbone NH...OC bridges in the helical as well as the
sheeted conformations of proteins. If, in a thought experiment,
the dipole moment of one bridge is completely switched off, one
would need to deliver much more energy than 3.6 kcal/mol in order
to overcome the static dipole-dipole interaction energy. If, however, the dipole moment change of 3.7 Debye is distributed over
several neighbouring H-bridges one could reduce greatly the otherwise strong influence of the dipole interaction. Clearly, further
experimental evidence is needed to support this rather indirectly
argued effect.

The microscopic nature of the relaxations 1 and 2 (see Fig.7)
is more unclear. The high relaxation frequencies of these processes indicate also a protonic effect and then, due to the same
arguments as above, direct the attention again to the backbone
H-bridges. If this could be accepted, we would arrive at dipole
moment changes of $\mu \simeq 0.6$ Debye for relaxation 2 and $\mu \simeq 0.2$ Debye
for relaxation 1. These relaxations perhaps involve dynamical
jumps perpendicular to the NH...OC bridge.

When discussing hydrogen bridges in connection with biomolecules like proteins one is immediately faced with the problem of
H_2O bound to these molecules. This question prompted our experiment on hydrated lysozyme /19/, a protein which is particularly
stable for adsorption and desorption of water. It has been concluded that completely dehydrated LY contains only 3 to 4 tightly
bound water molecules per LY molecule /15/ which alone cannot be
responsible for the abserved dielectric relaxation effect (Fig.3)
In fact, hydration shows a clear increase of the absorption above
150 K due to an additional relaxation process which is approximately proportional, by weight, to the adsorbed water (Fig.9) and

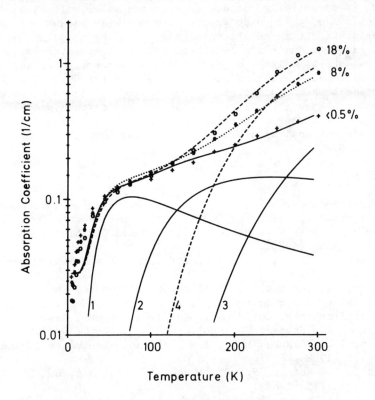

Fig.9: Absorption coefficient of lysozyme at 50 GHz for three
hydration levels. The dielectric function of the dried
material is composed of three relaxation processes (solid
lines 1-3) according to Eq.(7). The contribution of bound
water is represented by a further relaxation, which is
shown for 18% as the dashed line 4. (Taken from /19/).

which is nearly frequency independent. This last fact shows in
view of Eq.(4) that we are dealing with the case $\nu^2 \gg \nu_T^2$ which,
with Eq.(2) and the observed increase of α with T, does not ne-

cessarily need to assume an asymmetric double-well model. Under
this assumption, a symmetric double-well potential follows with
a barrier of 2.2 kcal/mol (1100 K)and a relaxation frequency
which cannot exceed 10 GHz at room temperature, corresponding to
a relaxation time of $\tau > 16$ ps. The dipole moment change μ is
then found to be 1.4 Debye if all the adsorbed water molecules
are considered to be in the same symmetric double-well potential,
which is surely an oversimplified assumption.

From this experiment, we should like to conclude that the
dielectrical mm-wave absorption, observed for the anhydreous materials, is essentially not due to adsorbed water molecules. Also
contributions of polar side groups seem unlikely because of the
short relaxation time and because $(Ala)_n$, having no polar side
groups, exhibits quite similar frequency and temperature dependence of α at the same level of magnitude. Consequently our assignment to processes involving the proton within the NH...OC group of
the peptide backbone seems rather conclusive. If our assignment
is correct, then such fast relaxation processes would greatly contribute to the number of possible substates of proteins /16/,
their conformational flexibility /17-18/ and thus the activity of
enzymes.

REFERENCES

1. Llewellyn-Jones, D.T., Knight, R.J., Moffat, P.H., Gebbie, H.A.
   "New Method of Measuring Low Values of Dielelctric Loss in the
   Near Millimeter Wave Region Using Untuned Cavities", Proc. IEE
   PtA, p. 535, (1980).
2. Kremer, F., Izatt, J.R., Int.J.of Infrared and Millemeter Waves,
   2, 675 (1981).
3. Kremer, F., Genzel, L., "Application of Untuned Cavities for
   Millimeter Wave Spectroscopy", Proc. 6th Int.Conf.on Infrared
   and Millimeter Waves, Miami, Dec. 1981, paper T 1-4.
4. Izatt, J.R., Kremer, F., Appl.Optics 20, 2555 (1981).
5. Birch, J.R., Clarke, R.N., The Radio and Electronic Engineer,
   52, 565 (1982).
6. Genzel, L., Kremer, F., Poglitsch, A., Bechtold, G., Biopolymers 22, 1715 (1983).
7. Genzel, L. Kremer, F., Poglitsch, A., Bechtold, G., "Millimeter wave and Far-infrared Spectroscopy on Biological Macromolecules", in "Coherent Excitations in Biological Systems",
   p. 58, Eds. H.Fröhlich and F.Kremer, Springer Verlag, Berlin,
   Heidelberg 1983.
8. Hunklinger, S., v.Schickfus, M., in "Amorphous solids low-temperature properties", Topics in Current Physics, Ed.W.A.Phillips,
   Springer Verlag, Berlin 1981.
9. Bilz, H., Genzel, L., Happ, H., Z.Physik 160, 535 (1960)

10. Stolen, R., Dransfeld, K., Phys.Rev.139, (4A), 1295 (1965)
11. Genzel, L., "Optische Absorption von Festkörpern durch Gitter-
    schwinungen". in Festkörperprobleme VI (Advances in Solid
    State Physics), Ed. O.Madelung, Vieweg, Braunschweig 1967.
12. Fröhlich, H., Theory of Dielectrics, Clarendon Press, Oxford
    1958.
13. Böttcher, C.J.F., Bordewijk, P., Theory of Electric Polariza-
    tion, Elserier, Amsterdam 1973.
14. Pethig, R., Dielectric and Electronic Properties of Biologi-
    cal Materials, John Wiley and Sons, Chichester 1979.
15. Imoto, T., Johnson, L.N., North, A.C.T., Phillips, D.C.,
    Rupley, J.A., in"The Enzymes", p.665, Ed. P.D. Boyer, Academic
    Press, New York 1972.
16. Fraunfelder, H., Petsko, G.A., Tsernoglou, D., Nature 280, 558,
    (1979).
17. Mc-Cammon, J.A., Gelin, B.R., Karplus, M., Nature 262, 325
    (1976).
18. Huber, R., Bennet, W.S., Biopolymers 22, 261 (1983)
19. Poglitsch, A., Kremer, F., Genzel, L., Submitted to J.Mol.Biol.

AUTHOR INDEX

Abrahamson, E.W. 257,385
Alix, A.J.P. 113
Allen's model 53
Alshuth, T. 329
Arbusov reaction 270
Atkinson, G.H. 606

Bamberg, E. 488
Bazzaz, M.B. 425
Bateman, J. 389
Bechtold, G. 621
Becker, R.S. 476,478
Bertoluzza, A. 91, 191
Birge, R.R. 457,473
Bixon, M. 36
Bjerrum fault 24
Borys, T.J. 385
Bowman, M.K. 441
Braiman, M. 309,319,321
Brereton, R.G. 419
Bridoux, M. 606
Brown, J.S. 421
Bull, T.E. 518
Busath, D. 504
Busch, G.E. 373,380
Buchet, R. 365,369

Chinsky, L. 115
Chothia, C. 69
Christensen 478
Claiborne, R. 548
Clementi, E. 10
Copan, W. 257
Cotton, T.M. 440,452
Coulson, C.A. 39
Curry, B. 310,311,316

Daly, M.C. 202
Day, R.S. 2, 6
Del Bene, J. 27
Delhaye, M. 114,587

Del Re, G. 15,36
Deshpande, S. 385
Diercksen, G.H.F. 46
Dogonadze, R.R. 36
Dörnemann, D. 420

Ebrey, Th.G. 367
Eigen, M. 23
Eisenman, G. 524
Elrod, J.P. 127
Emeis, D. 541
Emerson 426
Eyring, H. 19,529

Fabry Perot 605
Fairclough, D.P. 115
Fang, H.L. 469
Fateley, W.G. 571
Fermi level 11
Fock operator 4
Forsen, S. 518
Franck-Condon principle 56
Fraser, M.J. 175
Fritzsche, H. 180
Fry, R.C. 571

Geistefer, A. 385
Genzel, L. 609,621
Ghomi, M. 171
Graham, J.A. 571
Gramlich, V. 87
Green, P.K. 380
Gupta, P.T. 385

Hadzi, D. 39,61
Harbison, G.S. 284
Hargrave, P.A. 392
Harris, R.A. 464
Hartree-Fock 7,9
Hasser, R. 27,36
Haydon, D.A. 488,490

SUBJECT INDEX

Ab initio 1,3,67
Absorption coefficient 625
Absorption spectra 338,373
Acholeplasma laidlawii B. 557
Actinomycine D 115
Activation energy 48
Adenine 100
Adenosine monophosphate 106
Adenine residue 164
Adenosine 106
A-DNA 154,157,171,292,293
Adriamycine 116,126
Adsorbed water 172
A-genus polynucleotide 154
L-alanine 609,621,627
Albumin 615
Alkali metal ions 138
Alkaline earth ions 138
Alkylating agents 91
Allophycocyanins 427
All-trans retinal 267,311,329, 377
Amide 63,487,566
Amino-acids 609,611
Anisotropic fluorescence 431
Anesthetic 487
Anesthetic action 563
Angular momentum 248
Anharmonic coupling 53
Anion vs cation selectivity 531
Anisotropy 55
Anthocyanin Raman spectrum 600
15-anti configuration 269
Anti conformation 182
Antitumour compounds 113,295
Antiviral agents 115
Aperiodic DNA 5,11
ApG 87
Arginine 75
Argon Inductively Coupled Plasma 87

Asymmetry parameter 55
Asymmetric double-well 629
Atmospheric pressure 571
Atomic Emission 571
ATP 304
Aziridinyl ring 125

B A transition 173,176
B Z transition 158,173,183
Ba 292
Bacterial plasma, FT-IR spectrum 556
Bacteria 435
Bacteriohodopsin 81,303,329,340, 347,469
Base line 118
Base pairs 70,71
Band width 53
Bathorhodopsin 373
Bchl b 419
B-DNA 9,154,157,171,185,292,293
B-genous 159
B geometry 177
Bifurcated bonds 65
Bilipeptide 413
Biliproteins 409
Bilirubins 433
Binding constants 514
Binding sites 137,497
Biological target 113
Biomembranes 550,556
Biomolecules 609
Biosynthesis of tetrapyrroles 421
Biphotonic events 380
Bleomycine 115
BR-570 317,329,333,355,577

11-cis retinal 267,377
13-cis retinal 267,315,329,335
C2'-endo sugar pucker 173

Date Due